Amino Acids, Peptides and Proteins

Volume 34

A Specialist Periodical Report

Amino Acids, Peptides and Proteins

Volume 34

A Review of the Literature Published
during 2001

Senior Reporters
G.C. Barrett, *Oxford, UK*
J.S. Davies, *University of Wales, Swansea, UK*

Reporters
D.T. Elmore, *Oxfordshire, UK*
B. Penke, *University Szeged, Hungary*
G. Tóth, *University Szeged, Hungary*
G. Váradi, *University Szeged, Hungary*

advancing the chemical sciences

NEW FROM 2003

If you buy this title on standing order, you will be given FREE access to the chapters online. Please contact sales@rsc.org with proof of purchase to arrange access to be set up.

Thank you.

ISBN 0-85404-242-3

ISSN 1361-5904

A catalogue record for this book is available from the British Library

Published by The Royal Society of Chemistry,
Thomas Graham House, Science Park, Milton Road, Cambridge CB4 0WF, UK

Registered Charity Number 207890

For further information see our web site at www.rsc.org

Typeset by Vision Typesetting, Manchester, UK
Printed by Athenaeum Press Ltd, Gateshead, Tyne and Wear, UK

Preface

The literature on amino acids, peptides and proteins arrives in a variety of ways at the desktops of those maintaining current awareness of these topic areas. Whether our readers choose to read the online version of the chapters in this volume or have the traditional book format in front of them, we offer this latest volume on the 2001 literature, with the hope that new leads and further clarity will be conveyed through browsing these chapters. During the production of this volume, proceedings[1,2] of two major international symposia were published, reflecting the wealth and spread of work currently being researched in this area. We wait with interest for the contents of these proceedings to work their way to edited journal presentations for further review.

The policy for Specialist Periodical Reports published by the Royal Society of Chemistry continues to require in-depth and critical coverage of the literature. The endeavours of the team of authors are focussed by this policy on to major trends in current research. Although an approach to comprehensive coverage of narrow topics can be seen here and there, in this volume the subject areas are now vast and have called for compression and selectivity, particularly in the proteins area.

For this volume we welcome new authors for a wide ranging chapter on analogue and conformational studies on peptides and peptide hormones, now under the authorship of Botond Penke, Gábor Tóth and Györgyi Váradi from the Medical University of Szeged, Hungary. We offer a warm welcome to them and look forward to having their company in future volumes. Authors that have been involved with previous volumes in this series have penned the other chapters. So we are very grateful for the unstinting efforts of Donald Elmore, Graham Barrett and John Davies, and for their commitment to the continuation of this series of reports. Unfortunately the usual chapter on amino acids failed to make the deadline for publication, but will appear in the next volume as a 2-year compilation.

In order to preserve good practice in the use of abbreviations in peptide science, we once again reproduce by permission from J. Wiley and Sons the publishers, John Jones's 'Short Guide to Abbreviations' as published in *J. Peptide Science*, 1999, **5**, 465-471.

---

[1] Peptides 2002 Proceedings of the 27th European Peptide Symposium, Sorrento, Italy. Eds. E. Benedetti and C. Pedone, Edizioni Ziino, S.a.s, 2002, 1057pp.

[2] Peptides: The wave of the Future. Proceedings of the 2nd International and 27th American Peptide Symposium, San Diego, USA. Eds. M. Lebl and R.A. Houghten, American Peptide Symposium, 2001, 1111pp.

As always, the reporters are sincerely grateful to RSC personnel who have overseen the project, and who have ensured that their manuscripts are transcribed efficiently to the document you are now reading.

Graham Barrett   John Davies
Oxford   Swansea

Contents

Chapter 1 Peptide Synthesis   **1**
By Donald T. Elmore

   1 Introduction   1

   2 Methods   1
      2.1 Amino-group Protection   1
      2.2 Carboxy-group Protection   3
      2.3 Side-chain Protection   4
      2.4 Disulfide Bond Formation   5
      2.5 Peptide Bond Formation   6
      2.6 Peptide Synthesis on Macromolecular Supports
           and Methods of Combinatorial Synthesis   10
      2.7 Enzyme-mediated Synthesis and Semisynthesis   13
      2.8 Miscellaneous Reactions Related to Peptide
           Synthesis   16

   3 Appendix: A List of Syntheses in 2001   18
      3.1 Natural Peptides, Proteins and Partial Sequences   19
      3.2 Sequential Oligo- and Poly-peptides   23
      3.3 Enzyme Substrates and Inhibitors   23
      3.4 Conformations of Synthetic Peptides   25
      3.5 Glycopeptides   26
      3.6 Phosphopeptides and Related Compounds   27
      3.7 Immunogenic Peptides   27
      3.8 Nucleopeptides   28
      3.9 Miscellaneous Peptides   29
      3.10 Purification Methods   31

      References   31

Chapter 2   Analogue and Conformational Studies on Peptides, Hormones
**   and Other Biologically Active Peptides   55**
                  By Botond Penke, Gábor Tóth and Györgyi Váradi

         1   Introduction   55

         2   Peptide Backbone Modifications and Peptide Mimetics   55
             2.1   Aza, Oxazole, Oxazolidine and Tetrazole Peptides   55
             2.2 $\psi[CH_2CH_2]$, $\psi[Z\text{-}CF=CH]$, retro-$\psi[NHCH(CF_3)]$,
                   $\psi[epoxy]$, $\psi[CONHNR]$, $\psi[CSNH]$, $\psi[oxetane]$, ψ
                   $[CH_2NH]$ $\psi[CHOH\text{-}cyclopropyl\text{-}CONH]$,
                   $\psi[PO_2RCH]$ 56
             2.3   Rigid Amino Acid, Peptide and Turn Mimetics   57

         3   Cyclic Peptides   62

         4   Biologically Active Peptides   65
             4.1   Peptides Involved in Alzheimer's Disease   65
             4.2   Antimicrobial Peptides   68
             4.3   ACTH Peptides   74
             4.4   Angiotensin II Analogues and Non-peptide
                   Angiotensin II Receptor Ligands   74
             4.5   Bombesin/Neuromedin Analogues   74
             4.6   Bradykinin Analogues   75
             4.7   Cholecystokinin Analogues, Growth Hormone-
                   releasing Peptide and Analogues   77
             4.8   Integrin-related Peptide and Non-peptide Analogues   78
             4.9   LHRH and GnRH Analogues   82
             4.10 α-MSH Analogues   83
             4.11 MHC Class I and II Analogues   83
             4.12 Neuropeptide Y (NPY) Analogues   85
             4.13 Opioid (Neuropeptide FF, Enkephalin, Nociceptin,
                   Deltorphin and Dynorphin) Peptides   85
             4.14 Somatostatin Analogues   89
             4.15 Tachykinin (Substance P and Neurokinins)
                   Analogues   90
             4.16 Vasopressin and Oxytocin Analogues   92
             4.17 Insulin and Chemokines   93
             4.18 Peptide Toxins   93
             4.19 Miscellaneous   95

         5   Enzyme Inhibitors   97
             5.1   Aminopeptidase and Deformylase Inhibitors   97
             5.2   Calpain Inhibitors   98
             5.3   Carboxypeptidase Inhibitors   99
             5.4   Caspase Inhibitors   100

5.5   Cathepsin and Other Cysteine Protease Inhibitors   101
5.6   Cytomegalovirus and Rhinovirus 3C Protease
      Inhibitors   102
5.7   Converting Enzymes and Their Inhibitors   103
5.8   Elastase Inhibitors   103
5.9   Farnesyltransferase Inhibitors   105
5.10 HIV-Protease Inhibitors   107
5.11 Matrix Metalloproteinase Inhibitors   110
5.12 NO-Synthase Inhibitors   113
5.13 Proteasome Inhibitors   114
5.14 Protein Phosphatase Inhibitors   116
5.15 Renin and Other Aspartyl Proteinase Inhibitors   117
5.16 Thrombin and Factor-Xa Inhibitors   120
5.17 Tyrpsin and Other Serine Protease Inhibitors   122
5.18 tRNA Synthetase Inhibitors   123
5.19 Miscellaneous   123

6   Phage Library Leads   125

7   Protein–Protein Interaction Inhibitors   126
7.1   SH2 and SH3 Domain Ligands   126

8   Advances in Formulation/Delivery Technology   127

References   128

Chapter 3   Cyclic, Modified and Conjugated Peptides   149
           By John S. Davies

1   Introduction   149

2   Cyclic Peptides   150
2.1   General Considerations   150
2.2   Cyclic Dipeptides (Including Dioxopiperazines)   153
2.3   Cyclotripeptides   158
2.4   Cyclotetrapeptides   158
2.5   Cyclopentapeptides   160
2.6   Cyclohexapeptides   163
2.7   Cycloheptapeptides   166
2.8   Cyclooctapeptides/Cyclononapeptides   167
2.9   Cyclodecapeptides and Higher Cyclic Peptides   169
2.10 Peptides Containing Thiazole/Oxazole Rings   172
2.11 Cyclodepsipeptides   178

3   Modified and Conjugated Peptides   188
3.1   Phosphopeptides   188

3.2   Glycopeptide Antibiotics   189
3.3   Glycopeptides   194
3.4   Lipopeptides   203

4   Miscellaneous Structures   205

5   References   206

Chapter 4   Proteins   219
By Graham C. Barrett

1   Introduction   219

2   Structure of This Chapter   220
2.1   Cross-referencing in This Chapter   220

3   Textbooks and Monographs   220
3.1   Literature Searching in Protein Science   221
3.2   Protein Nomenclature   221

4   Structure Determination of Proteins   221
4.1   Proteomics and Genomics   221
4.2   Mass Spectrometry   222
4.3   NMR Spectroscopy   222
4.4   Structures of Proteins Determined using Physical
      Methods in Combination with Structural
      Derivatization   225
4.5   Molecular Modelling   226
4.6   Other Structural Features Established for Proteins
      by Physical Methods: Crosslinks   226
4.7   X-Ray Crystallographic Studies of Proteins   227
4.8   Structural Information for Proteins from Other
      Physical Techniques   240
4.9   Circular Dichroism Spectroscopy   243

5   Folding and Conformational Studies   243
5.1   Background to Protein Folding Studies   243
5.2   Effects of Metal Complexation on Protein Structure   247
5.3   Membrane Proteins and Channel Proteins   248
5.4   Prion Proteins   252
5.5   Rare Folding Motifs within Proteins   257

6   Adhesion and Binding Studies   257
6.1   Textbooks and Monographs   257
6.2   General Results   257
6.3   New Results from Binding Studies   258

6.4 Protein–Protein Interactions Involving Chaperones 258
6.5 Proteins Complexed with Non-protein Species 263

7 Enzyme Studies 271
7.1 Textbooks and Monographs 271
7.2 Mechanistic Studies 271

8 Proteins as Enzyme Inhibitors 315

9 Enzyme Inhibition by Non-protein Species 315

10 Other Biological Functions for Proteins 318

11 Other Protein Studies 318
11.1 Assisted Biosynthesis of Proteins 318
11.2 Disulfide Interchange 319

12 Other Protein Properties 319

References 321

A Short Guide to Abbreviations and Their Use in Peptide Science

Abbreviations, acronyms and symbolic representations are very much part of the language of peptide science – in conversational communication as much as in its literature. They are not only a convenience, either – they enable the necessary but distracting complexities of long chemical names and technical terms to be pushed into the background so the wood can be seen among the trees. Many of the abbreviations in use are so much in currency that they need no explanation. The main purpose of this editorial is to identify them and free authors from the hitherto tiresome requirement to define them in every paper. Those in the tables that follow – which will be updated from time to time – may in future be used in this Journal without explanation.

All other abbreviations should be defined. Previously published usage should be followed unless it is manifestly clumsy or inappropriate. Where it is necessary to devise new abbreviations and symbols, the general principles behind established examples should be followed. Thus, new amino-acid symbols should be of form Abc, with due thought for possible ambiguities (Dap might be obvious for diaminoproprionic acid, for example, but what about diaminopimelic acid?).

Where alternatives are indicated below, the first is preferred.

Amino Acids

Proteinogenic Amino Acids

Ala	Alanine	A
Arg	Arginine	R
Asn	Asparagine	N
Asp	Aspartic acid	D
Asx	Asn *or* Asp	
Cys	Cysteine	C
Gln	Glutamine	Q
Glu	Glutamic acid	E
Glx	Gln *or* Glu	
Gly	Glycine	G
His	Histidine	H
Ile	Isoleucine	I
Leu	Leucine	L

Lys	Lysine	K
Met	Methionine	M
Phe	Phenylalanine	F
Pro	Proline	P
Ser	Serine	S
Thr	Threonine	T
Trp	Tryptophan	W
Tyr	Tyrosine	Y
Val	Valine	V

Other Amino Acids

Aad	α-Aminoadipic acid
βAad	β-Aminoadipic acid
Abu	α-Aminobutyric acid
Aib	α-Aminoisobutyric acid; α-methylalanine
βAla	β-Alanine; 3-aminopropionic acid (avoid Bal)
Asu	α-Aminosuberic acid
Aze	Azetidine-2-carboxylic acid
Cha	β-cyclohexylalanine
Cit	Citrulline; 2-amino-5-ureidovaleric acid
Dha	Dehydroalanine (also ΔAla)
Gla	χ-Carboxyglutamic acid
Glp	pyroglutamic acid; 5-oxoproline (also pGlu)
Hph	Homophenylalanine (Hse = homoserine, and so on). Caution is necessary over the use of the use of the prefix homo in relation to α-amino-acid names and the symbols for homo-analogues. When the term first became current, it was applied to analogues in which a side-chain CH_2 extension had been introduced. Thus homoserine has a side-chain CH_2CH_2OH, homoarginine $CH_2CH_2CH_2NHC(=NH)NH_2$, and so on. In such cases, the convention is that a new three-letter symbol for the analogue is derived from the parent, by taking H for homo and combining it with the first two characters of the parental symbol – hence, Hse, Har and so on. Now, however, there is a considerable literature on β-amino acids which are analogues of α-amino acids in which a CH_2 group has been inserted between the α-carbon and carboxyl group. These analogues have also been called homo-analogues, and there are instances for example not only of 'homophenylalanine', $NH_2CH(CH_2CH_2Ph)CO_2H$, abbreviated Hph, but also 'homophenylalanine', $NH_2CH(CH_2Ph)CH_2CO_2H$ abbreviated Hph. Further, members of the analogue class with CH_2 interpolated between the α-carbon and the carboxyl group of the parent α-amino acid structure have been called both 'α-homo'- and 'β-homo'. Clearly great care is essential, and abbreviations for 'homo' analogues ought to be fully defined on every occasion. The term 'β-homo' seems preferable for backbone extension (emphasizing as it does that the

residue has become a β-amino acid residue), with abbreviated symbolism as illustrated by βHph for $NH_2CH(CH_2Ph)CH_2CO_2H$.

Hyl	δ-Hydroxylysine
Hyp	4-Hydroxyproline
αIle	*allo*-Isoleucine; 2S, 3R in the λ-series
Lan	Lanthionine; S-(2-amino-2-carboxyethyl)cysteine
MeAla	*N*-Methylalanine (MeVal = *N*-methylvaline, and so on). This style should not be used for α-methyl residues, for which either a separate unique symbol (such as Aib for α-methylalanine) should be used, or the position of the methyl group should be made explicit as in αMeTyr for α-methyltyrosine.
Nle	Norleucine; α-aminocaproic acid
Orn	Ornithine; 2,5-diaminopentanoic acid
Phg	Phenylglycine; 2-aminophenylacetic acid
Pip	Pipecolic acid; piperidine-s-carboxylic acid
Sar	Sarcosine; *N*-methylglycine
Sta	Statine; (3S, 4S)-4-amino-3-hydroxy-6-methyl-heptanoic acid
Thi	β-Thienylalanine
Tic	1,2,3,4-Tetrahydroisoquinoline-3-carboxylic acid
αThr	*allo*-Threonine; 2S, 3S in the λ-series
Thz	Thiazolidine-4-carboxylic acid, thiaproline
Xaa	Unknown or unspecified (also Aaa)

The three-letter symbols should be used in accord with the IUPAC-IUB conventions, which have been published in many places (e.g. *European J. Biochem.* 1984; **138**: 9–37), and which are (May 1999) also available with other relevant documents at: http://www.chem.qnw.ac.uk/iubmb/iubmb.html#03

It would be superfluous to attempt to repeat all the detail which can be found at the above address, and the ramifications are extensive, but a few remarks focussing on common misuses and confusions may assist. The three-letter symbol standing alone represents the unmodified intact amino acid, of the λ-configuration unless otherwise stated (but the λ-configuration may be indicated if desired for emphasis: e.g. λ-Ala). The same three-letter symbol, however, also stands for the corresponding amino acid *residue*. The symbols can thus be used to represent peptides (e.g. AlaAla or Ala-Ala = alanylalanine). When nothing is shown attached to either side of the three-letter symbol it is meant to be understood that the amino group (always understood to be on the left) or carboxyl group is unmodified, but this can be emphasized, so AlaAla = H-AlaAla-OH. Note however that indicating free termini by presenting the terminal group in full is wrong; $NH_2AlaAlaCO_2H$ implies a hydrazino group at one end and an α-keto acid derivative at the other. Representation of a free terminal carboxyl group by writing H on the right is also wrong because that implies a terminal aldehyde.

Side chains are understood to be unsubstituted if nothing is shown, but a substituent can be indicated by use of brackets or attachment by a vertical bond up or down. Thus an *O*-methylserine residue could be shown as **1**, **2**, or **3**.

$$—Ser(Me)—\quad \textbf{1}$$

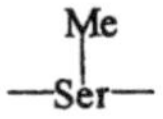

$$\begin{array}{c} Me \\ | \\ —Ser— \end{array}\quad \textbf{2}$$

$$\begin{array}{c} —Ser— \\ | \\ Me \end{array}\quad \textbf{3}$$

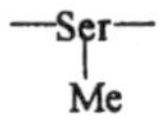

Note that the oxygen atom is not shown: it is contained in the three-letter symbol – showing it, as in Ser(OMe), would imply that a peroxy group was present. Bonds up or down should be used only for indicating side-chain substitution. Confusions may creep in if the three-letter symbols are used thoughtlessly in representations of cyclic peptides. Consider by way of example the hypothetical cyclopeptide threonylalanylalanylglutamic acid. It might be thought that this compound could be economically represented **4**.

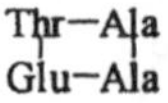

$$\begin{array}{c} Thr—Ala \\ Glu—Ala \end{array}\quad \textbf{4}$$

But this is wrong because the left hand vertical bond implies an ester link between the two side chains, and strictly speaking if the right hand vertical bond means anything it means that the two Ala α-carbons are linked by a CH_2CH_2 bridge. This objection could be circumvented by writing the structure as in **5**.

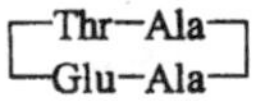

$$\begin{array}{c} \lceil Thr—Ala \rceil \\ \lfloor Glu—Ala \rfloor \end{array}\quad \textbf{5}$$

But this is now ambiguous because the convention that the symbols are to be read as having the amino nitrogen to the left cannot be imposed on both lines. The direction of the peptide bond needs to be shown with an arrow pointing from CO to N, as in **6**.

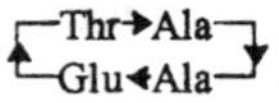

$$\begin{array}{c} \lceil Thr{\rightarrow}Ala \rceil \\ \lfloor Glu{\leftarrow}Ala \rfloor \end{array}\quad \textbf{6}$$

Actually the simplest representation is on one line, as in **7**.

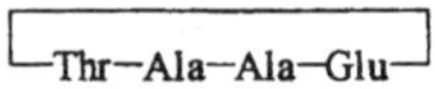

$$\lceil\overline{\quad\quad\quad\quad}\rceil \\ \lfloor Thr—Ala—Ala—Glu \rfloor \quad \textbf{7}$$

Substituents and Protecting Groups

Ac	Acetyl
Acm	Acetamidomethyl
Adoc	1-Adamantyloxycarbonyl
Alloc	Allyloxycarbonyl
Boc	*t*-Butoxycarbonyl
Bom	π-Benzyloxymethyl
Bpoc	2-(4-Biphenylyl)isopropoxycarbonyl

Btm	Benzylthiomethyl
Bum	π-t-Butoxymethyl
Bu^i	i-Butyl
Bu^n	n-Butyl
Bu^t	t-Butyl
Bz	Benzoyl
Bzl	Benzyl (also Bn); Bzl(OMe) = 4-methoxybenzyl and so on
Cha	Cyclohexylammonium salt
Clt	2-Chlorotrityl
Dcha	Dicyclohexylammonium salt
Dde	1-(4,4-Dimethyl-2,6-dioxocyclohex-1-ylidene)ethyl
Ddz	2-(3,5-Dimethoxyphenyl)-isopropoxycarbonyl
Dnp	2,4-Dinitrophenyl
Dpp	Diphenylphosphinyl
Et	Ethyl
Fmoc	9-Fluorenylmethoxycarbonyl
For	Formyl
Mbh	4,4′-Dimethoxydiphenylmethyl, 4,4′-Dimethoxybenzhydryl
Mbs	4-Methoxybenzenesulphonyl
Me	Methyl
Mob	4-Methoxybenzyl
Mtr	2,3,6-Trimethyl,4-methoxybenzenesulphonyl
Nps	2-Nitrophenylsulphenyl
OAll	Allyl ester
OBt	1-Benzotriazolyl ester
OcHx	Cyclohexyl ester
ONp	4-Nitrophenyl ester
OPcp	Pentachlorophenyl ester
OPfp	Pentafluorophenyl ester
OSu	Succinimido ester
OTce	2,2,2-Trichloroethyl ester
OTcp	2,4,5-Trichlorophenyl ester
Tmob	2,4,5-Trimethoxybenzyl
Mtt	4-Methyltrityl
Pac	Phenacyl, $PhCOCH_2$ (care! Pac also = $PhCH_2CO$)
Ph	Phenyl
Pht	Phthaloyl
Scm	Methoxycarbonylsulphenyl
Pmc	2,2,5,7,8-Pentamethylchroman-6-sulphonyl
Pr^i	i-Propyl
Pr^n	n-Propyl
Tfa	Trifluoroacetyl
Tos	4-Toluenesulphonyl (also Ts)
Troc	2,2,2-Trichloroethoxycarbonyl
Trt	Trityl, triphenylmethyl
Xan	9-Xanthydryl

Z   Benzyloxycarbonyl (also Cbz). Z(2Cl)=2-chlorobenzyloxycarbonyl
            and so on

Amino Acid Derivatives
DKP   Diketopiperazine
NCA   *N*-Carboxyanhydride
PTH   Phenylthiohydantoin
UNCA   Urethane *N*-carboxyanhydride

Reagents and Solvents
BOP   1-Benzotriazolyloxy-tris-dimethylamino-phosphonium hexafluoro-
            phosphate
CDI   Carbonyldiimidazole
DBU   Diazabicyclo[5.4.0]-undec-7-ene
DCCI   Dicyclohexylcarbodiimide (also DCC)
DCHU   Dicyclohexylurea (also DCU)
DCM   Dichloromethane
DEAD   Diethyl azodicarboxylate (DMAD=the dimethyl analogue)
DIPCI   Diisopropylcarbodiimide (also DIC)
DIPEA   Diisopropylethylamine (also DIEA)
DMA   Dimethylacetamide
DMAP   4-Dimethylaminopyridine
DMF   Dimethylformamide
DMS   Dimethylsulphide
DMSO   Dimethylsulphoxide
DPAA   Diphenylphosphoryl azide
EEDQ   2-Ethoxy-1-ethoxycarbonyl-1,2-dihydroquinoline
HATU   This is the acronym for the 'uronium' coupling reagent derived from
            HOAt, which was originally thought to have the structure **8**, the
            *H*exafluorophosphate salt of the *O*-(7-*A*zabenzotriazol-lyl)-*T*et-
            ramethyl *U*ronium cation.

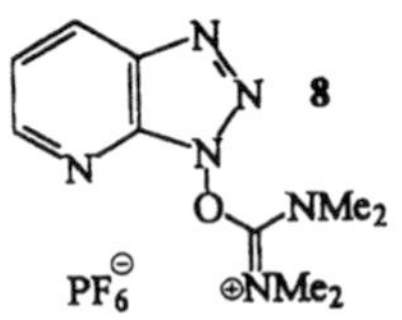

In fact this reagent has the isomeric *N*-oxide structure **9** in the
crystalline state, the unwieldy correct name of which does not con-
form logically with the acronym, but the acronym continues in use.

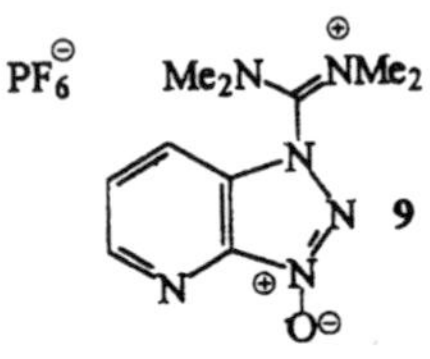

Similarly, the corresponding reagent derived from HOBt has the firmly attached label HBTU (the tetrafluoroborate salt is also used: TBTU), despite the fact that it is not actually a uronium salt.

HMP	Hexamethylphosphoric triamide (also HMPA, HMPTA)
HOAt	1-Hydroxy-7-azabenzotriazole
HOBt	1-Hydroxybenzotriazole
HOCt	1-Hydroxy-4-ethoxycarbonyl-1,2,3-triazole
NDMBA	*N,N'*-Dimethylbarbituric acid
NMM	*N*-Methylmorpholine
PAM	Phenylacetamidomethyl resin
PEG	Polyethylene glycol
PtBOP	1-Benzotriazolyloxy-tris-pyrrolidinophosphonium hexafluorophosphate
SDS	Sodium dodecyl sulphate
TBAF	Tetrabutylammonium fluoride
TBTU	See remarks under HATU above
TEA	Triethylamine
TFA	Trifluoroacetic acid
TFE	Trifluoroethanol
TFMSA	Trifluoromethanesulphonic acid
THF	Tetrahydrofuran
WSCI	Water soluble carbodiimide: 1-ethyl-3-(3'-dimethylaminopropyl)-carbodiimide hydrochloride (also EDC)

Techniques

CD	Circular dichroism
COSY	Correlated spectroscopy
CZE	Capillary zone electrophoresis
ELISA	Enzyme-linked immunosorbent assay
ESI	Electrospray ionization
ESR	Electron spin resonance
FAB	Fast atom bombardment
FT	Fourier transform
GLC	Gas liquid chromatography
hplc	High performance liquid chromatography
IR	Infra red
MALDI	Matrix-assisted laser desorption ionization
MS	Mass spectrometry
NMR	Nuclear magnetic resonance
nOe	Nuclear Overhauser effect
NOESY	Nuclear Overhauser enhanced spectroscopy
ORD	Optical rotatory dispersion
PAGE	Polyacrylamide gel electrophoresis
RIA	Radioimmunoassay
ROESY	Rotating frame nuclear Overhauser enhanced spectroscopy
RP	Reversed phase

SPPS	Solid phase peptide synthesis
TLC	Thin layer chromatography
TOCSY	Total correlation spectroscopy
TOF	Time of flight
UV	Ultraviolet

Miscellaneous

Ab	Antibody
ACE	Angiotensin-converting enzyme
ACTH	Adrenocorticotropic hormone
Ag	Antigen
AIDS	Acquired immunodeficiency syndrome
ANP	Atrial natriuretic polypeptide
ATP	Adenosine triphosphate
BK	Bradykinin
BSA	Bovine serum albumin
CCK	Cholecystokinin
DNA	Deoxyribonucleic acid
FSH	Follicle stimulating hormone
GH	Growth hormone
HIV	Human immunodeficiency virus
LHRH	Luteinizing hormone releasing hormone
MAP	Multiple antigen peptide
NPY	Neuropeptide Y
OT	Oxytocin
PTH	Parathyroid hormone
QSAR	Quantitative structure–activity relationship
RNA	Ribonucleic acid
TASP	Template-assembled synthetic protein
TRH	Thyrotropin releasing hormone
VIP	Vasoactive intestinal peptide
VP	Vasopressin

J. H. Jones

1
Peptide Synthesis

BY DONALD T. ELMORE

1   Introduction

As in the previous Report,[1] many reviews have been published. Some[2-7] relate to several sections of this article whereas others are cognate to particular sections as follows: Section 2.1,[8] Section 2.3,[9] Section 2.4,[10] Section 2.5,[11-15] Section 2.6,[15-23] Section 2.7,[24-29] Section 3.1,[30-32] Section 3.2,[33] Section 3.3,[34,35] Section 3.4,[36] Section 3.5,[37-41] Section 3.6,[41-42] Section 3.8,[43,44] Section 3.9[45-51] and Section 3.9.[52,53].

2   Methods

2.1   Amino-group Protection. – It often happens in chemistry that a technique that has been regarded as passé for some years, is slightly modified and resumes its former status in the literature. Protection of the α-amino group by the phthaloyl group is a recent example. The need for rather vigorous conditions in order to effect removal with hydrazine led to discontinuance of its use. An improved method of phthaloylation using monomethyl phthalate[54] would hardly have stayed its abandonment, but the discovery that tetrachlorophthaloyl groups, while stable to piperidine and to acids, are removed with $N_2H_4/CHONMe_2$ (3:17) at 40°C for 1 hr[55] has offered a fresh orthogonal substituent. Moreover, preparation of tetrachlorophthaloyl amino acids under microwave irradiation[56] is straightforward. Microwave irradiation also permits rapid synthesis of phthaloyl derivatives with no loss of chiral purity,[57] but this development has probably come too late for modern peptide synthesis. New variations in the conditions for removal of the Boc group have been reported. A fast selective method uses 4M-HCl in dioxan.[58] Deprotection is complete in 30 min. at room temperature. The method is selective in the presence of Bu^t esters and alkyl esters and also Bu^t thioethers but not in the presnce of Bu^t phenolic ethers. Nitrolytic removal of Boc groups has been reported[59] and Z-groups and Bu^t esters are unaffected, but oxidation of susceptible groups is a hazard. Boc groups can be removed from substituted peptides on Wang resin using conc. H_2SO_4 in dioxan (1:9 v/v) at 8°C for 2 hr.[60] Cleavage of the peptide from the resin is quite limited. Lipidated peptides are often very labile to acids and bases and it is valuable that Boc groups can be removed using trimethylsilyl triflate in the presence of a tertiary base such as lutidine or $EtMe_2N$.[61] The use of silica gel under microwave irradiation for the detachment of acid-labile groups was disappointing.[62] Strong irradiation for long periods was required and this gave unacceptable amounts of byproducts. A precolumn method of preparation of Fmoc-amino acids and -peptides can be carried out with Fmoc-Cl after adsorption on silica gel.[63] Excess reagent is washed away with EtOAc. The Fmoc-derivatives are then eluted for analysis; much less byproduct is present than with earlier protocols and Fmoc-Cl is completely absent. Details for the synthesis on a large scale of Fmoc-4-aminomethyl benzoic acid, Fmoc-trans-4-(aminomethyl)cyclohexane carboxylic acid and Fmoc derivatives of cis-β-amino acids have been described.[64] Fmoc- and Z-derivatives of N-methylserine and N-methylthreonine can be accessed by the general method in Scheme 1.[65] Polymer-bound N-hydroxysuccinimide can be prepared from commercial styrene-

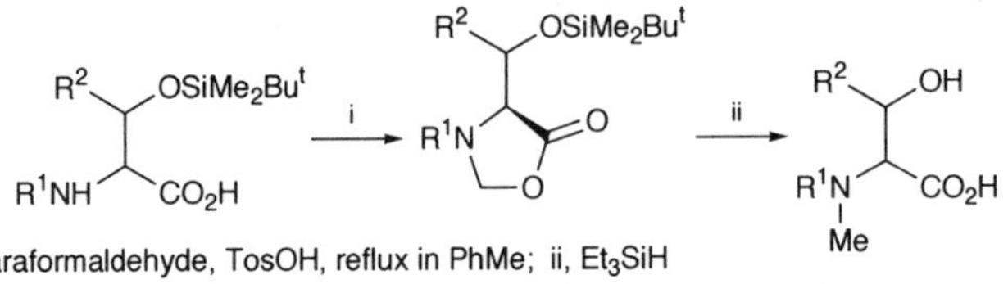

Reagents: i, paraformaldehyde, TosOH, reflux in PhMe; ii, Et_3SiH

Scheme 1

maleic anhydride copolymer by reaction with 50% w/v aqueous NH_2OH. Treatment of the product with Fmoc-Cl/aq. K_2CO_3 gives rise to polymer-bound Fmoc-OSu. This is a convenient reagent for the preparation of Fmoc amino acids and the unused and reacted polymer can be filtered off and reused.[66] Fmoc-Amino acids can also be directly synthesised[67] using organo-zinc chemistry. For example, Fmoc-3-iodoalanine-OBu^t (obtained in 7 steps from L-serine) is converted into the organo-zinc derivative and then into a substituted Phe derivative ready for SPPS after appropriate deprotection (Scheme 2). An inter-

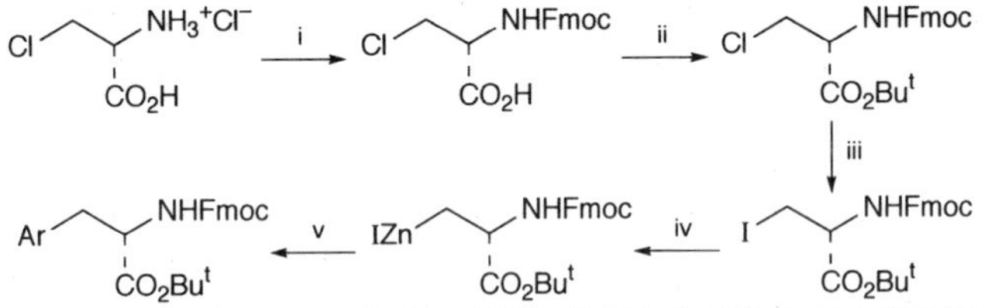

Reagents: i, Fmoc–OSu, $NaHCO_3$; ii, $Me_2C=CH_2$, H_2SO_4; iii, NaI, Me_2CO; iv, Zn*, $CHONMe_2$; v, Pd_2dba_3, $P(o\text{-tol})_3$, ArI, $CHONMe_2$
Pd_2dba_3 = tris(dibenzylideneacetone)dipalladium
$P(o\text{-tol})_3$ = tri-2-tolylphosphine
Zn* is prepared from Zn dust using Me_3SiCl in $CHONMe_2$

Scheme 2

esting method of preparing Fmoc- and Z-derivatives of amino acids uses the chloroformate at neutral pH in presence of activated Zn powder.[68] An improved method for the acidolytic removal of protecting groups from thionopeptides used aqueous CF_3CO_2H (<80% w/v) for about 2h.[69] This procedure minimises the acid-catalysed cleavage at the peptide bond immediately following the -CSNH- moiety. The (2-nitrofluoren-9-yl)methoxycarbonyl group offers interesting possibilities.[70] Not only does it possess enhanced lability to bases, but it also undergoes photo-chemical cleavage under appropriate conditions. α-Fmoc groups have been employed *inter alia* for the synthesis of cyclic peptides involving ring closure on a specially synthesised backbone amino acid.[71] This procedure has been recommended for combinatorial synthesis of potential peptide drug candidates. The Alloc group can be used to protect α-amino groups in the synthesis of lipopeptides since Pd(0) rather than acid can be used for deprotection as an alternative to that mentioned above.[72] The same paper recommends the use of Bu^t groups for the protection of carboxy and thiol groups. Neutral conditions can be used to release protected amino groups if the 2-nitrobenzyl group is used.[73] A low cathodic potential converts the nitro group into a hydroxyamino group that undergoes ring closure to benzisoxazolone with release of the amino group. The 2-[phenyl-(methyl)sulfonio]ethoxycarbonyl group ensures that, after protection of the amino group, the product will be soluble in water and Leu enkephalin has been synthesised using it.[74] This could be useful in syntheses employing enzyme-catalysed steps and it could also simplify the production of crystals for X-ray diffraction studies. A base-labile group, 2-(2,4-dinitrophenylsulfonyl)ethoxycarbonyl, has been recommended for solid-phase peptide synthesis, but does not appear to offer any special advantages.[75] A novel side reaction was discovered when an O-silyl group was removed from protected pyroglutaminol.[76] The expected product was formed admixed with an isomer in which a Boc group had migrated (Scheme 3). The use of enzymes in peptide

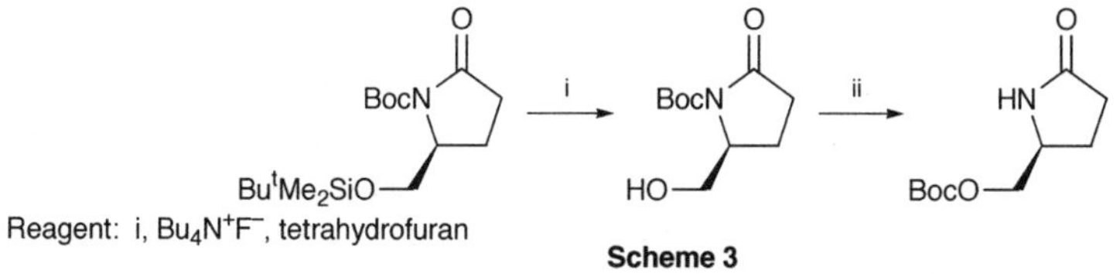

Reagent: i, Bu$_4$N$^+$F$^-$, tetrahydrofuran

Scheme 3

synthesis is still limited to a few workers. Immobilised penicillin G acylase in toluene has been used to acylate amino acids,[77] but attempts to effect esterification and trans-esterification failed. It is not surprising that an enzyme will not operate on both ends of an amino acid molecule. If it did, a jumble of polymers would be the outcome. The established PhAcOZ group, which is enzyme-labile, has been used to synthesise acid- and base-labile nucleo-peptides.[78]

2.2 Carboxy-group Protection. – Although trifluoromethyl esters are not much used in peptide synthesis, it may be useful to know that these derivatives of *N*-acylamino acids are accessible using catalysis by 4-dimethylamino pyridine.[79] 2,2,2-Trichloro-t-butyl (Tcb) esters are of interest because they are deprotected in slightly acidic or neutral conditions in the presence Zn^{2+} or Cd^{2+} ions with the super nucleophile cobalt(I) phthalocyanine.[80] Preparation of Tcb esters so far described involve the use of acyl chlorides, but unsymmetrical anhydrides might be a suitable alternative starting material. The esterification of amino acids can be effected in the presence of 'triphosgene' (CCl₃OCOOCCl₃) in tetrahydrofuran at 55-60°C.[81] Amino acids have been esterified with MeOH under pressure and in the presence of an ultrastable zeolite as catalyst.[82] There has been a further report for the esterification of acylamino acids with 3,4,5-tris(octadecyloxy)-benzyl alcohol before peptide synthesis.[83] The product is purified by size-exclusion chromatography on Sephadex LH-20. Using Fmoc amino acids, subsequent deprotection could be effected using 4M-HCl in EtOAc. This could alternatively be achieved under milder conditions using a lipase for deprotection in the presence of activated charcoal to retain the liberated alcohol with filtration or centrifugation to isolate the product. Treatment of Boc-Ala-OBuᵗ with CeCl₃.7H₂O/NaI removed both protecting groups.[84] If the CeCl₃.7H₂O were heated in MeCN under reflux, then cooled to room temperature and added to the substrate, the ester group was removed but the Boc group survived. Obviously further work is required to delineate conditions for selective deprotection. The β-CO₂H of Asp in peptides prepared by SPPS can be isotopically enriched by hydrolysis with Na¹⁷OH in MeOH/CH₂Cl₂.[85] The peptide is detached from the support using acid.

2.3   Side-chain Protection. – An efficient synthesis of *N*-Boc-*O*-cyclohexyl-L-tyrosine has been described.[86] Boc-Tyr-OH was treated with NaH in CHONMe₂ and then with 3-bromocyclohexene. The product was hydrogenated over PtO₂. The highest yield was obtained when a byproduct from the previous step was *not* removed because the yield then was almost quantitative. A problem has been encountered when the hydroxyl group of Thr was protected by tosylation.[87] A mixture of *O*-tosyl- and dehydro-Thr was formed. The ratio of the two products depended strongly on the choice of groups for protecting the amino and carboxy groups. In the case of Fmoc-Thr-OBzl, only the unsaturated product was formed. In order to obtain fluorescent enzyme substrates, α-Fmoc-ε-[(7-methoxycoumarin-4-yl)acetyl]-L-lysine was used as starting material.[88] Strictly speaking, the fluorescent substituent is not a protecting group, since it can not be detached without destroying the peptide. Further examples of this type of fluorescent substrate are cited in Section 3.3. For protection of the thiol group in the side chain of an amino acid, 4-methoxytrityl chloride is appropriate.[89] The *S*-(4-methoxytrityl) group can be retained if the peptide is detached from the resin using CH₃CO₂H/CF₃CH₂OH/CH₂Cl₂ (1:2:7) for 15 min. at room temperature, but is cleaved with 1-5% CF₃CO₂H in CH₂Cl₂/Et₃Si (95:5). Allylic protection of cysteine is possible with the *S*-[*N*-{2,3,5,6-tetrafluoro-4-(phenylthio)-phenyl}-*N*-allyloxycarbonyl]-aminomethyl (Fsam) group (1)[90], which can be removed by oxidation with I₂ with concomitant formation of disulphide bonds or the allylic moiety can be removed with Pd(PPh₃)₄ in the presence of an allyl scavenger to give the peptide with free thiol groups.

The side-chain of Met, normally protected as the sulfoxide, can be regenerated by using Bu₄NBr in CF₃CO₂H as a reducing agent.[91] Nucleophilic attack of halide at the S atom of the protonated sulfoxide is proposed. The tetracovalent S intermediate undergoes release of water after further protonation to give a halosulfonium ion and this is the rate-determining step.

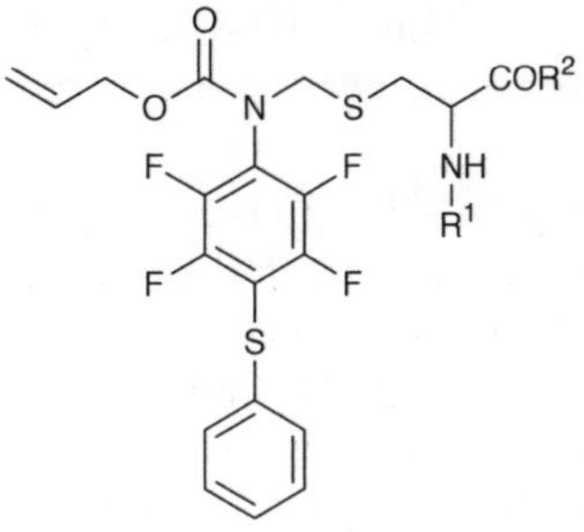

(1)

Protection of the imidazole ring of His with the Bum group has been re-examined.[92] Byproduct formation during deprotection could be diminished by scavenging the HCHO formed using MeONH$_2$.

2.4 Disulfide Bond Formation. – There has been a surprising report that a cyclic disulfide is formed from a peptide with cysteine at both termini in degassed water under reduced pressure.[93] Cyclisation occurred even when the amount of oxygen present in the sytem was only 1/16 of that theoretically required. The authors maintain that disulfide formation does not necessarily require an oxidant such as elementary oxygen or iodine. The report last year that exposure to trans-[Pt(en)$_2$Cl$_2$] effects intramolecular disulfide bond formation from peptides containing two cysteinyl residues has been amplified by the discovery that fully reduced α-conotoxin GI and SI are similarly converted into the oxidised form with 3 disulfide bonds in one step at pH values between 3 and 7.[94]

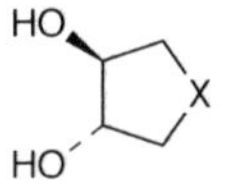

(2) X = Se, SeO, SeSe

Water-soluble reagents containing Se such as (2) are quite powerful oxidising agents, although these have not been applied to the formation of peptide cyclic disulfides.[95] Based on appropriate model experiments, it was shown that Fmoc-Cys-β-Ala-Cys-OMe, which was immobilised on a resin through a benzyl thioether link involving the side chain of the *C*-terminal cysteine residue, underwent detachment and intramolecular cyclisation in the presence of *N*-chlorosuccinimide (Scheme 4).[96] A library of amphipathic bicyclic peptides in which one of

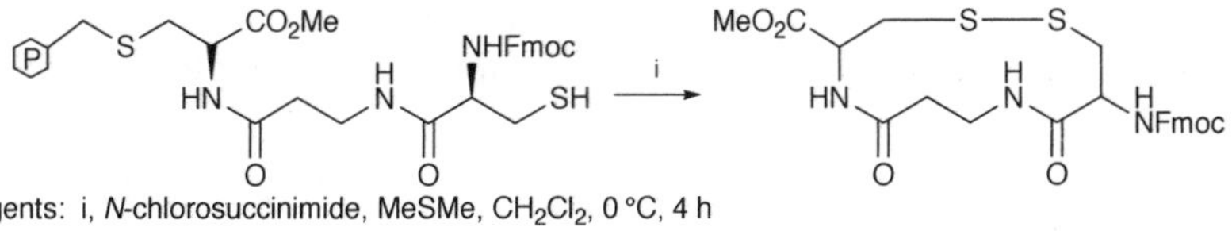

Reagents: i, *N*-chlorosuccinimide, MeSMe, CH$_2$Cl$_2$, 0 °C, 4 h

Scheme 4

the rings contained a disulfide has been assembled. The first cyclic structure was produced by intramolecular thioester ligation while attached to a solid support and the second ring was closed by oxidation with MeSOMe of the side chains of two cysteinyl residues after detachment from the resin.[97]

2.5   Peptide Bond Formation. – Boc_2O is commonly used in the unsymmetrical anhydride synthesis of peptides or for protecting -OH or -NH- groups. Imidazole or trifluoroethanol are convenient scavengers to react with excess reagent.[98] Pentafluoro-, 2,4,5-trichloro- and pentachloro-phenyl esters of Fmoc amino acids are conveniently prepared using a 2-phase system in which 3% $NaHCO_3$ solution is the aqueous phase. Good yields are obtained in 2-3 h.[99] Pentafluorophenyl 4-nitrobenzenesulfonate is a useful reagent for peptide coupling and HOBt strongly promotes synthesis.[100] $4\text{-}NO_2\text{-}C_6H_4\text{-}SO_2OBt$ is a probable intermediate. No loss of chiral purity was detected when Boc amino acids were coupled but not surprisingly there was extensive loss when Bz amino acids were used. *N*-Protected amino acid bromides have not previously been used for peptide synthesis because they are unstable and difficult to purify. They have now been obtained from *N*-protected amino acids using 1-bromo-*N*,*N*-2-trimethyl-1-propenylamine under mild and neutral conditions.[101] They are useful for coupling very hindered amino acids and are used in the presence of collidine. Carbodiimides have not been completely abandoned as coupling agents. Sparingly soluble protected peptides have been coupled in $CHCl_3/PhOH$ using $EtN=C=N(CH_2)_3NMe_2$ in the presence of 3,4-dihydro-3-hydroxy-4-oxo-1,2,3-benzotriazine (HOOBt) with no loss of chiral purity.[102] Similarly, Z-Gly-Phe-OH and H-Phe-OBzl were coupled in the same solvent but using $Pr^iN=C=NPr^i$ with a combination of HOOBt and its Bu_4N^+ salt as additive.[103] There has been another paper extolling the value of Cu(II) derivatives for retaining chiral purity during SPPS.[104] The use of the Cu(II) derivatives of HOBt and HOAt seem to be particularly effective. Another group has adduced further supporting evidence. For example, the use of HOBt and $CuCl_2$ minimises loss of chiral purity when PyBOP is the coupling agent.[105] Again, $CuCl_2$ and HOAt eliminates enantiomerisation in segment coupling reactions.[106] It really is quite surprising that this technique is not routinely used in the light of reports in recent years. It has been reported that HOBt, HOOBt and HOSu react with CH_2Cl_2 and $ClCH_2CH_2Cl$ to form respectively di- and mono-substituted derivatives.[107] In the light of the successful use of $CHCl_3/PhOH$ mentioned above, it would be useful to determine if $CHCl_3$ reacts with the above additives. Carpino has synthesised the 4,5- and 5,6-benzo derivatives of HOAt.[108] Both can be converted into coupling reagents with $(Me_2N)_2C^+ClPF_6^-$. The 4,5-derivative gives a uronium derivative (3) whereas the 5,6-derivative gives a guanidinium derivative (4). Although neither possesses any special advantage over HOAt, the former caused coupling faster than

(3)   (4)

the latter. Li and Xu have described a battery of onium-type peptide coupling agents.[109] Some of these (e.g. BOMI, BDMP, BEP, BEMT) have been reported by these workers before and have been mentioned in vols. 32 and 33 of this series of publications. Li and Xu have summarised their conclusions about the potential value of some of these and earlier reagents. Reagents based on HOBt and incorporating phosphonium and uronium structural components (e.g. BOP, HBTU) are regarded as satisfactory for syntheses involving coded amino acids. For segment coupling, however, where loss of chiral purity may occur, reagents based on HOBt and incorporating immonium structural components (e.g. BOMI, BDMP) would be a better choice, because of their ability to limit loss of chiral purity and on account of their reactivity. Although reagents based on HOAt are acknowledged as the strongest candidates for the synthesis of hindered peptides, Li and Xu cite the disadvantage of high expense of reagent. They suggest, however, that 2-halopyridinium (e.g. BEP,5) and 2-halothiazolium (e.g

(5)   (6)

BEMT,6) reagents are also suitable and can be synthesised from cheaper starting materials. The original paper should be consulted for a list of structures of other potential coupling reagents. Prop-2-ynyl triphenylphosphonium bromide is another onium-type coupling agent that has been used to synthesise a number of small peptides.[110] Microwave radiation has been used to promote the synthesis of Aib peptides using PyBOP or HBTU in CHONMe$_2$ as solvent.[111] A one-pot method for deprotecting Alloc amines and coupling to N-Boc or N-Fmoc amino acids uses Pd(PPh$_3$)$_4$ and 1,8-diazabicyclo[2,2,2]-cyclooctane (DABCO) as scavenger and a carbodiimide to effect coupling involves a reaction time of 10-20 min.[112] Although derivatives of 1,3,5-triazine have been used as coupling agents in the past,there has been a further development in this area.[113] When racemic N-protected amino acids were used in peptide synthesis in the presence of a chiral tertiary amine such as strychnine, brucine or sparteine and 2-chloro-4,6-dimethoxy-1,3,5-triazine, the coupling proceeded enantio-selectively. Kagan enantioselectivity parameters were derived and were in the range 1.6-195. Strychnine gave the best results ($s = 98$-195). The authors postulated that chiral triazinyl-ammonium chlorides were formed as intermediates. Another coupling method merits further assessment. Instead of using N-protected amino acids, α-azido acids, prepared by Wong's Cu^{2+}-catalysed diazo transfer method, were the starting material.[114] There was negligible loss of chiral purity and peptides that are prone to formation of diketopiperazines were obtained in good yield by a solid-phase method.

When synthesising peptides of Pro, the ratio of cis/trans conformations of the prolyl peptide bond is strongly influenced by the chirality of the acyl residue immediately preceding the Pro residue. Acyl moieties from (2S)-2,6-dimethyl-3-oxo-3,4-dihydro-2H-1,4-benzoxazine-2-carboxylic acid and (2R)-3-methoxy-2-

methyl-2-(4'-methyl-2'-nitrophenoxy)-3-oxopropionic acid in the acylproline moiety constrains the dihedral angle to the trans orientation.[115] These are promising chiral peptide mimetic units that strongly favour the trans structure. A special method is available for coupling a peptide with an *N*-terminal Ser residue, the amino group of which is unprotected.[116] Selective oxidation with $NaIO_4$ gave a α-*N*- glyoxylyl peptide and this was reductively aminated using 4,5-dimethoxy-2-tritylthiobenzylamine. After removal of the Trt group, the product was condensed with a peptide thioester. The remaining protecting group, 4,5-dimethoxy-2-mercaptobenzyl-, was removed with trifluoromethane-sulfonic acid. Note that the *N*-terminal Ser residue of the *C*-terminal fragment is converted into a Gly residue. Another special coupling method is exemplified by allowing the complex of Gly-Gly-OMe and $[(\eta^6\text{-}p\text{-cymol})RuCl_2]_2$ or $[(\eta^6\text{-}C_6Me_6)RuCl_2]_2$ to react with $NH_2(CH_2)_nCO_2Me$ to give the tripeptide complex.[117] A total synthesis of (-)-hemiasterlin has been described that used only the *N*-benzothiazole-2-sulfonyl (Bts) protecting group for the following reasons;- (a) Bts-Cl requires no additive for coupling, (b) purification of products was ex-pected to be easy and (c) Bts-amino acids should be easy to methylate and there are two *N*-methyl groups in the peptide.[118] During a reaction involving coupling with an oxazolone derivative and in the presence of an azacrown ether, the latter underwent ring opening (Scheme 5) and this led to the authors developing a

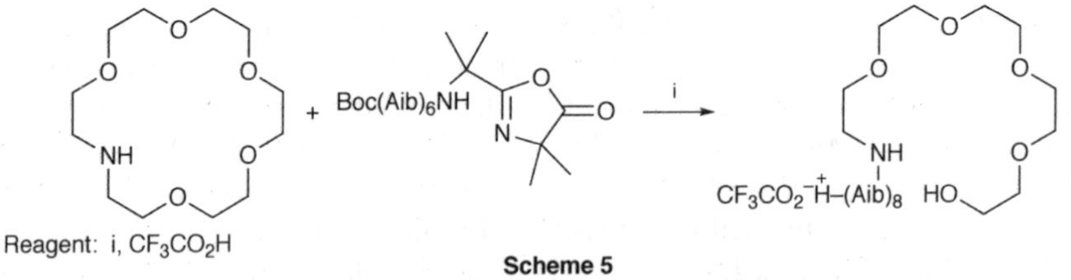

Reagent: i, CF_3CO_2H

Scheme 5

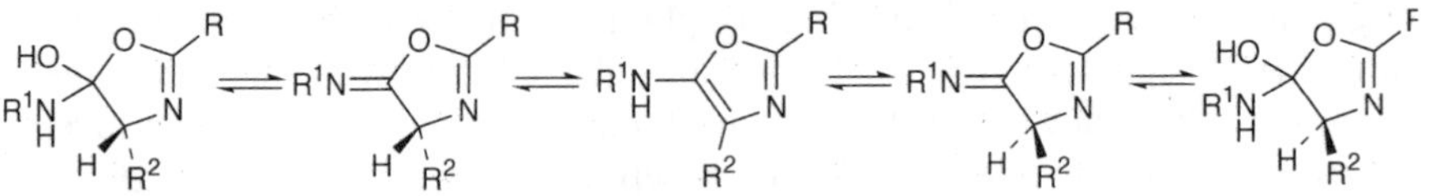

Scheme 6

theory (Scheme 6) for the loss of chiral purity of oxazolines during peptide synthesis.[119] During the synthesis of long peptides by fragment condensation changing the phase of synthesis from solid to liquid or vice versa was found to be beneficial in difficult couplings.[120] In the synthesis of monobiotinylated rat relaxin, Hmb protection of the B chain was found to prevent chain aggrega-tion.[121] In the synthesis of chaperonin 60.1 (195-217), ordinary SPPS was unsuc-cessful and replacement of one of the two Pro residues by $(\psi^{Me,Me})$Pro was also unsatisfactory. When both Pro residues were similarly replaced, synthesis pro-ceeded satisfactorily.[122] A different ploy has been introduced for synthesising long peptides.[123] Partial acetylation of amino groups in the resin decreased the capacity for interaction of chains and peptides containing >90 residues were satisfactorily assembled. Acylation with *N*-protected iminoacetic acid in solution can give mono- or di-substituted derivatives, but if the nucleophile is attached to an insoluble support, only monosubstituted products are formed.[124] Synthesis of amyloid peptides can be complicated by difficulty in removing Fmoc groups and a solution involves using diazabicyclo[5.4.0]-undec-7-ene (DBU)(2%w/v).[125] Because of the medical importance of amyloid proteins, there are further citations below (see refs. 139, 162, 163, 191, 210 and 513). Pyroglutamic acid can be readily introduced into a pseudopeptide as a replacement for Pro.[126] Benzyl (S)-pyroglutamate can be acylated with pentafluorophenyl esters in the presence of a strong base such as NaH or the Li derivative of hexamethyl-disilazide.

The method of native chemical ligation using peptide thioesters continues to attract much attention.[127,128] The alkane sulfonamide safety-catch resin is useful for the synthesis of *C*-terminal peptide thioesters and it has now been shown that addition of e.g. LiBr increases the swelling of the resin.[129] In addition, the yield is increased if cleavage of the peptide is carried out in 2M LiBr in tetrahydrofuran. The usual method for native chemical ligation involves a peptide thioester and a peptide possessing an *N*-terminal Cys residue. If desired, the latter can subsequently be converted into Ala by desulfurisation with Pd/Al_2O_3, $Pd/BaSO_4$ or Pd/C.[130] The method was illustrated by the synthesis of three proteins. If the desulfurisation was carried out in 6M guanidine at pH 7.5, a small amount of disulfide dimer was formed, but this complication can be avoided by working at pH 4.5. A variation uses a removable *N*-(1-phenyl-2-mercaptoethyl) auxiliary[131]

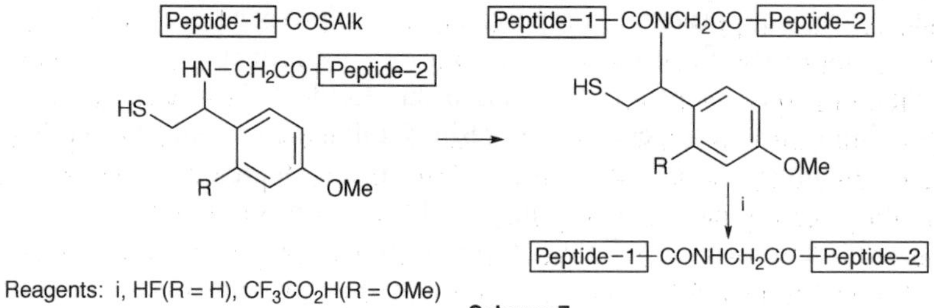

Reagents: i, HF(R = H), CF_3CO_2H(R = OMe)

Scheme 7

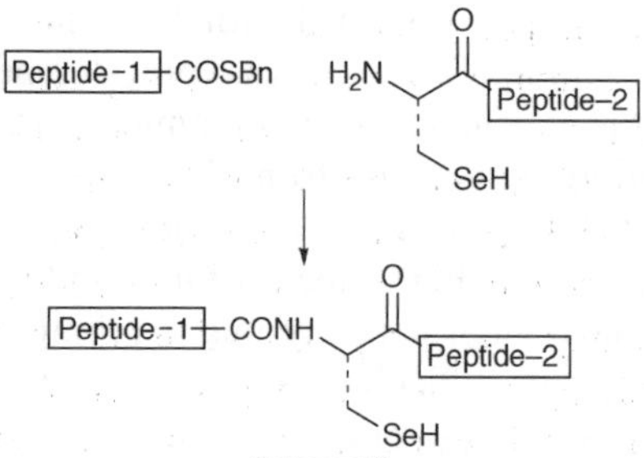

Scheme 8

(Scheme 7). This results in the replacement of what would have been Cys by Gly. Selenocysteine peptides can also participate in native chemical ligation reactions.[132,133] (Scheme 8) Glycopeptides can also be synthesised by native chemical ligation.[134,135] There is no doubt that this technique offers an important alternative to recombinant DNA methodology because any number of uncoded amino acids could be incorporated into appropriate synthetic targets.

2.6 Peptide Synthesis on Macromolecular Supports and Methods of Combinatorial Synthesis. – The chloromethyl groups of Merrifield resin can be converted into aminomethyl groups by reaction with a non-nucleophilic base such as lithium hexamethyldisilazide in high yield.[136] The use of polystyrene-based supports cross-linked with methylacrylate esters of diols, especially 1,4-butanediol, continues to be popular because the open structure markedly diminishes steric problems that can interfere with peptide syntheses. The latter employ well-known chemistry such as attachment of the *C*-terminal residue using the Cs salt[137] and *N*-protection with the Boc[138-140] or Fmoc[140] groups. As a token of the success possible with this kind of support, the synthesis of an extremely insoluble, hydrophobic antiparallel β-sheet peptide, LMVGGVVIA, the *C*-terminal fragment of β-amyloid peptide, is cited.[139] In a further development, this type of support has been fitted with a 2-nitrobenzyl group (7). The assembled peptide

(7)

can then be detached photolytically.[141] Several other new supports have been designed. A cheap support can be prepared from a cation-exchange resin bearing carboxy groups. Esterification of these followed by reduction to polymeric benzyl alcohols gives a product that has been tested in the synthesis of two peptides,[142] but the close structure of the resin is unlikely to permit the synthesis of long or difficult peptides. More promising is a development of the extended cross-linking moiety near the beginning of this section. Using *O,O'*-bis(acrylamidopropyl)-PEG as the crosslinker for polystyrene, a support was obtained on which the notorious (65-74) sequence of acyl carrier protein was successfully assembled.[143] It remains to be determined if this support offers any marked advantage over those supports obtained with 1,4-butanediol dimethylacrylate cross-linking. *N*-Fmoc aminoxy-2-chlorotrityl polystyrene has been simply prepared and used for the synthesis of hydroxamic acids.[144] The latter can be detached by mild acidolysis. Starting from Tris, an extended dendrimer was synthesised and repeatedly coupled to a core-shell type resin.[145] The dendrimer structure of the resin ensured a high capacity for peptide synthesis. This may or may not be an advantage; high capacity might promote physical interaction between chains resulting in difficult couplings. A new silyl linker has been designed for N→C assembly of peptides.[146] Chiral purity of product is at considerable risk when using this tactic and it is probably best avoided except under special circumstances e.g. when the residue at the *C*-terminus is itself the product of a long and/or difficult synthesis. Even then, the best ploy may be to synthesise most of the structure by the conventional C→N assembly and then to attach a small segment at the *C*-terminus by a method that precludes loss of chiral purity, e.g. enzyme-catalysed coupling, formation of a Gly-Xaa bond or native chemical ligation in one of its modes.

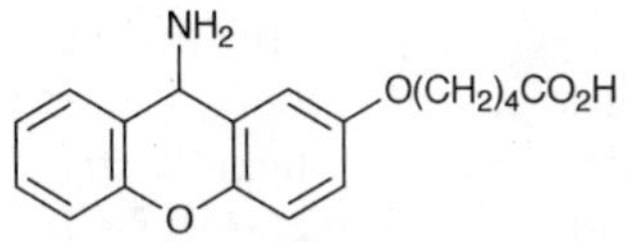

(8)

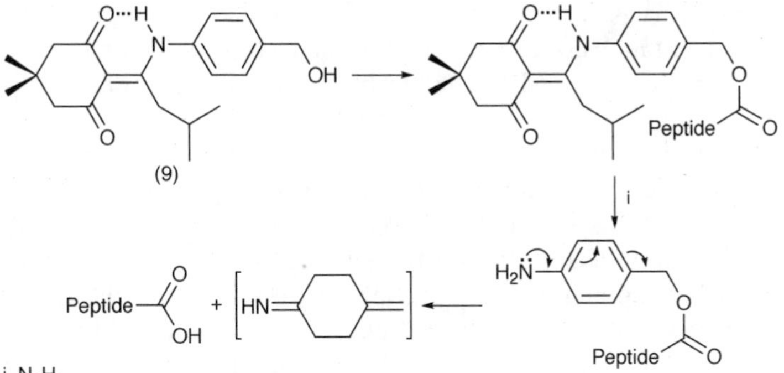

(9)

Reagent: i, N$_2$H$_4$

Scheme 9

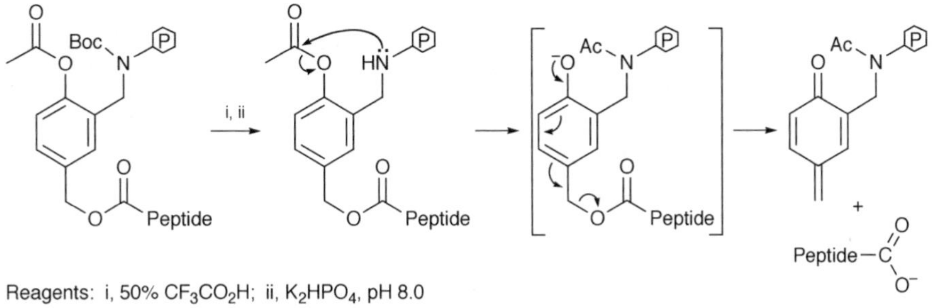

Reagents: i, 50% CF$_3$CO$_2$H; ii, K$_2$HPO$_4$, pH 8.0

Scheme 10

Several new linkers for the conventional C→N approach to SPPS have been described. 5-[(9-Aminoxanthen-2-yl)oxy]valeric acid (8) is an acid-labile linker. It has been used to synthesise CCK-8 sulfate and a phosphorylated fragment of the T-cell receptor complex. The products were detached with 50% CH$_3$CO$_2$H in CH$_2$Cl$_2$ in the presence of scavengers in 15 min. at room temperature.[147] A useful linker based on the Dde-derived protecting group (9) is stable to acid and base conditions but releases a synthetic peptide attached to the benzylic hydroxy group on treatment with hydrazine (Scheme 9)[148] Another linker that relies on a 1,6-elimination process to release the product is depicted in Scheme 10. In this case, the quinone moiety is left attached to the resin[149] Note that the N-Boc group acts as a safety catch. This method was used in the synthesis of the antibacterial squalamine. Another safety-catch method uses Boc chemistry to assemble libraries of cyclic peptides.[150] The C-terminal residue is attached to the resin as an ester and when the peptide has reached the appropriate length, removal of the Boc group causes the intramolecular formation of an amide group. The use of dimethyl 2,3-O-isopropylidene-D-tartrate attached to a peptide, after removal of the isopropylidene group and oxidation with periodate, can give rise to α-oxoaldehydes as potential proteinase inhibitors.[151] A backbone amide linker has been designed in order to construct an addition at the *C*-terminus.[152] It may prove to be a more flexible approach to attach a Lys residue at the ε-amino group and with protection of either the α-amino or carboxy group, further assembly could continue at either end. The 4-alkoxy-2-hy-droxybenzaldehyde linker has been used to synthesise a dimeric peptide via a N_2H_4 moiety.[153] A new avenue in SPPS has opened, is presently free from congestion, but may not retain this freedom from competition much longer. The conventional route for SPPS has hitherto involved immobilising the *N*-pro-tected *C*-terminal residue on a support by forming a σ-bond using a reactive derivative of the *C*-terminal carboxy group. *N*-Deprotection and coupling steps follow alternately until the peptide is detached from the support. The new approach[154] involves π-bonding between unsaturated moieties in substrate and support through an appropriate transition element. A convenient support is polystyrene containing triphenylphosphine groups. Hexacarbonyl-chromium(0) was complexed with an unsaturated group in the amino acid (e.g. Phe) to give (10) as an example. Reaction with polymer-bound phosphine under irradiation then afforded (11). Deprotection of the amino group and peptide coupling then

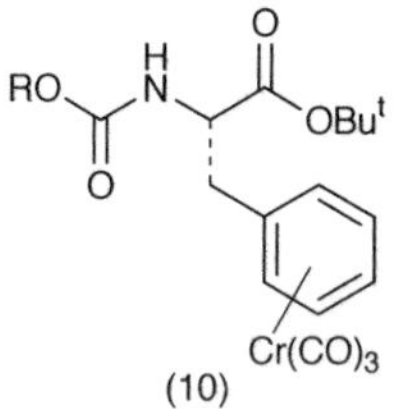

(10) 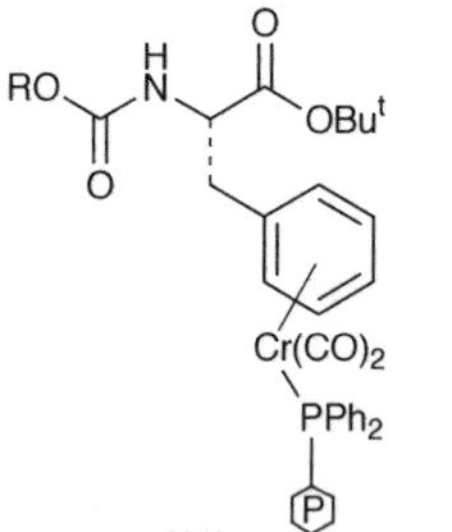   (11)

followed uneventfully and the protected peptide was detached from the support by aerial oxidation. Much remains to be done to test and assess the method as a possible viable alternative to conventional SPPS. For example, *N*-protection of the amino acids with aromatic groups (e.g.Fmoc) was shown to give rise to complexation at one of two sites. Again, the use of a polystyrene support instead of polyacrylamide might give rise to problems of adsorption of products. De-tachment of product by aerial oxidation could also cause difficulties with pep-tides containing Cys and Met. In the event that the use of enzymes increases in peptide synthesis, it would be essential to discover if the enzymes would be irreversibly adsorbed or even inactivated by the use of transition elements. Nevertheless, this new branch of peptide chemistry is to be welcomed. It might be useful in combinatorial syntheses. Even if it proves essential to have an aromatic amino acid at the *C*-terminus, this could be removed at the end of the synthesis by exposure to carboxypeptidase A.

This section concludes with a miscellany of preparative and analytical methods pertaining to SPPS. A detailed study of the cleavage of peptides from resins using $CF_3SO_3H/CF_3CO_2H/PhSMe$ has been made.[155] Not surprisingly, the ease of product detachment depends on several factors including the type of support used, the size and composition of the peptide and temperature. PAM resin and its attached peptide generally part company most readily. [Gly^8]-Angiotensin II was much more easily detached from any resin examined than was the [Phe^8]-derivative. All this work was carried out using Boc chemistry. Polymer-bound *N*-hydroxysuccinimide has been used as an additive for carbodiimide-mediated peptide synthesis.[156] For the synthesis of *C*-terminal thioesters of peptides, Fmoc chemistry can be used and the peptide can be detached using AlMe$_3$, the latter improvement leading to diminished production of by-products.[157] The *N*-terminal fragment (1-37) of bovine pancreatic trypsin inhibitor was assembled and this was used to synthesise the complete protein by native chemical ligation. Analytical construct technology has been applied to the single-bead analysis of a split-mix combinatorial library.[158] Cleavage at orthogonal sites permits biological assay of one fragment and electrospray mass spectrometry of an alternative fragment. Micro-reactors have been designed to permit SPPS to be carried out in a 4-compartment apparatus.[159] The four reservoirs are separated by microfrits. The first contains Fmoc amino acid, the second a carbodiimide coupling reagent, the third the amino component and the fourth CHONMe$_2$ to collect the product. Each reservoir contains a Pt electrode and an external voltage (usually 700v) is applied to induce electroosmotic flow of reagents. The progress of peptide assembly in SPPS can be monitored by time of flight static secondary ion mass spectrometry on individual beads.[160] If the linkage of the peptide to the resin is through an ester or amide bond, this is orthogonally cleaved in the bombardment and the main peptide structure survives for identification. The technique is sensitive at the femtomole level. A quite different method for monitoring continuous-flow SPPS involves measurement of pressure changes.[161] The latter are monitored with a resistance strain gauge attached to the inlet. This technique can detect structural changes in the peptide resin conjugate such as those that occur during aggregation.

Amylin is a 37-residue peptide formed in type 2 diabetes and it displays some homology with the insoluble peptide that is formed in Alzheimer's disease. Amylin deposits on the β-cells of the pancreas causing apoptosis. Two fragments of amylin (17-37) and (24-37) circulate in patients with type 2 diabetes and both are amyloidogenic. Both fragments have been synthesised[162] and pose problems of aggregation and insolubility similar to those found with ACP(65-74). 1,1,1,3,3,3-Hexafluoropropanol was found to be the best solvent. Di-β-peptoids have been synthesised using Fmoc chemistry.[163] *N*-Alkyl-β-amino acids can be used and these offer scope for exercises in the synthesis of insoluble peptides. An abbreviated prion protein has been synthesised.[164] Difficult sequences were synthesised using Boc chemistry and HATU as the coupling agent. Presumably, if the complete protein is to be made, native chemical ligation or enzyme-catalysed fragment condensatiion would be suitable routes avoiding undesirable decreases in solubility due to the use of hydrophobic protecting groups, a problem that is likely to occur ever more frequently.

2.7 Enzyme-mediated synthesis and Semisynthesis. – Esters of amino acids and peptide derivatives have long been accessible by enzyme catalysis. Aminoacyl derivatives of glycerol can be obtained using a variety of proteinases and lipases in glycerol containing 0-10% v/v of aqueous buffer at 50°C.[165] There was no regiospecificity displayed towards the hydroxy groups of glycerol, but with Asp and Glu derivatives, the α-carboxy group of the amino acid is selectively esterified. α-Methylglucose undergoes regioselective thermolysin-catalysed transesterification with vinyl esters of *N*-protected amino acids (Phe is best) to give esters at the 6-hydroxy group.[166] *N*-Protected amino acids or peptides can be esterified with αω-dihydroxyalkanes in presence of subtilisin and the product can undergo aminolysis with peptide alcohols using either papain or genetically modified subtilisin ('subtiligase').[167] Although no reaction occurred between IgG and cysteine in the presence of carboxypeptidase Y, when cysteine methyl ester and IgG methyl ester were used as substrates, the former was attached at the *C*-terminus of the protein.[168] Moreover, when cysteine isobutyl ester was used, both light and heavy chains of IgG were labelled. The use of V8 proteinase from *S. aureus* catalysed reaction at the β-carboxy group of Z-Asp or the γ-carboxy group of Z-Glu.[169]

The use of organic solvents in enzyme-catalysed peptide synthesis continues to attract considerable interest. Esters of Z-Aspartame were synthesised in almost quantitative yield in toluene at controlled activity of water using thermolysin adsorbed on Celite R-640.[170] It was suggested that the porous support controls the hydration of the enzyme. Isolated yields were >90% and no purification was required if equimolar amounts of substrates were used. The synthesis of Z-Ala-Ala-Leu-Phe-NHC$_6$H$_4$NO$_2$ from Z-Ala-Ala-Leu-OMe and H-Phe-NHC$_6$H$_4$NO$_2$ catalysed by subtilisin in ternary mixtures of CHONMe$_2$, MeCN and water has been studied in some detail.[171,172] Yields peaked at 95% when the concentration of CHONMe$_2$ was 60%v/v. In view of the work of Halling *et al.* reported below, the high yields could be attributed to the low solubility of the product. Similar results were obtained in ethanol. Longer peptides have been synthesised by attaching the *C*-terminal tripeptide to an insoluble support and using subtilisin in the presence of SDS to effect coupling.[173] Again, the reaction is presumably driven by the disappearance of the soluble acylating peptide from solution. A proteinase from *Clostridium thermo-hydrosulfuricam* has been used to catalyse the coupling of *N*-protected amino acids (especially Asp) with amino acid esters at pH 6.5 and 45°C in a solution saturated with EtOAc. Yields were only modest and may be attributed to the solubility of the product.[174] Z-Ala-Phe-OMe has been synthesised in a 2-phase membrane reactor and the transfer of product to the organic phase probably explains the high yields obtained.[175] Another example has been reported of the preferential formation of peptide containing D-amino acids using inverse substrates such as 4-amidinophenyl and 3- and 4-guanidinomethyl esters with trypsin as the catalyst.[176] The products, after removing basic moieties released during coupling, were resistant to hydrolysis by trypsin.

Early reports from about 1993 of the use of solid-to-solid enzymatic syntheses of peptides, in which hydrated salts provided the small amount of water for reaction, did not precipitate a massed foray into this field, but occasional applications have been reported since then. For example, some small peptides have been made recently by this method.[177] This situation is likely to change very significantly as a result of two papers published by Halling *et al.*[178,179] Practical studies on the choice of added salt, substrates, enzymes and physical conditions have been supplemented by considering factors that favour the reaction proceeding in a forward direction as a result of precipitation of the product. Only a brief account of these papers can be given here and a more detailed study is strongly recommended. The thermolysin-catalysed synthesis of Aspartame is greatly accelerated by the addition of $KHCO_3$ or K_2CO_3 and less so by addition of the corresponding Na salts. These effects are attributed to adjustment of acid-base equilibria, but the liquid phase was trapped between reactants and product and pH measurements were not possible. A theoretical treatment of the effects of reaction conditions that favour precipitation of product is advanced. It requires a comparison of the equilibrium constant, K_{eq}, and the saturated mass action ratio, z_{sat}. The latter can be predicted from a knowledge of the mps. of reactants and product and, in some cases, the pK_a values of ionisable groups. One feature of solid-to-solid systems is the high substrate concentration that is operative. High substrate concentration is also possible in systems that contain much more water. The coupling of Ac-Phe-OEt and H-Leu-NH_2 under chymotryptic catalysis has been intensively studied.[180] Water-in-oil emulsions were used as reaction media. The reaction performance was independent of the emulsion system and yields of 90-94% were obtained at high substrate concentration. α-Chymotrypsin showed superactivity in systems containing non-ionic detergents (0.2 − 0.8% w/v). High yields of Z-Tyr-Arg-NH_2, a kyotorphin precursor, were obtained with α-chymotrypsin as catalyst in eutectic mixtures of substrate and solvent.[181] Finally, in the study of medium effects on peptide synthesis, reactions catalysed by bovine trypsin and chymotrypsin were studied in aqueous ethanol, dioxan and MeCN.[182] The K_m decreased as the fraction of organic solvent was increased, but then increased as high solvent concentrations were used. Both enzymes had good stability in the presence of organic solvents.

The use of mutant forms of proteolytic enzymes has advanced to only a disappointing extent. A deletion mutant of tyrocidine synthetase comprising the adenylation domain as an independent unit that catalyses adenylate formation was used to investigate its possible ability to catalyse peptide bond formation. Several Phe dipeptides were synthesised.[183] Mutants of *B. lentus* subtilisin were earlier shown to catalyse the formation of both D- and L-peptides. It has now been demonstrated that these mutant enzymes catalyse syntheses from non-coded amino acids such as β-alanine and the β-homologue of phenylalanine.[184] The D189E mutant of trypsin catalysed syntheses using mimetics of trypsin substrates similar to the work described above,[176] but the mutant enzyme permitted the use of nucleophilic acyl acceptors such as derivatives of Lys or Arg without causing product hydrolysis.[185] For example, coupling of Bz-Gln-OGp and H-Ala-Ala-Arg-Ala-Gly-OH gave 69% of the expected hexapeptide. The site-selective glycosylation of residue 166, a component of the primary specificity pocket of *B.lentus* subtilisin, gave a mutant enzyme that is capable of catalysing the synthesis of both L- and D-peptides.[186]

Enzymes other than proteinases are useful in amino acid and peptide chemistry. Lipases from *Candida antarctica*, *Thermomyces lanugenosus* and *Pseudomonas alcaligenes* catalyse the enantio-selective ammonolysis of aminoacid esters.[187] This selectivity could be enhanced by lowering the temperature of reaction. Esters of Boc amino acid esters gave products of almost total chiral purity. Enantiomerically enriched D- and L-silylated alanines were obtained by deracemisation of DL-silylmethylated hydantoins.[188]

There is one example of the use of an abzyme, a disappointing outcome for a technique that looked so promising. *N*-Z-4-Phosphotyrosine as a hapten was conjugated on to bovine serum albumin. This was used to immunise mice and fusion of the spleen cells produced hybridomas that catalysed the hydrolysis of *O*-benzoyl-L-tyrosine derivatives in a reaction[189] that was acccelerated by a factor of 10^4.

As a final example of the use of enzymes in peptide synthesis, the chemoenzymatic attachment of mannose units to the α- and ε-amino groups is cited. 4-Isothiocyanato-β-D-gluco-pyranoside is linked to the reducing end of an oligosaccharide rich in mannose by the transglycosylation activity of endo-β-*N*-acetylglucosaminidase A.[190] The product can then be coupled via the isothiocyanato group to free amino groups in nonglycosylated proteins.

2.8   Miscellaneous Reactions Related to Peptide Synthesis. – Fluorescent labelling of peptides and proteins is an important technique. Fluorescein isothiocyanate has been used many times and it has recently been selected again for labelling peptides in solution.[191] With β-amyloid peptides, some of which tend to aggregate, a physical property that accounts for neuronal death in Alzheimer's disease and of pancreatic β-cells in type 2 diabetes, fluorescent labelling is carried out in a high concentration of Me_2SO to prevent this behaviour. 5(6)-Carboxy-fluorescein or 7-amino-4-methyl-3-coumarinylacetic acid were used. The introduction of guanidino groups at ω-amino groups has also received considerable attention before. A new reagent (12) has been designed;[192] it reacts quickly in tetrahydrofuran at room temperature. Conjugation of unprotected peptide to dauno- and doxo-rubicin has been effected (a) by coupling $BocNHOCH_2CO_2H$ to the *N*-terminus of the peptide, and (b) introducing a $>C=O$ into the drug.[193] After removal of the Boc group from the modified peptide, oxime formation occurs between thw $>C=O$ group in the drug and the peptide derivative. A new method for modification of Phe peptides in order to assemble combinatorial libraries has been proposed.[194] A Suzuki-Miyaura coupling reaction is carried out using a 4-iodo-Phe derivative and an aryl or hetero-cyclic boronic acid (Scheme 11). Michael addition of thiols, carbon nucleophile and amines to dehydropeptide derivatives is a suitable method for introducing e.g. an alkylthio group at the β-carbon atom of the unsaturated amino acid.[195] Amide formation with an α-hydroxy-β-aminoacid can give rise to a homobislactone (13) as a byproduct.[196] β-Lactams can be prepared by SPPS using a β-hydroxyacid and resin-bound hydroxylamine. Ring closure is effected under Mitsunobu conditions (Scheme 12). Reductive cleavage from the resin is achieved with SmI_2.[197] Peptide conjugates bearing thiofarnesyl or thioglycoside substituents can be

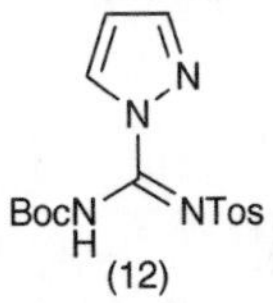

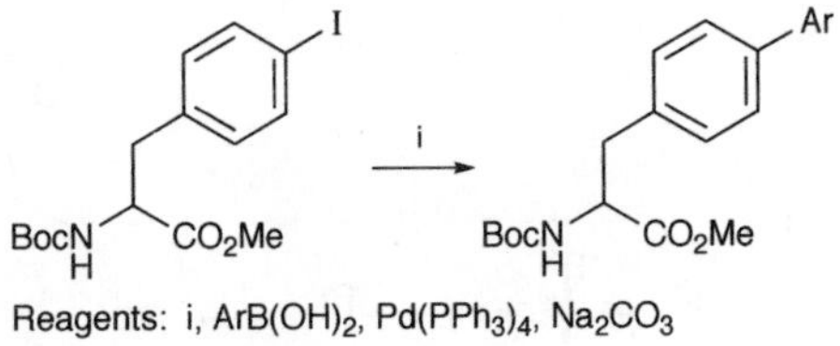

Reagents: i, ArB(OH)$_2$, Pd(PPh$_3$)$_4$, Na$_2$CO$_3$

Scheme 11

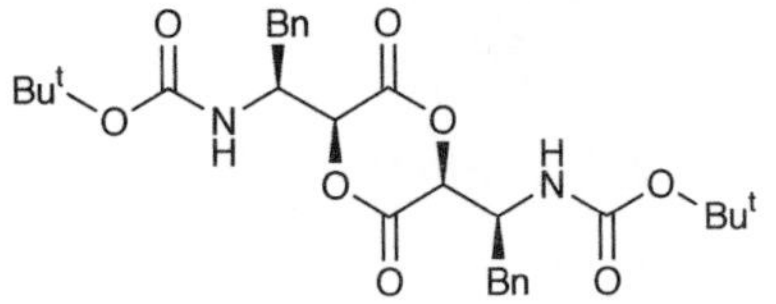

(13)

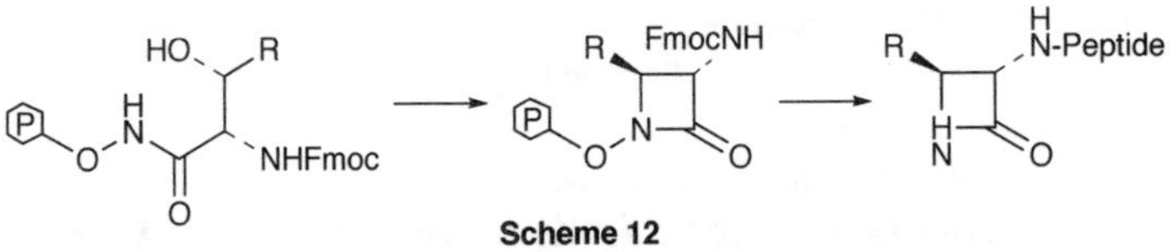

Scheme 12

obtained by first synthesising a peptide containing a dehydroalanine residue. Addition of the desired thiol on the alkene side chain produces the desired product.[198]. Using a small peptide containing a Gly residue, a xanthyl group can be introduced on the latter by irradiation of the peptide and *N*-bromosuccinimide with a 150w uv lamp, adding potassium xanthate and allowing to stand for several hours. The xanthate product could be subjected to a free radical reaction with e.g. allyl ethyl sulfone in the presence of dilauryl peroxide. The product is a peptide in which the Gly residue has been converted into a peptide containing 2-allylglycine.[199] α-Aminoketones can be obtained from *N*-protected amino acid chlorides by the Friedel-Craft reaction and these can be converted into peptidyl ketones by conventional peptide synthesis.[200] There has been a warning[201] about the possible occurrence of a side reaction when 2-amino-thiolane is used to cross link peptides. One of the products (14) is stable at acid pH but cyclises to (15)

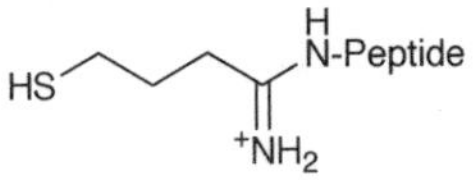

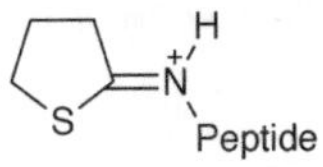

under basic conditions. The existence of the thiol (14) was demonstrated by reaction with *N*-ethylmaleimide. The feasibility of using the Diels-Alder reaction for covalently linking two peptides has been demonstrated[202] by the sequence of reactions outlined in Scheme 13. Presumably, it would be possible to synthesise cyclic peptides by attaching a diene at one end of a peptide and a dienophile at the other end. Intermolecular reaction might possibly be prevented by having the diene as a detachable linker on an insoluble support. There appears to be considerable possible mileage as an attractive journey to be made. A new amino

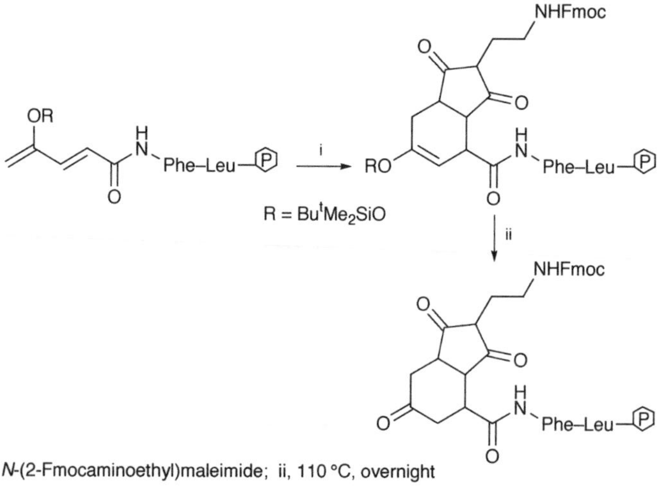

Reagents: i, *N*-(2-Fmocaminoethyl)maleimide; ii, 110 °C, overnight

Scheme 13

acid derivative with a diol side chain, L-2-amino-4,5-dihydroxypentanoic acid, has been prepared from L-allylglycine as the Fmoc derivative.[203] After synthesis of a peptide from this compound, the vicinal diol moiety can be oxidised with NaIO$_4$. Conversion into the corresponding oxime offers the possibility of producing new inhibitors of metalloproteinases. Peptides containing from 17 to 33 residues and terminating with a 4-iodobenzoyl moiety have been coupled to a trialkyne nucleus under Pd(0) catalysis in aqueous solution at pH values ranging from 5.0-7.5.[204] It would be interesting to know if all three peptide chains have to be identical or if there can be two or three different peptide chains. Finally, there is a report of an unexpected isomerisation (Scheme 14) of thioanilides of peptides with Aib at the *C*-terminus.[205] This reaction proceeds with retention of chiral purity.

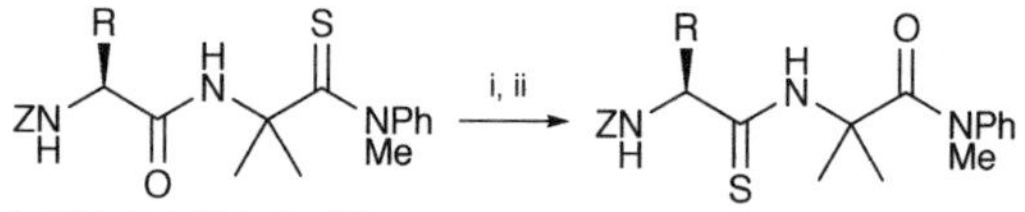

Reagents: i, ZnCl$_2$ in AcOH; ii, HCl in AcOH

Scheme 14

3   Appendix: A List of Syntheses in 2001

The syntheses in Section 3.1 are listed under the name of the peptide/protein to which they relate, but no arrangement is attempted under the subheading. In some cases, closely related peptides are listed together.

Peptide/Protein   *Ref.*

3.1 Natural Peptides, Proteins and Partial Sequences. –
Activin
    Analogues of human activin β_A (12-116) by SPPS   206
Adaptor protein Grb2
    Inhibitors of Grb2-SH2 domain   207
Albumin
    Analogues of *N*-terminus of bovine serum albumin and coordination
        with Cu(II) ions   208
Allium chemistry
    Se-alk(en)ylselenocysteines and γ-glutamyl derivatives   209
Amyloid proteins
    β-Amyloid precursor peptide and presenilin segments   210
    Simulation of the lipophilic and antigenic sites of Aβ(1-42) peptide
        of Alzheimer's disease   211
Angiotensin
    Analogues of angiotensin II   212,213
Anorectic peptide
    Cocaine- and amphetamine-regulated transcript (CART) (55-102)   214,215
Antibiotics
    Gramicidin S analogues   216-218
    Model of binding site of vancomycin   219
    Solid- and solution-phase synthesis of vancomycin   220
    Symmetrical dimeric form of temporin A   221
    Conjugate of temporin A and cecropin A   222
    Analogues of trichogin GAIV   223
    Antifungal analogue of echinocandin-like lipopeptide   224
    Antifungal macrocyclic lipopeptidolactones   225
    Antifungal analogues of pseudomycin   226
    Library of nikkomycin analogues   227
    Fragments of *Raphanus sativus* antifungal protein 2   228
    A 15-residue fragment of lactoferricin   229
    Cholesterol conjugates of analogues of distamycin   230
    Apicidin, an antimalarial fungal metabolite   231
    All-D-stereoisomer of leucocin A   232
    Analogues of mureidomycin   233
    Fragments of esculentin-1   234
    Peptaibolin and analogues   235
    Magainin 2 derivatives   236
    Antiviral activity of peptides of unnatural amino acids   237
    Peptides from subunit A of *E.coli* DNA gyrase   238
ATP synthase
    Subunit c of F_1F_2 ATP synthase   239
Bacterial peptides
    Pheromone of *Enterococcus faecalis* inducing gelatinase biosynthesis 240

Bombesin
    ^{99}Tc-chelated to a peptide derived from bombesin   241
Bradykinin
    Analogues   242,243
    Cyclic analogues   244
    Analogues containing 1-cycloalkane-1-carboxylic acid   245
Calcitonin
    Salmon calcitonin I   246
    Analogue of eel calcitonin (1-9)   247
Casein
    Analogues of human β-casein fragment (54-59)   248
Cecropin
    All-D-cecropin B   249
Chemotactic peptides
    fMLP analogues   250-252
    MIFL derivatives as partial agonists of human neutrophil
        formylpeptide receptors   253
    Chemotactic peptide containing a 1,2-dithiolane ring   254
Cholecystokinin and gastrin
    Fragment (26-33) analogues of CCK   255
    Porphyrin-CCK8 conjugate   256
    CCK receptor antagonists   257
    Human big gastrin II and an analogue   258
Collagen
    Collagen model containing Fe^{II}(bipyridine)$_3$ complex   259
    Collagen models labelled with ^{13}C and ^{15}N   260
    Collagen models having a pyrene group at the *N*-terminus   261
Didemnins
    Structure-activity relationships of didemnin derivatives   262
Elastin
    Crosslinked poly(KGGVG)   263
Endothelin
    Endothelin receptor antagonists   264-266
Enolase
    Peptides of small domain of *P. falciparum* enolase   267,268
Fertilin
    Dimyristoylated peptides as polyvalent fertilin mimics   269
Fibrinogen
    Analogues of γ-fragment containing RGD moiety   270
Fungal pathogens
    Alternariolide, phytotoxin affecting apple trees   271
Galanin
    Analogue of galanin (1-19)   272
Gla proteins
    Human matrix Gla protein   273
Glucagon

Cyclic analogue containing 24 amino-acid residues 274
GnRH/LHRH
    Long-acting analogue 275
    Antagonists 276
    Conjugate of [D-Lys^6]GnRH and emodin 277
Growth-hormone releasing factor
    Analogues of human GHRF 278
Heat-shock proteins
    Cyclic analogue of fragment of human protein 70 279
Immunosuppressants
    Total synthesis of stevastelin B 280
Insect peptides
    Analogues of cockroach tachykinin; SARP studies 281
    Neuropeptides containing phenylglycine derivatives 282
    Proctolin analogues containing phenylglycine 283
    Proctolin analogues containing oxazole ring 284
    Pheromone biosynthesis activating neuropeptide 285
    Analogues of oostatic peptide of *Neobellieria bullata* 286
Insulin
    The insulin-like 4 gene product 287
    2 Photoprobes of insulin-like factor (IGF) selective for photoaffinity
      labelling of proteins binding IGF 288
    B_{19}-Gly-B_{20} human insulin 289
    Rat insulin 3 and its biological activity 290
Integrins
    Analogue of a potent inhibitor of integrin $\alpha_4\beta_1$ 291
    Sulfonylated dipeptides as inhibitors of integrin VLA-4 292
    Integrin ligand conjugates 293
Ion-channel peptides
    Sequence dependence of peptides containing Aib 294
    N-Terminal modification of channel-forming peptide 295
    Antagonists of ATP-induced K^+-ion efflux from cells 296
Laminin
    Hydrophobic laminin-related peptides 297
    Peptide-PEG hybrids related to laminin 298
Lysozyme
    Conformationally restricted analogues of HEL(52-61) 299
Marine organism peptides
    Minalemine A from *Didemnum rodriguesi* 300
    SPPS of trunkamide A 301
Melanotropins
    Active cyclic peptides lacking a His residue 302
Motilin
    Cyclic peptides related to the *N*-terminus 303
Moult-inhibiting hormone (MIH)
    MIH of American crayfish (*Procambarus clarkii*) 304

Neuropeptides
    Antagonist of molluscan neuropeptide APGW 305
    Neuropeptide Y (NPY) receptor antagonists 306
    Tc-labelled NPY analogues for tumour imaging 307
Opioids, antinociceptive peptides and receptors
    SPPS of deltorphin 308
    Deltorphin analogues 309,310
    Glycopeptide derivative of enkephalin 311
    Enkephalin analogues 312-317
    Analogues of endomorphin 318,319
    Incorporation of (2S)-2-methyl-3-(2′,6′-dimethyl-4′-hydroxyphenyl)
        propionic acid into an opioid peptide 320
    Dynorphin A analogues 321
    Specifically tritiated endorphins 322
    Potent δ-opioid receptor antagonist 323
    Endomorphin-1 324
    Nociceptin/orphanin FQ receptor antagonist 325,326
Orexin
    Truncated analogues of orexin-A 327
Phytochelatins
    Cysteine-rich peptides as analogues of phytochelatins 328
Posterior pituitary hormones
    C-terminal tetrapeptide of oxytocin 329,330
    Analogues of oxytocin and D-Har-vasopressin 331
    Oxytocin analogues modified at position 2 332
    Oxytocin antagonists 333-335
    Conjugates of oxytocin analogues and penicillamine 336
    SPPS of tocinoic acid derivatives 337
RGD peptides
    Peptides containing the RGDS sequence 338
    RGD scaffolds and binding to integrin receptor 339
    Bivalent PEG hybrid containing RGD and PHSRN 340
    Peptides related to fibronectin fragments 341,342
    RGD analogue containing carnosine 343
    Arg-Gly-Asp-β-Ala-His 344
    126 Glycoside mimetics based on the RGD sequence 345
    Cyclic RGD analogue containing (E)-alkene dipeptide isostere 346
Sleep-inducing peptide
    ^{3}H-Labelled δ-sleep-inducing peptide 347
Somatostatin
    Heptapeptide analogues with *C*-terminal modifications 348
    Octapeptide analogues 349
    Receptor antagonists 350,351
    Cyclic peptide library of somatostatin analogues 352
    Analogue labelled with ^{99m}Tc 353
    Analogues that induce apoptosis in various tumour cells 354

Substance P
    Substance P receptor antagonists   355
    Conjugates with litorin   356
    Substance P bearing an oligosaccharide on Gln residue   357
    1,1'-Ferrocenophane lactam mimetic of substance P   358
Tachykinins
    [Aib^{16}]Scyliorhinin II and [Sar^{16}]scyliorhinin   359
Thrombin
    Analogues of SFLLR motif of thrombin receptor   360
    Thrombin receptor antagonists   361
    Analogues of the peptide tethered to thrombin receptor   362
Thymosin
    Thymosin α_1 363
Thyroid proteins
    Transthyretin via regioselective chemical ligation   364
Thyrotropin
    TRH analogues   365
Toxic peptides
    α-Conotoxins and analogues   366
    μ-Conotoxin analogues   367
    Peptides having high affinity for α-bungarotoxin   368
    Spider toxin from *Nephila madagascariencis*   369
Tuftsin
    Tuftsin receptor-binding peptide labelled with Tc(V)   370
Viral proteins
    The Tat(49-57) domain of HIV-1 protein and analogues   371
    Epitopic peptides from protein p24 of HIV-1   372
    Envelope protein fragments of human hepatitis B virus   373,374

3.2   Seqential Oligo-and Poly-peptides. –
    Anti-HIV activity of *N*- and *O*-sulfated Tyr oligomers   375,376
    Heparin-like activity of homopoly(*N*-acroyl)amino acids   377
    Conformation of peptides with repeated Xaa-Pro sequences   378
    Photoresponsive elastin-like polymer   379
    Elastomeric polypentapeptides   380
    Optically active poly(β-peptides)   381
    Poly(γ-benzyl-α-L-glutamate) of defined length 382
    Surface grafting of poly(L-glutamates)   383,384
    Poly(Asp-co-Lys) by thermal condensation   385
    α-*N*-Substituted hydrazinoacetic acid oligomers   386
    Polymer containing sulfonylated pyrrolidine residues   387
    Copolymer of Ala and 3-hydroxybutanoic acid   388
    Copolymer of L-Asp and L-Glu   389

3.3   Enzyme Substrates and Inhibitors. –
    Efficient synthesis of peptide 4-nitroanilides   390

Peptide α-keto-β-aldehyde inhibitors of Ser proteinases   391
SPPS of peptide aldehydes on acetal resin   392
Inhibitors of human neutrophil elastase   393
Epoxypeptidomimetics as inhibitors of thiol proteinases   394
Analogue of Bowman-Birk proteinase inhibitor   395
Bicyclic peptide inhibitor of trypsin from sunflower   396
Fragments of *Cucurbita maxima* trypsin inhibitor III   397
Eglin C mutants as inhibitors of kexin and furin   398
Fluorescent farnesylated *Ras* heptapeptide   399
Ras Farnesyltransferase inhibitors   400
Tripeptide derivative containing Phe-Arg isostere   401
S-Aminoacyl-*N*-acetylcysteamines thioesters are substrates for
  nonribosomal peptide synthases   402
Inhibition of human cytomegalovirus proteinase   403
Inhibitors of matrix metalloproteinases   404-406
Tyropeptins A and B, new proteasome inhibitors   407
Inhibitors of catalytic β-subunits of proteasomes   408
Tripeptide of N^G-nitroarginine as inhibitor of nitric oxide synthase   409
Dipeptide photoisomeric inhibitor of chymotrypsin   410
Benzamidine carboxamides containing phenylglycine as inhibitors of
  factor Xa   411
Potent inhibitors of factor Xa   412
Peptidyl phosphonate diphenyl ester inhibitors of serine proteinases
  from peptide amides   413
Fluorogenic substrates of cathepsin B   414
Tetrapeptide 4-nitroanilides containing dehydroamino acids and
  behaviour towards cathepsin C   415
New cathepsin k inhibitors   416
N-Nitrosoanilines as inhibitors of caspase-3 peptidases   417
α-Ketohydroxamate inhibitors of human calpain I   418
1,2-Benzothiazine-1,1-dioxide peptide inhibitors of calpain I   419
Chromogenic substrates of Glu,Asp proteinases   420
2-Substituted statines as Asp proteinase inhibitors   421
Unsymmetrical peptidyl ureas inhibitors of aspartic proteinases   422
Pepstatin analogue with 2 trifluoromethyl hydroxymethylene
  isosteres   423
Computer-designed inhibitors of aspartic proteinases   424
Inhibitors of tyrosine tRNA synthase   425,426
Proteinase substrate specificity using membrane-bound peptides   427
Long-lasting prodrugs that inhibit enkephalin-degrading enzymes   428
Thrombin inhibitors   429,430
Inhibitors of cysteine proteinase from hepatitis A   431
Inhibitors of proteinase from hepatitis C virus   432
Peptidyl diazomethylketones as cysteine proteinase inhibitors   433
Inhibitors of recombinant cysteine proteinase from *Leishmania*
  mexicana   434

Specificity of cysteine proteinase of *L. mexicana*   435
Cyclic alkoxyketones as inhibitors of cathepsin K   436,437
Inhibitors of HIV proteinase   438-441
Conjugate of HIV proteinase inhibitor and nucleoside reverse
    transcriptase inhibitor   442
Inhibitors of HIV proteinase containing pyrrolidin-3-ol or
    pyrrolidin-3-one moieties   443
Inhibitors of HIV proteinase containing alophenylnorstatine   444
Water-soluble prodrugs of HIV proteinase inhibitors   445
Inhibitors of Zn β-lactamase of *B. cereus*   446
Inhibitors of trypanothione reductase   447,448
Inhibitors of guinea-pig liver transglutaminase   449,450
Inhibitors of folylpoly-γ-glutamate synthase   451
Intramolecularly quenched substrates for cathepsin D   452
Internally quenched fluorogenic substrate of γ-glutamyl hydrolase   453
Dipeptide amide inhibitors of neuronal NO synthase   454
Dimer formed from 25-residue fragment of bovine pancreatic trypsin
    inhibitor   455

3.4   Conformations of Synthetic Peptides. –
Conformationally constrained reverse turn mimetics   456
Oligomers of *trans*-(4S,5R)-4-carboxybenzyl-5-methyl-
    oxazolidin-2-one   457
Conformation of peptides containing an α-ethylated-αα-disubstituted
    amino acid   458
Conformation of a peptide linked between *i* and *i* + 4 positions   459
Helical peptides with cavity for fluorescent ligand   460
Rate of helix-coil transition in 19-residue peptide   461
Helical peptides containing Asn and Lys that bind to
    A-T base pairs of oligonucleotides   462
α-Helical peptide library   463
Amphipathic control of $3_{10}/\alpha$-helical equilibrium   464
10-Helices in oxetane β-amino acid hexamers   465
Oligomers of oxetane-2-carboxylate   466
Copper centres in 4-helix-bundle proteins   467
Coiled-coil formation by peptides containing unnatural hydrophobic
    amino acids   468
Self-sorting of coiled-coils containing Leu or hexafluoroleucine cores   469
β-Hairpin peptides   470,471
β-Hairpin peptide containing a thioamide group   472
An azatripeptide containing a βVI-like turn   473
Imidazoline pseudodipeptides as reverse turn mimics   474
Phenylisoserine in β-turn constructs   475
Bicyclic β-turn peptidomimetics   476,477
Coupling helices and hairpins in peptides   478
Peptides containing an α-tetrasubstituted amino acid   479

Tetracosapeptide containing 8 different amino acids   480
β-Heptapeptide with a 3_{14}-helix stabilised by a salt bridge   481
Medium-dependent self-assembly of amphiphilic peptide   482
Peptides of 1-aminocyclodecane-1-carboxylic acid   483
Peptides of 1-aminocycloundecane-1-carboxylic acid   484
βII'-type turn from *N*-acyl-γ-dipeptide amides   485
Triply-templated artificial β-sheet   486
Homotrimeric ββα peptide   487
Solvent-induced transition from β- to helical structure   488
Conformation of peptides containing APP repeats   489
Conformation of short Ser peptides   490
Helical dehydropeptides with β-substituents   491
Increased helical content of PrP^c mutated protein in Creutzfeld-Jakob
    disease   492
Turn-forming peptides from α-tetrasubstituted amino acids   493
Conformation of di-*N*-propylglycine residue in a Gly-rich sequence   494
Peptidomimetics containing a carbocyclic 1,3-diacid   495
β-Barrels with interior His residues having ion-channel and esterase
    activity   496
Conformation of peptides containing 2,2,6,6-tetramethylpiperidine-
    4-amino-4-carboxylic acid   497

3.5   Glycopeptides. –
Fluorobenzoyl groups in glycopeptide synthesis   498
Protecting group for hydroxyl groups in glycopeptide synthesis   499
Galactosyl fluoride for trans-glycosylation   500
Glycosyl-α-amino acid derivatives   501
Antitumour activity of muramoylpeptide-acridine and -acridone
    conjugates   502
Multiple muramoyldipeptide:4-branched structure on Lys   503
Glycosides of muramoyldipeptides and chromone aglycones   504
Synthesis of glycopeptides from Asn linked to free oligosaccharides;
    isolation by Et_2O precipitation   505
Galactosyl derivatives of Ser and Thr   506,507
SPPS of an antifreeze glycoprotein   508
SPPS of glycopeptides using a novel silyl linker   509
20(S)-Camptothecin glycoconjugates   510
Glycosylation at a thiol group with a bivalent glycan   511
Combinatorial SPPS of cyclic neoglycopeptides   512
Protected chitobiose-asparagine; amyloidogenic peptides   513
Enzymatic galactosylation of *C*-glycoside analogues   514
Chemoenzymatic synthesis of neoglycopeptides   515
Chemoenzymatic synthesis of sialylated glycopeptides from mucins
    and T-cell stimulating peptides   516
Cyclic sialyl Lewis X mimetic inhibits P-selectin   517
Stable helical conjugate of a peptide, cyclodextrin and cholic acid   518

4-α-Helix bundle assembled on a galactopyranoside template 519
Glycopeptide mimetics from Asn surrogates by chemoselective
    ligation 520
Automated synthesis of glycoproteins 521
Antifreeze glycoprotein analogues 522
Effect of glycosylation on conformation of folded oligopeptides 523
Use of hexafluoroacetone in synthesis of *N*-linked glycopeptoids 524
Carbopeptoids incorporating the anomeric centre of mannopyranose 525
C-terminal glycopeptide of human salivary mucin 526
Ig domain I of Emmprin linked to chitobiose 527
Ester-linked carbohydrate-peptide conjugates 528
Solid-phase glycosylation of peptide templates 529
SPPS of glycoconjugate biomolecules 530
Gel-phase synthesis of 43-mer peptide from CD-4 binding domain
    of HIV envelope glycoprotein 531
Glycopeptide fragments of serglycin 532
Chemoenzymatic synthesis of a 2,3-sialyl-threonine and the
    N-terminal sequence of leukosialin 533
Enzymatic synthesis of glycopeptides using endo-α-*N*-
    acetylgalactosaminidase from *Streptomyces sp.* 534

3.6 Phosphopeptides and Related Compounds. –
SPPS of a *O*-phosphopeptide 535
Phosphoramidite as an *O*-phosphitylation agent 536
O-Ethyl 1-azidoalkylphosphonic acids for synthesising protected
    phosphonamidate peptides 537
Coupling of phosphopeptides with dichlorotriphenyl-phosphorane 538
SPPS of α-hydroxyphosphonates and hydroxystatine amides 539
SPPS of phosphopeptides using ammonium t-Bu phosphonate 540
Phosphorylated p21Max protein 541
Phosphopeptide derived from an expressed *E.coli* peptide 542
Peptides from CF_2- or CHF-substituted phosphoamino acids 543
Protected phosphinic pseudopeptide blocks for SPPS 544
Staudinger ligation of phosphinothioester and azide to form a
    peptide 545
Phosphinodipeptide analogues 546
Reactivity of peptidyl phosphinic esters 547
Phosphinic dipeptide as inhibitor of human cyclophilin 548
N-Phosphonomethylglycine dipeptides 549
Phosphonopeptides from pyridylomethylphosphonate diphenyl
    esters 550
Phosphonopeptides containing uracil or thymine 551

3.7 Immunogenic and Immunosuppressant Peptides. –
Mannosylated peptide constructs with 3 branched epitopes 552
Peptide immunogens related to sperm tail protein type-1 553

Immunodominant regions of hepatitis C viral proteins   554
Antibody hypervariable loop mimetics   555
Plasmodium falciparum antigenic dendrimers   556
Antigenic properties of analogues of a melanoma protein   557
Diepitopic sequential oligopeptide carrier mimic of the Sm
   autoantigen synthesis   558
Peptide that stimulates T-cells and carries tumour-associated
   carbohydrate antigens   559
Peptides from V3 region DNA-mutants of AIDS patients   560
Peptide containing epitopes of B- and T-cell epitopes   561
Immunogenicity and antigenicity of glycopeptide fragments of
   tick-borne encephalitis virus   562
Peptides containing up to 4 copies of a T cell epitope   563
Different methods for coupling peptides to carriers   564
Antamanide analogues as immunosuppressants   565
Conjugate of N-terminus of β_2GPI and anti-β_2GPI antibodies   566

3.8   Nucleopeptides. –
Peptide nucleic acids (PNAs) containing a psoralen unit   567
SPPS of a PNA monomer   568
PNA oligomers using Fmoc/acyl-protected monomers   569
Novel peptide nucleic acid monomers   570-573
Thymine mimetics in PNAs   574
Polycationic analogue of oligothymidylic acid   575
Flavin-tethered peptide nucleic acid   576
Aminoethylprolyl PNA: PNA/DNA duplex formation   577
2-Stranded helical peptides containing L-α-nucleobase amino
   acids   578
SPPS of DNA-peptide conjugates   579,580
Sequence fidelity of template-directed PNA-ligation   581
SPPS of peptide-oligonucleotide phosphorothioate conjugates 582
Large scale synthesis of oligonucleotide-peptide conjugates 583
Native ligation for peptide-oligonucleotide conjugation   584,585
PNA-related oligonucleotide mimics   586
Backbone modifications of aromatic PNA monomers   587
Boc/acyl chemistry in PNA synthesis   588
(3R,6R) and (3S,6R)-piperidone PNA   589
N-Boc and -Fmoc dipeptoids as PNA monomers   590
PNA monomers using the Mitsunobu reaction   591
Chiral PNA using Fmoc chemistry   592
PNA fragments containing 5-methylcytosine derivatives   593
Liver cells specifically take up lactose-PNA conjugates   594
Analogues of peptides from HIV-1 rev containing nucleobase amino
   acids   595
A PNA containing hydrophilic Ser residues   596
Peptide-PNA conjugates containing anthracene probe   597

PNA-DNA duplexes containing 3-nitropyrrole 598
PNA monomers containing fluoroaromatics 599
Cr coordination by peptide-oligonucleotides 600
Pyrrolidinyl PNA containing β-amino acid spacers 601
PNA containing cis-4-adeninyl-L-prolinol unit 602
PNA synthesis via oxime and thiazolidine formation 603
DNA-3′-PNA chimeras using Bhoc/Fmoc PNA monomers 604
Loss of chiral purity during SPPS of PNA analogues 605
5′-Aminoacylated oligouridylic acids 606

3.9   Miscellaneous Peptides. –
Synthetic technology of Aspartame 607
L-Asp-D-Ala-fenchyl esters 608
H-L-Ala-Gln-OH 609
SPPS of peptides containing αβ-dehydroamino acids 610
Peptidocalix[4]arenes 611,612
Oligourea peptidomimetics 613
Peptide conjugates of 2,6-dimethoxyhydroquinone-3-mercaptoacetic
    acid 614,615
Antibiotic and cytotoxic conjugates of indolecarbazole and peptides 616
Anthraquinone peptides as inhibitors of AP-1 transcription factor 617
Peptides containing β-alanine: self assembly 618
3-Aminopiperid-2,5-dione as a surrogate of -Ala-Gly- 619
Di- and tri-peptides containing two quaternary N-groups 620
Solid-phase synthesis of chemokines 621
Tat-peptides as chelates for Tc^{99m} and Re 622
Peptide labelling with Tc^{99m} in a diphosphine ligand 623
Boronated peptides for boron neutron capture therapy 624
Synthesis and properties of ferrocenyl-peptides 625-627
Redox properties of ferrocenoyl-dipeptides 628
Peptides containing a fullerine amino acid 629
Pseudopeptides containing sulfoximines in the backbone 630
β-Peptides containing guanidino groups 631
Interaction of ferrocenyl dipeptides and 3-amino-pyrazole
    derivatives 632
Peptides labelled with In^{115} 633
Peptide diazeniumdiolate conjugates as prodrugs 634
Polyamides containing pyrrole and imidazole amino acids 635
Imide and lactam derivatives of N-benzyl pyroglutamyl-Phe 636
SPPS of peptide hydroxamic acids 637
Retro-ψ[NHCH(CF$_3$)]peptidyl hydroxamates 638
Functional *Ras* lipoproteins and fluorescent derivatives 639
Fluorescent bahaviour of tripeptides containing 9-amino-fluorene-9-
    carboxylic acid 640
Amphiphilic β-structure of a peptide-porphyrin conjugate 641
Multiporphyrinic helices: light-harvesting possibilities 642

Peptides containing ruthenium tris(pyridine) complexes   643
Metal complexing by tripeptides containing α,α-substituted Gly with
    pyridine rings   644
Co(II) complexation by plastocyanin peptides   645
Peptides containing nucleobases that bind RNA   646
Peptides that bind to DNA   647,648
Incorporation of L-4-[sulfono(difluoromethyl)phenylalanine into a
    peptide   649
Peptides with rigid or unnatural structures   650
Boc-(Leu-Leu-Lac)₃-Leu-Leu-OEt   651
Triazine conjugates for multiple specific-site labelling of peptides   652
Peptides of N-Fmoc-N,N'-bis-Boc-7-guanyl-1,2,3,4-tetrahydro-
    isoquinoline-3-carboxylic acid   653
2H-Azirin-3-amine as a dipeptide synthon (Aib-Hyp)   654
Antitumour prodrug of cytarabine linked to peptide   655
Anthracycline-peptide conjugates as antitumour drugs   656
2,3-Disubstituted poly(β-peptides)   657
9-Amino-4,5-diazafluorene-9-carboxylic acid (Daf) and
    peptides thereof   658
Synthetic chlorophyll proteins   659
Inhibitor of primitive haematopoietic cell division   660
Peptides containing DOPA   661
Antimicrobial peptides of β-amino acids   662
Lipidated fragment of endothelial NO-synthase   663
Dipeptide mimics based on α-ketotetrazoles   664
Diketopiperazine-based receptors for small peptides   665
Receptors for acylated amino acids and peptides   666
'Carbadipeptides' with basic amino acid at C-terminus   667
Synthetic peptides that inhibit HIV infection   668,669
DNA cleavage by tripeptide-cytotoxic drug conjugates   670
31-residue peptide with oxaloacetate decarboxylase activity   671
Fragment (106-126) of prion peptide and its properties   672
Solution-phase synthesis of H-Arg₈-OH, a molecular transporter   673
Peptide derivatives of (S)-4-oxoazetidine-2-carboxylate   674
Peptides containing cyclohexylether δ-amino acids   675
Organisation of peptide nanotubes into bundles   676
Antibacterial conjugates of peptides and 1,4,5,8-naphthalene-
    tetracarboxylic diimide   677
Phe mimetics by alkylation of 6-benzylpiperazine-2,3,5-trione   678
Conjugate of albumin and doxorubicin active against murine renal
    cell carcinoma   679
β-Ala-Leu-Ala-Leu-doxorubicin: a tumour-activated drug   680
Glycoprotein III/IIIa receptor antagonists   681
Solution-phase synthesis of tripeptide derivatives   682
Large-scale automatic synthesis of small peptides   683
SPPS of peptide amides from unnatural amino acids   684

Ferrocenes bearing podand dipeptide chains 685
Fragment (649-654) of retinoblastoma as a dendrite 686

3.10 Purification Methods. –
Peptide separation by solid-phase precipitation and extraction 687
SDS-polyacrylamide gel electrophoresis of small peptides 688
Chiral stationary phases and their resolution efficiency 689
N-Hydroxysuccinimidyl-fluorescein-*O*-acetate for labelling peptides
    before liquid chromatography 690
Covalent capture: tool for purification of polypeptides 691
Isolation and purification of thymosin α_1 692

References

1. D. T. Elmore, Specialist Periodical Report, *Amino Acids, Peptides and Proteins*, 2002, **33**, 83.
2. V. V. S. Babu, *Resonance*, 2001, **6**, 68.
3. Y. Okada, *Curr. Org. Chem.*, 2001, **5**, 1.
4. S. Aimoto, *Curr. Org. Chem.*, 2001, **5**, 45.
5. G. Pang, Q. Wang and Q. Chen, *Shipin Kexue (Beijing)*, 2001, **22**, 80.
6. G. G. Kochendoerfer, *Curr. Opin. Drug Discovery Dev.*, 2001, **4**, 205.
7. M. Goodman, C. Zapf and Y. Rew, *Biopolymers*, 2001, **60**, 229.
8. J. D. Wade, in *Solid-Phase Synthesis*, eds. S. A. Kates and F. Albericio, Marcel Dekker, Inc., New York, 2000, p. 103.
9. A. L. Doherty-Kirby and G. A. Lajoie, in *Solid-Phase Synthesis*, eds. S. A. Kates and F. Albericio, Marcel Dekker, Inc., New York, 2000, p. 129.
10. D. Andreu and E. Nicolas, in *Solid-Phase Synthesis*, eds. S. A. Kates and F. Albericio, Marcel Dekker, Inc., New York, 2000, p. 365.
11. F. Albericio, R. Chinchilla, D. J. Dodsworth and C. Najera, *Org. Prep. Proced. Int.*, 2001, **33**, 203.
12. S. Aimoto, *Tanpakushitsu Kakusan Koso*, 2001, **46**, 1471.
13. J. P. Tam, J. Xu and K. D. Eom, *Biopolymers*, 2001, **60**, 194.
14. O. Seitz, *Nachr. Chem.*, 2001, **49**, 488.
15. B. C. Hamper, J. J. Owen, K. D. Jerome, A. S. Scates, S. A. Kolodziej, R. C. Chott and A. S. Kesselring, *ACS Symp. Ser.*, 2001, **774**, 180.
16. P. Lloyd-Williams and E. Giralt, in *Solid-Phase Synthesis*, eds. S. A. Kates and F. Albericio, Marcel Dekker, Inc., New York, 2000, p. 377.
17. C. Blackburn, in *Solid-Phase Synthesis*, eds. S. A. Kates and F. Albericio, Marcel Dekker, Inc., New York, 2000, p. 197.
18. F. Albericio and S. A. Kates, in *Solid-Phase Synthesis*, eds. S. A. Kates and F. Albericio, Marcel Dekker, Inc., New York, 2000, p. 275.
19. Y. Lee and R. B. Silvermann, *Proc. ECSOC-3, Proc. ECSOC-4, 1999, 2000*, 1999-2000, p. 1290.
20. T. Yokum and G. Barany, in *Solid-Phase Synthesis*, eds. S. A. Kates and F. Albericio, Marcel Dekker, Inc., New York, 2000, p. 79.
21. R. Reents, D. A. Jeyaraj and H. Waldmann, *Adv. Synth. Catal.*, 2001, **343**, 501.

22.   C. Enjabal, S. Subra, R. Combarieu, J. Martinez and J.-L. Aubagnac, *Recent Res. Dev. Org. Chem.*, 2000, **4**, 29.
23.   G. Guichard, in *Solid-Phase Synthesis*, eds. S. A. Kates and F. Albericio, Marcel Dekker, Inc., New York, 2000, p. 649.
24.   Z.-z. Chen, Y.-m. Li, G. Zhao and Y.-f. Zhao, *Youji Huaxue*, 2001, **21**, 273.
25.   Z.-h. Peng, W.-t. Lin, J.-l. Feng, M.-h. Zong and R.-h. Yao, *Fenzi Cuihua*, 2000, **14**, 461.
26.   V. Rolland and R. Lazaro, *Methods Biotechnol.*, 2001, **15**, 357.
27.   Y. Asano, *J. Microbiol. Biotechnol.*, 2000, **10**, 573.
28.   S.-T. Chen, B. Sookkheo, S. Phutrahul and K.-T. Wang, *Methods Biotechnol.*, 2001, **15**, 373.
29.   P. Liu, G.-L. Tian and Y.-H. Ye, *Gaodeng Xuexiao Huaxue*, 2001, **22**, 1342.
30.   M. Amblard, J. A. Fehrentz, I. Daffix, G. Berge, C. Devin, M. Llinares, C. Genu-Dellac, Ph. Bedos, P. Dodey, D. Pruneau, J. L. Paquet, J. M. Luccarini, P. Belichard, F. Bellamy, S. Halazy, W. Riquet, F. Colpaert and J. Martinez, *Actual. Chim. Ther.*, 1999, **25**, 207.
31.   D. Yang, Y. Yang, Q. Xiao and Y. Zhong, *Huaxi Yaoxue*, 2001, **16**, 440.
32.   T. Shioiri and Y. Hamada, *Synlett*, 2001, 184.
33.   L. Fang and T. Tan, *Huagong Jinzhan*, 2001, **20**, 24.
34.   Q.-W. Zhang, M.-H. Zhou and H.-L. Shi, *Zhongguo Yiyao Gongyi Zazhi*, 2000, **81**, 248.
35.   J. V. Potetinova, E. I. Milgotina, V. A. Makarov and T. L. Voyushina, *Russ. J. Bioorg. Chem.*, 2001, **27**, 151.
36.   K. Burgess, *Acc. Chem. Res.*, 2001, **34**, 826.
37.   M. Mamoru, *Trends Glycosci. Glycotechnol.*, 2001, **13**, 11.
38.   P. M. St. Hilaire and M. Meldal, *Comb. Chem.*, 1999, 291.
39.   A. Bernardi and L. Manzoni, *Chemtracts*, 2000, **13**, 588.
40.   K. Ajisaka, *Trends Glycosci. Glycotechnol.*, 2001, **13**, 305.
41.   C. García-Echeverría, in *Solid-Phase Synthesis*, eds. S. A. Kates and F. Albericio, Marcel Dekker, Inc., New York, 2000, p. 419.
42.   J. S. McMurray, D. R. Coleman, W. Wang, and M. L. Campbell, *Biopolymers*, 2001, **60**, 3.
43.   M. Egholm and R. A. Casale, in *Solid-Phase Synthesis*, eds. S. A. Kates and F. Albericio, Marcel Dekker, Inc., New York, 2000, p. 549.
44.   R. Eritja, in *Solid-Phase Synthesis*, eds. S. A. Kates and F. Albericio, Marcel Dekker, Inc., New York, 2000, p. 529.
45.   T. Okamura and K. Ueyama, *Kobunshi Kako*, 2000, **49**, 261.
46.   A. Bianco, T. Da Ros, M. Prato and C. Toniolo, *J. Pept. Sci.*, 2001, **7**, 208.
47.   O. Seitz, I. Heineman, A. Mattes and H. Waldmann, *Tetrahedron*, 2001, **57**, 2247.
48.   H.-B. Kraatz and M. Galka, *Met. Ions Biol. Syst.*, 2001, **38**, 385.
49.   A. Golebiowski, S. R. Klopfenstein and D. E. Portlock, *Curr. Opin. Drug Discovery Dev.*, 2001, **4**, 428.
50.   Y. Shigeri, Y. Tatsu and N. Yumoto, *Pharmacol. Ther.*, 2001, **91**, 85.
51.   P. Bravo, L. Bruche, C. Pesenti, F. Viani, A. Volonterio and M. Zanda, *J. Fluorine Chem.*, 2001, **112**, 153.
52.   J. S. McMurray, in *Solid-Phase Synthesis*, eds. S. A. Kates and F. Albericio, Marcel Dekker, Inc., New York, 2000, p. 735.
53.   N. P. Ambulos, L. Bibbs, L. F. Bonewald, S. A. Kates, A. Ashok, K. F. Medzih-radszky and S. T. Weintraub, in *Solid-Phase Synthesis*, eds. S. A. Kates and F. Albericio, Marcel Dekker, Inc., New York, 2000, p. 751.

54. J. R. Casimir, G. Guichard, D. Tourwé and J.-P. Briand, *Synthesis*, 2001, 1985.
55. E. Cros, M. Planas, X. Mejías and E. Bardají, *Tetrahedron Lett.*, 2001, **42**, 6105.
56. E. Cros, M. Planas and E. Bardají, *Synthesis*, 2001, 1313.
57. A. R. Rajipour, S. E. Mallakpour and G. Imanzadeh, *Indian J. Chem.*, 2001, **40B**, 250.
58. G. Han, M. Tamaki and V. J. Hruby, *J. Pept. Res.*, 2001, **58**, 338.
59. P. Strazzolini, T. Melloni and A. G. Giumanini, *Tetrahedron*, 2001, **57**, 9033.
60. H. S. Trivedi, M. Anson, P. G. Steel and J. Worley, *Synlett*, 2001, 1932.
61. D. Kadereit, P. Deck, I. Heinemann and R. Waldmann, *Chem. – Eur. J.* 2001, **7**, 1184.
62. J. Sebestik, J. Hlavacek and I. Stibor, *Chem. Listy*, 2001, **95**, 365.
63. D. Shangguan, Y. Zhao, H. Han, R. Zhao and G. Liu, *Anal. Chem.*, 2001, **73**, 2054.
64. J. M. Dener, P. P. Fantauzzi, T. A. Kshirsagar, D. E. Kelly and A. B. Wolfe, *Org. Process Res. Dev.*, 2001, **5**, 445.
65. Y. Luo, G. Evindar, D. Fishlock and G. A. Lajoie, *Tetrahedron Lett.*, 2001, **42**, 3807.
66. R. Chinchilla, D. J. Dodsworth, C. Nájera and J. M. Soriano, *Tetrahedron Lett.*, 2001, **42**, 7579.
67. H. J. C. Deboves, A. G. N. Christian and R. F. W. Jackson, *J. Chem. Soc., Perkin Trans 1*, 2001, 1876.
68. V. Vommina, S. Babu and K. Ananda, *Indian J. Chem.*, 2001, **40B**, 70.
69. J. H. Miwa, L. A. Margarida and A. E. Meyer, *Tetrahedron Lett.*, 2001, **42**, 7189.
70. B. Henkel and E. Bayer, *J. Pept. Sci.*, 2001, **7**, 152.
71. G. Gellerman, A. Elgavi, Y. Salitra and M. Kramer, *J. Pept. Res.*, 2001, **57**, 277.
72. D. Kadereit and H. Waldmann, *ChemBioChem.*, 2000, **1**, 200.
73. A. Chibani and Y. Bendaoud, *J. Alger. Soc. Chim.*, 2001, **11**, 17.
74. K. Hojo, M. Maeda and K. Kawasaki, *J. Pept. Sci.*, 2001, **7**, 615.
75. J. Gurnani, C. K. Narang and M. R. K. Sherwani, *Orient. J. Chem.*, 2000, **16**, 469.
76. L. Bunch, P.-O. Norrby, K. Frydenvang, P. Krogsgaard-Larsen and U. Madsen, *Org. Lett.*, 2001, **3**, 433.
77. A. Basso, S. Biffi, L. De Martin, L. Gardossi and P. Linda, *Croat. Chem. Acta.*, 2001, **74**, 757.
78. D. A. Jeyaraj and H. Waldmann, *Tetrahedron Lett.*, 2001, **42**, 835.
79. P. Liu, A. Yan and Y. Ye, *Huaxue Tongbao*, 2001, 777.
80. L. Kovacs, P. Forgo and Z. Kele, *Proc. ECSOC-3, Proc. ECSOC-4*, 1999, 2000, 1015.
81. I. A. Rivero, S. Heredia and A. Ochoa, *Synth. Commun.*, 2001, **31**, 2169.
82. M. A. Wegman, J. M. Elzinga, E. Neeleman, F. van Rantwijk and R. A. Sheldon, *Green Chem.*, 2001, **3**, 61.
83. H. Tamiaki, T. Obata, Y. Azefu and K. Toma, *Bull. Chem. Soc. Jpn.*, 2001, **74**, 733.
84. E. Marcantoni, M. Massaccesi, E. Torregiani, G. Bartoli, M. Bosco and L. Sambri, *J. Org. Chem.*, 2001, **66**, 4430.
85. V. Theodorou-Kassioumis, N. Biris, C. Sakarellos and V, Tsikaris, *Tetrahedron Lett.*, 2001, **42**, 7703.
86. Y. Nishiyama, S. Ishizuka, S. Shikama and K. Kurita, *Chem. Pharm. Bull.*, 2001, **49**, 233.
87. C. Somlai, S. Lovas, P. Forgó, R. F. Murphy and B. Penke, *Synth. Commun.*, 2001, **31**, 3633.
88. N. B. Malkar and G. B. Fields, *Lett. Pept. Sci.*, 2000, **7**, 263.
89. S. Mourtas, D. Gatos, V. Kalaitzi, C. Katakalou and K. Barlos, *Tetrahedron Lett.*, 2001, **42**, 6965.

90. P. Gomez-Martinez, F. Guibé and F. Albericio, *Lett. Pept. Sci.*, 2000, **7**, 187.

91. L. Taboada, E. Nicolás and E. Giralt, *Tetrahedron Lett.*, 2001, **42**, 1891.

92. M. Mergler, F. Dick, B. Sax, J. Schwindling and Th. Vorherr, *J. Pept. Sci.*, 2001, **7**, 502.

93. L. M. Berezhkovskiy and S. V. Deshpande, *Biophys. Chem.*, 2001, **91**, 319.

94. T. Shi and D. L. Rabenstein, *Tetrahedron Lett.*, 2001, **42**, 7203.

95. M. Iwaoka, T. Takahashi and S. Tomoda, *Heteroat. Chem.*, 2001, **12**, 293.

96. T. Zoller, J.-B. Ducep, T. Tahtaoui and M. Hibert, *Tetrahedron Lett.*, 2000, **41**, 9989.

97. Y. Sun, G. Lu and J. P. Tam, *Org. Lett.*, 2001, **3**, 1681.

98. Y. Basel and A. Hassner, *Synthesis*, 2001, 550.

99. V. V. S. Babu, K. Ananda and R. I. Mathad, *Lett. Pept. Sci.*, 2000, **7**, 239.

100. K. Pudhom and T. Vilaivan, *Synth. Commun.*, 2001, **31**, 61.

101. A. DalPozzo, R. Bergonzi and M. Ni, *Tetrahedron Lett.*, 2001, **42**, 3925.

102. T. Inui, H. Nishio, Y. Nishiuchi and T. Kimura, *Pept. Sci.*, 2001, 31.

103. N. Mihala, J. Bodi, A. Gomory and H. Suli-Vargha, *J. Pept. Sci.*, 2001, **7**, 565.

104. W. Van Den Nest, S. Yuval and F. Albericio, *J. Pept. Sci.*, 2001, **7**, 115.

105. S. Ishizuka, K. Kurita and Y. Nishiyama, *Pept. Sci.*, 2001, 39.

106. Y. Nishiyama, S. Ishizuka and K. Kurita, *Tetrahedron Lett.*, 2001, **42**, 8789.

107. J.-G. Ji, D.-Y. Zhang, J. Jian, Y.-H. Ye and Q.-Y. Xing, *Huaxue Xuebao*, 2001, **59**, 1740.

108. L. A. Carpino and F. J. Ferrer, *Org. Lett.*, 2001, **3**, 2793.

109. P. Li and J. C. Xu, *J. Pept. Res*, 2001, **58**, 129.

110. M. Y. Ali, H. N. Roy, M. S. Zaman, M. A. Islam and A. B. M. H. Haque, *Indian J. Chem.*, 2000, **39B**, 769.

111. V. Santagada, F. Fiorino, E. Perisutti, B. Severino, V. De Filippis, B. Vivenzio and G. Caliendo, *Tetrahedron Lett.*, 2001, **42**, 5171.

112. C. Zorn, F. Gnad, S. Salmen, T. Herpin and O. Reiser, *Tetrahedron Lett.*, 2001, **42**, 7049.

113. Z. Kamiński, B. Kolesińska, J. E. Kamińska and J. Góra, *J. Org. Chem.*, 2001, **66**, 6276.

114. J. T. Lundquist and J. C. Pelletier, *Org. Lett.*, 2001, 781.

115. M. Breznik, S. G. Grdadolnik, G. Giester, I. Leban and D. Kikelj, *J. Org. Chem.*, 2001, **66**, 7044.

116. T. Kawakami, K. Akaji and S. Aimoto, *Org. Lett.*, 2001, **3**, 1403.

117. K. Haas and W. Beck, *Eur. J. Inorg. Chem.*, 2001, 2485.

118. E. Vedejs and C. Kongkittingam, *J. Org. Chem.*, 2001, **66**, 7355.

119. A. M. Belostotskii, E. Genizi and A. Hassner, *Chem. Commun.*, 2001, 1960.

120. D. Gatos and C. Tzavara, *J. Pept. Res.*, 2001, **57**, 168.

121. M. N. Mathieu, J. D. Wade, B. Catimel, C. P. Bond, E. C. Nice, R. J. Summers, L. Otvos and G. W. Tregear, *J. Pept. Res.*, 2001, **57**, 374.

122. M. Keller and A. D. Miller, *Bioorg. Med. Chem. Lett.*, 2001, **11**, 857.

123. J. Kawaguchi, T. Matsuoka and N. Serizawa, *Pept. Sci.*, 2001, 27.

124. Sh. N. Khattab, A. El-Faham, A. M. El-Massry, E. M. E. Mansour and M. M. A. El-Rahman, *Lett. Pept. Sci.*, 2001, **7**, 331.

125. A. K. Tickler, C. J. Barrow and J. D. Wade, *J. Pept. Sci.*, 2001, **7**, 488.

126. C. Tomasini and M. Villa, *Tetrahedron Lett.*, 2001, **42**, 5211.

127. H. Hojo, *Pept. Sci.*, 2001, 5.

128. S. Biancalana, D. Hudson, M. F. Songster and S. A. Thompson, *Lett. Pept. Sci.*, 2000, **7**, 291.

129. R. Quaderer and D. Hilvert, *Org. Lett.*, 2001, **3**, 3181.

130. L. Z. Yan and P. E. Dawson, *J. Amer. Chem. Soc.*, 2001, **123**, 526.
131. P. Botti, M. R. Carrasco and S. B. H. Kent, *Tetrahedron Lett.*, 2001, **42**, 1831.
132. M. D. Gieselman, L. Xie and W. A. van der Donk, *Org. Lett.*, 2001, **3**, 1331.
133. R. Quaderer, A. Sewing and D. Hilvert, *Helv. Chim. Acta*, 2001, **84**, 1197.
134. L. A. Marcaurelle, L. S. Mizoue, J. Wilken, L. Oldham, S. B. H. Kent, T. M. Handel and C. R. Bertozzi, *Chem. –Eur. J.*, 2001, **7**, 1129.
135. P. Durieux, J. Fernandez-Carneado and G. Tuchscherer, *Tetrahedron Lett.*, 2001, **42**, 2297.
136. D. F. J. Sampson, R. G. Simmonds and M. Bradley, *Tetrahedron Lett.*, 2001, **42**, 5517.
137. M. Roice and V. N. R. Pillai, *Protein Pept. Lett.*, 2000, **7**, 365.
138. I. M. Krishnakumar and B. Mathew, *Lett. Pept. Sci.*, 2001, **7**, 317.
139. P. K. Ajikumar and K. S. Davaky, *Lett. Pept. Sci.*, 2000, **7**, 207.
140. P. K. Ajikumar and K. S. Devaky, *J. Pept. Sci.*, 2001, **7**, 641.
141. K. S. Kumar, M. Roice and V. N. R. Pillai, *Tetrahedron*, 2001, **57**, 3151.
142. E. Lattmann, D. C. Billington, P. Arayarat, H. Singh and M. Offel, *Science Asia*, 1999, **25**, 107.
143. S. Leena and K. S. Kumar, *J. Pept. Res.*, 2001, **58**, 117.
144. W. C. Chan, S. L. Mellor, G. E. Atkinson, E. L. Fritzen and D. J. Staples, *Solid-Phase Org. Synth.*, 2001, **1**, 85.
145. J. K. Cho, D.-W. Kim, J. Namgung and Y.-S. Lee, *Tetrahedron Lett.*, 2001, **42**, 7443.
146. B. H. Lipshutz and Y.-J. Shin, *Tetrahedron Lett.*, 2001, **42**, 5629.
147. C. Somlai, P. Hegyes, R. Fenyo, G. K. Toth, B. Penke and W. Voelter, *Z. Naturforsch., B: Chem. Sci.*, 2001, **56**, 526.
148. S. R. Chhabra, H. Parekh, A. N. Khan, B. W. Bycroft and B. Kellam, *Tetrahedron Lett.*, 2001, **42**, 2189.
149. B. Chitkul, B. Atrash and M. Bradley, *Tetrahedron Lett.*, 2001, **42**, 6211.
150. G. T. Bourne, S. W. Golding, R. P. McGeary, W. D. F. Meutermans, A. Jones, G. R. Marshall, P. F. Alewood and M. L. Smythe, *J. Org. Chem.*, 2001, **66**, 7706.
151. O. Melnyk, J.-S. Fruchart, C. Grandjean and H. Gras-Masse, *J. Org. Chem.*, 2001, **66**, 4153.
152. J. Alsina, K. J. Jensen, M. F. Songster, J. Vagner, F. Albericio, G. Barany, J. Flygare and M. Fernandez, *Solid-Phase Org. Synth.*, 2001, **1**, 121.
153. T. Okayama and V. J. Hruby, *Pept. Sci.*, 2001, 35.
154. A. C. Comely, S. E. Gibson, N. J. Hales and M. A. Peplow, *J. Chem. Soc., Perkin Trans. 1*, 2001, 2526.
155. G. N. Jubilut, E. M. Cilli, M. Tominaga, A. Miranda, Y. Okada and C. R. Nakaie, *Chem. Pharm. Bull.*, 2001, **49**, 1089.
156. R. Chinchilla, D. J. Dodsworth, C. Nájera and J. M. Soriano, *Tetrahedron Lett.*, 2001, **42**, 4487.
157. A. Sweing and D. Hilvert, *Angew. Chem., Int. Ed.*, 2001, **40**, 3395.
158. O. Lorthioir, R. A. E. Carr, M. S. Congreve, M. H. Geysen, C. Kay, P. Marshall, S. C. McKeown, N. J. Parr, J. J. Scicinski and S. P. Watson, *Anal. Chem.*, 2001, **73**, 963.
159. P. Watts, C. Wiles, S. J. Haswell, E. Pombo-Villar and P. Styring, *Chem. Commun.*, 2001, 990.
160. D. Maux, C. Enjalbal, J. Martinez, J.-L. Aubagnac and R. Combarieu, *J. Amer. Soc. Mass Spectrum*, 2001, **12**, 1099.
161. M. B. Baru, L. G. Mustaeva, I. V. Vagenina, E. Yu. Gorbunova and V. V. Cherskii, *J. Pept. Res.*, 2001, **57**, 193.
162. M. R. Nilsson, L. L. Nguyen and D. P. Raleigh, *Anal. Biochem.*, 2001, **288**, 76.

163. B. C. Hamper, A. S. Kesselring, M. H. Parker and J. A. Turner, *Solid-Phase Synthesis*, 2000, **1**, 55.
164. H. L. Ball, D. S. King, F. E. Cohen, S. B. Prusiner and M. A. Baldwin, *J. Pept. Res.*, 2001, **58**, 357.
165. C. Morán, M. R. Infante and P. Clapés, *J. Chem. Soc., Perkin Trans. 1*, 2001, 2063.
166. V. Boyer, M. Stanchev, A. J. Fairbanks and B. G. Davis, *Chem. Commun.*, 2001, 1908.
167. C.-F. Liu and J. P. Tam, *Org. Lett.*, 2001, **3**, 4157.
168. S. Lin and C. R. Lowe, *Anal. Biochem.*, 2000, **285**, 127.
169. N. Wehofsky, F. Bordusa and M. Alisch, *Chem. Commun.*, 2001, 1602.
170. L. De Martin, C. Ebert, L. Gardossi and P. Linda, *Tetrahedron Lett.*, 2001, **42**, 3395.
171. A. V. Bacheva, I. Y. Filippova, E. N. Lysogorskaya and E. N. Oksenoit, *J. Mol. Catal. B: Enzym*, 2000, **11**, 89.
172. I. V. Getun, I. Y. Filippova, E. N. Lysogorskaya and E. S. Oksenoit, *J. Mol. Catal., B: Enzym.*, 2001, **15**, 105.
173. S. V. Kolobanova, I. Yu. Filippova, E. N. Lysogorskaya, *Russ. J. Bioorg. Chem.*, 2001, **27**, 306.
174. J. S. Yadav, H. M. Meshram, A. R. Prasad, Y. S. S. Ganesh, A. B. Rao, G. Seenayya, M. V. Swamy and M. G. Reddy, *Tetrahedron: Asymmetry*, 2001, **12**, 2505.
175. A. Trusek-Holownia and A. Naworyta, *Chem. Pap.*, 2000, **54**, 442.
176. H. Sekizaki, K. Itoh, E. Toyota and K. Tanizawa, *Amino Acids*, 2001, **21**, 175.
177. C. Zhou, G.-L. Tian, H.-Y. Shen and Y.-H. Ye, *Huaxue Xuebao*, 2001, **59**, 1707.
178. M. Erbeldinger, X. Ni and P. J. Halling, *Biotechnol. Bioeng.*, 2001, **72**, 69.
179. R. V. Ulijn, A. E. M. Janssen, B. D. Moore and P. J. Halling, *Chem.–Eur. J.*, 2001, **7**, 2089.
180. P. Clapés, L. Espelt, M. A. Navarro and C. Solans, *J. Chem. Soc., Perkin Trans. 2*, 2001, 1394.
181. C. Kim, I. K. Lee, J. E. Ahn and C. S. Shin, *Biotechnol. Lett.*, 2001, **23**, 1423.
182. L. M. Simon, M. Kotorman, G. Garab and I Laczko, *Biochem. Biophys. Res. Commun.*, 2001, **280**, 1367.
183. R. Dieckmann, T. Neuhof, M. Pavela-Vrancic and H. von Dohren, *FEBS Lett.*, 2001, **498**, 42.
184. K. Khumtaveeporn, A. Ullmann, K. Matsumoto, B. G. Davis and J. B. Jones, *Tetrahedron: Asymmetry*, 2001, **12**, 249.
185. S. Xu, K. Rall and F. Bordusa, *J. Org. Chem.*, 2001, **66**, 1627.
186. K. Matsumoto, B. G. Davis and J. B. Jones, *Chem. Commun.*, 2001, 903.
187. P. López-Serrano, M. A. Wegman, F. van Rantwijk and R. A. Sheldon, *Tetrahedron: Asymmetry*, 2001, **12**, 235.
188. R. J. Smith, M. Pietzsch, T. Waniek, C. Syldatk and S. Bienz, *Tetrahedron: Asymmetry*, 2001, **12**, 157.
189. F. Benedetti, F. Berti, A. Colombatti, M. Flego, L. Gardossi, P. Linda and S Peressini, *Chem. Commun.*, 2001, 715.
190. K. Fujita and K. Takegawa, *Biochem. Biophys. Res. Commun.*, 2001, **282**, 678.
191. L. Fulop, B. Penke and M. Zarandi, *J. Pept. Sci.*, 2001, **7**, 397.
192. Y. Zhang and A. J. Kennan, *Org. Lett.*, 2001, **3**, 2341.
193. P. Ingallinella, A. Di Marco, M. Taliani, D. Fattori and A. Pessi, *Bioorg. Med. Chem. Lett.*, 2001, **11**, 1343.
194. S. Kotha and K. Lahiri, *Bioorg. Med. Chem. Lett.*, 2001, **11**, 2887.
195. P. M. T. Ferreira, H. L. S. Maia, L. S. Monteiro and J. Sacramento, *J. Chem. Soc., Perkin Trans. 1*, 2001, 3167.

196. Y. Hayashi, Y. Kinoshita, K. Hidaka, A. Kiso, H. Uchibori, T. Kimura and Y. Kiso, *J. Org. Chem.*, 2001, **66**, 5537.
197. M. M. Meloni and M. Taddei, *Org. Lett.*, 2001, **3**, 337.
198. Y. Zhu and W. A. van der Donk, *Org. Lett.*, 2001, **3**, 1189.
199. P. Blakskjaer, L. Pedersen and T. Skrydstrup, *J. Chem. Soc., Perkin Trans. 1*, 2001, 910.
200. M. L. Di Gioia, A. Leggio, A. Liguori, A. Napoli, C. Siciliano and G. Sindona, *J. Org. Chem.*, 2001, **66**, 7002.
201. M. Mokotoff, Y. M. Mocarski, B. L. Gentsch, M. R. Miller, J.-H. Zhou, J. Chen and E. D. Ball, *J. Pept. Res.*, 2001, **57**, 383.
202. A. Graven and M. Meldal, *J. Chem. Soc., Perkin Trans. 1*, 2001, 3198.
203. J. C. Spetzler and T. Hoeg-Jensen, *J. Pept. Sci.*, 2001, **7**, 537.
204. D. T. Bong and M. R. Ghadiri, *Org. Lett.*, 2001, **3**, 2509.
205. R. A. Breitenmoser and H. Heimgartner, *Helv. Chim. Acta*, 2001, **84**, 786.
206. H. H. Keah, N. Allen, R. Clay, R. I. Boysen, T. Warner and M. T. W. Hearn, *Biopolymers*, 2001, **60**, 279.
207. J. Schoepfer, B. Gay, N. End, E. Muller, G. Scheffel, G. Caravatti and P. Furet, *Bioorg. Med. Chem. Lett.*, 2001, **11**, 1201.
208. E. K. Quagraine, H.-B. Kraatz and R. S. Reid, *J. Inorg. Chem.*, 2001, **85**, 23.
209. E. Block, M. Birringer, W. Jiang, T. Nakahodo, H. J. Thompson, P. J. Toscano, H. Uzar, X. Zhang and Z. Zhu, *J. Agric. Food Chem.*, 2001, **49**, 458.
210. L. Balaspiri and U. Langel, *J. Pept. Sci.*, 2001, **7**, 58.
211. P. P. Mager and K. Fischer, *Mol. Simul.*, 2001, **27**, 237.
212. M. Tominaga, S. R. Barbosa, E. F. Poletti. J. Zukerman- Schpector, R. Marchetto, S. Schrier, A. C. M. M. Paiva and C. R. Nakaie, *Chem. Pharm. Bull.*, 2001, **49**, 1027.
213. S. Lindman, G. Lindeberg, A. Gogoll, F. Nyberg, A. Karlen and A. Hallberg, *Bioorg. Med. Chem.*, 2001, **9**, 763.
214. T. Abiko and R. Fujimura, *Protein Pept. Lett.*, 2001, **8**, 147.
215. T. Abiko and R. Ogawa, *Protein Pept. Lett.*, 2001, **8**, 289.
216. R. Akasaka, K. Yamada, T. Kawaguchi, R. Tanaka, H. Yamamura, S. Araki, K. Saito, F. Kato and M. Kawai, *Pept. Sci.*, 2001, 429.
217. A. Yoshida, H. Fujishima, S. Ando, H. Nishikawa and H. Takiguchi, *Pept. Sci.*, 2001, 185.
218. K. Yamaguchi, T. Ueda, T. Yoshida, Y. Soejima, H. Aoyagi and N. Izumiya, *Saga Daigaku Nogakubu Iho*, 2000, **85**, 123.
219. M. C. F. Monnee, A. J. Brouwer, L. M. Verbeek, A. M. A. van Wageningen and R. M. J. Liskamp, *Bioorg. Med. Chem. Lett.*, 2001, **11**, 1521.
220. K. C. Nicolaou, S. Y. Cho, R. Hughes, N. Winssinger, C. Smethurst, H. Labis-chinski and R. Endermann, *Chem. –Eur. J.*, 2001, **7**, 3798.
221. H. Hujakka, J. Ratilainen, T. Korjamo, H. Lankinen, P. Kuusela, H. Santa. R. Laatikainen and A. Narvanen, *Bioorg. Med. Chem.*, 2001, **9**, 1601.
222. D. Wade, T. Bergman, J. Silberring and H. Lankinen, *Protein Pept. Lett.*, 2001, **8**, 443.
223. C. Peggion, V. Moretto, F. Formaggio, M. Crisma, C. Toniolo, J. Kamphuis, B. Kaptein and Q. B. Broxterman, *J. Pept. Res.*, 2001, **58**, 317.
224. A. Fujie, T. Iwamoto, B. Sato, H. Muramatsu, C. Kasahara, T. Furuta, Y. Hori, M. Hino and S. Hashimoto, *Bioorg. Med. Chem. Lett.*, 2001, **11**, 399.
225. D. Barrett, A. Tanaka, K. Harada, H. Ohki, E. Watabe, K. Maki and F. Ikeda, *Bioorg. Med. Chem. Lett.*, 2001, **11**, 479.
226. Y.-Z. Zhang, X. Sun, D. J. Zeckner, R. K. Sachs, W. L. Current, J. Gidda, M.

Rodriguez and S.-H. Chen, *Bioorg. Med. Chem. Lett.*, 2001, **11**, 903.

227. A. Suda, A. Ohta, M. Sudoh, T. Tsukuda and N. Shimma, *Heterocycles*, 2001, **55**, 1023.

228. W. M. M. Schaaper, G. A. Posthuma, H. H. Plasman, L. Sijtsma, F. Fant, F. A. M. Borremans, K. Thevissen, W. F. Broekaert, R. H. Meloen and A. Van Amerongen, *J. Pept. Res.*, 2001, **57**, 409.

229. B. E. Haug and J. S. Svendsen, *J. Pept. Sci.*, 2001, **7**, 190.

230. S. Bhattacharya and M. Thomas, *Tetrahedron Lett.*, 2001, **42**, 3499.

231. P. J. Murray, M. Kranz, M. Ladlow, S. Taylor, F. Berst, A. B. Holmes, K. N. Keavey, A. Jaxa-Chamiec, P. W. Seale, P. Stead, R. J. Upton, S. L. Croft, W. Clegg and M. R. J. Elsegood, *Bioorg. Med. Chem. Lett.*, 2001, **11**, 773.

232. L. Z. Yan, A. C. Gibbs, M. E. Stiles, D. S. Wishart and J. C. Vederas, *J. Med. Chem.*, 2000, **43**, 4579.

233. A. Bozzoli, W. Kazmierski, G. Kennedy, A. Pasquarello and A. Pecunioso, *Bioorg. Med. Chem. Lett.*, 2000, **10**, 2759.

234. M. Roice, G. Suma, K. S. Kumar and V. N. R. Pillai, *J. Protein Chem.*, 2001, **29**, 25.

235. M. Crisma, A. Barazza, F. Formaggio, B. Kaptein, Q. B. Broxterman, J. Kamphuis and C. Toniolo, *Tetrahedron*, 2001, **57**, 2813.

236. H. Kanegae, S. Okumura, T. Akao, K. Harata and H. Morii, *Pept. Sci.*, 2001, 377.

237. M. Saito and T. Chiba, *Akita Kogyo Koto Senmon Gakko Kenkyu Kiyo*, 2001, **36**, 40.

238. R. Marchetto, E. Nicolas, N. Castillo, J. Bacardit, M. Navia, J. Vila and E. Giralt, *J. Pept. Sci.*, 2001, **7**, 27.

239. T. Sato, T. Kawakami, K. Akaji, H. Konishi, K. Mochizuki, Fujiwara, H. Akutsu and S. Aimoto, *Pept. Sci.*, 2001, 29.

240. J. Nakayima, Y. Cao, T. Horii, S. Sakuda and H. Nagasawa, *Biosci., Biotechnol., Biochem.*, 2001, **65**, 2322.

241. J. F. Valliant, R. W. Riddoch, D. W. Hughes, D. G. Roe, T. K. Fauconnier and J. R. Thornback, *Inorg. Chim. Acta*, 2001, **325**, 155.

242. L. Biondi, F. Filira, M. Gobbo, B. Scolaro, R. Rocchi, R. Galeazzi, M. Orena, A. Zeegers and T. Piek, *J. Pept. Sci.*, 2001, **7**, 626.

243. I. Derdowska, A. Prahl, K. Neubert, B. Hartrodt, A. Kania, D. Dobrowolski, S. Melhem, H. I. Trzeciak, T. Wierzba and B. Lammek, *J. Pept. Res.*, 2001, **57**, 11.

244. M. Gobbo, L. Blondi, F. Filira, R. Rocchi and T. Pick, *Lett. Pept. Sci.*, 2000, **7**, 171.

245. P. Rovero, M. Pellegrini, A. Di Fenza, S. Meini, L. Quartara, C. A. Maggi, F. Formaggio, C. Toniolo and D. F. Mierke, *J. Med. Chem.*, 2001, **44**, 274.

246. D. Gatos and C. Tzavara, *J. Pept. Res.*, 2001, **57**, 168.

247. V. Gut, V. Cerovsky, M. Zertova, E. Korblova, P. Malon, H. Stocker and E. Wunsch, *Amino Acids*, 2001, **21**, 255.

248. J. Rysz, A. Redlinski, J. Mudyna, M. Luciak and Z. J. Kaminski, *Acta Pol. Pharm.*, 2000, **57**(Suppl.), 11.

249. J. M. Bland, A. J. De Lucca, T. J. Jacks and C. B. Vigo, *Mol. Cell. Biochem.*, 2001, **218**, 105.

250. G. Cavicchioni, M. Turchetti and S. Spisani, *Bioorg. Med. Chem. Lett.*, 2001, **11**, 3157.

251. M. Yoshiki, D. Asai, H. Kodama, M. Miyazaki, I. Fujita, Y. Hamasaki, M. Yuhei, S. Miyazaki and M. Kondo, *Pept. Sci.*, 2001, 171.

252. I. Torrini, M. Nalli, M. P. Paradisi, G. P. Zecchini, G. Lucente and S. Spisani, *J. Pept. Res.*, 2001, **58**, 56.

253. A. Dalpiaz, A. Scatturin, G. Vertuani, R. Pecoraro, P. A. Borea, K. Varani, S.

Traniello and S. Spisani, *Eur. J. Pharmacol.*, 2001, **41**, 327.
254. E. Morera, G. Lucente, G. Ortar, M. Nalli, F. Mazza, E. Gavuzzo and S. Spisani, *Bioorg. Med. Chem.*, 2001, **10**, 147.
255. T. Abiko and N. Nakatsubo, *Protein Pept. Lett.*, 2001, **8**, 153.
256. S. De Luca, D. Tesauro, P. Di Lello, R. Fattorusso, M. Saviano, C. Pedone and G. Morelli, *J. Pept. Sci.*, 2001, **7**, 386.
257. J. M. Bartolomé-Nebreda, R. Patiño-Molina, M. Martín-Martínez, I. Gómez-Monterrey, M. T. García-López, R. González-Muñiz, E. Cenarruzabeitia, M. Lattore, J. Del Río and R. Herranz, *J. Med. Chem.*, 2001, **44**, 2219.
258. K. Kitagawa, C. Aida, H. Fujiwara, T. Yagami and S. Futaki, *Chem. Pharm. Bull.*, 2001, **49**, 958.
259. T. Koide, M. Yuguchi and Y. Ito, *Pept. Sci.*, 2001, 369.
260. M. Doi, T. Nakazawa, K. Watanabe, S. Uchiyama, T. Ohkubo and Y. Kobayashi, *Pept. Sci.*, 2001, 341.
261. Y. Tanaka, K. Ueda, E. Ebisuzaki, T. Akanuma, T. Kato and N. Nishino, *Pept. Sci.*, 2001, 355.
262. X. Ding, M. D. Vera, B. Liang, Y. Zhao, M. S. Leonard and M. M. Joullié, *Bioorg. Med. Chem. Lett.*, 2001, **11**, 231.
263. M. Martino and A. M. Tamburro, *Biopolymers*, 2001, **59**, 29.
264. W. Yu, Y. Liang, K. Liu, Y. Zhao, G. Fei and H. Wang, *Junshi Yixue Kexueyuan Yuankan*, 2001, **25**, 190.
265. Y. Liang, W. Yu, K. Liu, Y. Zhao, G. Fei and H Wang, *Zhongguo Yaowu Huaxue Zazhi*, 2001, **11**, 63.
266. F. Dasgupta, N. Gangadhar, M. Bruhaspathy, A. K. Verma, S. Sarin and A. K. Mukherjee, *Bioorg. Med. Chem. Lett.*, 2001, **11**, 555.
267. R. Nonaka, H. Oku, K. Sato, S. Kano, M. Suzuki and R. Katakai, *Pept. Sci.*, 2001, 301.
268. T. Ishiguro, H. Oku, K. Yamada and R. Katakai, *Pept. Sci.*, 2001, 293.
269. S. Gupta and N. S. Sampson, *Org. Lett.*, 2001, **3**, 3333.
270. L. T. Vezenkov, S. A. Malamov, N. St. Pencheva, N. M. Penev and A. I. Mladenova, *Bulg. Chem. Commun.*, 2001, **33**, 35.
271. C. Morita, T. Okuno, K. Hashimoto and H. Shirahama, *Heterocycles*, 2001, **54**, 917.
272. I. M. K. Kumar and B. Mathew, *J. Pept. Sci.*, 2001, **7**, 606.
273. T. M. Hackeng, J. Rosing, H. M. H. Spronk and C. Vermeer, *Protein Sci.*, 2001, **10**, 864.
274. P. Grieco, P. M. Gitu and V. J. Hruby, *J. Pept. Res.*, 2001, **57**, 250.
275. S. Rahimipour, N. Ben-Aroya, M. Fridkin and Y. Koch, *J. Med. Chem.*, 2001, **44**, 3645.
276. G. Jiang, J. Stalewski, R. Galyean, J. Dykert, C. Schteingart, P. Broqua, A. Aebi, M. L. Aubert, G. Semple, P. Robson, K. Akinsanya, R. Haigh, P. Riviere, J. Trojnar, J. L. Junien and J. E. Rivier, *J. Med. Chem.*, 2001, **44**, 453.
277. S. Rahimipour, I. Bilkis, V. Peron, G. Gescheidt, F. Barbosa, Y. Mazur, Y. Koch, L. Weiner and M. Fridkin, *Photochem. Photobiol.* , 2001, **74**, 226.
278. E. Witkowska, A. Orlowska, B. Sagan, M. Smoluch and J. Izdebski, *J. Pept. Sci.*, 2001, **7**, 166.
279. B. Karawajczyk, I. Wirkus-Romanowska, J. Wysocki, K. Rolka, Z. Mackiewicz, R. Glosnicka and G. Kupryszewski, *Pol. J. Chem.*, 2001, **75**, 265.
280. N. Kohyama and Y. Yamamoto, *Synlett*, 2001, 694.
281. D. Lombarska-Sliwinska, D. R. Naessel and D. Konapinska, *Pestycydy*, 2001, 11.

282. W. Szeszel-Fedorowicz, G. Rosinski, J. Issberner, R. Osborne, I. Janssen, A. De Loof and D. Konopinska, *Acta Pol. Pharm.*, 2000, **57**(Suppl.), 88.

283. W. Szeszel-Fedorowicz, M. Lisowski, G. Rosinski, J. Issberner, R. Osborne and D. Konopinska, *Pol. J. Chem.*, 2001, **75**, 411.

284. A. Plant, F. Stieber, J. Scherkenbeck, P. Lösel and H. Dyker, *Org. Lett.*, 2001, **3**, 3427.

285. I. Zeltser, O. Ben-Aziz, I. Schefler, K. Bhargava, M. Altstein and C. Gilon, *J. Pept. Res.*, 2001, **58**, 275.

286. J. Mařík, B. Bennettová, R. Tykva, M. Buděšínský and J. Hlaváček, *J. Pept. Res.*, 2001, **57**, 401.

287. E. E. Büllesbach and C. Schwabe, *J. Pept. Res.*, 2001, **57**, 77.

288. M. J. Horney, C. A. Evangelista and S. A. Rosenzweig, *J. Biol. Chem.*, 2001, **276**, 2880.

289. L. Chen and Y. Sun, *Zhongguo Shenghua Yaowu Zazhi*, 2001, **22**, 234.

290. K. J. Smith, J. D. Wade, A. A. Claasz, L. Otvos, C. Temelcos, Y. Kubota, J. M. Hutson, G. W. Tregear and R. A. Bathgate, *J. Pept. Sci.*, 2001, **7**, 495.

291. H. Li, X. Jiang and M. Goodman, *J. Pept. Sci.*, 2001, **7**, 82.

292. W. K. Hagmann, P. L. Durette, T. Lanza, N. J. Kevin, S. E. de Laszlo, I. E. Kopka, D. Young, P. A. Magriotis, B. Li, L. S. Lin, G. Yang, T. Kamenecka, L. L. Chang, J. Wilson, M. MacCoss, S. G. Mills, G. Van Riper, E. McCauley, L. A. Egger, U. Kidambi, K. Lyons, S. Vincent, R. Stearns, A. Colletti, J. Teffera, S. Tong, J. Fenyk-Melody, K. Owens, D. Levorse, P. Kim, J. A. Schmidt and R. A. Mumford, *Bioorg. Med. Chem. Lett.*, 2001, **11**, 2709.

293. D. Boturyn and P. Dumy, *Tetrahedron Lett.*, 2001, **42**, 2787.

294. H. Yamaguchi, H. Kodama, I. Fujita, F. Kato and M. Kondo, *Pept. Sci.*, 2001, 175.

295. J. R. Broughman, K. E. Mitchell, R. L. Sedlacek, T. Iwamoto, J. M. Tomich and B. D. Schultz, *Amer. J. Physiol.*, 2001, **280**, C451.

296. R. G. Ravi, S. B. Kertesy, G. R. Dubyak and J. A. Jacobson, *Drug Develop. Res.*, 2001, **54**, 75.

297. F. Reig, I. Haro, D. Polo, P. Sospedra and M. A. Alsina, *J. Colloid Interface Sci.*, 2001, **239**, 64.

298. M. Maeda, H. Kamada, K. Hojo, Y. Yamamoto, S. Nakagawa, T. J. Smith, T. Mayumi and K. Kawasaki, *Chem. Pharm. Bull.*, 2001, **49**, 488.

299. J. R. Casimir, K. Iterbeke, W. Van Den Nest, M.-C. Trescol- Biemont, H. Dumortier, S. Muller, D. Gerlier, C. Rabourdin- Combe, D. Tourwe and J. Paris, *J. Pept. Res.*, 2001, **56**, 398.

300. A. Expósito, M. Fernández-Suárez, T. Inglesias, L. Muñoz and R. Riguera, *J. Org. Chem.*, 2001, **66**, 4206.

301. J. M. Caba, I. M. Rodriguez, I. Manzanares, E. Giralt and F. Albericio, *J. Org. Chem.*, 2001, **66**, 7568.

302. P. Prusis, R. Muceniece, I. Mutule, F. Mutulis and J. E. S. Wikberg, *Eur. J. Med. Chem.*, 2001, **36**, 137.

303. M. Haramura, A. Okamachi, K. Tsuzuki, K. Yogo, M. Ikuta, T. Kozono, H. Takanashi and E. Murayama, *Chem. Pharm. Bull.*, 2001, **49**, 40.

304. T. Kawakami, C. Toda, K. Akaji, T. Nishimura, T. Nakatsuji, K. Ueno, M. Sonobe and S. Aimoto, *Pept. Sci.*, 2001, 77.

305. M. Ohtani and S. Aimoto, *Pept. Sci.*, 2001, 179.

306. A. Balasubramaniam, V. C. Dhawan, D. E. Mullins, W. T. Chance, S. Sheriff, M. Guzzi, M. Prabhakaran and E. M. Parker, *J. Med. Chem.*, 2001, **44**, 1479.

307. M. Langer, R. La Bella, E. Garcia-Garayoa and A. G. Beck- Sickinger, *Bioconju-*

gate Chem., 2001, **12**, 1028.

308. Q. Jia, *Sichuan Shifan Daxue Xuebao, Ziran Kexueban*, 2000, **23**, 643.
309. S. E. Schullery, D. W. Rodgers, S. Tripathy, D. E. Jayamaha, M. D. Sanvordekar, K. Renganathan, C. Mousigian and D. L. Heyl, *Bioorg. Med. Chem.*, 2001, **9**, 2633.
310. S. Liao, M. Shenderovich, K. E. Kövér, Z. Zhang, K. Hosohata, P. Davis, F. Porreca, H. I. Yamamura and V. J. Hruby, *J. Pept. Res.*, 2001, **57**, 257.
311. S. A. Mitchell, M. R. Pratt, V. J. Hruby and R. Polt, *J. Org. Chem.*, 2001, **66**, 2327.
312. F. Gosselin, D. Tourwe, M. Ceusters, T. Meert, L. Heylen, M. Jurzak and W. D. Lubell, *J. Pept. Res.*, 2001, **57**, 337.
313. Y. Sasaki, A. Ambo, H. Murase, M. Hirabuki, H. Ouchi and Y. Yamamoto, *Pept. Sci.*, 2001, 117.
314. C. Grison, S. Genève, E. Halbin and P. Coutrot, *Tetrahedron*, 2001, **57**, 4903.
315. N. St. Pencheva, L. T. Vezenkov, E. D. Naydenova, P. B. Milanov, T. I. Pajpanova, S. A. Malamov and L. St. Bojkova, *Bulg. Chem. Commun.*, 2001, **33**, 47.
316. Y. Sasaki, M. Hirabuki, A. Ambo, H. Ouchi and Y. Yamamoto, *Bioorg. Med. Chem. Lett.*, 2001, **11**, 327.
317. D. Pawlak, M. Oleszezuk, J. Wojcik, M. Pachulska, N. N. Chung, P. W. Schiller and J. Izdebski, *J. Pept. Sci.*, 2001, **7**, 128.
318. X.-F. Huo, N. Wu, W.-H. Ren and A. S. C. Chen, *Gaodeng Xuexiao Huaxue Xuebao*, 2001, **22**, 1157.
319. M. Keller, C. Boissard, L. Patiny, N. N. Chung, C. Lemieux, M. Mutter and P. W. Schiller, *J. Med. Chem.*, 2001, **44**, 3896.
320. Y. Lu, G. Weltrowska, C. Lemieux, N. N. Chung and P. W. Schiller, *Bioorg. Med. Chem. Lett.*, 2001, **11**, 323.
321. J. V. Aldrich, Q. Zheng and T. F. Murray, *Chirality*, 2001, **13**, 125.
322. C. Tomboly, R. Dixit, I. Lengyel, A Borsodi and G. Toth, *J. Labelled Compd. Radiopharm.*, 2001, **44**, 355.
323. V. Santagada, G. Balboni, G. Caliendo, R. Guerrini, S. Salvadori, C. Bianchi, S. D. Bryant and L. H. Lazarus, *Bioorg. Med. Chem. Lett.*, 2000, **10**, 2745.
324. L.-m. Wu, J.-h. Chen, X.-q. Mao, Q.-h. Xu and X.-n. Zhao, *Zhongguo Shengwu Huaxue Yu Fenzi Shengwu Xuebao*, 2001, **17**, 215.
325. R. Guerrini, G. Caló, R. Bigoni, D. Rizzi, A. Rizzi, M. Zucchini, K. Varani, E. Hashiba, D. G. Lambert, G. Toth, P. A. Borea, S. Salvadori and D. Regoli, *J. Med. Chem.*, 2001, **44**, 3956.
326. R. Guerrini, G. Calo, R. Bigoni, D. Rizzi, D. Regoli and S. Salvadori, *J. Pept. Res.*, 2001, **57**, 215.
327. J. G. Darker, R. A. Porter, D. S. Eggleston, D. Smart, S. J. Brough, C. Sabido-David and J. C. Jerman, *Bioorg. Med. Chem. Lett.*, 2001, **11**, 737.
328. H. Satofuka, T. Fukui, M. Takagi, H. Atomi and T. Imanaka, *J. Inorg. Biochem.*, 2001, **86**, 595.
329. A. K. Ivanov, A. Antonov and I. A. Donetskii, *Khim. Prir. Soedin*, 1992, 393.
330. A. K. Ivanov, E. E. Grigor'eva, A. A. Antonov and I. A. Donetskii, *Khim. Prir. Soedin*, 1992, 400.
331. Z. Prochazka, M. Zertova and J. Slaninová, *Lett. Pept. Sci.*, 2000, **7**, 179.
332. J. Slaninová, L. Maletinska, J. Vondrasek and Z. Prochazka, *J. Pept. Sci.*, 2001, **7**, 413.
333. M. Manning, S. Stoev, C. L. Cheng, N. C. Wo and W. Y. Chan, *J. Pept. Sci.*, 2001, **7**, 449.
334. E. Carnazzi, A. Aumelas, B. Mouillac, C. Breton, L. Guillou, C. Barberis and R. Seyer, *J. Med. Chem.*, 2001, **44**, 3022.

335. J. Havass, K. Bakos, G. Toth, F. Fulop and G. Falkay, *Acta Pharm. Hung.*, 2000, **70**, 168.

336. L. Bélec, L. Maletinska, J. Slaninová and W. D. Lubell, *J. Pept. Res.*, 2001, **58**, 263.

337. J. Hlavacek and U. Ragnarsson, *J. Pept. Sci.*, 2001, **7**, 349.

338. Y. Hirano, T. Iuchi, M. Kayahara, K. Sato, M. Oka and T. Hayashi, *Pept. Sci.*, 2001, 333.

339. M. Royo, W. Van Den Nest, M. del Fresno, A. Frieden, D. Yahalom, M. Rosenblatt, M. Chorev and F. Albericio, *Tetrahedron Lett.*, 2001, **42**, 7387.

340. K. Hojo, Y. Susuki, M. Maeda, I. Okazaki, M. Nomizu, H. Kamada, Y. Yamamoto, S. Nakagawa, T. Mayumi and K. Kawasaki, *Bioorg. Med. Chem. Lett.*, 2001, **11**, 1429.

341. Y. Susuki, K. Hojo, M. Maeda, M. Nomizu, I. Okazaki, N. Nishi, H. Kamada, Y. Yamamoto, S. Nakagawa, T. Mayumi and K. Kawasaki, *Pept. Sci.*, 2001, 193.

342. T. Isamoto, N. Kiyosawa, T. Yamada and S. Sofuku, *Pept. Sci.*, 2001, 113.

343. M. L. Mondeshki, L. T. Vezenkov and N. St. Pencheva, *Bulg. Chem. Commun.*, 2001, **33**, 29.

344. M. L. Mondeshki, L. T. Vezenkov and N. St. Pencheva, *Bulg. Chem. Commun.*, 2001, **33**, 23.

345. N. Moitessier, S. Dufour, F. Chretien, J. P. Thiery, B. Maigret and Y. Chapleur, *Bioorg. Med. Chem.*, 2001, **9**, 511.

346. S. Oishi, T. Kamano, A. Niida, M. Kawaguchi, R. Hosatani, M. Imamura, N. Yawata, K. Ajito, H. Tamamura, A. Otaka and N. Fujii, *Pept. Sci.*, 2000, 249.

347. M. Giraud, J.-L. Morgat, F. Cavelier and J. Martinez, *J. Labelled Compd. Radiopharm.*, 2001, **44**, 501.

348. A. Janecka and M. Zubrzycka, *Endocr. Regul.*, 2001, **35**, 75.

349. A. Janecka and M. Zubrzycka, *Pol. J. Chem.*, 2001, **75**, 1877.

350. B. A. Hay, B. M. Cole, F. M. DiCapua, G. W. Kirk, M. C. Murray, R. A. Nardone, D. J. Pelletier, A. P. Ricketts, A. S. Robertson and T. W. Siegel, *Bioorg. Med. Chem. Lett.*, 2001, **9**, 2557.

351. W. G. Rajeswaran, S. J. Hocart, W. A. Murphy, J. E. Taylor and D. H. Coy, *J. Med. Chem.*, 2001, **44**, 1305.

352. G. T. Bourne, S. W. Golding, W. D. F. Meutermans and M. L. Smythe, *Lett. Pept. Sci.*, 2001, **7**, 311.

353. J. K. Amartey, R. S. Parhar and I. Al-Jammaz, *Nucl. Med. Biol.*, 2001, **28**, 225.

354. S. A. W. Gruner, G. Kéri, R. Schwab, A. Venetianer and H. Kessler, *Org. Lett.*, 2001, **3**, 3723.

355. R. Millet, L. Goossens, K. Bertrand-Caumont, R. Houssin, B. Rigo, J.-F. Goossens and J.-P. Henichart, *Lett. Pept. Sci.*, 2000, **7**, 269.

356. E. N. Galyuk, S. V. Egorova, E. B. Gurina, *Khim. Prir. Soedin*, 1992, 112.

357. K. Haneda, T. Inazu, M. Mizuno, R. Iguchi, H. Tanabe, K. Fujimori, K. Yamamoto, H. Kumagai, K. Tsumori and E. Munekata, *Biochim. Biophys. Acta*, 2001, **1526**, 242.

358. S. Maricic, A. Ritzén, U. Berg and T. Frejd, *Tetrahedron*, 2001, **57**, 6523.

359. S. Rodziewicz-Motowidto, A. Lesner, A. Legowska, C. Czaplewski, A. Liwo, K. Rolka, R. Patacchini and L. Quartara, *J. Pept. Res.*, 2001, **58**, 159.

360. K. Alexopoulos, D. Panagiotopoulos, T. Mavromoustakos, P. Fatseas, M. C. Paredes-Carbajal, D. Mascher, S. Mihailescu and J. Matsoukas, *J. Med. Chem.*, 2001, **44**, 328.

361. H.-C. Zhang, D. F. McComsey, K. B. White, M. F. Addo, P. Andrade-Gordon, C. K. Derian, D. Oksenberg and B. E. Maryanoff, *Bioorg. Med. Chem. Lett.*, 2001, **11**,

2105.
362. A. Matsushima, T. Fujita, K. Okada, N. Shirasu, T. Nose and Y. Shimohigashi, *Bull. Chem. Soc. Jpn.*, 2000, **73**, 2531.
363. Q. Li and W. Wu, *Yaowu Shengwu Jishu*, 2001, **8**, 227.
364. J. A. Wilce, S. G. Love, S. J. Richardson, P. F. Alewood and D. J. Craik, *J. Biol. Chem.*, 2001, **276**, 25997.
365. R. Jain, J. Singh, J. H. Perlman and M. C. Gershengorn, *Bioorg. Med. Chem.*, 2001, **10**, 189.
366. M. N. Zhmak, I. E. Kasheverov, Yu. N. Utkin, V. I. Tsetlin, O. M. Volpina and V. T. Ivanov, *Russ. J. Bioorg. Chem.*, 2001, **27**, 67.
367. M. Nakamura, Y. Ishida, T. Kohno, K. Sato and H. Nakamura, *Pept. Sci.*, 2001, 85.
368. R. Kasher, M. Balass, T. Scherf, M. Fridkin, S. Fuchs and E. Katchalski-Katzir, *Chem. Biol.*, 2001, **8**, 147.
369. T. Wakamiya, A. Yamamoto, K. Kawaguchi, T. Kinoshita, J. Yamaguchi, Y. Itagaki, H. Naoki and T. Nakajima, *Bull. Chem. Soc. Jpn.*, 2001, **74**, 1743.
370. E. Wong, S. Bennett, B. Lawrence, T. Fauconnier, L. F. L. Lu, R. A. Bell, J. R. Thornback and D. Eshima, *Inorg. Chem.*, 2001, **40**, 5695.
371. P. Wender, D. J. Mitchell, K. Pattabiraman, E. T. Pelkey, L. Steinman and J. B. Rothbard, *Proc. Natl. Acad. Sci., U. S. A.*, 2000, **97**, 13003.
372. G. Tonarelli, J. Lottersberger, J. L. Salvetti, S. Jacchieri, R. A. Silva-Lucca and L. M. Beltramini, *Lett. Pept. Sci.*, 2000, **7**, 217.
373. S. De Falco, M. Ruvo, A. Verdoliva, A. Scarallo, D. Raimondo, A. Raucci and G. Fassina, *J. Pept. Res.*, 2001, **57**, 390.
374. I. L. Kuranova, S. I. Churkina, A. V. Os'mak, E. B. Filonova, V. L. Lyudmirova and O. V. Noskova, *Khim. Prir. Soedin*, 1992, 406.
375. M. Ueki, S. Watanabe, T. Saitoh, H. Nakashima, N. Yamamoto and H. Ogawara, *Bioorg. Med. Chem.*, 2001, **9**, 487.
376. M. Ueki, S. Watanabe, Y. Ishii, O. Okunaka, K. Uchino, T. Saitoh, K. Higashi, H. Nakashima, N. Yamamoto and H. Ogawara, *Bioorg. Med. Chem.*, 2001, **9**, 477.
377. A. Bentolila, I. Vlodavsky, C. Haloun and A. J. Domb, *Polym. Adv. Technol.*, 2000, **11**, 377.
378. Y. Hirano, M. Shimoda, M. Hattori, M. Oka and T. Hayashi, *Pept. Sci.*, 2001, 337.
379. M. Alonso, V. Reboto, L. Guiscardo, V. Mate and J. C. Rodriguez-Cabello, *Macromolecules*, 2001, **34**, 8072.
380. J. Lee, C. W. Macosko and D. W. Urry, *Biomacromolecules*, 2001, **2**, 170.
381. J. Cheng and T. J. Deming, *Macromolecules*, 2001, **34**, 5169.
382. X. Wang, W. H. Daly, P. Russo and M. Ngu-Schwernlein, *Biomacromolecules*, 2001, **2**, 1214.
383. R. H. Wieringa, E. A. Siesling, P. J. Werkman, E. J. Vorenkamp and A. J. Schouten, *Langmuir*, 2001, **17**, 6491.
384. R. H. Wieringa, E. A. Siesling, P. F. M. Geurts, P. J. Werkman, E. J. Vorenkamp, V. Erb, M. Stamm and A. J. Schouten, *Langmuir*, 2001, **17**, 6477.
385. H. L. Jiang, G. P. Tang and K. J. Zhu, *Macromol. Biosci.*, 2001, **1**, 266.
386. A. Cheguillaume, A. Salaün, S. Sinbandhit, M. Potel, P. Gall, M. Baudy-Floc'h and P. Le Grel, *J. Org. Chem.*, 2001, **66**, 4923.
387. H.-S. Lee, F. A. Syud, X. Wang and S. H. Gellman, *J. Amer. Chem. Soc.*, 2001, **123**, 7721.
388. Z. G. Arkin, J. Rydz, G. Adamus and M. Kowalczuk, *J. Biomater. Sci., Polym. Ed.*, 2001, **12**, 297.
389. G. Tang, Y. Zhu, X. Xie and Q. Wu, *Shengwu Yixue Gongehengxue Zazhi*, 2001, **18**,

337.

390. G. Abbenante, D. Leung, T. Bond and D. P. Fairlie, *Lett. Pept. Sci.*, 2000, **7**, 347.

391. J. F. Lynas, S. L. Martin and B. Walker, *J. Pharm. Pharmacol.*, 2001, **53**, 473.

392. W. Yao and H. Y. Xu, *Tetrahedron Lett.*, 2001, **42**, 2549.

393. K. Ohmoto, M. Okuma, T. Yamamoto, H. Kijima, T. Sekioka, K. Kitagawa, S. Yamamoto, K. Tanaka, K. Kawabata, A. Sakata, H. Imawaka, H. Nakai and M. Toda, *Bioorg. Med. Chem.*, 2001, **9**, 1307.

394. M. Demarcus, M. L. Ganadu, G. M. Mura, A. Porcheddu, L. Quaranta, G. Reginato and M. Taddei, *J. Org. Chem.*, 2001, **66**, 697.

395. D. Scarpi, E. G. Occhiato, A. Trabocchi, R. J. Leatherbarrow, A. B. E. Brauer, M. Nievo and A. Guarna, *Bioorg. Med. Chem.*, 2001, **9**, 1625.

396. Y.-Q. Long, S.-L. Lee, C.-Y. Lin, I. J. Enyedy, S. Wang, P. Li, R. B. Dickson and P. P. Roller, *Bioorg. Med. Chem. Lett.*, 2001, **11**, 2515.

397. P. Fraszczak, K. Lis, A. Jaskiewicz, H. Miecznikowska, K. Rolka and K. Kupryszewski, *Pol. J. Chem.*, 2001, **75**, 1863.

398. H. Fei, M. Luo, Y. Ye, D. Ding and Z. Qi, *Shengwu Huaxue Yu Shengwu Wuli Xuebao*, 2001, **33**, 591.

399. Z. Xia and C. D. Smith, *J. Org. Chem.*, 2001, **66**, 5241.

400. M. C. Hillier, J. P. Davidson and S. F. Martin, *J. Org. Chem.*, 2001, **66**, 1657.

401. M. Brewer and D. H. Rich, *Org. Lett.*, 2001, **3**, 945.

402. D. E. Ehmann, J. W. Trauger, T. Stachelhaus and C. T. Walsh, *Chem. Biol.*, 2000, **7**, 765.

403. P. Xu and W. Lin, *Zhongguo Yaowu Huaxue Zazhi*, 2001, **11**, 32.

404. T. Fujisawa, S.-I. Katakura, S. Odake, Y. Morita, J. Yasuda, I. Yasumatsu and T. Morikawa, *Chem. Pharm. Bull.*, 2001, **49**, 1272.

405. D. L. Musso, M. W. Andersen, R. C. Andrews, R. Austin, E. J. Beaudet, J. D. Becherer, D. G. Bubacz, D. M. Bickett, J. H. Chan, J. G. Conway, D. J. Cowan, M. D. Gaul, K. C. Glennon, K. M. Hedeen, M. H. Lambert, M. A. Leesnitzer, D. L. McDougald, J. L. Mitchell, M. L. Moss, M. H. Rabinowitz, M. C. Rizzolio, L. T. Schaller, J. B. Stanford, T. K. Tippin, J. R. Warner, L. G. Whitesell and R. W. Wiethe, *Bioorg. Med. Chem. Lett.*, 2001, **11**, 2147.

406. T. Fujisawa, S. Odake, J. Yasuda and M. Morikawa, *Pept. Sci.*, 2001, 181.

407. I. Momose, R. Sekizawa, S. Hirosawa, D. Ikeda, H. Naganawa, H. Iinuma and T. Takeuchi, *J. Antibiotics*, 2001, **54**, 1004.

408. B. M. Kessler, D. Tortorella, M. Altun, A. F. Kisselev, E. Fiebiger, B. G. Hekking, H. L. Ploegh and H. S. Overkleeft, *Chem. Biol.*, 2001, **8**, 913.

409. J. Yang, S. Wu, K. Zhou, S. Peng, S. Meng, Z. Wu, Y. Ruan and F. Dong, *Zhongguo Yaowu Huaxue Zazhi*, 2001, **11**, 13.

410. A. J. Harvey and A. D. Abell, *Bioorg. Med. Chem. Lett.*, 2001, **11**, 2441.

411. S. D. Jones, J. W. Liebeschuetz, P. J. Morgan, C. W. Murray, A. D. Rimmer, J. M. E. Roscoe, B. Waszkowycz, P. M. Welsh, W. A. Wylie, S. C. Young, H. Martin, J. Mahler, L. Brady and K. Wilkinson, *Bioorg. Med. Chem. Lett.*, 2001, **11**, 733.

412. T. Su, J. Wu, B. Doughan, Z. J. Jia, J. Woolfrey, B. Huang, P. Wong, G. Park, U. Sinha, R. M. Scarborough and B.-y. Zhu, *Bioorg. Med. Chem. Lett.*, 2001, **11**, 2947.

413. M. Umezaki, F. Kondo, T. Fujii, T. Yoshimura and S. Ono, *Pept. Sci.*, 2001, 33.

414. R. L. Melo, R. C. Barbosa Pozzo, L. C. Alves, E. Perisutti, G. Caliendo, V. Santagada, L. Juliano and M. A. Juliano, *Biochim. Biophys. Acta.*, 2001, **1547**, 82.

415. M. Makowski, M. Pawelczak, R. Latajka, K. Nowak and P. Kafarski, *J. Pept. Sci.*, 2001, **7**, 141.

416. R. A. Smith, A. Bhargava, C. Browe, J. Chen, J. Dumas, H. Hatoum-Mokdad and

R. Romero, *Bioorg. Med. Chem. Lett.*, 2001, **11**, 2951.

417.   Z. Guo, M. Xian, W. Zhang, A. McGill and P. G. Wang, *Bioorg. Med. Chem.*, 2001, **9**, 99.
418.   K. A. Josef, F. K. Kauer and R. Bihovsky, *Bioorg. Med. Chem. Lett.*, 2001, **11**, 2615.
419.   G. J. Wells, M. Tao, K. A. Josef and R. Bihovsky, *J. Med. Chem.*, 2001, **44**, 3488.
420.   E. I. Milgotina, A. S. Shcheglov, G. B. Lapa, G. G. Chestukhina and T. L. Voyushina, *J. Pept. Res.*, 2001, **58**, 12.
421.   J. M. Travins, M. G. Bursavich, D. F. Veber and D. H. Rich, *Org. Lett.*, 2001, **3**, 2725.
422.   N. A. Dales, R. S. Bohacek, K. A. Satyshur and D. H. Rich, *Org. Lett.*, 2001, **3**, 2313.
423.   C. Pesenti, A. Arnone, S. Bellosta, P. Bravo, M. Canavesi, E. Corradi, M. Frigerio, S. V. Meille, M. Monetti, W. Panzeri, F. Viani, R. Venturini and M. Zanda, *Tetrahedron*, 2001, **57**, 6511.
424.   A. S. Ripka, K. A. Satyshur, R. S. Bohacek and D. H. Rich, *Org. Lett.*, 2001, **3**, 2309.
425.   R. L. Jarvest, J. M. Berge, P. Brown, D. W. Hamprecht, D. J. McNair, L. Mensah, P. J. O'Hanlon and A. J. Pope, *Bioorg. Med. Chem. Lett.*, 2001, **11**, 715.
426.   P. Brown, D. S. Eggleston, R. C. Haltiwanger, R. L. Jarvest, L. Mensah, P. J. O'Hanlon and A. J. Pope, *Bioorg. Med. Chem. Lett.*, 2001, **11**, 711.
427.   R. A. Laursen, C. Zhu and Y. Duan in *High-Throughput Synthesis*, ed. M. Sucholeiki, pub. M. Dekker,Inc. New York, 2001, pp. 109-115.
428.   H. Chen, F. Noble, B. P. Roques and M.-C. Fournié-Zaluski, *J. Med. Chem.*, 2001, **44**, 3523.
429.   K. Lee, W.-H. Jung, M. Kang and S.-H. Lee, *Bioorg. Med. Chem. Lett.*, 2001, **10**, 2775.
430.   J. Cossy and D. Belotti, *Bioorg. Med. Chem. Lett.*, 2001, **11**, 1989.
431.   Y. K. Ramtohul, N. I. Martin, L. Silkin, M. N. G. James and J. C. Vederas, *Chem. Commun.*, 2001, 2740.
432.   M.-A. Poupart, D. R. Cameron, C. Chabot, E. Ghiro, N. Goudreau, S. Goulet, M. Poirier and Y. S. Tsantrizos, *J. Org. Chem.*, 2001, **66**, 4743.
433.   F. Kasprzykowski, P. Kania, K. Plucinska, A. Grubb, M. Abrahamson and C. Schalen, *Pol. J. Chem.*, 2001, **75**, 831.
434.   A. Graven, P. M. St. Hilaire, S. J. Sanderson, J. C. Mottram, G. H. Coombs and M. Meldal, *J. Comb. Chem.*, 2001, **3**, 441.
435.   L. C. Alves, R. L. Melo, S. J. Sanderson, J. C. Mottram, G. H. Coombs, G. Caliendo, V. Santagada, L. Juliano and M. A. Juliano, *Eur. J. Biochem.*, 2001, **268**, 1206.
436.   A. E. Fenwick, B. Garnier, A. D. Gribble, R. J. Ife, A. D. Rawlings and J. Witherington, *Bioorg. Med. Chem. Lett.*, 2001, **11**, 195.
437.   A. E. Fenwick, A. D. Gribble, R. J. Ife, N. Stevens and J. Witherington, *Bioorg. Med. Chem. Lett.*, 2001, **11**, 199.
438.   M. Marastoni, M. Bazzaro, S. Salvadori, F. Bortolotti and R. Tomatis, *Bioorg. Med. Chem.*, 2001, **9**, 939.
439.   H. Matsumoto, T. Matsuda, S. Nakata, T. Mitoguchi, T. Kimura, Y. Hayashi and Y. Kiso, *Bioorg. Med. Chem.*, 2001, **9**, 417.
440.   Y. Konda, Y. Takahashi, S. Arima, N. Sato, K. Takeda, K. Dobashi, M. Baba and Y. Harigaya, *Tetrahedron*, 2001, **57**, 4311.
441.   M.-c. Song, S. Rajesh, Y. Hayashi and K. Kiso, *Bioorg. Med. Chem. Lett.*, 2001, **11**, 2465.
442.   H. Matsumoto, T. Kimura, T. Hamawaki, A. Kumagai, T. Goto, K. Sano, Y. Hayashi and Y. Kiso, *Bioorg. Med. Chem.*, 2001, **9**, 1589.
443.   J. Courcambeck, F. Bihel, C. De Michelis, G. Quéléver and J. L. Kraus, *J. Chem.*

Soc., Perkin Trans 1, 2001, 1421.

444. V.-D. Le, C. C. Mak, Y.-C. Lin,. J. H. Elder and C.-H. Wong, *Bioorg. Med. Chem.*, 2001, **9**, 1185.
445. H. Matsumoto, Y. Sohma, D. Shuto, T. Hamawaki, T. Kimura, Y. Hayashi and Y. Kiso, *Pept. Sci.*, 2001, 237.
446. S. Bounaga, M. Galleni, A. Laws and M. I. Page, *Bioorg. Med. Chem.*, 2001, **9**, 503.
447. D. Orain and M. Bradley, *Tetrahedron Lett.*, 2001, **42**, 515.
448. J. H. McKie, J. Garforth, R. Jaouhari, C. Chan, H. Yin, T. Besheya, A. H. Fairlamb and K. T. Douglas, *Amino Acids*, 2001, **20**, 145.
449. C. Marrano, P. de Macedo, P. Gagnon, D. Lappiere, C. Gravel and J. W. Keillor, *Bioorg. Med. Chem.*, 2001, **9**, 3231.
450. C. Marrano, P. de Macedo and J. W. Keillor, *Bioorg. Med. Chem.*, 2001, **9**, 1923.
451. N. Valiaeva, D. Bartley, T. Konno and J. K. Coward, *J. Org. Chem.*, 2001, **66**, 5146.
452. K. Paschalidou and C. Tzougraki, *Lett. Pept. Sci.*, 2000, **7**, 249.
453. J. J. Pankuch and J. K. Coward, *Bioorg. Med. Chem. Lett.*, 2001, **11**, 1561.
454. H. Huang, P. Martasek, L. J. Roman and R. B. Silverman, *J. Enzyme Inhib.*, 2001, **16**, 233.
455. N. Carulla, C. Woodward and G. Barany, *Bioconjugate Chem*, 2001, **12**, 726.
456. W. Qiu, X. Gu, V. A. Soloshonok, M. D. Carducci and V. J. Hruby, *Tetrahedron Lett.*, 2001, **42**, 145.
457. S. Lucarini and C. Tomasini, *J. Org. Chem.*, 2001, **66**, 727.
458. M. Tanaka, M. Oba, N. Imawaka, Y. Tanaka, M. Kurihara and H. Suemune, *Helv. Chim. Acta*, 2001, **84**, 32.
459. L. M. A. McNamara, M. J. I. Andrews, F. Mitzel, G. Siligardi and A. B. Tabor, *Tetrahedron Lett.*, 2001, **42**, 1591.
460. I. Obataya, S. Sakamoto, A. Ueno and H. Mihara, *Biopolymers*, 2001, **59**, 65.
461. C.-Y. Huang, Z. Getahun, T. Wang, W. F. DeGrado and F. Gai, *J. Amer. Chem. Soc.*, 2001, **123**, 12111.
462. H. Matsuno, K. Niikura and Y. Okahata, *Biochemistry*, 2001, **40**, 3615.
463. I. Fujii, Y. Takaoka, K. Suzuki and T. Kanaka, *Tetrahedron Lett.*, 2001, **42**, 3323.
464. L. G. J. Hammarstrom, T. J. Gauthier, R. P. Hammer and M. L. McLaughlin, *J. Pept. Res.*, 2001, **58**, 108.
465. T. D. W. Claridge, J. M. Goodman, A. Moreno, D. Angus, S. F. Barker, C. Taillefumier, M. P. Watterson and G. W. J. Fleet, *Tetrahedron Lett.*, 2001, **42**, 4251.
466. S. F. Barker, D. Angus, C. Taillefumier, M. R. Probert, D. J. Watkin, M. P. Watterson, T. D. W. Claridge, N. L. Hungerford and G. W. J. Fleet, *Tetrahedron Lett.*, 2001, **42**, 4247.
467. R. Schnepf, P. Hörth, E. Bill, K. Wieghardt, P. Hildebrandt and W. Haehnel, *J. Amer. Chem. Soc.*, 2001, **123**, 2186.
468. N. A. Schnarr and A. J. Kennan, *J. Amer. Chem. Soc.*, 2001, **123**, 11081.
469. B. Bilgicer, X. Xing and K. Kumar, *J. Amer. Chem. Soc.*, 2001, **123**, 11815.
470. C. Das, G. A. Naganagowda, I. L. Karle and P. Balaram, *Biopolymers*, 2001, **58**, 335.
471. B. Aguilera, G. Siegal, H. S. Overkleeft, N. J. Meeuwenoord, F. P. J. T. Rutjes, J. C. M. Van Hest, H. E. Schoemaker, G. A. van der Marel, J. H. van Boom and M. Overhand, *Eur. J. Org. Chem.*, 2001, 1541.
472. J. H. Miwa, A. K. Patel, N. Vivatrat, S. M. Popek and A. M. Meyer, *Org. Lett.*, 2001, **3**, 3373.
473. C. Hemmerlin, M. T. Cung and G. Boussard, *Tetrahedron Lett.*, 2001, **42**, 5009.
474. R. C. F. Jones and J. Dickson, *J. Pept. Sci.*, 2001, **7**, 220.
475. I. A. Motorina, C. Huel, E. Quiniou, J. Mispelter, E. Adjadj and D. S. Grierson, *J.*

Amer. Chem. Soc., 2001, **123**, 8.
476. W. D. Kohn and L. Zhang, *Tetrahedron Lett.*, 2001, **42**, 4453.
477. M. Eguchi, M. S. Lee, M. Stasiak and M. Kahn, *Tetrahedron Lett.*, 2001, **42**, 1283.
478. C. Das, S. C. Channaveerappa and P. Balaram, *Chem. –Eur. J.*, 2001, **7**, 840.
479. F. Formaggio, C. Peggion, M. Crisma, C. Toniolo, L. Tchertanov, J. Guilhem, J.-P. Mazaleyrat, Y. Goubard, A. Gaucher and M. Wakselman, *Helv. Chim. Acta*, 2001, **84**, 481.
480. D. Seebach, J. V. Schreiber, P. I. Arvidsson and J. Frackenpohl, *Helv. Chim. Acta*, 2001, **84**, 271.
481. P. I. Arvidsson, M. Rueping and P. Seebach, *Chem. Commun.*, 2001, 649.
482. E. T. Powers and J. W. Kelly, *J. Amer. Chem. Soc.*, 2001, **123**, 775.
483. A. Moretto, F. Formaggio, M. Crisma, C. Toniolo, M. Saviano, R. Iacovino, R. M. Vitale and E. Benedetti, *J. Pept. Res.*, 2001, **57**, 307.
484. M. Saviano, R. Iacovino, E. Benedetti, V. Moretto, A. Banzato, F. Formaggio, M. Crisma and C. Yoniolo, *J. Pept. Sci.* , 2000, **6**, 571.
485. M. Brenner and D. Seebach, *Helv. Chim. Acta*, 2001, **84**, 2155.
486. J. S. Nowick, J. M. Cary and J. H. Tsai, *J. Amer. Chem. Soc.*, 2001, **123**, 5176.
487. A. R. Mezo, R. P. Cheng and B. Imperiali, *J. Amer. Chem. Soc.*, 2001, **123**, 3885.
488. S. K. Awasthyi, S. C. Shankaramma, S. Raghothama and P. Balaram, *Biopolymers*, 2001, **58**, 465.
489. T. Nakamura, M. Wakahara, M. Oka, T. Hayashi, M. Hattori and Y. Hirano, *Pept. Sci.*, 2001, 321.
490. Y. Ohnuma, M. Wakahara, M. Oka, T. Hayashi, S. Nishimura, M. Hattori and Y. Hirano, *Pept. Sci.*, 2001, 317.
491. Y. Inai and T. Hirabayashi, *Biopolymers*, 2001, **59**, 356.
492. A. J. Thompson, K. J. Barnham, R. S. Norton and C. J. Barrow, *Biochim. Biophys. Acta*, 2001, **1544**, 242.
493. B. Kaptein, Q. B. Broxterman, H. E. Schoemaker, F. P. J. T. Rutjes, J. J. N. Veerman, J. Kamphuis, C. Peggion, F. Formaggio and C. Toniolo, *Tetrahedron*, 2001, **57**, 6567.
494. K. Bhargava and R. B. Rao, *Protein Pept. Lett.*, 2001, **8**, 307.
495. T. K. Chakraborty, A. Ghosh, R. Nagaraj, A. R. Sankar and A. C. Kunwar, *Tetrahedron*, 2001, **57**, 9169.
496. B. Baumeister, N. Sakai and S. Matile, *Org. Lett.*, 2001, **3**, 4229.
497. L. Martin, A. Ivancich, C. Vita, F. Formaggio and C. Toniolo, *J. Pept. Res.*, 2001, **58**, 424.
498. P. Sjölin and J. Kihlberg, *J. Org. Chem.*, 2001, **66**, 2957.
499. D. Kumagai, M. Miyazaki and S.-I. Nishimura, *Tetrahedron Lett.*, 2001, **42**, 1953.
500. K. Fukase, T. Yasukochi and S. Kusumoto, *Bull. Chem. Soc. Jpn.*, 2001, **74**, 1123.
501. C. Grison, F. Coutrot and P. Coutrot, *Tetrahedron*, 2001, **57**, 6215.
502. K. Dzierzbicka, A. M. Kołodziejczyk, B. Wysocka-Skrzela, A. Myśliwski and D. Sosnowska, *J. Med. Chem.*, 2001, **44**, 3606.
503. S. De Zhang, G. Liu and S. Q. Xia, *Chin. Chem. Lett.*, 2001, **12**, 887.
504. V. O. Kur'yanov, T. A. Chupakhina, A. E. Zemlyakov, V. Ya. Chirva, V. V. Ishchenko and V. P. Khilya, *Chem. Nat. Compd.*, 2001, **37**, 39.
505. S. Wen and Z. Guo, *Org. Lett.*, 2001, **3**, 3773.
506. J.-j. Zhou, X.-z. Wang, Y.-z. Wu, Z.-c. Jia, W. Zhou and J. Bian, *Di-San Junyi Daxue Xuebao*, 2000, **22**, 958.
507. J. W. Lane and R. L. Halcomb, *Tetrahedron*, 2001, **57**, 6531.
508. P. H. Tseng, W.-T. Jiaang, M.-Y. Chang and S. T. Chen, *Chem.- Eur. J.*, 2001, **7**, 585.

509. A. Ishii, H. Hojo, Y. Nakahara, Y. Nakahara and Y. Ito, *Pept. Sci.*, 2001, 43.
510. H.-G. Lerchen and K. Von dem Bruch, *J. Prakt. Chem.*, 2001, **342**, 753.
511. B. G. Davis, *Chem. Commun.*, 2001, 351.
512. V. Wittmann and S. Seeberger, *Angew. Chem., Int. Ed.*, 2000, **39**, 4348.
513. C. J. Bosques, V. W.-F. Tai and B. Imperiali, *Tetrahedron Lett.*, 2001, **42**, 7207.
514. L. Tarantini, D. Monti, L. Panza, D. Prosperi and S. Riva, *J. Mol. Catal. B: Enzym.*, 2001, **11**, 343.
515. D. Ramos, P. Rollin and W. Klaffke, *J. Org. Chem.*, 2001, **66**, 2948.
516. S. K. George, T. Schwientek, B. Holm, C. A. Reis, H. Clausen and J. Kihlberg, *J. Amer. Chem. Soc.*, 2001, **123**, 11117.
517. C.-Y. Tsai, X. Huang and C.-H. Wong, *Tetrahedron Lett.*, 2000, **41**, 9499.
518. K. Hamasaki, R. Suzuki, H. Mihara and A. Ueno, *Macromol. Rapid Commun.*, 2001, **22**, 262.
519. J. Brask and K. J. Jensen, *Bioorg. Med. Chem. Lett.*, 2001, **11**, 697.
520. S. Peluso and B. Imperiali, *Tetrahedron Lett.*, 2001, **42**, 2085.
521. P. Sears and C.-H. Wong, *Science*, 2001, **291**, 2344.
522. A. Eniade and R. N. Ben, *Biomacromolecules*, 2001, **2**, 557.
523. M. Gobbo, A. Nicotra, R. Rocchi, M. Crisma and C. Toniolo, *Tetrahedron*, 2001, **57**, 2433.
524. K. Burger, C. Böttcher, G. Radics and L. Hennig, *Tetrahedron Lett.*, 2001, **42**, 3061.
525. D. D. Long, R. J. Tennant-Eyles, J. C. Estevez, M. R. Wormald, R. A. Dwek, M. D. Smith and G. W. J. Fleet, *J. Chem. Soc., Perkin Trans. 1*, 2001, 807.
526. J. Satyanarayana, T. L. Gururaja, S. Narasimhamurthy, G. A. Naganagowda and M. J. Levine, *Biopolymers*, 2001, **58**, 500.
527. H. Hojo, J. Watabe, Y. Nakahara, Y. Nakahara, Y. Ito and K. Nabeshima, *Pept. Sci.*, 2001, 47.
528. I. Jerić and S. Horvat, *Eur. J. Org. Chem.*, 2001, 1533.
529. K. M. Halkes, C. H. Gotfredsen, M. Grøtli, L. P. Miranda, J. O. Duus and M. Meldal, *Chem. –Eur. J.* 2001, **7**, 3584.
530. G. Di Fabio, A. De Capua, L. De Napoli, D. Montesarchio, G. Piccialli, F. Rossi and E. Benedetti, *Synlett*, 2001, 341.
531. M. Roice and V. N. R. Pillai, *Lett. Pept. Sci.*, 2000, **7**, 281.
532. Y. Nakahara, S. Ando, Y. Ito, H. Hojo and Y. Nakahara, *Biosci., Biotechnol., Biochem.*, 2001, **65**, 1358.
533. N. Bézay, G. Dudziak, A. Liese and H. Kunz, *Angew. Chem., Int. Ed.*, 2001, **40**, 2292.
534. K. Ajisaka, M. Miyasato and I. Ishii-Karakasa, *Biosci. Biotechnol. Biochem.*, 2001, **65**, 1240.
535. S.-B. Chen, Y.-M. Li, G. Zhao, S.-Z. Luo and Y.-F. Zhao, *Gaodeng Xuexiao Huaxue Xuebao*, 2001, **22**, 106.
536. S.-B. Chen, Y.-M. Li, S.-Z. Luo, G. Zhao, B. Tan and Y.-F. Zhao, *Phosphorus, Sulfur Silicon Relat. Elem.*, 2000, **164**, 277.
537. D. Sikora, T. Nonas and T. Gajda, *Tetrahedron*, 2001, **57**, 1619.
538. S.-Z. Dong, H. Fu and Y.-F. Zhao, *Synth. Commun.*, 2001, **31**, 2067.
539. R. E. Dolle, T. F. Herpin and Y. C. Shimshock, *Tetrahedron Lett.*, 2001, **42**, 1855.
540. Z. Kupihár, Z. Kele and G. K. Tóth, *Org. Lett.*, 2001, **3**, 1033.
541. T. Kawakami, K. Hasegawa, K. Teruya, K. Akaji, M. Horiuchi, F. Inagaki, Y. Kurihara, S. Uesugi and S. Aimoto, *J. Pept. Sci.*, 2001, **7**, 474.
542. T. Kawakami, K. Teruya, K. Hasegawa, K. Akaji, M. Horiuchi, F. Inagaki, Y. Kurihara, S. Uesugi and S. Aimoto, *Pept. Sci.*, 2001, 21.
543. A. Otaka, E. Mitsuyama, H. Watanabe, H. Tamamura, T. Hayashi, K. Nakao and

N. Fujii, *Pept. Sci.*, 2001, 17.
544. D. Georgiadis, M. Matziari and A. Yiotakis, *Tetrahedron*, 2001, **57**, 3471.
545. B. L. Nilsson, L. L. Kiessling and R. T. Raines, *Org. Lett.*, 2001, **3**, 9.
546. H. J. Cristau, A. Coulombeau, A. Genevoie-Borella and J. L. Pirat, *Tetrahedron Lett.*, 2001, **42**, 4491.
547. D. Georgiadis, V. Dive and A. Yiotakis, *J. Org. Chem.*, 2001, **66**, 6604.
548. L. Demange and C. Dugave, *Tetrahedron Lett.*, 2001, **42**, 6295.
549. Z. Ziora, P. Kafarski, J. Holband and G. Wojcik, *J. Pept. Sci.*, 2001, **7**, 466.
550. B. Boduszek, *Phosphorus, Sulfur, Silicon Relat. Elem.*, 2001, **176**, 119.
551. X.-J. Liu and R.-Y. Chen, *Phosphorus, Sulfur, Silicon Relat. Elem.*, 2001, **176**, 19.
552. G. Kragol and L. Otvos, *Tetrahedron*, 2001, **57**, 957.
553. H. H. Keah, M. K. O'Bryan, D. M. De Kretser and M. T. W. Hearn, *J. Pept. Res.*, 2001, **57**, 1.
554. K. S. Kumar, M. Roice, P. G. Sasikumar, C. D. Poduri, V. S. Sugunan, V. N. R. Pillai and M. R. Das, *J. Pept. Res.*, 2001, **57**, 140.
555. M. Favre, K. Mohle and J. A. Robinson, *Proc. ECSOC-3, Proc. ECSOC-4*, 1999, 2000, 1582.
556. F. Chaves, J. C. Calvo, C. Carvajal, Z. Rivera, L. Ramirez, M. Pinto, M. Trujillo, F. Guzman and M. E. Patarroyo, *J. Pept. Res.*, 2001, **58**, 307.
557. A. Quesnel, A. Zerbib, F. Connan, J.-G. Guillet, J.-P. Briand and J. Choppin, *J. Pept. Sci.*, 2001, **7**, 157.
558. C. Alexopoulos, V. Tsikaris, C. Rizou, E. Panou-Pomonis, M. Sakarellos-Daitsiotis, C. Sakarellos, P. G. Vlachoyiannopoulos and H. M. Moutsopoulos, *J. Pept. Sci.*, 2001, **7**, 105.
559. S. K. George, B. Holm, C. A. Reis, T. Schwientek, H. Clausen and J. Kihlberg, *J. Chem. Soc., Perkin Trans. 1*, 2001, 880.
560. K. Nokihara, S. Shimizu, R. Pipkorn, T. Yasuhara and T. Shioda, *Pept. Sci.*, 2001, 13.
561. L. J. Cruz, R. Padron, L. J. Gonzalez, J. C. Aguilar, E. Iglesias, H. E. Garay, V. Falcon, E. Rodriguez and O. Reyes, *Lett. Pept. Sci.*, 2000, **7**, 229.
562. T. D. Volkova, O. M. Vol'pina, M. N. Zhmak, V. T. Ivanov, M. F. Vorovich and M. F. Timoveev, *Russ. J. Bioorg. Chem.*, 2001, **27**, 151.
563. J. Mack, K. Falk, O. Rotzschke, T. Walk, J. L. Strominger and G. Jung, *J. Pept. Sci.*, 2001, **7**, 338.
564. L. J. Cruz, E. Iglesias, J. C. Aguilar, D. Quintana, H. E. Garay, C. Duarte and O. Reyes, *J. Pept. Sci.*, 2001, **7**, 511.
565. P. Zubrzak, K. Kaczmarek, M. L. Kowalski, B. Szkudlinska and J. Zabrocki, *Pol. J. Chem.*, 2001, **75**, 1869.
566. D. S. Jones, K. A. Cockerill, C. A. Gamino, J. R. Hammaker, M. S. Hayag, G. M. Iverson, M. D. Linnik, P. A. McNeeley, M. E. Tedder, H.-T. Ton-Nu and E. J. Victoria, *Bioconjugate Chem.*, 2001, **12**, 1012.
567. A. Okamoto, K. Tanabe and I. Saito, *Org. Lett.*, 2001, **3**, 925.
568. R. D. Viirre and R. H. E. Hudson, *Org. Lett.*, 2001, **3**, 3931.
569. G. Kovics, Z. Timar, Z. Kele and L. Kovacs, *Proc. ECSOC-3, Proc. ECSOC-4*, 1999, 2000, 1240.
570. Y. Wu and J.-C. Xu, *Huaxue Xuebao*, 2001, **59**, 1660.
571. J.-Q. Bai, Y. Li and K.-L. Liu, *Chin. J. Chem.*, 2001, **19**, 276.
572. B. Folkiewicz, W. Wisniowski, A. S. Kolodziejczyk and K. Wisniewski, *Nucleosides, Nucleotides Nucleic Acids*, 2001, **20**, 1393.
573. A. Slaitas and E. Yeheskiely, *Nucleosides, Nucleotides Nucleic Acids*, 2001, **20**, 1377.

574. A. B. Eldrup, B. B. Nielsen, G. Haaima, H. Rasmussen, J. S. Kastrup, C. Christensen and P. E. Nielsen, *Eur. J. Org. Chem.*, 2001, 1781.

575. C. K. Kwan, *Vestn. Mosk. Univ., Ser. 2: Khim*, 2001, **42**, 128.

576. H. Ikeda, K. Yoshida, M. Ozeki and I. Saito, *Tetrahedron Lett.*, 2001, **42**, 2529.

577. M. D'Costa, V. Kumar and K. N. Ganesh, *Org. Lett.*, 2001, **3**, 1281.

578. S. Matsumura, A. Ueno and H. Mihara, *Pept. Sci.*, 2001, 363.

579. T. Kubo, K. Dubey and M. Fujii, *Pept. Sci.*, 2001, 23.

580. T. Kubo, K. Dubey and M. Fujii, *Nucleosides, Nucleotides Nucleic Acids*, 2001, **20**, 1321.

581. A. Mattes and O. Seitz, *Chem. Commun.*, 2001, 2050.

582. M. Antopolsky and A. Azhayev, *Nucleosides, Nucleotides, Nucleic Acids*, 2001, **20**, 539.

583. S. O. Doronina, A. P. Guzaev and M. Manoharan, *Nucleosides, Nucleotides and Nucleic Acids*, 2001, **20**, 1007.

584. D. A. Stetsenko and M. J. Gait, *Nucleosides, Nucleotides Nucleic Acids*, 2001, **20**, 801.

585. M. C. De Koning, D. V. Filippov, N. Meeuwenoord, M. Overhand, G. A. van der Marel and J. H. van Boom, *Synlett*, 2001, 1516.

586. V. Efimov, M. Choob, A. Buryakova, D. Phelan and O. Chakhmakhcheva, *Nucleosides, Nucleotides Nucleic Acids*, 2001, **20**, 419.

587. L. D. Fader, M. Boyd and Y. S. Tsantrizos, *J. Org. Chem.*, 2001, **66**, 3372.

588. T. Kofoed, H. F. Hansen, H. Orum and T. Koch, *J. Pept. Sci.*, 2001, **7**, 402.

589. A. Püschl, T. Boesen, T. Tedeschi, O. Dahl and P. E. Nielsen, *J. Chem. Soc., Perkin Trans. 1*, 2001, 2757.

590. Y. Wu, J.-C. Xu, J. Liu and Y.-X. Jin, *Tetrahedron*, 2001, **57**, 3373.

591. B. Falkiewicz, A. S. Kołodziejczyk, B. Liberek and K. Wiśniewski, *Tetrahedron*, 2001, **57**, 7909.

592. Y. Wu and and J.-C. Xu, *Tetrahedron*, 2001, **57**, 8107.

593. E. Ferrer and R. Eritja, *Lett. Pept. Sci.*, 2000, **7**, 195.

594. X. Zhang, C. G. Simmons and D. R. Corey, *Bioorg. Med. Chem. Lett.*, 2001, **11**, 1269.

595. T. Takahashi, K. Hamasaki, A. Ueno and H. Mihara, *Bioorg. Med. Chem.*, 2001, **9**, 991.

596. T. Vilaivan, C. Khongdeesameor, W. Wiriyawaree, W. Wansawat, M. S. Westwell and G. Lowe, ScienceAsia, 2001, **27**, 113.

597. G. Balasundaram, T. Takahashi, A. Ueno and H. Mihara, *Bioorg. Med. Chem.*, 2001, **9**, 1115.

598. P. Zhang, M. Egholm, N. Paul, M. Pingle and D. E. Bergstrom, *Methods*, 2001, **23**, 132.

599. N. Shibata, B. K. Das, H. Honjo and Y. Takeuchi, *J. Chem. Soc., Perkin Trans. 1*, 2001, 1605.

600. E. R. Civitello, R. G. Leniek, K. A. Hossler, K. Haebe and D. M. Stearns, *Bioconjugate Chem.*, 2001, **12**, 459.

601. T. Vilaivan, C. Suparpprom, P. Harnyuttanakorn and G. Lowe, *Tetrahedron Lett.*, 2001, **42**, 5533.

602. M. Shigeyasu M. Kuwahara, M. Sisido and T. Ishikawa, *Chem. Lett.*, 2001, 634.

603. D. Forget, D. Boturyn, E. Defrancq, J. Lhomme and P. Dumy, *Chem. –Eur. J.*, 2001, **7**, 3976.

604. D. Capasso, L. De Napoli, G. Di Fabio, A. Messere, D. Montesarchio, C. Pedone, G. Piccialli and M. Saviano, *Tetrahedron*, 2001, **57**, 9481.

605. R. Corradini, S. Sforza, A. Dossena, G. Palla, R. Rocchi, F. Filira, F. Nastri and R. Marchelli, *J. Chem. Soc., Perkin Trans. 1*, 2001, 2690.
606. C. N. Tetzlaff and C. Richert, *Tetrahedron Lett.*, 2001, **42**, 5681.
607. L. Chen, Y. Xu, F. Xu, D. Wang and Y. Meng, *Huaxue Tongbao*, 2001, 428.
608. Y. Yuasa, A. Nagakura and H. Tsuruta, *J. Agric. Food Chem.*, 2001, **49**, 5013.
609. Y. Li, J. Qin and S. Zhao, *Di-San Junyi Daxue Xuebao*, 2001, **23**, 571.
610. M. Royo, J. C. Jiménez, A. López-Macià, E. Giralt and F. Albericio, *Eur. J. Org. Chem.*, 2001, 45.
611. S. B. Shuker, J. Esterbrook and J. Gonzalez, *Synlett*, 2001, 210.
612. M. Lazzarotto, F. Sansone, L. Baldini, A. Casnati, P. Cozzini and R. Ungaro, *Eur. J. Org. Chem.*, 2001, 595.
613. A. Boeijen, J. van Ameijde and R. M. J. Liskamp, *J. Org. Chem.*, 2001, **66**, 8454.
614. Y. F. Song and P. Yang, *Prep. Biochem. Biotechnol.*, 2001, **31**, 411.
615. Y.-F. Song and P. Yang, *Austr. J. Chem.*, 2001, **54**, 253.
616. P. Moreau, M. Sancelme, C. Bailly, S. Leonce, A. Pierre, J. Hickman, B. Pfeiffer and M. Prudhomme, *Eur. J. Med. Chem.*, 2001, **36**, 887.
617. T. Ijaz, P. Tran, K. C. Ruparelia, P. H. Teesdale-Spittle, S. Orr and L. H. Patterson, *Bioorg. Med. Chem. Lett.*, 2001, **11**, 351.
618. T. K.-K. Mong, A. Niu, H.-F. Chow, C. Wu, L. Li and R. Chen, *Chem.-Eur. J.*, 2001, **7**, 686.
619. M. A. Estiarte, A. Diez, M. Rubiralta and R. F. W. Jackson, *Tetrahedron*, 2001, **57**, 157.
620. L. H. D. J. Booij, L. A. G. M. van der Broek, W. Caulfield, B. M. G. Dommerholt-Caris, J. K. Clark, J. van Egmond, R. McGuire, A. W. Muir, H. C. J. Ottenheijm and D. C. Rees, *J. Med. Chem.*, 2000, **43**, 4822.
621. J. F. McNamara, H. Lombardo, S. K. Pillai, I. Jensen, F. Albericio and S. A. Kates, *J. Pept. Sci.*, 2000, **6**, 512.
622. V. Polyakov, V. Sharma, J. L. Dahlheimer, C. M. Pica, G. D. Luker and D. Piwnica-Worms, *Bioconjugate Chem.*, 2000, **11**, 762.
623. A. Boschi, C. Bolzati, E. Benini, E. Malago, L. Uccelli, A. Duatti, A. Piffanelli, F. Refosco and F. Tisato, *Bioconjugate Chem.*, 2001, **12**, 1035.
624. M. Takagaki, W. Powell, A. Sood, B. F. Spielvogel, N. S. Hosmane, M. Kirihata, K. Ono, S.-I. Masunaga, Y. Kinashi, S.-I. Miyatake and N. Hashimoto, *Radiation Res.*, 2001, **156**, 118.
625. Y. Xu, P. Saweczko and H.-B. Kraatz, *J. Organomet. Chem.*, 2001, **637-639**, 335.
626. J. Sehnert, A. Hess and N. Metzler-Nolte, *J. Organomet. Chem.*, 2001, **637-639**, 349.
627. M. Pietraszkiewicz, A. Wieckowska, R. Bilewicz, A. Misicka, L. Piela and K. Bajdor, *Mater. Sci. Eng.*, *C*, 2001, **18**, 121.
628. M. V. Baker, H.-B. Kraatz and Q. J. Wilson, *New J. Chem.*, 2001, **25**, 427.
629. F. Pellarini, D. Pantarotto, T. Da Ros, A. Giangaspero, A. Tossi and M. Prato, *Org. Lett.*, 2001, **3**, 1845.
630. C. Bolm, G. Moll and J. D. Kahmann, *Chem. –Eur. J.*, 2001, **7**, 1118.
631. K. Wen, H. Han, T. Z. Hoffman, K. D. Janda and L. E. Orgel, *Bioorg. Med. Chem. Lett.*, 2001, **11**, 689.
632. P. Saweczko, G. D. Enright and H.-B. Kraatz, *Inorg. Chem.*, 2001, **40**, 4409.
633. I. B. Nagy, I. Varga and F. Hudecz, *Anal. Biochem.*, 2000, **287**, 17.
634. X. Tang, M. Xian, M. Trikha, K. V. Honn and P. G. Wang, *Tetrahedron Lett.*, 2001, **42**, 2625.
635. N. R. Wurtz, J. M. Turner, E. E. Baird and P. B. Dervan, *Org. Lett.*, 2001, **3**, 1201.
636. J. W. Tilley, G. Kaplan, K. Rowan, V. Schwinge and B. Wolitzky, *Bioorg. Med.*

Chem. Lett., 2001, **11**, 1.

637.   W. Zhang, L. Zhang, X. Li, J. A. Weigel, S. E. Hall and J. P. Mayer, *J. Comb. Chem.*, 2001, **3**, 151.

638.   A. Volonterio, P. Bravo and M. Zanda, *Tetrahedron Lett.*, 2001, **42**, 3141.

639.   K. Kuhn, D. J. Owen, B. Bader, A. Wittinghofer, J. Kuhlmann and H. Waldmann, *J. Amer. Chem. Soc.*, 2001, **123**, 1023.

640.   T. Yamada, K. Hanada, K. Makihira, R. Yanagihara and T. Miyazawa, *Pept. Sci.*, 2001, 417.

641.   T. Arai, M. Inudo and N. Nishino, *Pept. Sci.*, 2001, 359.

642.   N. Solladie, A. Hamel and M. Gross, *Chirality*, 2001, **13**, 736.

643.   H. Ishida, D. Hesek, F. Aoki and Y. Inoue, *Pept. Sci.*, 2001, 353.

644.   T. Yamada, T. Ichino, R. Yanagihara and T. Miyazawa, *Pept. Sci.*, 2001, 305.

645.   N. Abe, H. Oku, K. Yamada and R. Katakai, *Pept. Sci.*, 2001, 289.

646.   T. Takahashi, I. Kumagai, K. Hamasaki, A. Ueno and H. Mihara, *Pept. Sci.*, 2001, 81.

647.   C. Behrens, N. Harrit and P. E. Nielsen, *Bioconjugate Chem.*, 2001, **12**, 1021.

648.   M. Fujii, K. Yokoyama, T. Kubo, R. Ueki, S. Abe, K. Goto, T. Niidome, H. Aoyagi, K. Iwakuma, S. Ando and S. Ono, *Pept. Sci.*, 2001, 109.

649.   S. Liu, C. Dockendorf and S. D. Taylor, *Org. Lett.*, 2001, **3**, 1571.

650.   H. Ishida and Y. Inoue, *Biopolymers.*, 2001, **55**, 469.

651.   T. Ohyama, H. Oku, M. Yoshida and R. Katakai, *Biopolymers*, 2001, **58**, 636.

652.   J. A. Zerkowski and D. A. Sheftner, *Protein Pept. Lett.*, 2001, **8**, 123.

653.   V. Santagada, F. Fiorino, B. Severino, S. Salvadori, L. H. Lazarus, S. D. Bryant and G. Caliendo, *Tetrahedron Lett.*, 2001, **42**, 3507.

654.   R. A. Breitenmoser, T. R. Hirt, R. T. N. Luykx and H. Heimgartner, *Helv. Chim. Acta.*, 2001, **84**, 972.

655.   G. Cavallaro, G. Pitarresi, M. Licciardi and G. Giammona, *Bioconjugate Chem.*, 2001, **12**, 143.

656.   M. Langer, F. Kratz, B. Rothen-Rutishauser, H. Wunderli-Allenspach and A. G. Beck-Sickinger, *J. Med. Chem.*, 2001, **44**, 1341.

657.   M. de Garcia-Martin, M. Violante de Banez, M. Garcia- Alvarez, S. Munoz-Guerra and J. A. Galbis, *Macromolecules*, 2001, **34**, 5042.

658.   J.-P. Mazaleyrat, K. Wright, M. Wakselman, F. Formaggio, M. Crisma and C. Toniolo, *Eur. J. Org. Chem.*, 2001, 1821.

659.   H. K. Rau, H. Snigula, A. Struck, B. Robert, H. Scheer and W. Haehnel, *Eur. J. Biochem.*, 2001, **268**, 3284.

660.   J. Thierry, C. Grillon, S. Gaudron, P. Potier, A. Riches and Wdzieczak-Bakala, *J. Pept. Sci.*, 2001, **7**, 284.

661.   M. J. Sever and J. J. Wilker, *Tetrahedron*, 2001, **57**, 6139.

662.   D. Liu and W. F. DeGrado, *J. Amer. Chem. Soc.*, 2001, **123**, 7553.

663.   R. Machauer and H. Waldmann, *Chem. –Eur. J.*, 2001, **7**, 2940.

664.   B. C. H. May and A. D. Abell, *Tetrahedron Lett.*, 2001, **42**, 5641.

665.   H. Wennemers, M. Conza, M. Nold and P. Krattiger, *Chem. – Eur. J.* , 2001, **7**, 3342.

666.   E. Botana, S. Ongeri, R. Arienzo, M. Demarcus, J. G. Frey, U. Piarulli, D. Potenza, J. D. Kilburn and C. Gennari, *Eur. J. Org. Chem.*, 2001, 4625.

667.   A. S. Kende, H.-Q. Dong, A. W. Mazur and F. H. Ebetino, *Tetrahedron Lett.*, 2001, **42**, 6015.

668.   H. Tamamura, A. Omagari, K. Hiramatsu, T. Kanamoto, K. Gotoh, K. Kanbara, N. Yamamoto, H. Nakashima, A. Otaka and N. Fujii, *Bioorg. Med. Chem.*, 2001, **9**,

2179.

669. K. Ikeda, K. Konishi, M. Sato, H. Hoshino and K. Tanaka, *Bioorg. Med. Chem. Lett.*, 2001, **11**, 2607.
670. Y. F. Song and P. Yang, *Chin. Chem. Lett.*, 2001, **12**, 697.
671. S. E. Taylor, T. J. Rutherford and R. K. Allemann, *Bioorg. Med. Chem. Lett.*, 2001, **11**, 2631.
672. M. F. Jobling, X. Huang, L. R. Stewart, K. J. Barnham, C. Curtain, I. Volitakis, M. Perugini, A. R. White, R. A. Cherny, C. L. Masters, C. J. Barrow, S. J. Collins, A. I. Bush and R. Cappai, *Biochemistry*, 2001, **40**, 8073.
673. P. A. Wender, T. C. Jessop, K. Pattabiraman, E. T. Pelkey and C. L. VanDeusen, *Org. Lett.*, 2001, **3**, 3229.
674. M. Schneider and H.-H. Otto, *Arch. Pharm.*, 2001, **334**, 167.
675. H.-D. Arndt, A. Knoll and U. Koert, *Angew. Chem., Int. Ed.*, 2001, **40**, 2076.
676. H. Matsui and G. E. Douberly, *Langmuir*, 2001, **17**, 7918.
677. C. T. Miller, R. Weragoda, E. Izbicka and B. L. Iverson, *Bioorg. Med. Chem.*, 2001, **9**, 2015.
678. P. D. Bailey, N. Bannister, M. Bernad, S. Blanchard and A. N. Boa, *J. Chem. Soc., Perkin Trans. 1*, 2001, 3245.
679. F. Kratz, J. Drevs, G. Bing, C. Stockmar, K. Scheuermann, P. Lazar and C. Unger, *Bioorg. Med. Chem. Lett.*, 2001, **11**, 2001.
680. A.-M. Fernandez, K. Van derpoorten, L. Dasnois, K. Lebtahi, V. Dubois, T. J. Lobl, S. Gangwar, C. Oliyai, E. R. Lewis, D. Shochat and A. Trouet, *J. Med. Chem.*, 2001, **44**, 3750.
681. T. M. Sielecki, J. Liu, S. A. Mousa, A. L. Racanelli, E. A. Hausner, R. R. Wexler and R. E. Olson, *Bioorg. Med. Chem. Lett.*, 2001, **11**, 2201.
682. N. Kuroda, T. Hattori, C. Kitada and T. Sugawara, *Chem. Pharm. Bull.*, 2001, **49**, 1138.
683. N. Kuroda, T. Hattori, Y. Fujioka, D. G. Cork, C. Kitada and T. Sugawara, *Chem. Pharm. Bull.*, 2001, **49**, 1147.
684. W. L. Scott, F. Delgado, K. Lobb, R. S. Pottorf and M. J. O'Donnell, *Tetrahedron Lett.*, 2001, **42**, 2073.
685. T. Moriuchi, A. Nomoto, K. Yoshida, A. Ogawa and T. Hirao, *J. Amer. Chem. Soc.*, 2001, **123**, 68.
686. V. Cavallaro, P. Thompson and M. Hearn, *J. Pept. Sci.*, 2001, **7**, 262.
687. A. G. Bonner, L. M. Udell, W. A. Creasey, S. R. Duly and R. A. Laursen, *J. Pept. Res.*, 2001, **57**, 48.
688. J. Shi, Y. Zhao, Y. Zhang, Z. Yan, W. Han and J. Wang, *Shengwu Gongcheng Jinzhan*, 2001, **21**, 38.
689. D. Kontrec, A. Abatangelo, V. Vinkovic and V. Sunjic, *Chirality*, 2001, **13**, 294.
690. H. Wang, J. Li, T.-X. Yang and H.-S. Zhang, *J. Chromatogr. Sci.*, 2001, **39**, 365.
691. M. Villain, J. Vizzavona and K. Rose, *Chem. Biol.*, 2001, **8**, 673.
692. Z. Huang, *Yaowu Shengwu Jishu*, 2001, **8**, 207.

2

Analogue and Conformational Studies on Peptides, Hormones and Other Biologically Active Peptides

BY BOTOND PENKE, GÁBOR TÓTH AND GYÖRGYI VÁRADI

1   Introduction

The subject matter included this year is similar to that of last year, despite a change of authorship: the basic structure of the chapter remains unchanged compared to last year's chapter in Volume 33[1]. Scientific papers published during 2001 have been sourced mainly from CA Selects[2] on Amino Acids Peptides and Proteins and in increasing number from the Web of Science databases[3,4] on the Internet. Some sub-chapters are new (*e.g.* proteasome inhibitors, NO-synthase inhibitors) and the sub-chapter 'Advances in Delivery Technology' returns.

Papers from Conference Proceedings were cited sporadically but no patents have been used as source material.

2   Peptide Backbone Modifications and Peptide Mimetics

2.1   Aza, Oxazole, Oxazolidine and Tetrazole Peptides. – A general method to azadepsipeptides, a novel class of peptidomimetics, has been established and was applied to the synthesis of analogues of an antiparasitic cyclooctadepsipeptide[5](**1**). X-ray crystallography showed conservation of the 3D structure of the natural compound. Linear azadepsipeptides were prepared by reaction of a hydrazide with 3-t-butoxycarbonylphenyl chlorocarbonate and hydrolysis of the product was examined[6]. Aza-β^3-peptides (**2**), a new class of pseudopeptides with pyramidal nitrogen atoms as chiral centers, were assembled from N^α-substituted hydrazine acetic acid monomers[7]. Oxazole analogues of the insect neuropeptide proctolin were prepared and tested for myotropic activity[8]. The trimer and tetramer of *trans*-(4*S*,5*R*)-4-carboxybenzyl 5-methyl oxazolidin-2-one as a new class of pseudoprolines which fully control the formation of a Xaa(i-1)-Pro(i) peptide bond formation in the *trans* conformation, have been synthesized[9]. Monophenyl and diphenyl pseudooxazolone derivatives of glycine and alanine (**3**) were prepared and found to be cysteine protease inhibitors[10]. Insertion of the tetrazole ring in the tetraalanine sequence led to a very effective peptide chelating agent towards Cu(II) ions[11]. The synthesis and crystal structure of Boc-Phe-

Amino Acids, Peptides and Proteins, Volume 34
© The Royal Society of Chemistry, 2003

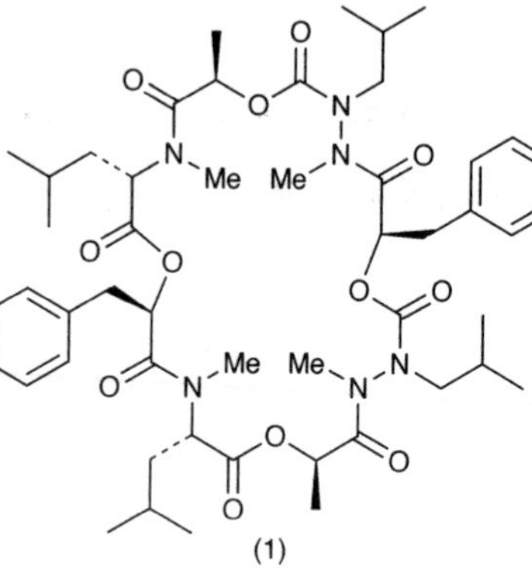

(1)

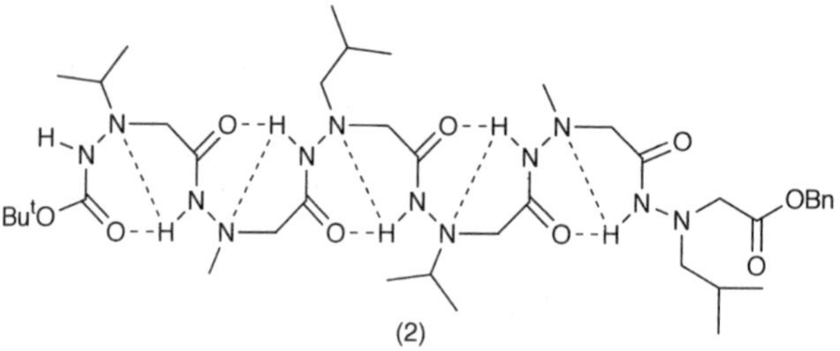

(2)

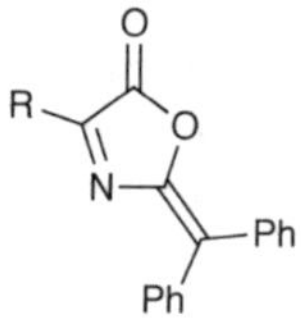

(3) R = H glycine derivative
R = Me alanine derivative

[COCN$_4$]-Gly-OBn, a novel *cis*-dipeptide mimic containing the non-hydrolys-able α-keto tetrazole isostere has been reported[12].

2.2   Ψ[CH$_2$CH$_2$],   Ψ[Z-CF = CH],   retro-Ψ[NHCH(CF$_3$)],   Ψ[epoxy], Ψ[CONHNR], Ψ[CSNH], Ψ[oxetane], Ψ[CH$_2$NH], Ψ[CHOH-cyclopropyl-CONH], Ψ[PO$_2$R-CH]. – A new approach to 'carba' dipeptides through a malonate intermediate resulted in the first synthesis of enantiopure Boc-D-Phe-Ψ[CH$_2$CH$_2$]-L-Arg(NO$_2$)-OH and Boc-D-Phe-Ψ[CH$_2$CH$_2$]-D-Arg(NO$_2$)-OH[13]. A new route to α-substituted fluoroalkene dipeptide isosteres using reduction-oxidative alkylation reactions with organocopper reagents was developed[14]. The synthesis of a novel family of partially modified retropeptidyl hydroxamates incorporating a [CH(CF$_3$)CH$_2$CO] unit (4) was achieved by using the Michael-type addition of free or polymer bound α-amino hydroxamates to 3-(*E*-enoyl)-1,3-oxazolidin-2-ones[15]. Dipeptide analogues containing an oxirane ring in the place of the peptide bond were prepared and used on solid-phase to generate a small library of epoxy peptidomimetics[16]. Some of the products were found to be cysteine protease inhibitors. Hydrazino peptides in which the β-carbon atoms in the β-amino acids of β-peptides are replaced by nitrogen atoms have been the subject of a 113-reference review[17]. Conformational analysis of thiopeptides provided force field parameters for sp^2 sulfur and showed that thio-substitution restricted the conformations of the amino acids in peptides[18,19]. Synthesis of *cis*- and *trans*-3-azido-oxetane-2-carboxylates as a new family of foldamers and of hexamers of *cis*-β-amino acids containing the oxetane structural motif was reported[20]. The key steps in the syntheses of oxetane amino acids are efficient

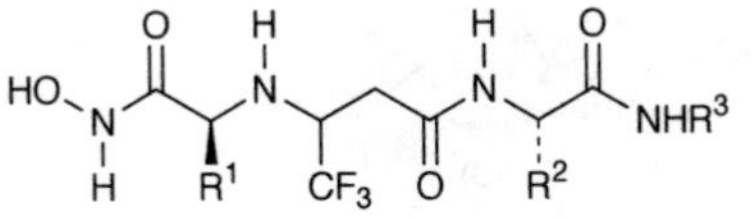

(4)

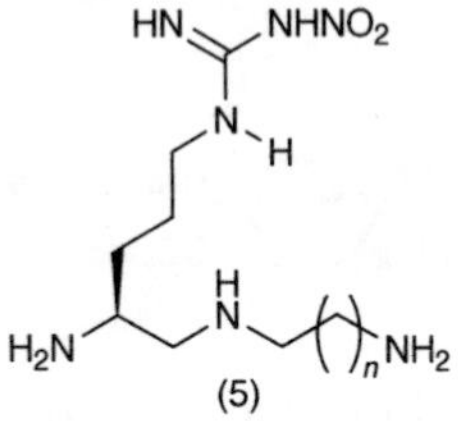

(5)

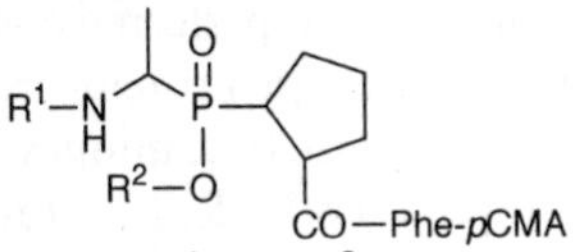

(6) R^1 = Ac, R^2 = H
R^1 = Ac, R^2 = Me

nucleophilic substitutions of 3-O-triflates of oxetanes by trifluoroacetate and azide. Detailed NMR study of oxetane β-amino acid hexamers revealed a well-defined left-handed helical structure stabilized by 10-membered hydrogen-bonded rings[21]. A series of fully reduced linear oligolysines has been synthesized on solid-phase and proved to form tight complexes with plasmid DNA making these molecules potential candidates for DNA carriers in gene delivery[22]. The deletion of the carbonyl group from the amide bond of neuronal Nitric Oxide Synthase inhibitors (5) greatly improved the selectivity over the endothelial and inducible variants of the enzyme[23]. Synthesis of cyclopropane-derived pep-tidomimetics using a new method for introducing side chains onto the C-terminal amino acid of the peptides *via* the opening of an N-Boc-aziridine with an organocuprate was reported[24]. All of these pseudopeptides were less potent inhibitors of Ras farnesyltransferase than the tetrapeptide parent suggesting the importance of the amide linkage. Phosphinic alanyl-proline surrogates (6) have been prepared *via* three- or four-step procedures and evaluated as inhibitors of the human cyclophilin hCyp-18 peptidyl-prolyl isomerase[25].

2.3 Rigid Amino Acid, Peptide and Turn Mimetics. – Structure-based rational design of nonpeptide ligands for peptide receptors has been reviewed in the context of drug delivery[26]. High-throughput synthesis of peptide mimetics has been the subject of a review of the literature from 1999 to March 2001 demon-strating the ability to produce a large number of compounds rapidly[27]. Com-pounds (7) and (8), illustrating a novel concept for constraining torsion angles in glycopeptides, have been synthesized through a convergent strategy and fully characterized[28]. Solution and solid-phase synthesis of hetarylene-carbopeptoids, a new class of compounds for combinatorial chemistry in glycopeptidomimetic design has been reported[29]. Optically pure 4,5-dihydro-3(2H)-pyridazinones (9) have been prepared by the asymmetric γ-alkylation of α,β-unsaturated glutamic acid derivatives, followed by cyclization, as a novel entry to these conformation-ally restricted glutamic acid derivatives[30]. Synthesis of optically active 4,5-dihydro-1,2,4-triazin-6(1H)-ones as a new series of small conformationally con-strained peptides have been reported[31].

Combinatorial turn mimetic synthesis and scaffold design, in addition to the

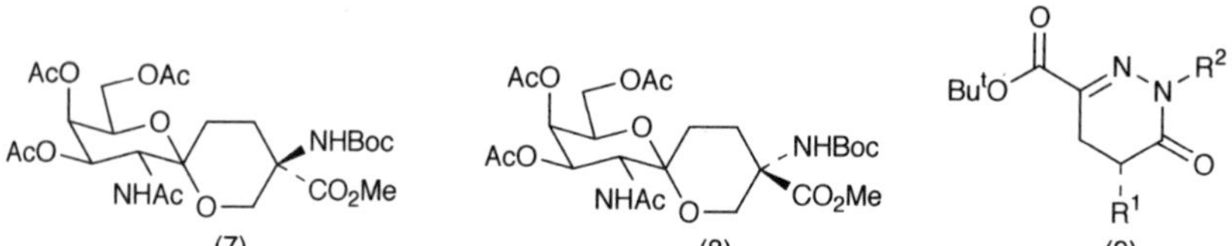

(7)   (8)   (9)

screening of these libraries towards biological targets have been reviewed in 47 references[32]. Solid-phase syntheses and preferred conformations of β-turn mimics that are designed to mimic or disrupt protein-protein interactions have been reviewed[33]. The importance of dimeric turn mimics and some new approaches to these have been also described. An 80-reference review discusses tethering strategies that lead to the development of various mimetics and antagonists, and also the application of aminocaproic acid as a tether in cyclic peptide β-turn mimics[34]. Strong intramolecular forces stabilize the conformation of a new type of spiro β-lactams (10) with a geometry very close to the ideal type II β-turns[35]. A library of urea type 6,6-bicyclic β-turn mimetics (11) has been successfully generated by solid-phase chemistry, with diversity at the i position, in good yield and purity[36]. A tripeptide-derived, 6,6-fused bicyclic scaffold (12) has been prepared and incorporated into peptides, through condensation of an aldehyde to form a thiazinone ring[37]. The bicyclic moiety represents a constrained β-turn mimetic in linear and macrocyclic peptide-heterocycle hybrids. A conformationally restricted Ala-Gly analogue (13) has been synthesized in multigram quantities[38]. The

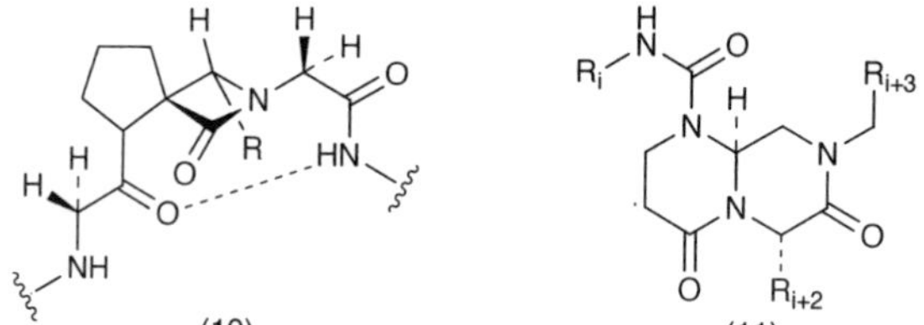

(10)   (11)

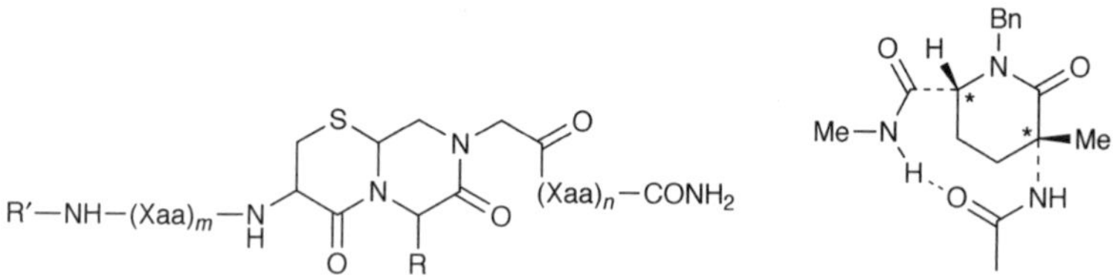

(12)   (13)

β-turn characteristics of the molecule were examined by molecular modeling and NMR spectroscopy. Aza-amino acid residues have been used to induce β-turn structure in model peptides. The dipeptide Ac-Aib-azaGly-NH$_2$ has been designed[39] to adopt specific torsion angles of the β-I turn and an azaamino acid residue induced[40] β-II turn conformation in the tetrapeptide Boc-Ala-Phe-azaLeu-Ala-OMe verified by IR, NMR and molecular modeling techniques. The aza-tripeptide Boc-Ala-AzaPip-Ala-NHiPr has been synthesized in seven steps and proved to adopt a β-IV-like turn around the N-terminal sequence[41]. An efficient approach to the enantiopure synthesis of β-turn mimetic azabicyclo[X.Y.0] alkane amino acids **(14-17)** has been developed[42]. Novel RGD mimetics (1-aza-bicyclo[5.2.0]nonan-2-ones) as scaffolds for β-turn structures have been prepared by photochemical rearrangement of oxaziridines and nitrones[43]. A new molecular entity of an all-*cis* 4,5-dihydroxy-1,3-cyclopentanedicarboxylic acid induced β-turn-like structures in compounds **(18)** and **(19)**[44].

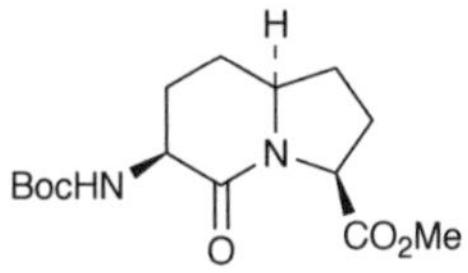

(14) 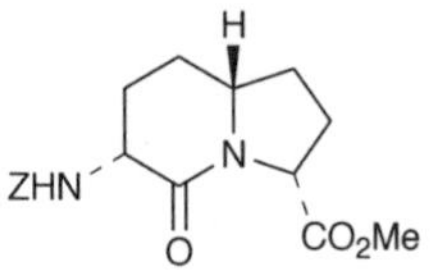   (15) 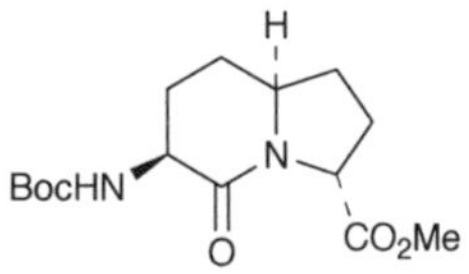   (16)

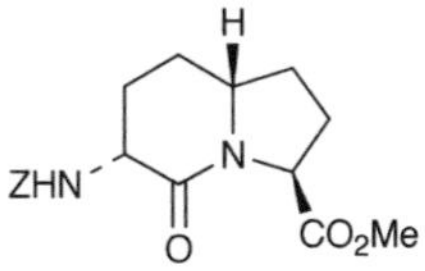

(17) 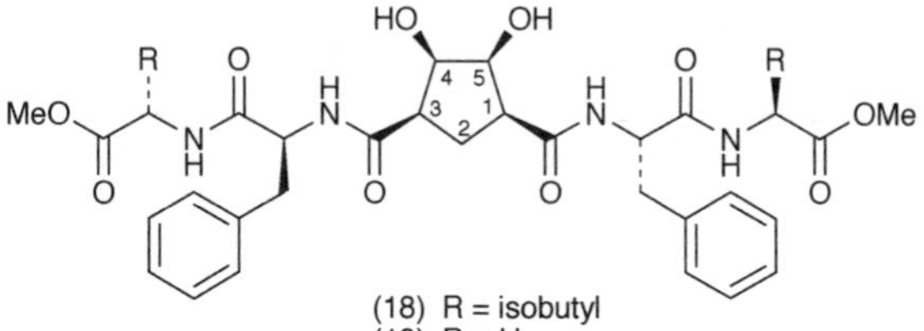   (18) R = isobutyl
         (19) R = H

Compounds **(20)** and **(21)** have been designed and synthesized as new Gly-Pro turn mimetics[45]. Both of the enantiomers (2R, 7aR) and (2S, 7aS) were shown to be potentially useful reverse turn mimics. An imidazoline dipeptide (a 4,5-dihydroimidazole-carboxylic acid) was prepared and reported to display an intramolecular hydrogen-bond consistent with a turn-conformation in solution[46]. A five-step synthesis of [5.5]-bicyclic reverse turn dipeptide mimetics **(22)** with side chain functional groups at the i+1 and i+2 positions has been developed[47]. Dihedral angles Ψ_2 and Φ_3 were greatly restricted in the bicyclic structure. A convenient method for the synthesis of both of the *cis*- and *trans*-fused stereoisomers of N-Boc-L-octahydroindole-2-carboxylic acid methyl ester **(23)** and **(24)**, as reverse turn mimetic precursors, has been reported[48]. This novel route, that has as the key step the ring-closing metathesis of a diallylated proline derivative, provides the first reported preparation of the *trans*-fused isomer. The new dipeptide isostere 7-endo-BtA **(25)** induced a reverse turn in an 11-residue peptide derived from the Bowman Birk Inhibitor class of serine protease inhibitors[49].

Novel constrained L-phenylalanine analogs have been synthesized[50]. The benzazepinone analogue **(26)** was comparable in potency to the unconstrained analogue **(27)** as a VCAM/VLA-4 antagonist. However, the spirocyclic aza-pinone derivative **(28)** showed 100-fold less potency than the unconstrained compound **(29)** in the same ELISA assay. The large-scale asymmetric synthesis of novel sterically constrained tyrosine and phenylalanine derivatives **(30)** and **(31)** has been reported[51]. A series of 5-vinyl substituted proline derivatives has been prepared using a hydroboration-Swern oxidation sequence[52]. None of the novel non-proteinogenic amino acids **(32)** and **(33)** showed activity on the glycine site of the NMDA receptor complex[53]. Selective alkylation of **(34)** at N^4, C^6 and N^1 led to a range of conformationally constrained phenylalanine deriva-

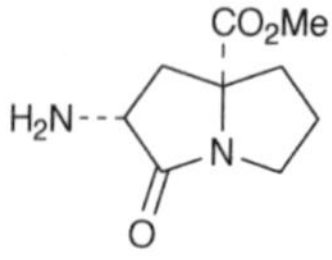

(20) (2*R*, 7a*R*) 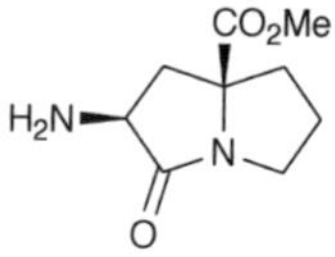   (21) (2*S*, 7a*S*) 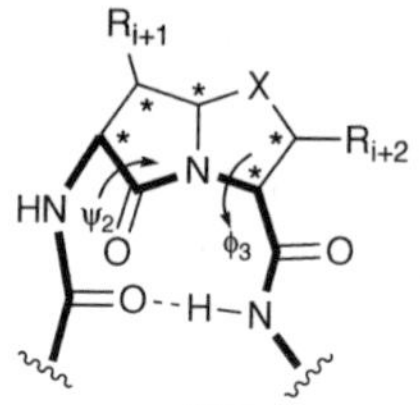   (22)

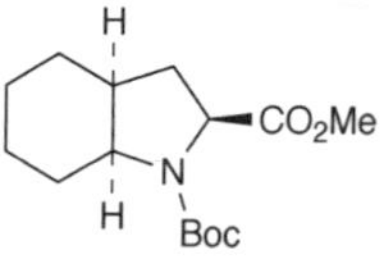

(23) *cis* 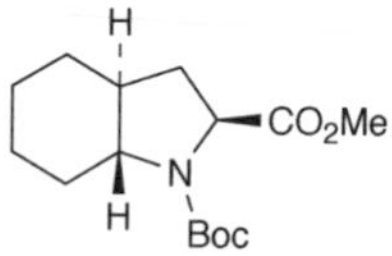   (24) *trans* 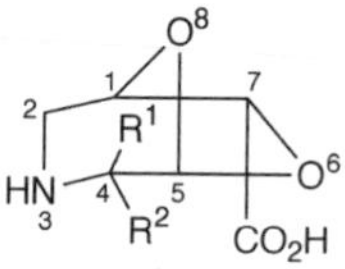   (25)

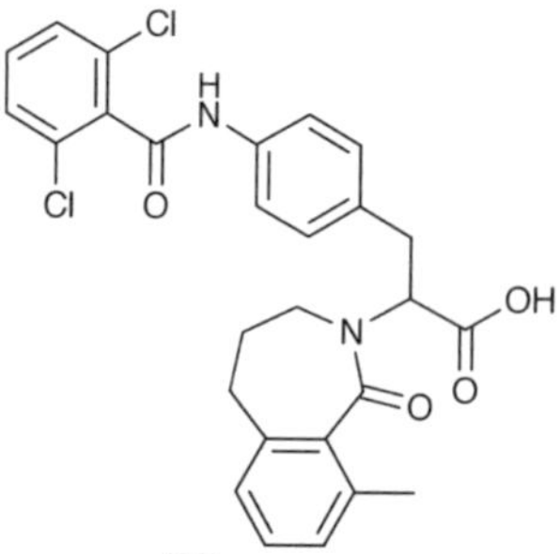

(26) 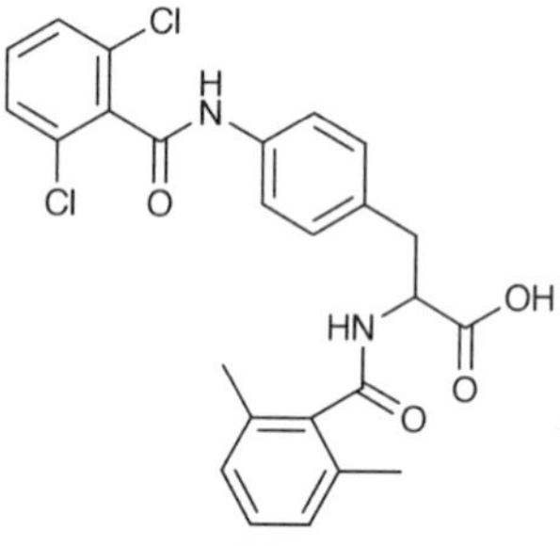   (27)

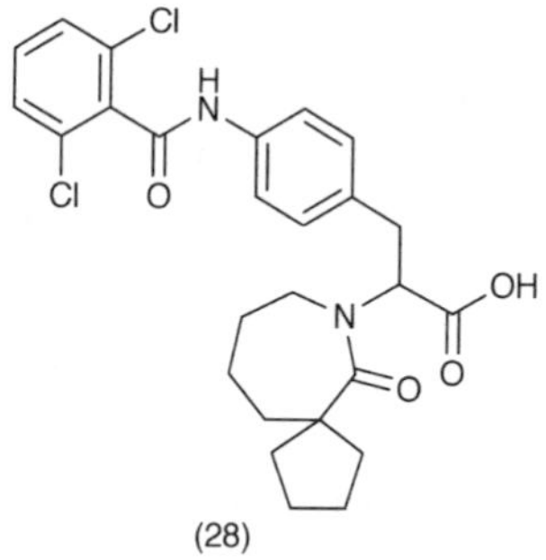

(28) 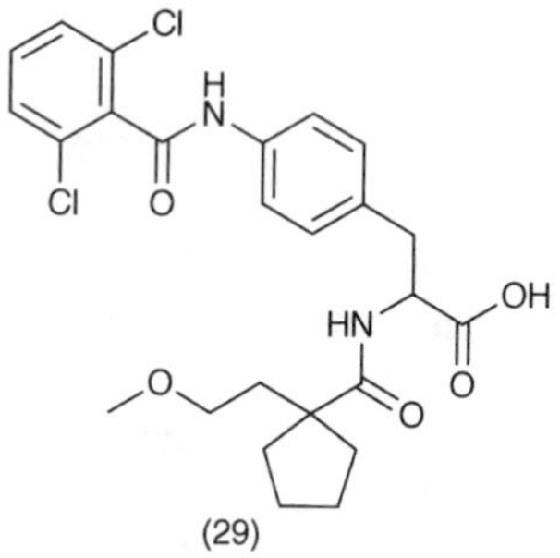   (29) 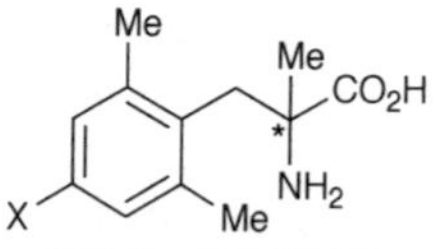   (30) X = H (α-TMP)
                            (31) X = OH (α-TMT)

tives[54]. In addition, constrained analogues of the dipeptides Phe-Gly (35, 36, 37, 38) and Phe-Ala (39) were prepared.

Enantiopure 7-[3-azidopropyl]indolizidin-2-one amino acid (40) as an Ala-Lys dipeptide mimetic has been synthesized from its 7-hydroxypropyl counterpart (41) with sodium azide and subsequent ester hydrolysis[55]. Indolizidin-9-one amino acid (42) has been used successfully as a constrained surrogate of the Gly-Gly dipeptide in Leu-enkephalin[56]. Organozinc chemistry was used to prepare 3-aminopiperidin-2,5-dione, a new Ala-Gly dipeptide mimic[57]. A 1,1'-ferrocenophane lactam Phe-Phe dipeptide mimetic (43) has been synthesized and incorporated into Substance P, giving a conformationally constrained organometallic analogue[58].

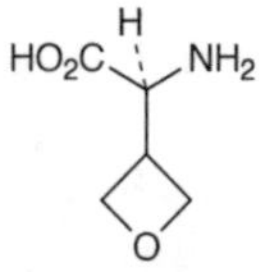

(32) 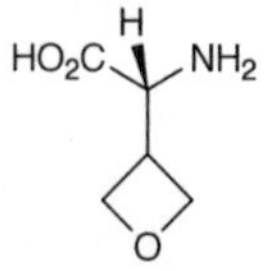   (33) 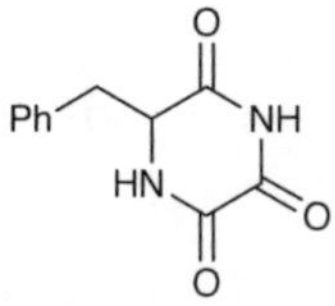   (34) 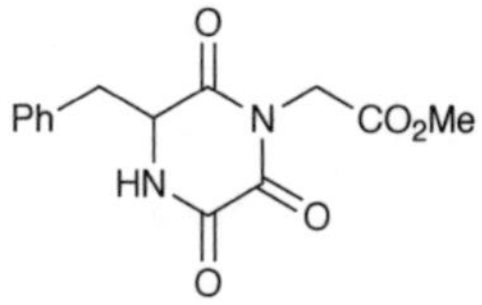   (35)

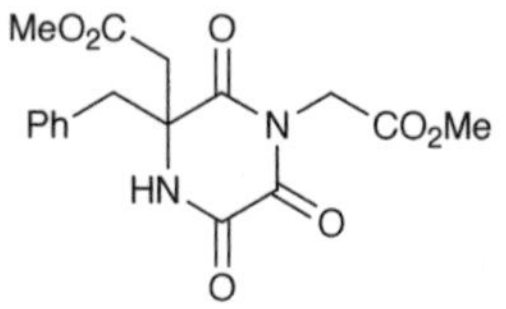

(36) 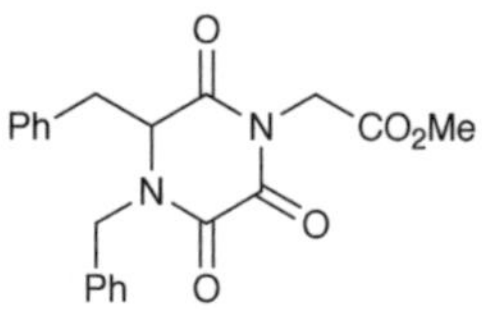   (37)

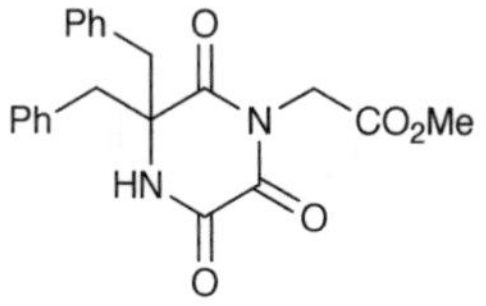

(38) 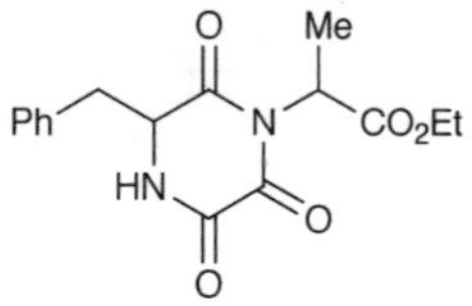   (39)

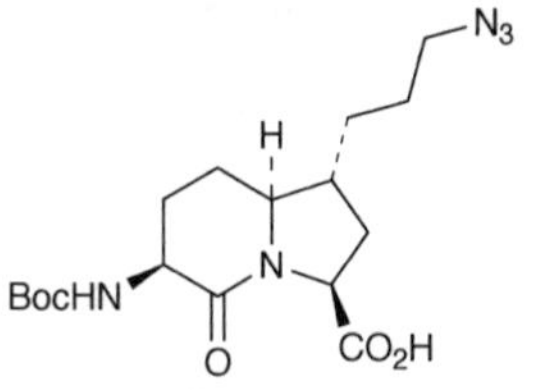

(40) 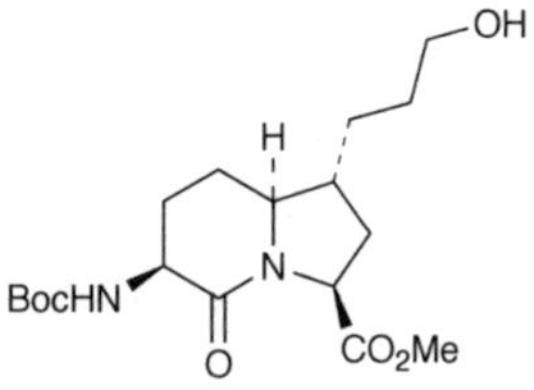   (41)

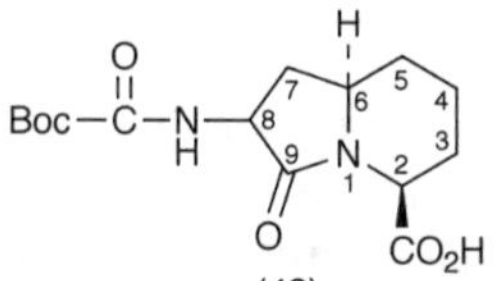

(42) 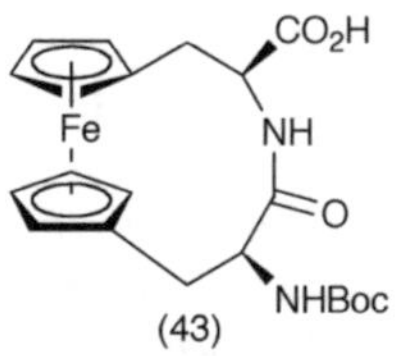   (43)

The terminally blocked homotrimer of (2S,4R)-4-amino-5-oxopyrrolidine-2-carboxylic acid has been characterized by a unique, alternating *cis-trans* amide sequence[59]. A versatile procedure **(Scheme 1)** to prepare both 6,5- [routes (a) and (b)] and 7,5- [route (c)] bicyclic lactam peptidomimetics based on a chemoselective reduction of the pyroglutamate ring carbonyl, followed by an intramolecular trapping of an N-acyliminium ion by side-chain hydroxyl groups has been developed[60]. The structure-activity relationship of a novel class of farnesyltransferase inhibitors based on the benzophenone scaffold **(44)** has been described[61]. A combinatorial library of 3-alkylated-1,4-benzodiazepines was synthesized on solid-phase and tested for CCK-B receptor binding[62]. A novel, secondary amide-linked resin-based solid-phase synthesis has been developed for the construction of a library of indole-based peptide mimetics **(45)** as PAR-1 antagonists[63]. Screening of the library led to the identification of a potent and

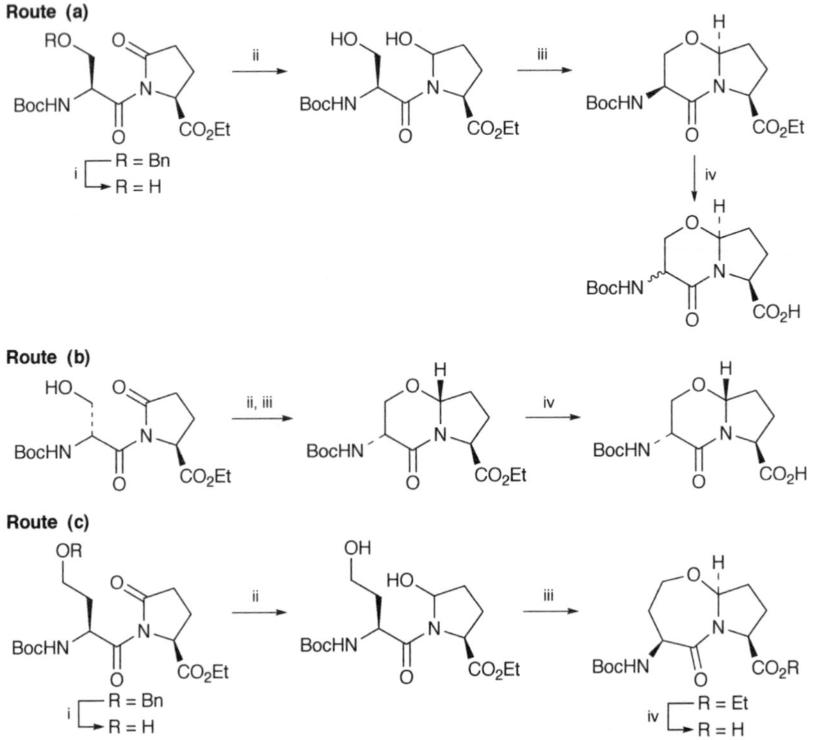

Route (a)

Route (b)

Route (c)

Reagents: i, H$_2$, Pd/C, MeOH; ii, LiBEt$_3$H, THF, −78 °C; iii, TFA(cat.), CH$_2$Cl$_2$, 50 °C; iv, LiOH, H$_2$O/THF

Scheme 1

selective PAR-1 antagonist. Three conformationally constrained didemnin B analogues and their total syntheses have been published by a research group[64,65]. While the isostatine residue in one of them was replaced by 2-hydroxy-3-cyclohexenecarboxylic acid as in compound (**46**), the two other analogues (**47, 48**) contain L-1,2,3,4-tetrahydroisoquinoline or L-1,2,3,4-tetrahydro-7-methoxyisoquinoline instead of N,O-dimethyltyrosine. In order to better understand the relationships between conformations and functional properties of β-homo-amino acids, crystal-state conformation of peptides containing C$^{\alpha,\alpha}$-dialkylated and β-homo-amino acid residues has been investigated[66].

3 Cyclic Peptides

As in last year's chapter in volume 33, the comprehensive coverage of this topic can be seen in Chapter 4. In addition to the discussion of new synthetic routes, this sub-section is confined to cyclic pseudopeptides and peptidomimetics.

A new safety-catch linker has been developed for the Boc solid-phase synthesis of libraries of cyclic peptides using a protected catechol derivative (**49**) in which one of the hydroxyl groups is masked with a benzyl group during the synthesis making the linker deactivated to aminolysis[67]. After the synthesis of the linear

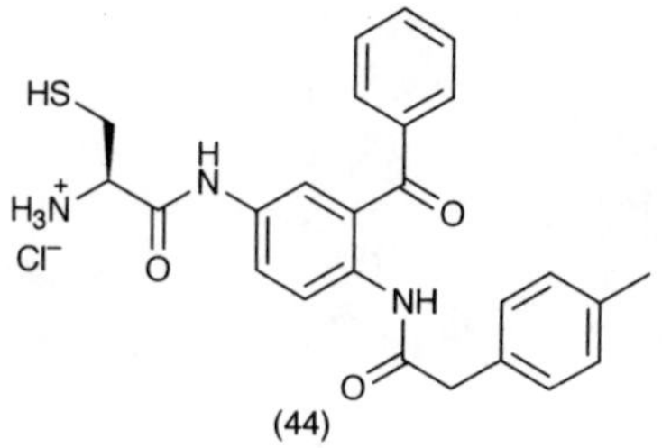

(44)

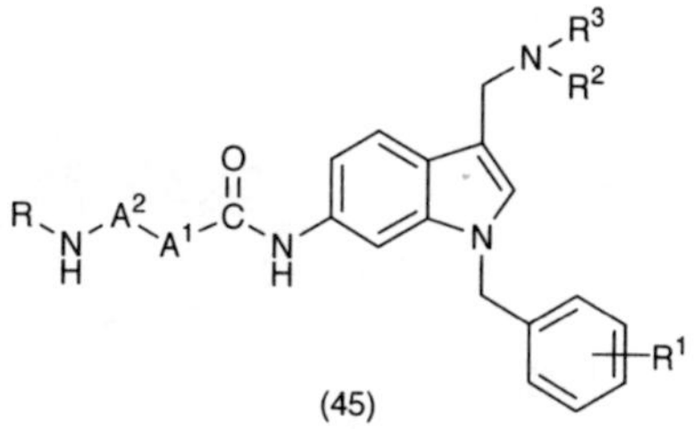

(45)

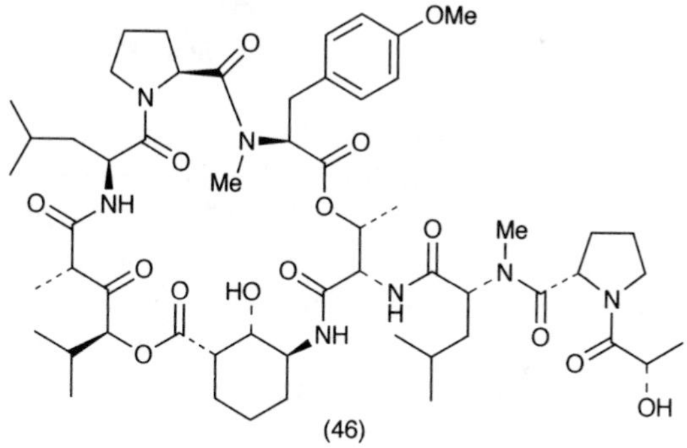

(46)

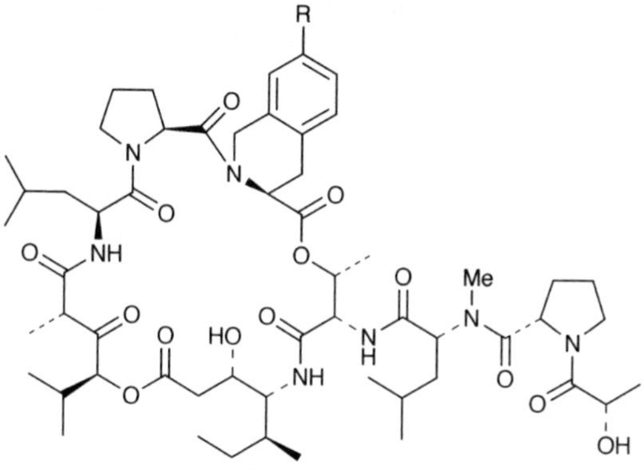

(47) R = H
(48) R = OMe

peptide, the linker is activated and the peptide deprotected. Neutralization of the N-terminal amine results in cyclization and cleavage from the resin. A library containing over 400 cyclic peptides was successfully prepared by the method. A general and rapid synthesis of a number of cyclic peptides, in which two cysteine sides chains are joined by an alkyl linker, was reported[68]. A Weinreb amine was introduced to the C-terminus of the peptides and transformed to an activated carbonyl function late in the synthetic sequence in order to avoid epimerization at the corresponding α-position. A triply orthogonal protecting group was used in the novel synthesis of cyclic peptides constrained by an aliphatic bridge between the α-carbons of the i and i+4 residues[69]. Contrary to the expected helical conformation, the model alanine-rich peptide adopted a β-turn conformation as shown by CD spectroscopy. Ring-closing metathesis was used to

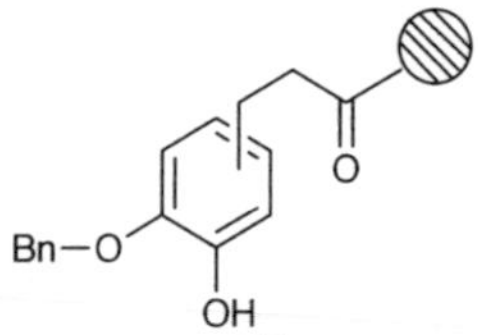

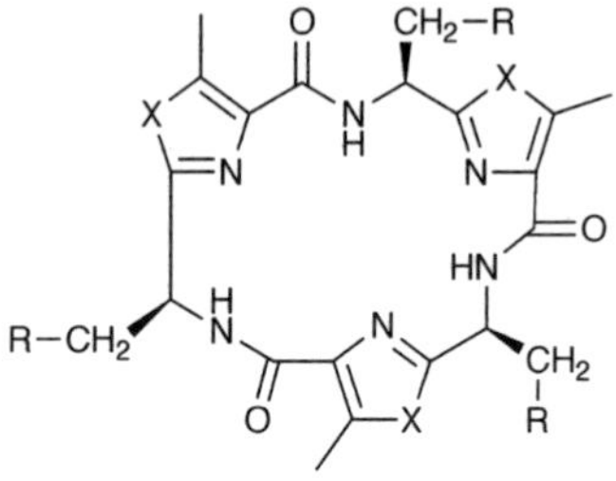

R = CO₂H or OH, X = O or S

Figure 1

prepare an eight-membered cyclic dipeptide for peptidomimetic research[70]. New molecular platforms with marine cyclopeptide related structure **(Figure 1)** have been synthesized on a gram scale[71]. Their rigid, almost planar heterocyclic ring-system and functionalized side-chains make these molecules ideal candidates for studies in molecular recognition and combinatorial chemistry. Total synthesis of dendroamide A **(50)**, that reverses multiple drug resistance in tumor cells, has been accomplished[72]. The method can be applied to the synthesis of further analogues whose small size, hydrophobic character and lack of charge make them cell permeable. Furthermore, the oxazole- and thiazole rings in the place of the peptide bonds and the unnatural D-stereochemistry of the amino acids used in the synthesis protect them from proteases. Total synthesis of trunkamide A **(51)**, a doubly prenylated cyclic peptide of marine origin, has been published[73]. The thiazoline ring was produced from the heptapeptide thioamide at the end of the synthesis, followed by macrocyclization that led to the cyclopeptide structure. The cyclic tetrapeptide cyclo[Lys-Trp-Lys-Ahx] (where Ahx is 6-aminohexanoic acid) has been found to be a novel DNA nicking agent with a higher binding constant and a nicking rate than the parent linear peptide[74]. A selenated alanine as an anchoring residue was designed and prepared for the solid-phase synthesis of the cyclic dehydrodepsipeptide, AM-toxin II **(52)**[75]. The synthesis of a novel class of $N^\alpha(\omega$-thioalkyl) amino acid building units **(53, 54** and **55)** and their incorporation into backbone cyclic peptides have been presented[76]. The general method was reductive alkylation of the appropriate amino acids with ω-protected thioaldehydes. NMR spectroscopy coupled with computational analysis of a novel cyclic angiotensin II analogue cyclo(3,5)-[Sar¹-Lys³-Glu⁵-Ile⁸] ANG II revealed that π^*-π^* interactions in the C-terminal aromatic residue are essential for agonist activity[77]. Novel cyclic peptide analogues based upon the minimal structural motif which has been found to exhibit biological activity at the thrombin receptor have been designed and synthesized[78]. Structure-activity studies showed that the Phe and Arg residues along with a primary amino group form an active recognition motif. This proposition was confirmed by the fact that the compound, cyclo(Phe-Leu-Leu-Arg-εLys-Dap) in which the Phe and Arg residues are in spatial proximity, has been found to be more active than the D-Phe analogue. Four nonpeptide thrombin receptor mimics were designed based upon these results using a piperazine template. Compounds **(56)** chosen from a small library of cyclic pseudopeptides led to two of the most active

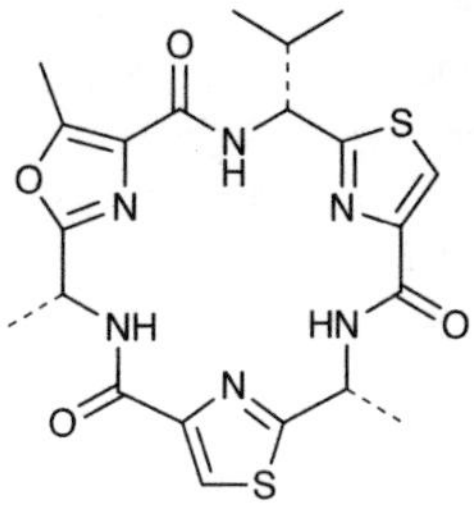

(50)

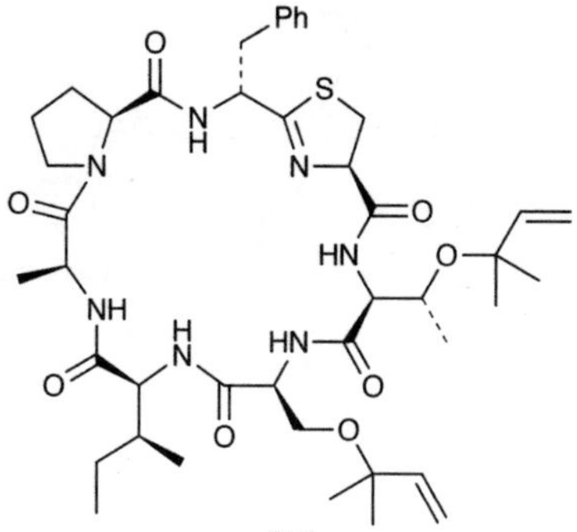

(51)

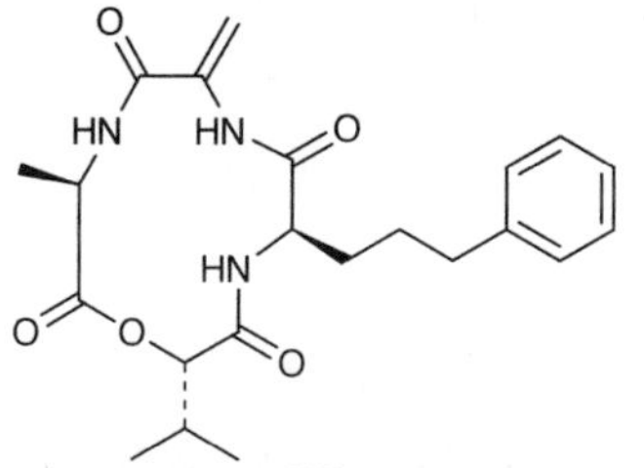

(52)

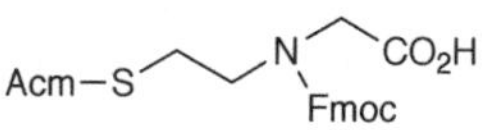

(53)

(54)

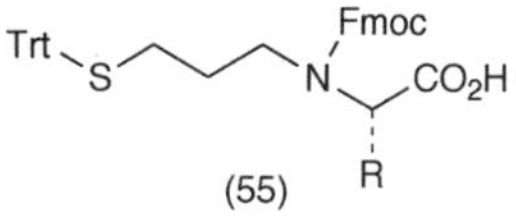

(55)

known inhibitors of the $\alpha_v\beta_3$ receptor[79]. A new approach for the design and synthesis of pheromone biosynthesis activating neuropeptide (PBAN) cyclic agonists and antagonists has been described[80]. Two backbone cyclic peptide libraries based on either the active C-terminal hexapeptide sequence Tyr-Phe-Ser-Pro-Arg-Leu-NH_2 of PBAN (1-33)NH_2 or the PBAN linear antagonist Arg-Tyr-Phe-Dphe-Pro-Arg-Leu-NH_2 were synthesized **(Figure 2)**. Several agonists and antagonists of the insect neuropeptide have been discovered by testing these libraries for biological activity. A family of cyclic pentapeptides **(57)** which adopt a type-II' β-turn conformation has been designed and synthesized as antagonists of the human bradykinin B_2 receptor[81]. Despite the structural homogeneity confirmed by NMR, the binding affinity of ten analogues of the series was strongly influenced by the actual side-chains in positions i and i + 3.

4   Biologically Active Peptides

4.1   Peptides Involved in Alzheimer's Disease. – The β-amyloid (Aβ), a 39-43 residue peptide, seems to play a central role by initiating neuronal cell death during the long period of the Alzheimer's disease. The chemical synthesis of this peptide on solid phase is rather tedious owing to solubilization problems inside

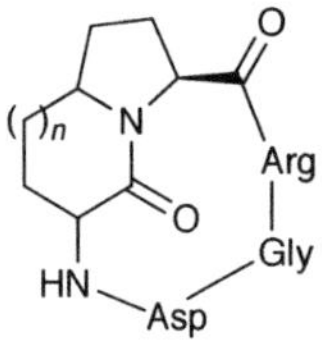

(56) $n = 1$ or 2

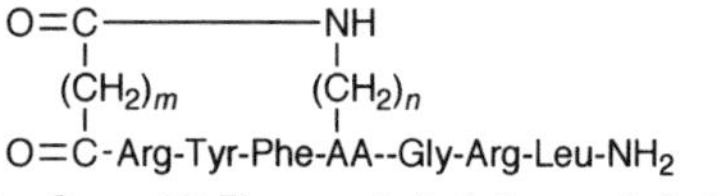

$O=C\text{———}NH$
$(CH_2)_m \qquad (CH_2)_n$
$O=C\text{-Arg-Tyr-Phe-AA--Gly-Arg-Leu-NH}_2$
AA = Ser or (D)-Phe, $n = 2, 3, 4, 6$, $m = 2, 3, 4$

Figure 2

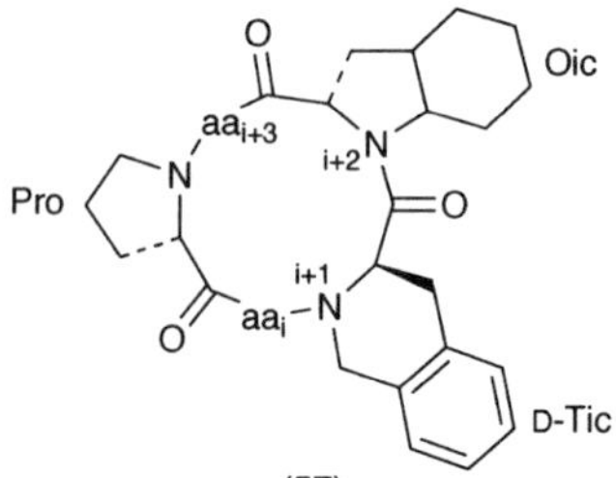

(57)

aa$_i$ = Ala, Dap, Dab(TFA), Orn(TFA), Phe, Ser, Thi
aa$_{i+3}$ = Ala, Arg(TFA), Dap(TFA), Phe, Ser

the solid phase and aggregation of the growing peptide chains ('on-resin aggregation'). Use of a stonger and more efficient base, DBU, at a concentration of 2% in DMF for N^α-Fmoc deprotection allowed substitutionally improved continuous flow solid phase synthesis of Aβ 29-40/42 fragments[82]. (Amylin, another difficult peptide, which also shows spontaneous aggregation, was synthesized on solid phase using NMP as the main solvent and hexafluoro-isopropanol for solubilizing the crude peptides[83].) Different segments of 10 to 40 amino acids of amyloid precursor protein (APP) were synthesized on solid phase by Boc chemistry[84]. Fluorescent labelling of Aβ 25-35, 1-40 and some short fragments was performed using 5(6)-carboxyfluorescein and 7-amino-4-methyl-3-coumarinylacetic acid[85]. At the root of the Alzheimer's disease we found changes in Aβ conformation where normal innocuous polypeptides transform to insoluble amyloid fibrils and deposit in the brain. The aggregation process has been reviewed with the focus on advancements made in elucidating the molecular structures of the Aβ amyloid fibril and alternate aggregation of the Aβ peptide formed during fibrillogenesis[86]. The most probable molecular interaction site of Aβ peptide was identified by using a fluorescence assay and was found to be the Aβ 16-20 sequence, KLVFF[87]. A new fluorescent method, time-resolved anisotropy measurement (TRAMS) has been used to study the aggregation of Aβ peptide. This sensitive technique detects the presence of preformed 'seed' particles in freshly prepared solutions of Aβ[88]. The method is suitable for early detection of Aβ aggregation before complexation becomes apparent in more conventional methods such as thio-flavin T fluorescence assay. A heptapeptide dimer (ALEQKLA)₂ was designed and synthesized, this compound undergoes a self-initiated structural transition from an α-helix to a β-sheet conformation and self-assembly into the amyloid fibrils[89]. An amyloidogenic peptide fragment of the protein amphoterin is homologous with Aβ and forms amyloid fibrils *in vitro* as well as binding to Aβ[90]. As the complete description of Aβ self-association kinetics require identification of the oligomeric species present at the pathway of association, a mathematical model of the kinetics of Aβ fibril formation was worked out[91]. The utility of the model for identifying toxic Aβ oligomers was demonstrated and might be useful for designing β-sheet breaker (BSB) compounds. The aluminium-hypothesis returned: it was found that Al^{3+} induced conformational changes in Aβ (α-helix → β-sheet) and enhanced its aggregation *in vitro*[92]. The mechanism of Aβ aggregation was studied by molecular modelling and found, that all Aβ domains (N-domain: 1-16; median domain: 17-22 and C-domain: 29-40/42) were crucial for intermolecular interaction and aggregation[93]. Aβ 1-40 fibrillogenesis can be inhibited by short Aβ congeners containing N-methyl amino acids at alternate residues, the most potent of these inhibitors is H_2N-K(Me-L)V(Me-F)F(Me-A)E-$CONH_2$[94]. This inhibitor appears to act by binding to growth sites of Aβ nuclei and/or fibrils and preventing propagation of the network of H-bonds that is essential for the formation of an extended β-sheet fibril. Two short reviews summarize the significance of Aβ aggregation in the physiopathology of Alzheimer's disease with special emphasis on physiological and synthetic, peptidic and non-peptidic Aβ-aggregation inhibitors[95,96]. Another publication deals with structure-function relationships for inhibitors of Aβ aggregation containing the recognition sequence KLVFF. Both the anionic (KLVFFEEEE) and cationic (KLVFFKKKK) peptides were effective by inhibiting fibril formation, while the neutral polar peptide (KLVFFSSSS) was ineffective[97]. Curcuninoid compounds like calebin-A proved to be neuroprotective against Aβ 25-35 (*in vitro* experiments, on pheocromocytoma and human neuroblastoma cells)[98]. One nanosecond molecular dynamics simulation was used to examine the structural stability of Aβ 25-35 and its analogue, $(Ala^{31,32,34,35})$ Aβ 25-35; calculations were performed using the modified GROMOS-87 force field. Bivalent cations (Ca^{2+}, Mg^{2+}, Zn^{2+}) stabilized the helix structure[99].

The mechanism of neurotoxicity of Aβ peptides has been intensively studied. Aβ peptides are capable of activating microglial cells *in vitro* and *in vivo*. Fibrillar amyloid peptide could maintain a chronic microglial activation, ultimately leading to the progressive neurodegeneration associated with Alzheimer's disease[100]. According to another hypothesis the interactions between Aβ-aggregates and cell membranes mediate Aβ toxicity. Membrane fluidity measurements show that exposed hydrophobic patches on the surface of Aβ aggregates interact with the hydrophobic core of the lipid bilayer leading to a reduction in membrane fluidity[101]. Decreases in membrane fluidity could hamper functioning of cell membrane receptors and ion channel proteins, such decreases might cause cellular toxicity. Reduction in cholesterol and sialic acid content protect cells against the toxic affect of Aβ-peptides: these results indicate the importance of interaction between Aβ-aggregates and cell membranes in the mechanisms of Aβ-toxicity[102].

Multiple caspases are involved in Aβ 25-35-induced neuronal apoptosis[103]. Ionic homeostasis is an important apoptotic effector and Aβ causes disruption of ionic homeostasis. Disruption of ionic balance by aggregated Aβ correlates with the toxic potential of the peptide[104].

Induction of learning and memory disfunction caused by different Aβ peptides (Aβ 25-35[105] and Aβ 1-42[106]) has been studied in rats. For a better understanding of the pathophysiology of Alzheimer's disease, APP transgenic mice were created that overexpress APP[107]. These transgenic mouse lines exhibit a deficit in spatial reference and working memory. For the early diagnosis of Alzheimer's disease in the future, a ligand was developed for imaging amyloid plaques in the brain[108]. The thioflavin derivative compound IBOX (2-(4-dimethylaminophenyl)-6-iodobenzoxarole was radioiodinated ([105]I) and has high binding potency for Aβ-plaques.

Formation of Aβ form APP can be decreased by inhibition of secretase enzymes. Memapsin 2 (β-secretase) can be controlled by an inhibitor OM99-2 (58), the X-ray crystal structure being critical for designing and synthesizing a series of potent β-secretase inhibitors[109]. Well known enzyme inhibitors like pepstatin A, peptide difluoroketone and peptide aldehydes inhibited Aβ 1-40 production in a dose-dependent fashion[110], probably by allosteric modulation of γ-secretase activity. Calcium ionophore A23187 specifically decreases the cleavage of APP by β-secretase in a caspase-dependent manner[111].

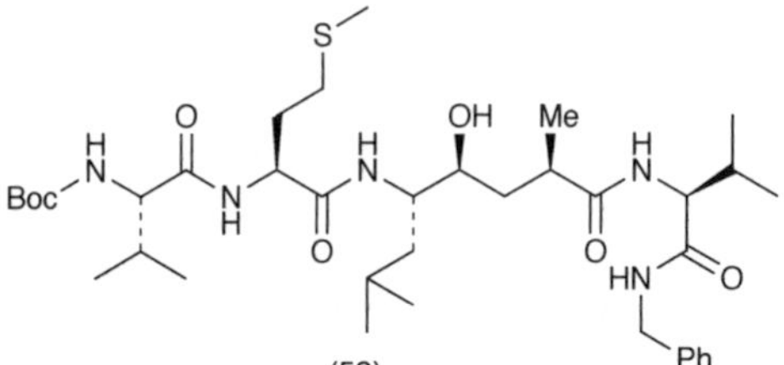

(58) 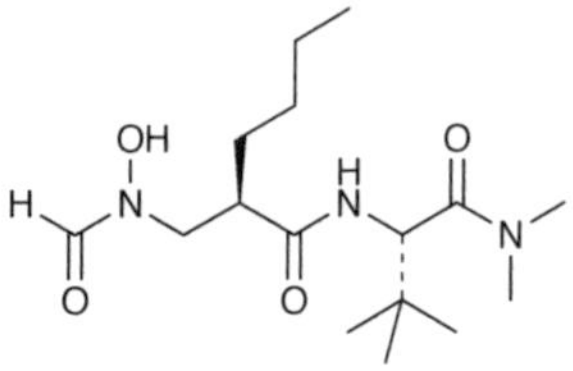   (59) (BB-3497)

4.2 Antimicrobial Peptides. – The functional and structural features of the naturally occurring antimicrobial peptides as well as their potential as therapeutic agents are reviewed in a 206-reference report[112]. The parameters that control activity and cell specificity of native antimicrobial peptides have been summarized[113]. Based on the 'carpet' mechanism and the role of the peptide oligomeric state, a novel class of diastereomeric antimicrobial peptides was developed. A 102-reference report has appeared on the molecular basis of peptide resistance, and the potential therapeutic applications of peptides in animal models of staphylococcal disease[114].

4.2.1 Antibacterial Peptides. Reviewing the cystine knot motif of toxins, the potential antibacterial applications of this uniquely stable structural feature have also been described[115]. Bacterial peptide deformylase (PDF), a novel target for the discovery of antibacterial drugs has been the subject of a 50-reference report[116]. Among others BB-3497 (59), a novel PDF inhibitor[117], an assay for detection of PDF inhibitors[118], and identification and characterization of novel deformylase inhibitors[119] have been discussed.

Six- and eight-residue cyclic D,L-α-peptides have been found to act preferentially on Gram-positive and/or Gram-negative bacterial membranes, increasing membrane permeability, collapsing transmembrane ion potentials, and causing rapid cell death[120]. The rapid action of these peptides and the fact that they act on membrane integrity rather than on vital biosynthetic processes suggests that resistance to these agents could be slower to develop. A new approach to antibacterial agents has emerged from the combination of a positively charged naphthalene tetracarboxylic diimide unit with various sequences of amino acids, generally lysine and glycine[121]. New generation cell-free systems were used for the biosynthetic production of antibacterial peptides such as the 31-residue cecropin P1[122]. Two-dimensional ^{1}H-NMR study of RTD-1, an antimicrobial defensin from Rhesus macaque leukocytes and its open chain analogue showed a similar extended β-hairpin structure with turns at one or both ends[123]. In contrast to many other antimicrobial peptides, RTD-1 did not display any amphiphilic character. Two novel insect defensins (Val-Thr-Cys-Asp-Leu-Leu-Ser-Phe-Glu-Ala-Lys-Gly-Phe-Ala-Ala-Asn-His-Ser-Ile-Cys-Ala-Ala-His-Cys-Leu-Ala-Ile-Gly-Arg-Lys-Gly-Gly-Ser-Cys-Gln-Asn-Gly-Val-Cys-Val-Cys-Arg-Asn   and Val-Thr-Cys-Asp-Leu-Leu-Ser-Phe-Glu-Ala-Lys-Gly-Phe-Ala-Ala-Asn-His-Ser-Ile-Cys-Ala-Ala-His-Cys-Leu-Val-Ile-Gly-Arg-Lys-Gly-Gly-Ala-Cys-Gln-Asn-Gly-Val-Cys-Val-Cys-Arg-Asn) were purified and characterized from larvae of the cupreous chafer, *Anomala cuprea* and showed slightly different activity against Gram-positive bacteria[124]. Cyclization of magainin 2 and melittin analogues altered their binding to phospholipid membranes[125]. Biological activity studies revealed that the linearity of the peptides is not essential for the disruption of the membrane, but rather provides the means to reach it. Modification of cationic charges from + 3 to + 7 of magainin peptides led to an optimized analog   (Gly-Ile-Gly-Lys-Phe-Ile-His-Ala-Val-Lys-Lys-Trp-Gly-Lys-Thr-Phe-Ile-Gly-Glu-Ile-Ala-Lys-Ser) whose selectivity is based on both reinforcement of activity against bacteria and reduction of the hemolytic effect[126]. Although all-D-cecropin B adopted identical α-helical structure compared to the natural peptide, the isomers differed in their interaction with lipopolysaccharide[127]. Pyrrhocoricin made of L-amino acids diminished the ATPase activity of recombinant DnaK and also reduced enzyme activities of alkaline phosphatase and β-galactosidase[128]. The pyrrhocoricin-binding site on *Escherichia coli* DnaK has also been determined.

Vancomycin analogues with activity against vancomycin-resistant bacteria have been synthesized in the solid- and solution-phase[129]. A number of *in vitro* highly potent antibacterial agents emerged from the biological evaluation of these libraries. Another effort in this area resulted in the synthesis of potent vancomycin or vancomycin derivative dimers using either disulfide formation or olefin metathesis ligation methods[130].

Substitution of Pro-9 in the α-helical antibiotic peptide P18 (Lys-Trp-Lys-Leu-Phe-Lys-Lys-Ile-Pro-Lys-Phe-Leu-His-Leu-Ala-Lys-Lys-Phe-NH$_2$) markedly reduced the antibacterial activity and increased hemolysis, suggesting that a proline kink in P18 serves as a hinge region to facilitate ion channel formation on bacterial cell membranes[131]. The presence of Arg residues at or near the N-terminus as well as a chain length exceeding 15 residues proved to be essential for the antibacterial activity of bactenecin 5, a 43mer peptide isolated from bovine neutrophils[132]. Structure-activity analysis of SMAP-29 (Arg-Glys-Leu-Arg-Arg-

Leu-Glly-Arg-Lys-Ile-Ala-His-Gly-Val-Lys-Lys-Tyr-Gly-Pro-Thr-Val-Leu-Arg-Ile-Ile-Arg-Ile-Ala-NH$_2$), a sheep leukocytes-derived antimicrobial peptide of the cathelicidin families, suggested that the N-terminal amphipathic α-helical region and the C-terminal hydrophobic region are responsible for antimicrobial and hemolytic acitivities, respectively, and the central Pro-19 in SMAP-29 plays a critical role in showing improved antibacterial activity[133]. The two Trp residues were found to be essential for antibacterial activity of a 15-residue fragment (Phe-Lys-Cys-Arg-Arg-Trp-Gln-Trp-Arg-Met-Lys-Lys-Leu-Gly-Ala) of bovine lactoferrin. Replacing Trp with natural and unnatural amino acids, the size, shape and aromatic character of Trp seemed to be the most important features for the activity[134]. In addition, it turned out that bulky aromatic amino acids increased the antibacterial activity of the 15-mer fragment, and that the peptides contained larger hydrophobic amino acids were generally more active against *Staphylococcus aureus* than against *Escherichia coli*[135]. Oligoalanyl substitution of polymyxin B nonapeptide (PMBN) **(60)**, the deacetylated amino derivative of polymyxin B did not effect most of PMBN's activities[136]. However, a hydrophobic aromatic substitution led to an analogue (Fmoc-PMBN) with high antibacterial activity and significantly reduced toxicity. Symmetrical dimeric derivative **(61)** of antimicrobial peptide temporin A (TA) has been synthesized using a novel

H$_2$N—FLPLIGRVLSGILAGG—NH

H$_2$N—FLPLIGRVLSGILAGG—NH

1 2   3   4   5   6   7   8   9
H-Thr-Dab-Dab-Dab-DPhe-Leu-Dab-Dab-Thr
    |
   CH$_2$—CH$_2$—NH—CO———┘

(60) (PMBN)   (61) (TAd)

branching unit 3-*N,N*-di(3-aminopropyl)amino propanoic acid[137]. TA and Tad (a dimeric analogue) showed similar antimicrobial effects against *Staphylococcus aureus*, a Gram-positive test strain. However, only the dimeric derivative had antimicrobial effect against *Escherichia coli*, the Gram-negative test strain. The synthesis and analyses of 11 new temporin A analogs, and a cecropin A-temporin A hybrid peptide has also been described[138]. A series of model amphipathic all L-amino acids peptides and their diastereomers with the sequence KX(3)KWX(2)KX(2)K, where X is Ala, Val, Ile, or Leu were synthesized and tested in order to investigate the initial stages leading to the binding and functioning of membrane-active polypeptides[139]. While hemolytic activity was drastically reduced, the antibacterial activity was preserved or increased upon substituting L-amino acids by their D-isomers. A correlation was found in the diastereomers between hydrophobicity and propensity to adopt helical/distorted helix structure and activity.

Four peptide fragments with bacteriocidal activity were isolated after proteolytic digestion of bovine β-lactoglobulin and their sequences were as follows: Val-Ala-Gly-Thr-Trp-Tyr, Ala-Ala-Ser-Asp-Ile-Ser-Leu-Leu-Asp-Ala-Gln-Ser-

Ala-Pro-Leu-Arg, Ile-Pro-Ala-Val-Phe-Lys and Val-Leu-Val-Leu-Asp-Thr-Asp-Tyr-Lys[140]. The peptides were synthesized and found to be effective against the Gram-positive bacteria only. Modifications of the last one yielded the peptide Val-Leu-Val-Leu-Asp-Thr-Arg-Tyr-Lys-Lys which enlarged the bacteriocidal activity spectrum to the Gram-negative bacteria *Escherichia coli* and *Bordetelle bronchiseptica*, and showed a high homology with the residues 55-64 of human blue-sensitive opsin. Beside antimicrobial lactoferrin fragments, a new peptide antibiotic corresponding to residues 63 to 117 of human casein-κ (Tyr-Gln-Arg-Arg-Pro-Ala-Ile-Ala-Ile-Asn-Asn-Pro-Tyr-Val-Pro-Arg-Tyr-Tyr-Tyr-Ala-Asn-Pro-Ala-Val-Val-Arg-Pro-His-Ala-Gln-Ile-Pro) was purified from human milk which inhibited the growth of Gram-positive, Gram-negative bacteria, and yeasts[141]. These results confirmed that antimicrobial peptides were liberated from human milk proteins during proteolysis and might play an important role in the host defense system of the newborn. Ser$(P)^{149}$κ-casein(106-169) designated as kappacin exhibited growth-inhibitory activity against the oral opportunistic pathogen *Streptococcus mutans*[142]. The nonphosphorylated peptide, however, did not inhibit growth at all. Peptides 5.11.1 (Asp-Ile-Gln-Ile-Pro-Gly-Ile-Lys-Lys-Pro-Thr-His-Arg-Asp-Ile-Ile-Ile-Pro-Asn-Trp-Asn-Pro-Asn-Val-Arg-Thr-Gln-Pro-Trp-Gln-Arg-Phe-Gly-Gly-Asn-Lys-Ser) and 8.4.1 (Glu-Asn-Phe-Phe-Lys-Glu-Ile-Glu-Arg-Ala-Gly-Gln-Arg-Ile-Arg-Asp-Ala-Ile-Ile-Ser-Ala-Ala-Pro-Ala-Val-Glu-Thr-Leu-Ala-Gln-Ala-Gln-Lys-Ile-Ile-Lys-Gly-Gly-Asp) have been purified from the hemolymph of the moth *Galleria mellonella* and also biochemically characterized[143]. While peptide 8.4.1 is a member of the cecropin family, peptide 5.11.1 is rich in proline and has a unique sequence in which the arrangement of proline residues does not fit to the consensus sequence of insect proline-rich peptides. Fifteen novel peptides, named ponericins, were isolated from the venom of the ant *Pachycondyla goeldii* and classified into three different families according to their sequence similarities[144]. Ten of these peptides were synthesized and further analysed. The high concentrations of each in the venom suggest that they may serve to protect against internal pathogens arising from alimentation, as in the case of peptides in some spider and snake venoms. Dicynthaurin, a novel homodimeric antimicrobial peptide from hemocytes of the solitary tunicate, *Halocynthia aurantium* showed broad-spectrum activity encompassing several Gram-positive and Gram-negative bacteria, but not *Candida albicans*, a fungus[145]. A new antimicrobial peptide, c(Thr-Ser-Tyr-Gly-Asn-Gly-Val-His-Cys-Asn-Lys-Ser-Lys-Cys-Trp-Ile-Asp-Val-Ser-Glu-Leu-Glu-Thr-Tyr-Lys-Ala-Gly-Thr-Val-Ser-Asn-Pro-Lys-Asp-Ile-Leu-Trp), referred to as MMFII, was purified from lactic acid bacteria *Lactococcus lactis*[146]. Molecular modeling showed three β-strands and an α-helical domain which did not exhibit an amphipathic structure. Seven peptides with antimicrobial activity were isolated from skin secretions of the diploid frog, *Xenopus tropicalis*[147]. The C-terminally amidated peptide XT-7 (Gly-Leu-Leu-Gly-Pro-Leu-Leu-Lys-Ile-Ala-Ala-Lys-Val-Gly-Ser-Asn-Leu-Leu-NH2) showed the lowest inhibitory concentrations against different microorganisms. The peptide was, however, hemolytic against human erythrocytes. The extract of the skin of the paradoxical frog *Pseudis paradoxa* was the source of four structurally related peptides and their

sequences were as follows: Gly-Leu-Asn-Thr-Leu-Lys-Lys-Val-Phe-Gln-Gly-
Leu-His-Glu-Ala-Ile-Lys-Leu-Ile-Asn-Asn-His-Val-Gln (pseudin-1), Gly-Leu-
Asn-Ala-Leu-Lys-Lys-Val-Phe-Gln-Gly-Leu-His-Glu-Ala-Ile-Lys-Leu-Ile-
Asn-Asn-His-Val-Gln (pseudin-2), Gly-Ile-Asn-Thr-Leu-Lys-Lys-Val-Ile-Gln-
Gly-Leu-His-Glu-Val-Ile-Lys-Leu-Val-Ser-Asn-His- Glu (pseudin-3), and Gly-
Ile-Asn-Thr-Leu-Lys-Lys-Val-Ile-Gln-Gly-Leu-His-Glu-Val-Ile-Lys-Leu-Val-
Ser-Asn-His-Ala (pseudin-4)[148]. Pseudins belong to the class of cationic, am-
phipathic α-helical antimicrobial peptides. Amino acid sequences of them did
not show similarity to any previously characterized peptides from frog skin. The
precursor of a novel peptide from a cDNA library prepared from pharyngeal
tissues of the tunicate, *Styela clava* has been cloned[149]. Its sequence predicted
clavaspirin, a histidine-rich amidated peptide that among other effects killed
Gram-positive as well as Gram-negative bacteria and permeabilized the outer
and inner membranes of *Escherichia coli*. The minimum inhibitory concentrations and mode of action of the 1-23 fragment of the α chain of bovine hemoglo-
bin towards *Micrococcus luteus* strain A270 were determined[150].

The interactions of the antimicrobial peptides magainin 1, melittin and the
C-terminally truncated melittin analog with hybrid bilayer membrane systems
were studied using surface plasmon resonance[151]. The results demonstrated that
these antimicrobial peptides bind to the lipids initially *via* electrostatic interactions which enhances the subsequent hydrophobic binding. Several methods
such as CD, FT-IR, ^{31}P-NMR spectroscopies have been used to investigate the
effect of cholesterol on the interactions of the antimicrobial peptide gramicidin S
with phosphatidylcholine and phosphatidylethanolamine model membrane systems[152]. The presence of cholesterol attenuated but did not abolish the interactions of gramicidin S with phospholipid bilayers.

4.2.2 Antifungal Peptides. A 53-reference review summarizes the development
of new generation azoles that are active against clinically relevant, drug-resistant
fungal pathogens, and that have advanced to late-stage clinical trials[153]. A study
reported that antibacterial peptides such as cecropin A and B from the silk moth
Cecropia as well as the porcine cecropin P1 are capable of inhibiting the growth
of and to kill yeast-phase *Candida albicans*[154]. Increasin the charge on an antimicrobial peptide (Lys-Lys-Val-Val-Phe-Lys-Val-Lys-Phe-Lys-NH$_2$) improved
antifungal activity without changing the antibacterial activity suggesting that the
net positive charge must play an important role in the specificity between
Candida albicans and Gram-positive bacteria[155].

The structure of a novel antifungal antibiotic, FR901469 (62), a 40-membered
macrocyclic lipopeptidolactone isolated from an unidentified fungus No. 11243
was determined[156]. A versatile synthesis of the lactam analogue (63) of (62)
including efficient formation of the ring by macrocyclization under high-dilution
conditions has also been published[157]. Besides, a series of acylated analogues of
FR901469 were designed and prepared, and several derivatives with comparable
in vivo antifungal efficacy but reduced hemolytic potential. were identified[158].

Conformational analysis of short-chain analogues of the lipopeptaibol antibiotic trichogin GA IV showed folded, but not helical structure[159]. Membrane

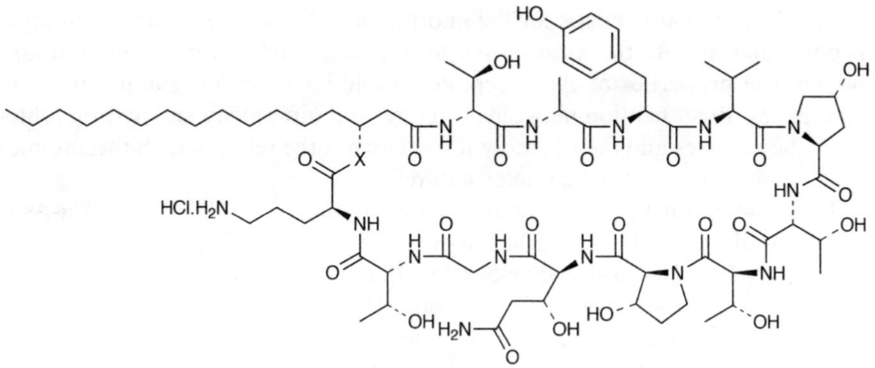

(62) (FR901469) X = O
(63) X = NH

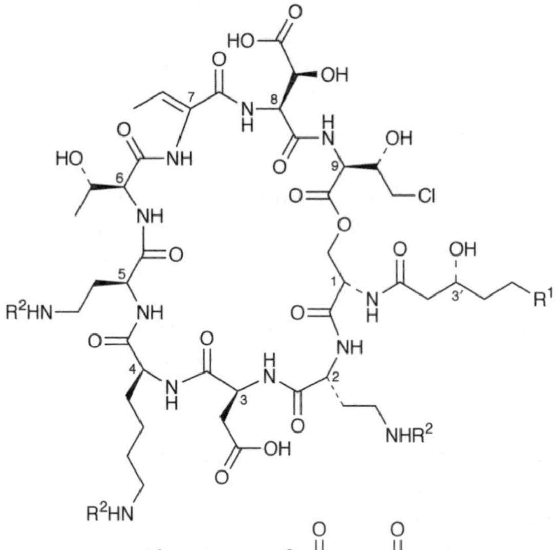

(64) R^1 = C$_9$H$_{19-n}$, R^2 = $\overset{O}{\overset{\|}{C}}OCH_2O\overset{O}{\overset{\|}{C}}$—Bu^t

(65) R^1 = C$_{11}$H$_{23-n}$, R^2 = $\overset{O}{\overset{\|}{C}}OCH_2O\overset{O}{\overset{\|}{C}}$—Bu^t

activity was found to increase from the shortest tetrapeptide up to the undecapeptide. The synthesis, bioconversion and antifungal activity of a series of N-acyloxymethyl carbamate linked triprodugs of pseudomycins have been described[160]. Two of these **(64** and **65)** showed excellent *in vivo* efficacy against systemic *Candidiasis*. Similar derivatives of 3-amido bearing pseudomycin analogues exhibited good *in vivo* efficacy against murine *Candidiasis*[161].

Secondary structure of PAFP-S, c(Ala-Gly-Cys-Ile-Lys-Asn-Gly-Gly-Arg-Cys-Asn-Ala-Ser-Ala-Gly-Pro-Pro-Tyr-Cys-Cys-Ser-Ser-Tyr-Cys-Phe-Gln-Ile-Ala-Gly-Gln-Ser-Tyr-Gly-Val-Cys-Lys-Asn-Arg), a peptide isolated from seeds of *Phytolacca Americana* and bearing a broad spectrum of antifungal activity has been examined by ^{1}H-NMR[162]. The results show that the molecular scaffold of PAFP-S features a triple-stranded β-sheet stabilized by a typical disulfide bridge motif which is characteristic for the knottin fold. PAFP-S is the first antifungal peptide that adopts the knottin-like fold. Investigation of conformation and biophysical properties of cyclic peptide antibiotics based on gramicidin S explained the direct relation between structure, amphipathicity and hydrophobicity of the peptides and their hemolytic activity but the relation with the antimicrobial activity is of a more complex nature[163].

Two novel antifungal peptides, designated α- and β-basrubrins (Gly-Ala-Asp-Phe-Gln-Glu-Cys-Met-Lys-Glu-His-Ser-Gln-Lys-Gln-His-Gln-His-Gln-Gly and Lys-Ile-Met-Ala-Lys-Pro-Ser-Lys-Phe-Tyr-Glu-Gln-Leu-Arg-Gly-Arg), were isolated from seeds of the Ceylon spinach *Basella rubra*[164]. After purification the antifungal potency of the peptides was determined. The antifungal activity and mechanism of action of PMAP-23, a 23-mer peptide with the sequence Arg-Ile-Ile-Asp-Leu-Leu-Trp-Arg-Val-Arg-Arg-Pro-Gln-Lys-Pro-Lys-Phe-Val-Thr-Val-Trp-Val-Arg-NH2 isolated from porcine myeloid was tested[165]. PMAP-23 prevented the regeneration of fungal cell walls and induced release of the fluorescent dye trapped in the membrane vesicles. Unguilin, a 18kDa cyclophilin-like protein was isolated[166] from seeds of the black-eyed pea (*Vigna unguiculata*) and its sequence is as follows: Phe-Asp-Met-Thr-Ala-Gly-Pro-Gln-Pro-Ala-Gly-Arg-Ile-Val-Phe-Glu-Gly-Phe-Ala-Asp-Met-Val-Gly-Arg-Thr-Ala-Val-Asn. Among several biological effects, unguilin exerted an antifungal activity toward fungi including *Coprinus comatus*, *Mycosphaerella arachidicola*, and *Botrytis cinerea*.

4.3   ACTH Peptides. – A new analogue of ACTH 4-10 (Met-Glu-His-Phe-Pro-Gly-Pro, Semax) was investigated as neuroprotective nitric oxide generating agent[167]. The peptide proved to be a novel nootropic substance.

4.4   Angiotensin II Analogues and Non-peptide Angiotensin II Receptor Ligands. – The amide-linked angiotensin II cyclic analogue, cyclo(3,5)-[Sar^1, Lys^2, Glu^5, Ile^8] angiotensin II has been synthesised[168]. This analogue was found to be an inhibitor of the parent peptide. NMR data and molecular modelling revealed structural similarities between losartan, EPRosartan, irbesartan and the above cyclic analogue. Cyclic 12-, 13- and 14-membered ring angiotensin II analogues have been synthesised[169]. For the ring closure, different Cys, Hcy and thioacetal (S-CH$_2$-S) containing analogues were prepared. Interestingly, compound **(66)**, a 13-membered analogue proved to be a full agonist. Conformational analysis suggested that some of these compounds adopted inverse γ-turn conformation. Several shorter fragments of angiotensin and their analogues have been prepared[170]. These short peptides were used for the determination of their binding affinity to AT4 receptor. Angiotensin derived peptides were also used to characterise the specificity of a lysosomal tripeptidyl peptidase I[171].

4.5   Bombesin/Neuromedin Analogues. – A potent and selective agonist (D-Phe-Gln-Trp-Ala-Val-ß-Ala-His-Phe-Nle-NH$_2$) of the gastrin releasing peptide preferring bombesin receptor (BB$_2$) and its truncated fragments have been synthesized[172]. In order to determine the key structural features, an alanine scan of

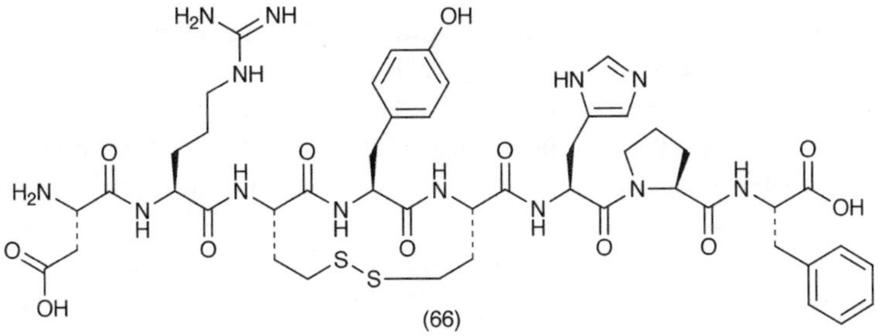

(66)

the above peptide was made. The D-Phe[6], Val[10], Phe[13] residues seemed to be the most responsible for the receptor selectivity. The mechanism of the G-protein mediated action of substance P analogues and bombesin was studied[173]. Orphan receptor (bombesin receptor subtype 3) selective bombesin analogues have been synthesized[174]. The incorporated conformationally constrained amino acids included: β-alanine (β-Ala), (R) and (S)-3-carboxypiperidine (Cpi), (R)- and (S)-*trans*-2-amino-1-cyclohexane-carboxylic acid (Achc), (R)- and (S)-*cis*-2-amino-1-cyclohexane-carboxylic acid (Achc), (R)- and (S)-3-amino-butyric acid (Aba), (R)- and (S)-3-amino-isobutyric acid (Aia), (R)- and (S)-3-amino-3-phenylpropionic acid (Apa), (R)-amino-3-(4-Cl-benzyl)-propionic acid (Acpb4), (R)-3-amino-3-(2-Cl-benzyl)-propionic acid)(Acpb2). The peptides Ac-[(R)-Aba[11],Phe[13],Nle[14]]Bn-(8-14) and Ac-[(S)-Aba[11],Phe[13],Nle[14]]Bn-(8-14) showed the highest selectivity for the hBRS-3 over the mammalian Bn receptors and did not interact with receptors for other gastrointestinal hormones/neurotransmitters. Molecular modelling demonstrated a unique β conformation at the 11th amino acid residue. A bombesin analogue (D-Phe-Gln-Trp-Ala-Val-ß-Ala-His-Phe-Nle-NH$_2$) was used to investigate the signal transduction mechanism of bombesin receptor subtype 3[175]. A ^{99}Tc chelate-bombesin conjugate (67) has been synthesised and used for ^{1}H-NMR investigations[176]. The above compound can be used as a receptor selective radiopharmaceutical. The activity of a bombesin antagonist

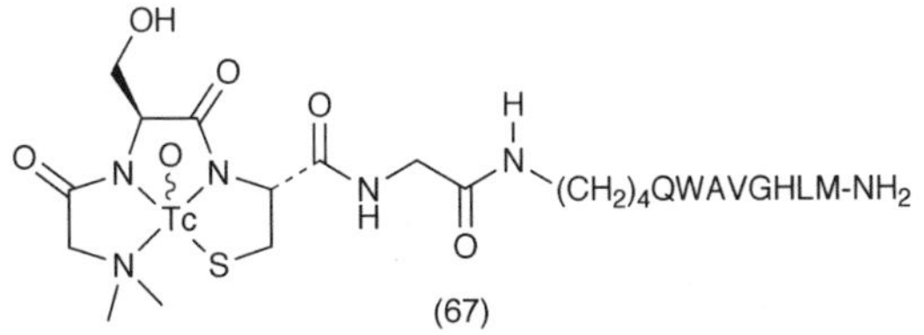

(67)

(D-Tpi[6]-Leu[13] ψ[CH$_2$NH]Leu[14] BN 6-14, RC 3095) and an LHRH antagonist (Ac-D-Nal[1]-D-Cpa[2]-D-Pal[3]-D-Cit[6]-D-Ala[10] LH-RH, Cetrorelix) against ovarian carcinoma was evaluated[177]. The synthesis of four bombesin analogues (H-Asn-Gln-Trp-Ala-Val-Gly-His-Cfa-Met-NH$_2$,H-Asn-Gln-Trp-Ala-Val-Sar-His-Leu-Met-NH$_2$, H-Asn-Gln-Trp-Ala-Val-ß-Ala-His-Leu-Met-NH$_2$, H-Asn-Gln-Nal-Ala-Val-Gly-His-Leu-Met-NH$_2$) and their effect on food intake has been described[178].

4.6   Bradykinin Analogues. – A series of cyclic pentapeptides based upon the C terminal fragment of the known B_2 kinin receptor antagonist (H-D-Arg-Arg-Pro-Hyp-Gly-Thi-Ser-D-Tic-Oic-Arg-OH) has been designed with the aim at obtaining a type-II' β-turn around the Tic and Oic residues[179]. The general structure of these peptides is shown in **Figure 3**. The most potent analogue was the c[Pro-Orn-D-Tic-Oic-Arg], which exhibited 86% inhibition. Two lipopeptide agonistic analogues of bradykinin (Pam-Lys-Arg-Pro-Pro-Gly-Phe-Ser-Pro-Phe-Arg-OH, PKD and Pam-Gly-Lys-Arg-Pro-Pro-Gly-Phe-Ser-Pro-Phe-Arg-OH, PGKD) were synthesized and characterised by NMR and binding to human B_2 receptor[180].

A series of analogues of desArg[9]-Lys-bradykinin, Lys-Arg-X-Ac$_n$c-X-Ser-Pro-Phe, in which the spacer X-Ac$_n$c-X replaces the central tetrapeptide Pro-Pro-Gly-Phe of bradykinin has been synthesized and characterized by NMR spectroscopy and binding to bradykinin B_1 receptor[181]. The ten 1-aminocycloalkane-1-carboxylic acid containing analogues are as follows:

H-Lys-Arg-Ado-Ser-Pro-Phe-OH, H-Lys-Arg-Ac$_7$c-Ser-Pro-Phe-OH, H-Lys-Arg-Gly-Ac$_7$c-Gly-Ser-Pro-Phe-OH, H-Lys-Arg-βAla-Ac$_7$c-βAla-Ser-Pro-Phe-OH, H-Lys-Arg-γAbu-Ac$_7$c-γAbu-Ser-Pro-Phe-OH, H-Lys-Arg-Gly-Ac$_6$c-Gly-Ser-Pro-Phe-OH, H-Lys-Arg-Gly-Ac$_8$c-Gly-Ser-Pro-Phe-OH, H-Lys-Arg-Gly-Ac$_9$c-Gly-Ser-Pro-Phe-OH, H-Lys-Arg-Gly-Ac$_{12}$c-Gly-Ser-Pro-Phe-OH, and H-Lys-Arg-Gly-Gly-Gly-Ser-Pro-Phe-OH.

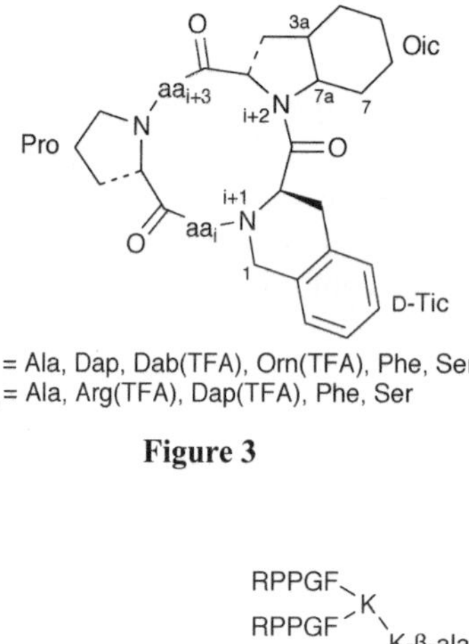

aa$_i$ = Ala, Dap, Dab(TFA), Orn(TFA), Phe, Ser, Thi
aa$_{i+3}$ = Ala, Arg(TFA), Dap(TFA), Phe, Ser

Figure 3

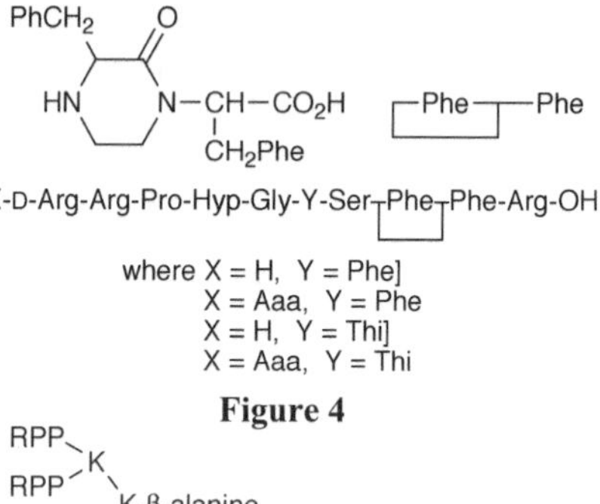

X-D-Arg-Arg-Pro-Hyp-Gly-Y-Ser-Phe-Phe-Arg-OH

where X = H, Y = Phe]
X = Aaa, Y = Phe
X = H, Y = Thi]
X = Aaa, Y = Thi

Figure 4

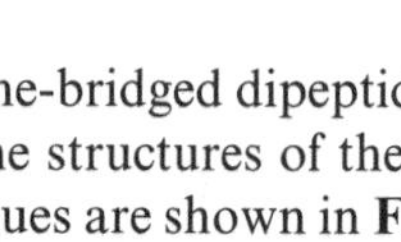

RPPGF-K
RPPGF-K
RPPGF-K-β-alanine
RPPGF-K
(68)

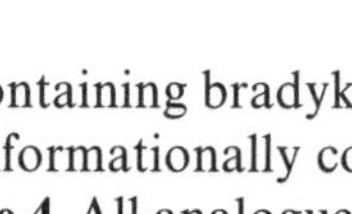

RPP-K
RPP-K
RPP-K-β-alanine
RPP-K
(69)

Eight new, ethylene-bridged dipeptide containing bradykinin analogues have been prepared[182]. The structures of the conformationally constrained dipeptide as well as the analogues are shown in **Figure 4**. All analogues were characterized by their antagonistic potencies (blood pressure, antiuterotonic and uterotonic activities).

Branched peptides (**68** and **69**) containing the N-terminal part of bradykinin were synthesized and investigated as thrombin inhibitors[183]. MAP4-RPPGF at

35.5 μM inhibits γ-thrombin- induced platelet aggregation 100% and α-thrombin-induced calcium mobilization in fibroblasts 84%. The stable metabolic fragment of bradykinin (RPPGF) was radioiodinated and used for radioimmunoassay[184]. Three linear Thr[6]-bradykinin analogues in which either one or both of the two phenylalanine residues in the sequence have been substituted by N-benzylglycine, and their head-to-tail cyclic analogues, were prepared and tested on isolated rat duodenum[185].

4.7   Cholecystokinin Analogues, Growth Hormone-Releasing Peptide and Analogues. – An improved solid-phase method was described for the synthesis of sulfated tyrosine containing peptides including CCK-12 (Ile-Ser-Asp-Arg-Asp-Tyr(SO$_3$H)-Met-Gly-Trp-Met-Asp-Phe-NH$_2$), mini-gastrin-II (H-Leu-Glu-Glu-Glu-Glu-Glu-Ala-Tyr(SO$_3$H)-Gly-Trp-Met-Asp-Phe-NH$_2$), little gastrin-II (Pyr-Gly-Pro-Trp-Leu-Glu-Glu-Glu-Glu-Glu-Ala-Tyr(SO$_3$H)-Gly-Trp-Met-Asp-Phe-NH$_2$), big gastrin-II (Pyr-Leu-Gly-Pro-Gln-Gly-Pro-Pro-His-Leu-Val-Ala-Asp-Pro-Ser-Lys-Lys-Gln-Gly-Pro-Trp-Leu-Glu-Glu-Glu-Glu-Glu-Ala-Tyr(SO$_3$H)-Gly-Trp-Met-Asp-Phe-NH$_2$), CCK-33 (H-Lys-Ala-Pro-Ser-Gly-Arg-Met-Ser-Ile-Val-Lys-Asn-Leu-Gln-Asn-Leu-Asp-Pro-Ser-His-Arg-Ile-Ser-Asp-Arg-Asp-Tyr(SO$_3$H)-Met-Gly-Trp-Met-Asp-Phe-NH$_2$) and CCK-39 (H-Tyr-Ile-Gln-Gln-Ala-Arg-Lys-Ala-Pro-Ser-Gly-Arg-Met-Ser-Ile-Val-Lys-Asn-Leu-Gln-Asn-Leu-Asp-Pro-Ser-His-Arg-Ile-Ser-Asp-Arg-Asp-Tyr(SO$_3$H)-Met-Gly-Trp-Met-Asp-Phe-NH$_2$)[186]. In this approach the growing peptide chain was assembled on 2-chlorotrityl resin using Fmoc-Tyr(SO$_3$Na)-OH as the key building block. The conditions for the final deprotection were optimised using trifluoroacetic acid. The synthesised CCK peptides were tested on isolated pancreatic islets for insulin release.

Analogues of the cholecystokinin 26-30 fragment (H-Asp-Tyr(SO$_3$H)-Met-Gly-Trp-Met-Asp-Phe-NH$_2$) were synthesised by the solid-phase method[187]. The phenylalanine in position 33 was replaced with phenylglycine, naphthylalanine, p-fluorophenylalanine, cyclohexylalanine and O-methyltyrosine. The peptides were tested *in vivo* for decreasing the food uptake of rats.

A porphyrin ring containing derivative of CCK-8 (CCK 26-33) and its indium complex have been synthesised[188]. The natural sequence of CCK-8 (without the sulfate moiety on the tyrosine) was elongated with a lysine residue. The N$^\varepsilon$ group of the completed peptide was acylated with a porphyrin derivative of its indium complex. The steric structure of the resulting molecules was investigated using NMR spectroscopy and molecular dynamics simulations. These investigations revealed that the original peptide conformation does not change after the attachment of the bulky heterocyclic ring, and the metal complexes could be used in different nuclear medicine techniques. In a project aimed at the search for selective CCK$_1$ receptor antagonists numerous peptidomimetic analogues have been synthesised based on the structure of **(70)** and **(71)**[189].

The steric structure of human ghrelin (H-Gly-Ser-Ser(octanoyl)-Phe-Leu-Ser-Pro-Glu-His-Gln-Arg-Val-Gln-Gln-Arg-Lys-Glu-Ser-Lys-Lys-Pro-Pro-Ala-Lys-Leu-Gln-Pro-Arg-OH) and its six truncated analogues (H-Gly-Ser-Ser(octanoyl)-Phe-Leu-Ser-Pro-Glu-His-Gln-NH$_2$,   H-Gly-Ser-Ser(octanoyl)-Phe-

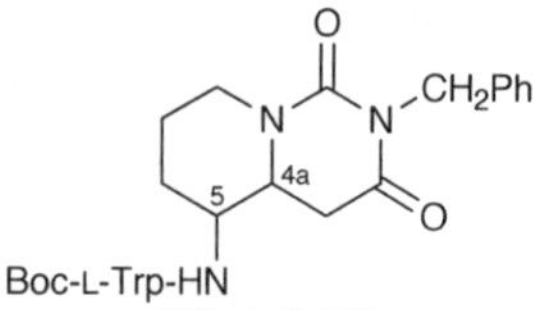

Boc-L-Trp-HN
(70) 4aS, 5R
(71) 4aR, 5S

Leu-Ser-Pro-Glu-His-Gln-Arg-Val-Gln-Gln-OH,   H-Gly-Ser-Ser(octanoyl)-
Phe-Leu-Ser-Pro-Glu-His-Gln-Arg-Val-Gln-Gln-Arg-Lys-Glu-Ser-NH$_2$,H-
Gly-Ser-Ser(octanoyl)-Phe-Leu-Ser-Pro-Glu-His-Gln-Arg-Val-Gln-Gln-Arg-
Lys-Glu-Ser-Lys-Lys-Pro-Pro-Ala-NH$_2$, H-Gly-Ser-Ser-Phe-Leu-Ser-Pro-Glu-
His-Gln-Arg-Val-Gln-Gln-Arg-Lys-Glu-Ser-Lys-Lys-Pro-Pro-Ala-Lys-Leu-
Gln-Pro-Arg-OH andH-Gly-Ser-Ser(octanoyl)-Phe-Leu-NH$_2$) has been inves-
tigated[190] by ^{1}H NMR measurements and CD spectroscopy. These investigations
proved that the presence or absence of the octanoyl group made no difference in
the structure of these peptides in solution.

Hexarelin (H-His-D-2MeTrp-Ala-Trp-D-Phe-Lys-NH$_2$), another GH releasing peptide and several analogues were synthesized and investigated for the
intestinal permeability[191] and on penile erection[192]. The analogue EP 91073
(Aib-D-Trp(2-Me)-D-Trp(2-Me)-Lys-NH$_2$) prevented the effect of EP 80661
(GAB-D-Trp(2-Me)-D-Trp(2-Me)-Lys-NH$_2$) and other EP peptides.

4.8   Integrin-related Peptide and Non-Peptide Analogues. – *4.8.1   IIb/IIIa Antagonists.* Conformationally constrained analogues of the GPIIb/IIIa antagonist
elarofiban (RWJ-53308) using the novel 1,2,4-triazolo[3,4-*a*]pyridine scaffold
have been synthesized[193]. Compounds **(72)** and **(73)** exhibited enhancement in

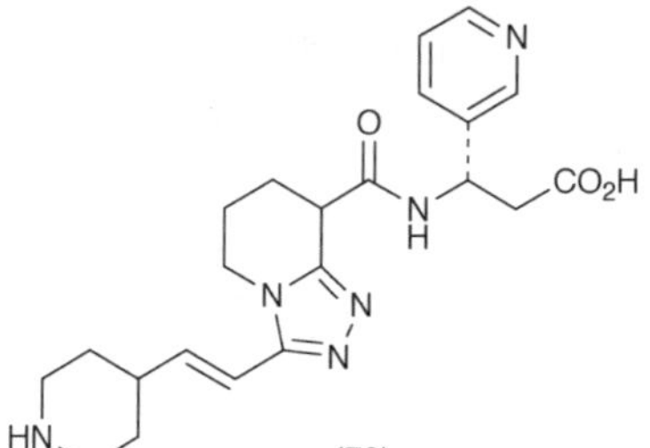

(72)

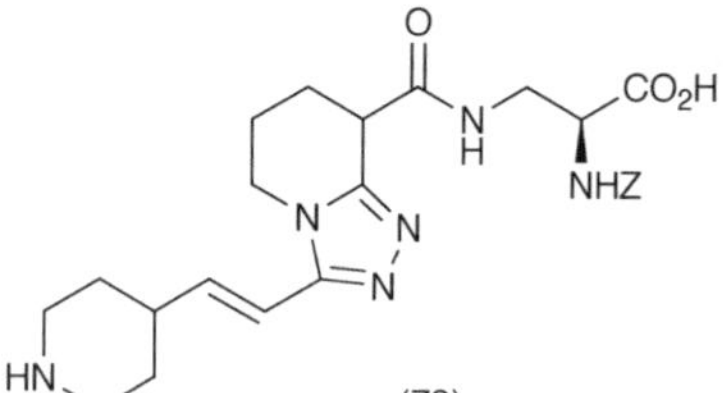

(73)

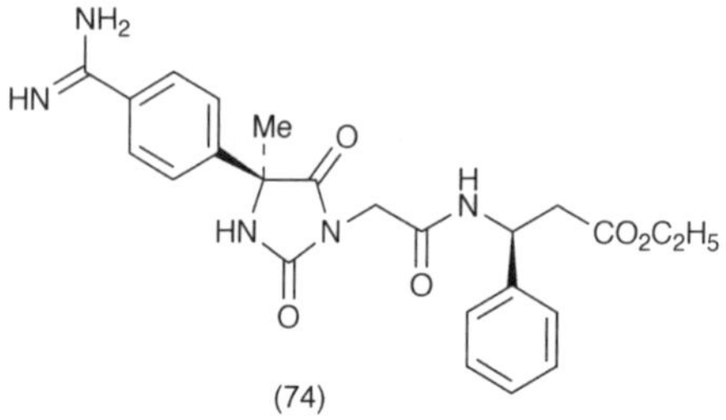

(74)

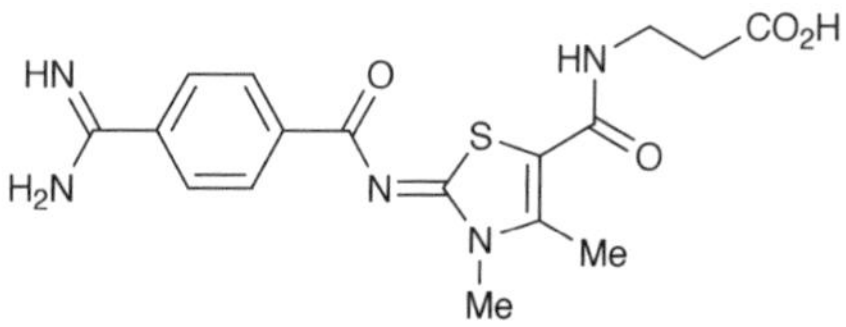

(75)

oral bioavailability, $t_{1/2}$, and duration of inhibition of adenosine-5'-diphosphate (ADP)-induced platelet aggregation relative to elarofiban. Structural modifications of the RGDS binding motif of fibrinogen led to the discovery of a non-peptide RGD mimetic GPIIb/IIIa antagonist (74) which inhibited platelet aggregation as well as ^{125}I-fibrinogen binding to ADP-activated human platelets and isolated GPIIb/IIIa with K_i values of 9 nM and 0.17 nM, respectively[194]. A series of the TAK-029 GPIIb/IIIa antagonists have been synthesized through modification on the glycine moiety[195]. The (3*S*, 2*S*)-4-methoxyphenylalanine derivative inhibited *in vitro* ADP-induced platelet aggregation with an IC_{50} value of 13 nM. Compound (75), a low molecular weight RGD mimic based on a novel sterically constrained 2-imino-3*H*-thiazoline scaffold proved to be a highly potent fibrinogen receptor antagonist[196]. Novel glycoprotein IIb-IIIa antagonists containing either the 3,9-diazaspiro[5.5]-undecane[197] or 3-azaspiro[5.5]undec-9-yl[198] nucleus have been described. However, none of these potent and specific antagonists showed good pharmacokinetic properties. As a continuation of the efforts made on the GPIIb/IIIa receptor antagonist XR299, a series of potent N-substituted benzamidine isoxazolines have been explored[199]. The substitution of the amidine presented a possibility for the modulation of duration of action within the sulfonamide series. Intravenous infusion of TAK-024 (76) at 1.6 μg/mL/min completely prevented arterial thrombus formation induced by endothelial injury in guinea pigs[200]. The dose of it that prolonged the bleeding time to three times the control value was 5.8 μg/mL/min. A series of RGD analogs have been synthesized as potential delivery vectors of pharmaceutical agents[201]. The analogs could tolerate side chain modifications fairly well as shown by platelet aggregation studies and confirmed by the fact that compound (77) with a side chain modification of poly(ethylene glycol) retained high affinity for glycoprotein IIb/IIIa ($IC_{50} = 150$ nM). A bivalent poly(ethylene glycol) hybrid con-

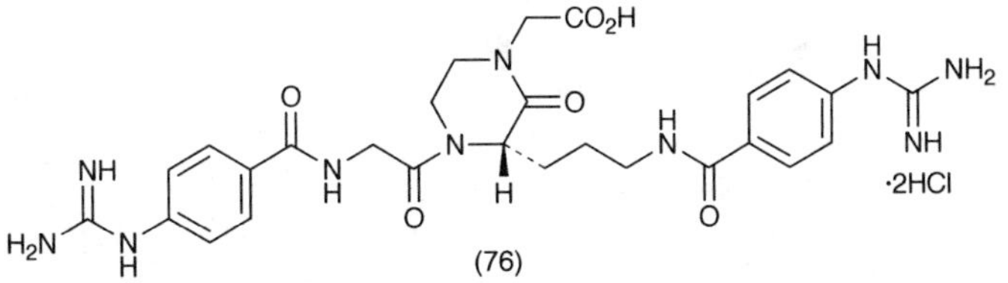

(76)

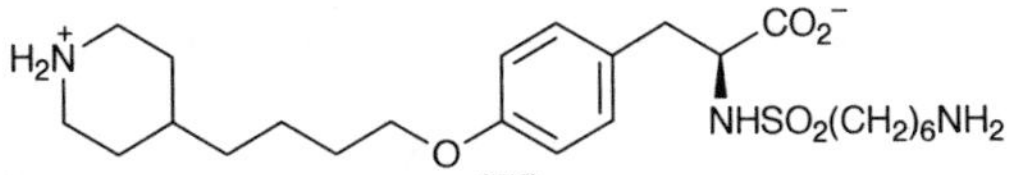

(77)

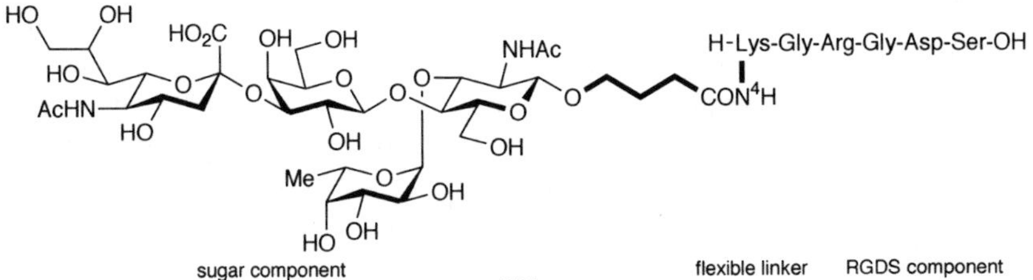

sugar component   flexible linker   RGDS component

(78)

taining an active site (RGD) and its synergistic site (PHSRN) of fibronectin has been prepared[202]. The PHSRN-aaPEG-RGD hybrid strongly promoted cell spreading activity compared with aaPEG-RGD suggesting that the PHSRN sequence synergically enhances the activity of the RGD sequence, and that the bivalent aaPEG hybrid method may be useful for conjugating functionally active pep-tides. An efficient synthesis of compound (78), a nonnatural glycopeptide that interacts with both selectins and integrins, has been described[203]. In addition to its high affinity with P-selectin, this compound also inhibited a specific interaction between human integrin β_1 and its monoclonal antibody. A novel synthetic route to SB214857, a potent GPIIb/IIIa receptor antagonist, has been developed using the CuI-catalysed coupling reaction of β-amino acids with aryl halides[204]. The reaction has been completed at 100 °C showing an accelerating effect of the β-amino acid for the Ullmann-type aryl amination reaction.

4.8.2 $\alpha_v\beta_3$ Antagonists. A new class of potent and selective $\alpha_v\beta_3$ antagonists based on a central thiophene scaffold and the acylguanidine Arg-mimetic has been discovered[205]. Compound (79), the adamantyl group-containing member of this series, showed a dose dependent effect in the *in vivo* TPTX rat model suggesting the possible application of these substances for the treatment of osteoporosis. Diketopiperazine- and hydantoin-based constrained RGD mi-metics were synthesized on the solid-phase[206]. The binding affinities to the $\alpha_v\beta_3$ receptor were in the low micromolar range for both families. The aza-RGD mimetic **80** was identified by an on-bead screening assay of a combinatorial library and showed good affinity and selectivity toward the $\alpha_v\beta_3$ integrin recep-tor (IC$_{50}$: 150 nM)[207]. Optimization of this compound by the substitution of an amide bond led to a library of highly active and selective RGD mimetics[208]. IC$_{50}$ values for the most potent 4-chloro substituted compounds (81 and 82) were 0.1 and 0.3 nM respectively. Novel malonamide derivatives as $\alpha_v\beta_3$ antagonists have been identified by modification of the glycine part of SC65811[209]. The most potent among them exhibited inhibitory activity with an IC$_{50}$ value of 0.42 nM. A convenient synthetic access to various conjugates of the cyclopeptide (-RGDfK-), a potent and selective $\alpha_v\beta_3/\alpha_v\beta_5$ ligand, through the regioselective derivatisation of lysine side chains, either in solution or on the solid support, was described[210]. Cyclo(-RGDf=V-) (83), a newly developed RGD peptide proved to be a potent inhibitor of tumor-induced angiogenesis *in vivo*, and therefore might be a useful agent preventing distant metastasis of the solid tumor[211]. A combina-torial library of stereodiverse RGD mimetics has been rationally constructed using molecular modeling[212]. The use of a single substituted carbohydrate scaf-fold proved to be an excellent way to mimic different conformations of the same peptide chain. The screening of small molecule libraries led to the identification of compounds which disrupt the MMM2-$\alpha_v\beta_3$ interaction so can serve as anti-angiogenic agents[213]. The synthesis of a photoreactive analogue of echistatin, a potent RGD-containing $\alpha_v\beta_3$ antagonist both *in vitro* and *in vivo* has been reported[214]. Cross-linking data achieved by using radiolabeled compounds sup-ported the hypothesis that the ligand-bound conformation of the integrin β_3 subunit differs from the known conformation of integrin domains.

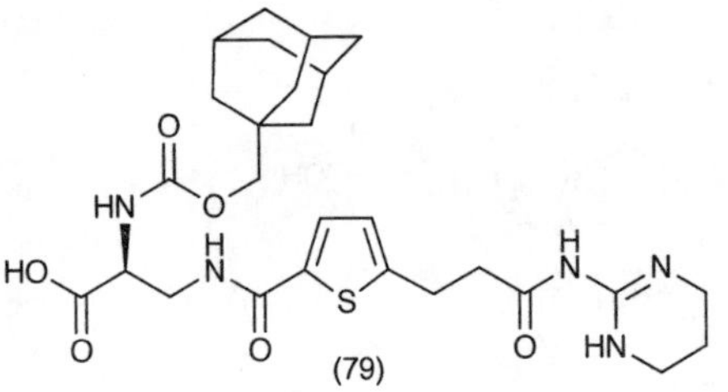

(79)

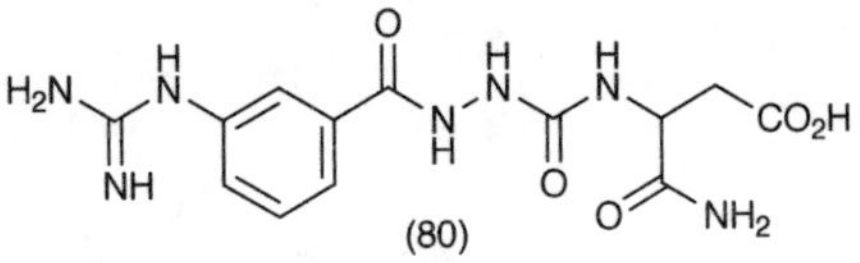

(80)

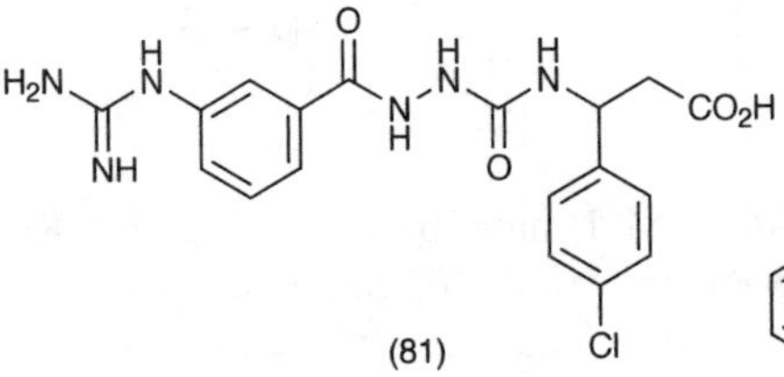

(81)

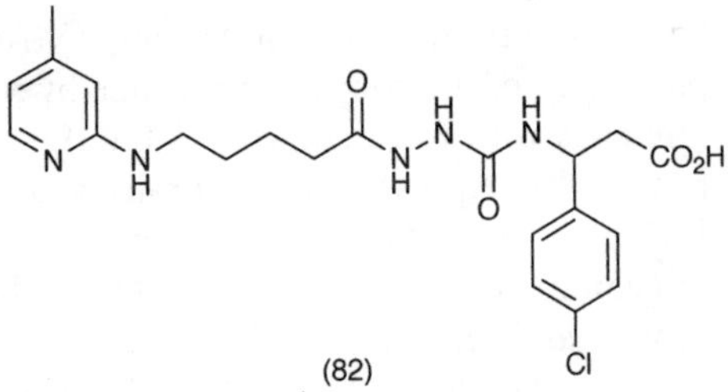

(82)

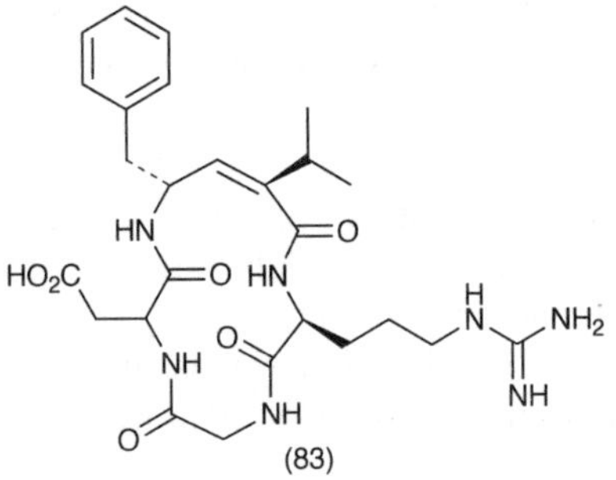

(83)

4.8.3 $\alpha_4\beta_1$, $\alpha_4\beta_7$ and $\alpha_5\beta_1$ Antagonists. Isoxazolyl, ozazolyl and thiazolyl-propionic acid derivatives derived from LDV were examined as $\alpha_4\beta_1$ antagonists[215]. Compound (84) showed nanomolar *in vitro* binding affinity and proved to be effective in an *in vivo* allergic mouse model. Selective inhibitors of VLA-4 based on the known bisarylurea series of LDV peptidomimetics that retain nanomolar potency, but are structurally simpler than the original leads have been generated[216]. Directed screening of a combinatorial library and subsequent optimization by solid-phase synthesis led to a series of sulfonylated dipeptide inhibitors of $\alpha_4\beta_1$[217]. Pharmacokinetic parameters determined for the most potent representatives which showed an IC_{50} value lower than 100 pM suggested rapid degradation ($t_{1/2} < 2.5$ min). Compound (85) was discovered as a selective and potent piperidinyl carboxylic acid-based inhibitor of the $\alpha_4\beta_1$ integrin and the vascular cell adhesion molecule 1 (VCAM-1) with a promising pharmacokinetic profile[218]. Several related peptides with the general formula cyclo(-Leu-Asp-Thr-Xaa-D-Pro-Xbb-), which include the LDT motif of the mucosal

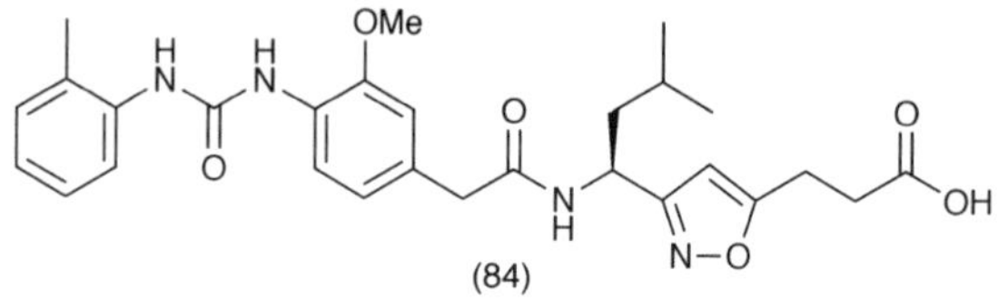

(84)

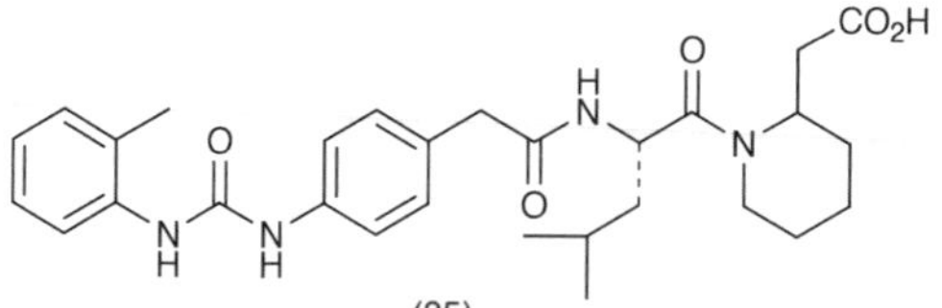

(85)

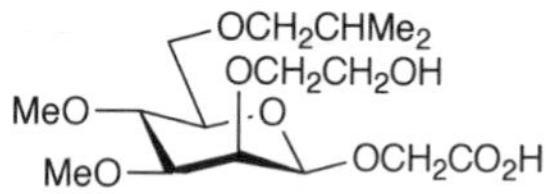

(86)

addressin cell adhesion molecule-1 (MAdCAM-1) have been developed[219]. Replacement of the peptidic backbone by a sugar scaffold led to a small library of mimetics with β-D-mannose as the rigid core[220]. This class of peptidomimetics has all the requirements for orally available drugs and compound (86) fulfills Lipinski's rules for bioavailability. A cellular solid-phase binding assay for screening $\alpha_4\beta_7$ integrin antagonists attached *via* photolinker to TentaGel macrobeads has been developed[221]. The resin bound compounds were identified by mass spectrometry. The $\alpha_4\beta_7$ – MAdCAM interaction was shown to be antagonized by an SVVYGLR-OH peptide which is a structural motif of osteopontin that binds integrins $\alpha_4\beta_1$ and $\alpha_9\beta_1$[222]. Inhibition studies revealed that Leu167 and the free C-terminal carboxylic acid of Arg168 have crucial roles in the interaction with α_4 integrins.

4.9   LHRH and GnRH Analogues. - Structurally new analogues of GnRH receptor antagonist cetrorelix (Ac-D-Nal(2)¹-D-Cpa²-D-Pal(3)³-D-Cit⁶-D-Ala¹⁰ GnRH) as well as conformationally constrained cyclized deca- or pentapeptides have been synthesized[223]. Receptor affinity measurements of these peptides were evaluated and the potential clinical use of them was discussed.

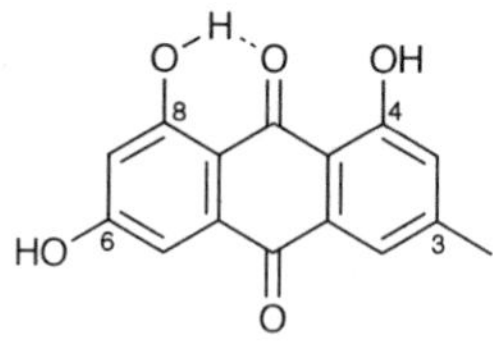

(87)

In an attempt to develop an efficient chemotherapeutic agent targeted at malignant cells that express receptors to GnRH, [D-Lys⁶]GnRH was coupled covalently to emodic acid to yield [D-Lys⁶(Emo)]GnRH (87)[224]. The synthesis, electron-transfer and photochemistry of (87) were demonstrated. In another article the biological evaluation of the upper derivative was discussed[225]. The prolonged activity of the [D-Lys⁶(Emo)] GnRH can be caused by the high binding affinity of the emodic acid moiety to serum proteins.

The inhibition of human ovarian carcinoma cell-line by cetrorelix was investigated[226]. A delivery system for a GnRH analogue (Leuprolide) [Des-Gly¹⁰,(D-

Leu⁶),Pro⁹]GnRH-ethylamide was prepared and evaluated[227]. This formulation resulted in a 120 day controlled release, which was potentially useful for testosterone suppression.

The low oral bioavailability, which is characteristic for GnRH and its analogues can be increased by protection of the peptides from their degradation by peptidases[228]. In this article the activity of different pancreatic endopeptidases towards GnRH was described. A GnRH analogue (des-Gly¹⁰,(D-Ala⁶)-LH-RH-ethylamide) was used to stimulate spermiation in paddlefish[229].

4.10 α-MSH Analogues. – Several analogues of α-MSH and its fragments have been synthesised[230]. The synthesised analogues were the following: Ser-Tyr-Ser-Ahx-Glu-His-Phe-Arg-Trp-Gly-Lys-Pro-Val, Ser-Tyr-Ser-Ahx-Glu-His-D-Phe-Arg-Trp-Gly-Lys-Pro-Val, Ahx-Asp-His-D-Phe-Arg-Trp-Lys, Ahx-Asp-Gln-D-Phe-Arg-Trp-Lys, Ahx-Asp-Trp-D-Phe-Arg-Trp-Lys, Ahx-Asp-Asn-D-Phe-Arg-Trp-Lys, Ahx-Asp-Arg-D-Phe-Arg-Trp-Lys, Ahx-Asp-Lys-D-Phe-Arg-Trp-Lys, Ahx-Asp-Tyr-D-Phe-Arg-Trp-Lys, Trp-D-Phe-Arg-Trp-Lys, and Trp-D-Phe-Arg-Trp. One of them (Ahx-Asp-Gln-D-Phe-Arg-Typ-Lys) proved to be 10 000 times less potent than the corresponding α-MSH fragment for the MC1R receptor subtype, while its potency on MC3R and MC4R was comparable to the parent peptide. In addition to the receptor binding measurements, NMR experiments and molecular modelling were performed.

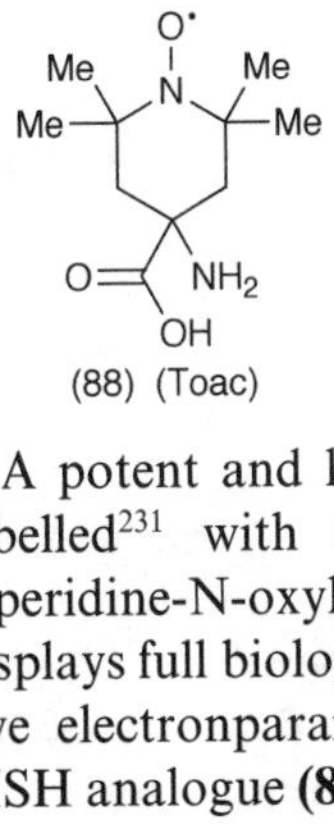

DPPE-(PEG)ₙ—N ... S ... Gly-[Nle⁴, D-Phe⁷]-α-MSH(4-10)

(88) (Toac) 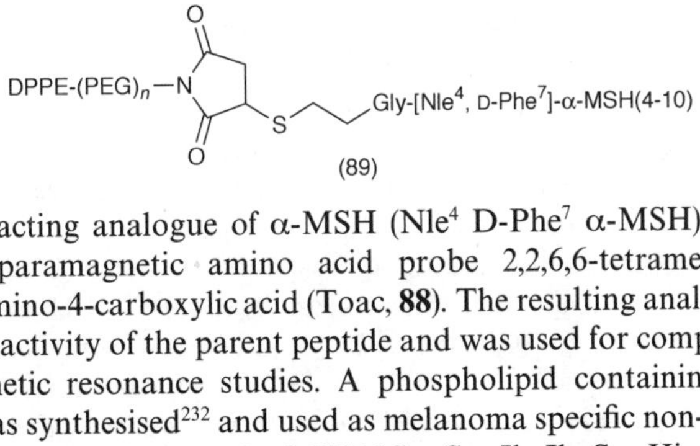   (89)

A potent and long acting analogue of α-MSH (Nle⁴ D-Phe⁷ α-MSH) was labelled[231] with the paramagnetic amino acid probe 2,2,6,6-tetramethyl-piperidine-N-oxyl-4-amino-4-carboxylic acid (Toac, **88**). The resulting analogue displays full biological activity of the parent peptide and was used for comparative electronparamagnetic resonance studies. A phospholipid containing α-MSH analogue **(89)** was synthesised[232] and used as melanoma specific non-viral gene delivery agent. Several analogues of α-MSH (Ser-Ser-Ile-Ile-Ser-His-Phe-Arg-Trp-Gly-Leu-Cys-Asp, Ser-Ser-Ile-Ile-Ser-His-Phe-Arg-Trp-Gly-Lys-Pro-Val, Ser-Tyr-Ser-Met-Glu-His-Phe-Arg-Trp-Gly-Leu-Cys-Asp) were synthesised[233] and used as highly selective MC₁ receptor selective ligands. Numerous selective antagonists of the human melanocortin receptor 4 have been synthesised **(Figure 5)**[234]. An *in vitro* system was used for the evaluation of their biological potency. The most interesting compound (cyclo(6β→10ε)-succinyl⁶-D-(2')Nal⁷-Arg⁸-Trp⁹-Lys¹⁰)-NH₂ showed 125-fold selectivity over hMC-3R and >300-fold selectivity over MC-1RB receptor.

4.11 MHC Class I and II Analogues. – Lipopeptides, which are currently being

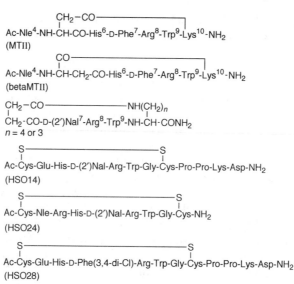

Figure 5

evaluated as candidate vaccines in human volunteers, provide a model system to define the pathways that lead exogenous proteins to associate with MHC class I molecules. One study investigated the presentation pathway of lipopeptides derived from an HLA-A2.1-restricted HIV-1 Reverse Transcriptase epitope in human dendritic cells[235].

HGP-30, a 30-amino acid peptide, homologous to a conserved region of HIV-1_{SF2}p17(86-115), is known to elicit both cellular and humoral immune responses when conjugated to KLH. Peptide conjugates consisting of a modified HGP-30 and the 38-50 region of human β-2-microglobulin, or a peptide from the MHC II β chain were evaluated in mice[236]. The new conjugates generated comparable or better immune responses to modified HGP-30 than KLH conjugates. In addition to that, minimal antibody responses were observed to the 'carrier' peptide while KLH conjugates induced significant anti-KLH antibody titers.

The peptide binding specificity of various common A2-supertype molecules has been investigated using single substitution analogue peptides or peptide libraries[237]. The data demonstrated that A2-supertype molecules recognize similar features at their peptide ligands, and also share largely overlapping peptide binding motifs.

A disulfide-linked octameric homodimer (90) that bears four copies of the influenza virus M2 protein ectodomain (M2) and two copies each of T-helper cell hemagglutinin epitopes, the I-E^d restricted S1 (S1) and the I-A^d restricted S2 (S2) fragments has been synthesized *via* intermolecular disulfide formation[238].

The C-saccharide analogue of the GalNAc (Tn epitope) has been covalently linked to a T cell epitope peptide using a chemoselective convergent synthetic approach and providing a non-hydrolysable synthetic vaccine[239]. The glycopeptide could bind to extracellular MHC molecules without internalization and processing and the C-glycoside part did not interfere with TCR recognition.

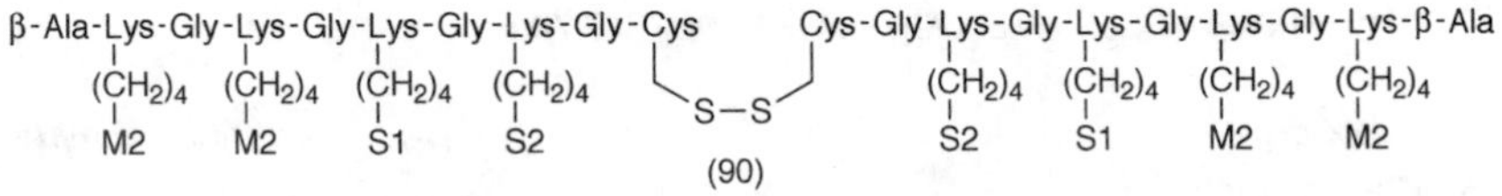

β-Ala-Lys-Gly-Lys-Gly-Lys-Gly-Lys-Gly-Cys   Cys-Gly-Lys-Gly-Lys-Gly-Lys-Gly-Lys-β-Ala

$(CH_2)_4$	$(CH_2)_4$	$(CH_2)_4$	$(CH_2)_4$		$(CH_2)_4$	$(CH_2)_4$	$(CH_2)_4$	$(CH_2)_4$
M2	M2	S1	S2	S—S	S2	S1	M2	M2

(90)

4.12 Neuropeptide Y (NPY) Analogues. – Nonapeptide fragments and their dimeric analogues have been synthesized based on the sequence of the 36-mer peptide, neuropeptide Y (NPY)[240]. A systematic investigation starting with [Pro^{30}, Tyr^{32}, Leu^{34}]NPY(28-36)-NH$_2$ (BW1911U90) has resulted in the development of highly selective and potent Y$_1$ receptor antagonists **(Figure 6)**.

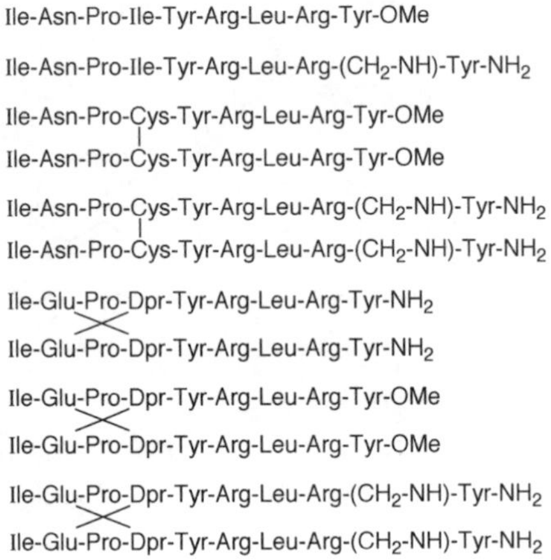

Ile-Asn-Pro-Ile-Tyr-Arg-Leu-Arg-Tyr-OMe

Ile-Asn-Pro-Ile-Tyr-Arg-Leu-Arg-(CH$_2$-NH)-Tyr-NH$_2$

Ile-Asn-Pro-Cys-Tyr-Arg-Leu-Arg-Tyr-OMe
$\qquad\qquad$|
Ile-Asn-Pro-Cys-Tyr-Arg-Leu-Arg-Tyr-OMe

Ile-Asn-Pro-Cys-Tyr-Arg-Leu-Arg-(CH$_2$-NH)-Tyr-NH$_2$
$\qquad\qquad$|
Ile-Asn-Pro-Cys-Tyr-Arg-Leu-Arg-(CH$_2$-NH)-Tyr-NH$_2$

Ile-Glu-Pro-Dpr-Tyr-Arg-Leu-Arg-Tyr-NH$_2$
$\qquad\times$
Ile-Glu-Pro-Dpr-Tyr-Arg-Leu-Arg-Tyr-NH$_2$

Ile-Glu-Pro-Dpr-Tyr-Arg-Leu-Arg-Tyr-OMe
$\qquad\times$
Ile-Glu-Pro-Dpr-Tyr-Arg-Leu-Arg-Tyr-OMe

Ile-Glu-Pro-Dpr-Tyr-Arg-Leu-Arg-(CH$_2$-NH)-Tyr-NH$_2$
$\qquad\times$
Ile-Glu-Pro-Dpr-Tyr-Arg-Leu-Arg-(CH$_2$-NH)-Tyr-NH$_2$

Figure 6

^{99m}Tc containing derivative of neuropeptide Y was synthesized as potential tumor imaging agent[241]. For complex formation, the 2-picolylamine-N,N-diacetic acid (PADA) containing neuropeptide derivatives were used. Both full length **(91)** and centrally truncated **(92)** analogs were synthesized in which His^{26} was replaced by alanine to avoid binding of ^{99m}Tc to the side chain of His. In another approach daunorubicin and doxorubicin containing neuropeptide Y analogues ([Cys^{15}]-NPY-Dauno-HYD **(93)**, [Cys^{15}]-NPY-Dauno-MBS **(94)** and [Cys^{15}]-NPY-Doxo-MBS **(95)**) were prepared and characterized[242]. According to biological investigations, only the biodegradable analogue **(93)** liberating the free antineoplastic agent is capable of displaying its cytotoxicity. A set of neuropep-tide Y analogues having 6-amino hexanoic acid moiety in position 8 were synthesized and used for the characterisation of the human, rat and guinea pig Y4 receptor[243].

4.13 Opioid (Neuropeptide FF, Enkephalin, Nociceptin, Deltorphin and Dynorphin) Peptides. – Four Leu-enkephalin analogues containing 2′,6′-dimethylphenylalanine in position 4 were prepared and tested for their receptor binding[244]. The Tyr-Gly-Gly-D-Dmp-Leu analogue was found to be an antagonist toward μ and δ opioid receptors with pA$_2$ values of 6.90 and 5.57, respectively.

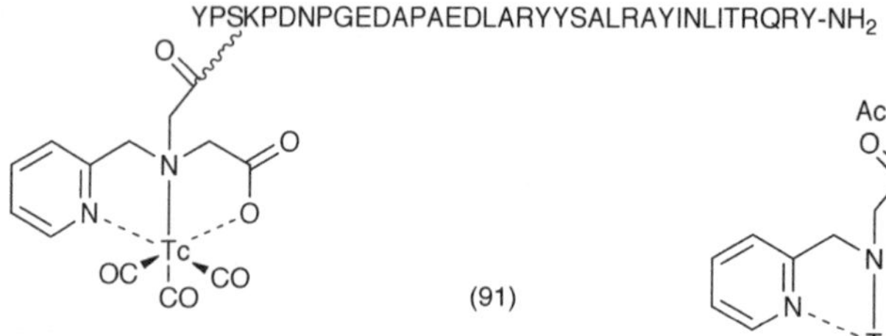

YPSKPDNPGEDAPAEDLARYYSALRAYINLITRQRY-NH₂

(91)

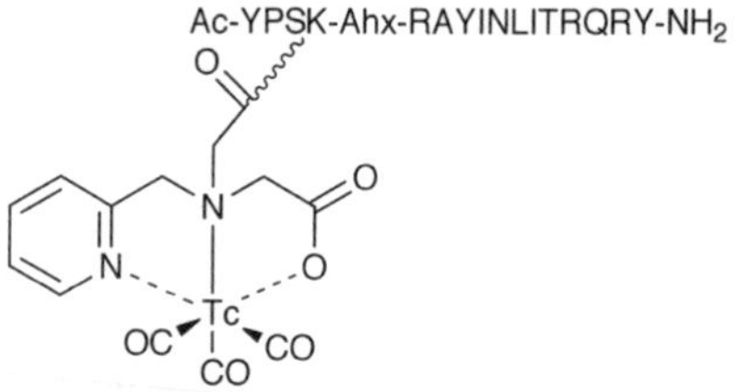

Ac-YPSK-Ahx-RAYINLITRQRY-NH₂

(92) Ahx = 6-aminohexanoic acid

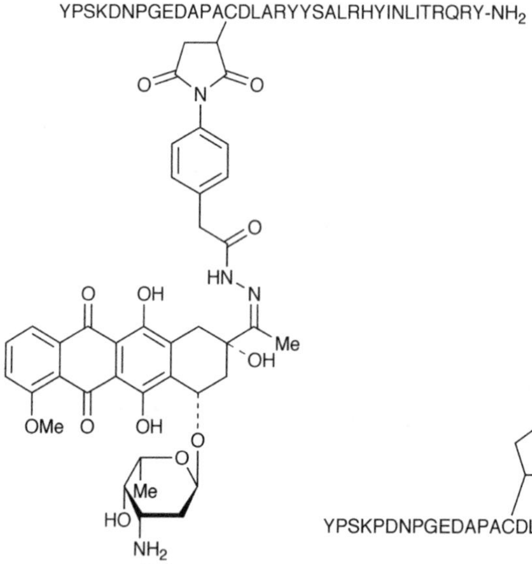

YPSKDNPGEDAPACDLARYYSALRHYINLITRQRY-NH₂

(93)

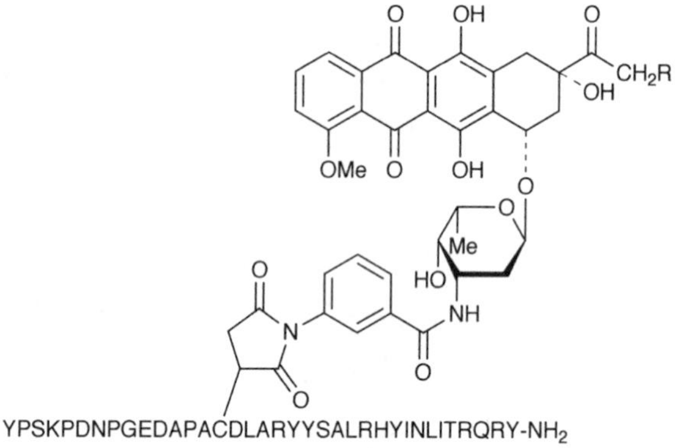

YPSKPDNPGEDAPACDLARYYSALRHYINLITRQRY-NH₂
(94) R = H
(95) R = OH

The enkephalin analogue (2S)-2-methyl-3-(2',6'-dimethyl-4'-hydroxyphenyl)-propionyl-D-Ala-Gly-Phe-Leu-NH₂ proved to be a quite potent δ opioid antagonist[245].

A theoretical conformational study has been performed on native enkephalins and their cyclic analogues[246]. The influence of solvents and Leu configuration on the conformation of cyclic enkephalin analogues has been studied by fluorescence decay of tryptophan[247], while the relationship between the conformation and biological activity of Leu-enkephalin has been studied using (2S,6R,8S)-9-oxo-8-N-(Boc)amino-1-azabicyclo[4.3.0]nonane-2-carboxylic acid as a constrained Gly²-Gly³ dipeptide surrogate[248]. Enkephalin analogue (96) exhibited low affinities for the μ and δ opioid receptors, while the duration of action was significantly enhanced indicating an increased metabolic stability.

An enzymatically stable analogue of YGGFMKKKFMR-Famide, a chimeric peptide of Met-enkephalin and FMRFa, was synthesised. The antinociceptive effects of intracerebroventricular injections of this analog – [D-Ala²]YAG-FMKKKFMRFamide - was then investigated[249]. According to these investigations the analogue causes modest to good antinociception effect in mice follow-

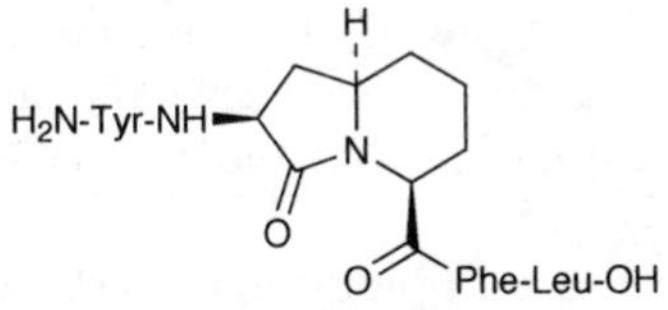

(96)

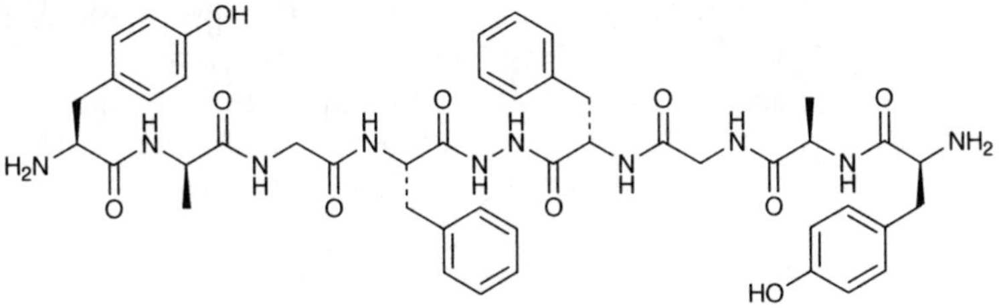

(97)

ing *i.c.v.* administration. Biphalin (**97**), the dimeric analogue of enkephalin showed high antinociceptive activity, while the side-effects were significantly descreased[250].

A set of constrained dipeptide analogues were synthesized and tested for binding to opioid receptors[251]. The general formula of the Dmt-Tic analogues is shown in **Figure 7**, where R is Ac, acetate; Me, methyl; All, allyl;Bn, benzyl; t-Bu, tert-butyl; Et, ethyl; adam, 1-adamantyl; Ph, phenyl. Some of these compounds proved to be potent and selective agonists/antagonists to various opioid receptors. Peptide analogues of the above pharmacophore display inhibition activity to the human multidrug resistance P-glycoprotein[252].

A set of pseudoproline containing analogues of morphiceptin and endomor-

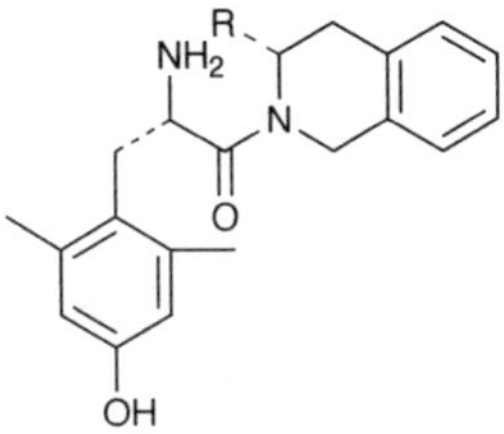

Figure 7

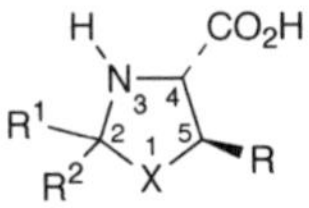

Cys-derived thiazolidines X = S, R = H
Ser-derived thiazolidines X = O, R = H
Thr-derived thiazolidines X = O, R = Me
R^1, R^2 = H, Me, aryl

Figure 8

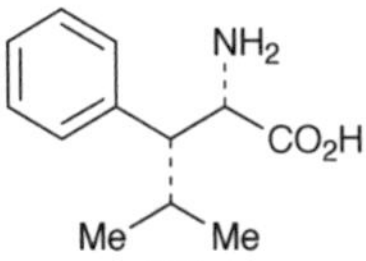

(98) 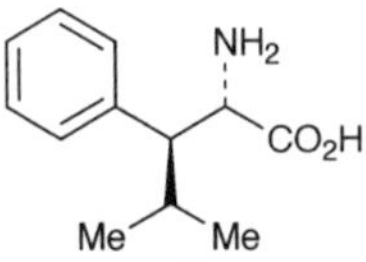　(99) 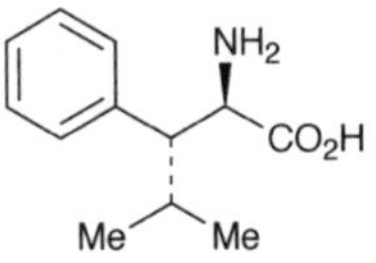　(100) 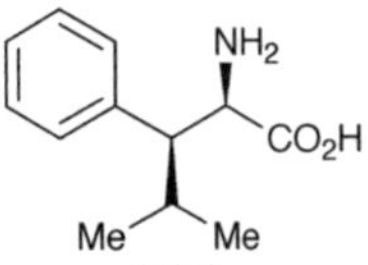　(101)

phin-2 was synthesized[253] **(Figure 8)**. The peptide analogues were used for conformational and binding studies, with several analogues showing good potency. Further proof was obtained that the *cis* Tyr-Pro bond is crucial for a bioactive conformation.

A highly potent and selective μ-opioid peptide (H-Dmt-D-Arg-Phe-Lys-NH$_2$) was pharmacologically characterised[254]. Four analogues of deltorphin I (Tyr-D-Ala-Phe-Asp-Val-Val-Gly-NH$_2$) were prepared using β-isopropyl phenylalanine isomers **(98, 99, 100** and **101)**[255]. The biological evaluation of these analogues resulted in exceptional selectivity. NMR conformational studies and molecular modeling suggested a model for the possible bioactive conformation. The role of backbone conformation in deltorphin II (Tyr-D-Ala-Phe-Gln-Val-Val-Gly-NH$_2$) was studied by QSAR method[256].

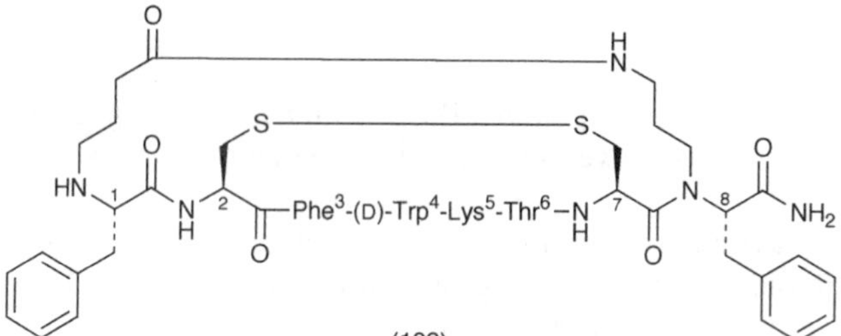

(102)

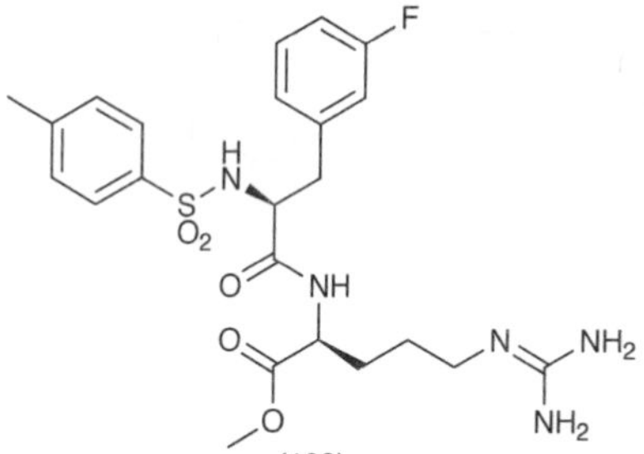

(103)

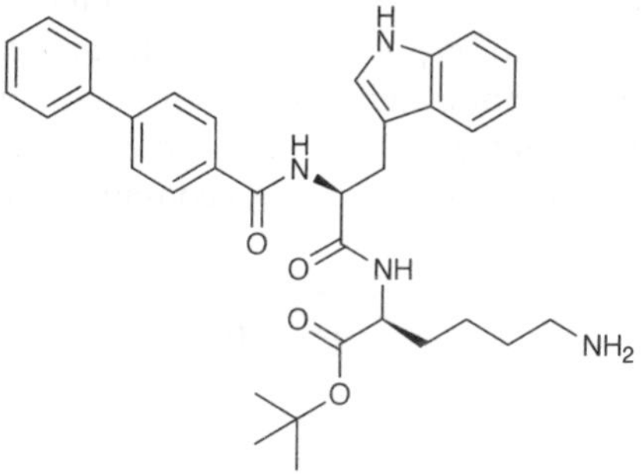

(104)

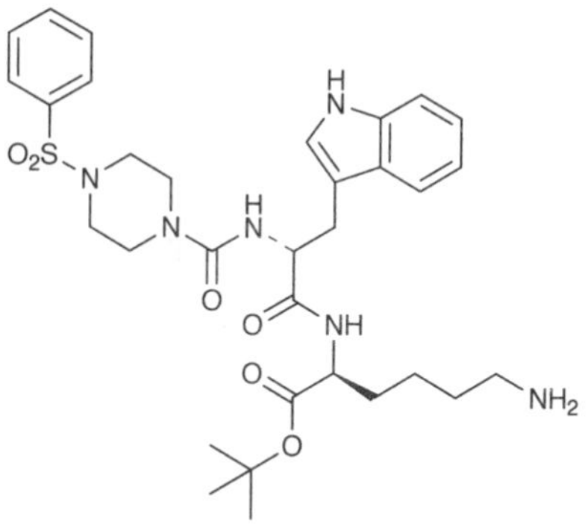

(105)

Orphanin/nociceptin 1-6 (Phe-Gly-Gly-Phe-Thr-Gly) and their C-terminally modified analogues have been tested[257]. The C-terminal glycine proved to be crucial for the hyperalgesic activity. A set of orphanin/nociceptin 1-13 analogues have been synthesized in order to investigate the role of the N-terminal amino acid in the biological activity which proved to be crucial[258]. Thirty two analogues of orphanin 1-13 were synthesized with substitution of the 4th amino acid (Phe) by varions unnatural aromatic amino acids[259]. Cyclic disulfide containing analogues of orphanin (Phe-Gly-Gly-Phe-Thr-Gly-Ala-Asn-Gln) were synthesized[260].

The role of the δ-opioid receptor in the antinociceptive actions was studied with a synthetic analogue of neuropeptide FF (D-Tyr-Leu-(Nme)Phe-Gln-Pro-Gln-Arg-Phe-NH$_2$)[261]. A structure-activity study was carried out to determine the importance of the C-terminal amino acids in neuropeptide FF[262]. The sequence of the C-terminal dipeptide seems to be responsible for the high affinity. A neuropeptide FF analogue, (D-Tyr^1-(NMe)Phe^3)neuropeptide FF, was used in three different studies to investigate the mechanism of the action of neuropeptide FF[263,264,265].

4.14 Somatostatin Analogues. – A bicyclic (backbone bridged and disulfide bridged) somatostatin analogue **(102)** was designed and synthesized[266] by Fmoc SPPS methodology. The 3D structure of this analogue was investigated by ^{1}HNMR spectroscopy. Biological assay revealed that this molecule displayed high selectivity to Hsst2 receptor subtype. Another Hsst2 receptor selective small molecule somatostatin analogue **(103)** based on the structure of **(104)** and **(105)** with good receptor binding affinities was prepared by Boc methodology[267]. Compound **(103)** antagonized the somatostatin action with IC$_{50}$ = 29 nM. The use of different agonistic and antagonistic analogues of somatostatin and their receptor binding to different somatostatin receptors (Hsst 1-5) have been reviewed[268]. Compounds **(106-110)** were considered as the most important analogues in the development of clinically useful compounds. In the same review several nonpeptide analogues of somatostatin and their potential use as chemotherapeutic agent was discussed. Another approach showed the possibility of the use of Tc-99m labelled somatostatin analogue (RC 160) as a peptide-based radiopharmaceutical[269].

Interesting and promising work has shown the possibility of the use of β-peptides as selective and potent ligands to Hsst4 receptor[270]. Molecular modelling and NMR investigations have revealed[271] the bioactive conformation of somatostatin analogues necessary for displaying high affinity to somatostatin receptor subtype Hsst2 and Hsst5. From an N-methylation scan of a potent somatostatin analogue (Cpa-cyclo(DCys-Pal-DTrp-Lys-Thr-Cys)-Nal-NH$_2$), several highly selective and potent antagonists were discovered[272].

The long time biokinetics and the possible therapeutic use of the radiolabelled somatostatin analogue (^{111}In-DTPA-D-Phe^1-ocreotide) have been investigated in mice transfected with the human carcinoid tumor, GOT1[273]. Lanreotine, an octapeptide analogue of somatostatin has been labelled with a commonly available radionuclide ^{99m}Tc[274].

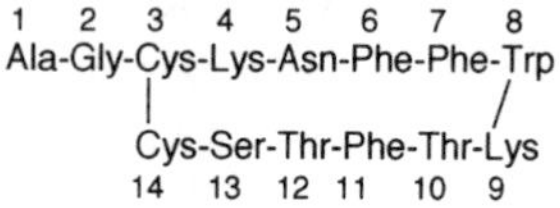

```
1   2   3   4   5   6   7   8              1   2   3   4   5   6   7    8
Ala-Gly-Cys-Lys-Asn-Phe-Phe-Trp           Ala-Gly-Cys-Lys-Asn-Phe-Phe-D-Trp
         |              /                           |               /
        Cys-Ser-Thr-Phe-Thr-Lys                    Cys-Ser-Thr-Phe-Thr-Lys
        14   13   12   11   10   9                  14   13   12   11   10   9
         (106) (SRIF-14)                             (107) [D-Trp⁸]SRIF
```

(106) (SRIF-14)

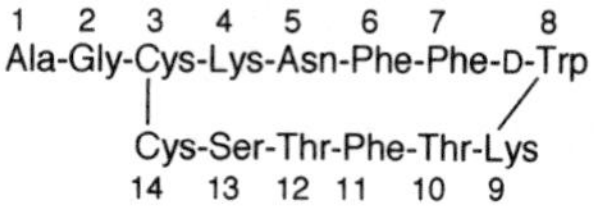

(107) [D-Trp[8]]SRIF

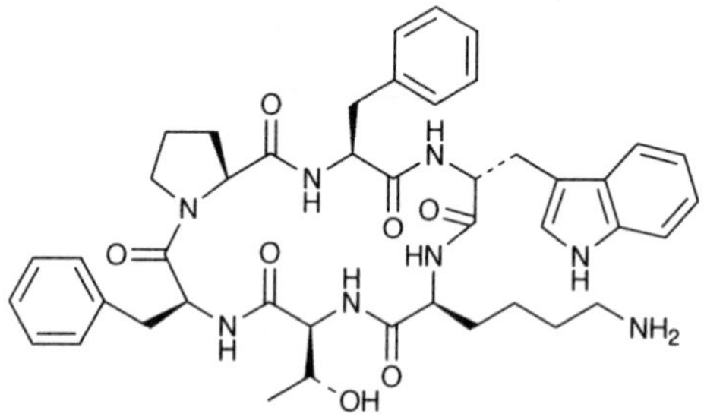

(108) (L-363,301)

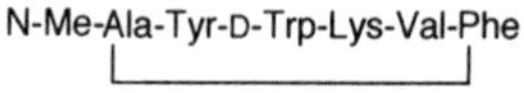

N-Me-Ala-Tyr-D-Trp-Lys-Val-Phe
|_____|

(109) (MK-678)

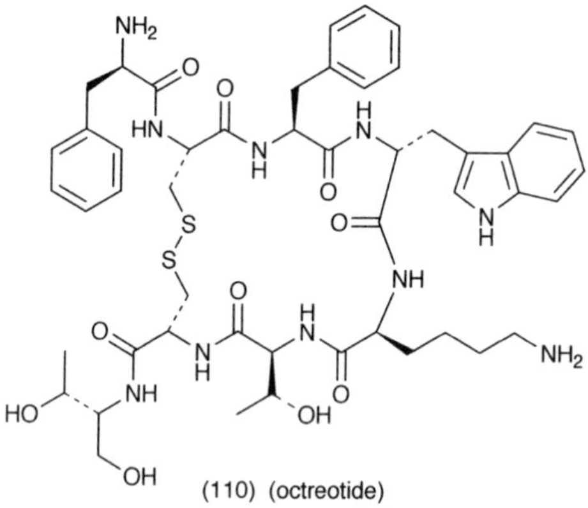

(110) (octreotide)

Cytotoxic somatostatin analogues (*e.g.* D-Phe-Cys-Tyr-D-Trp-Lys-Val-Cys-Thr-NH₂ covalently linked to 2-pyrrolinodoxorubicin) were investigated for the inhibition of the growth of small-cell lung a carcinoma[275]. The apoptotic effect of the somatostatin analogue TT-232 (D-Phe-Cys-Tyr-D-Trp-Lys-Cys-Thr-NH₂)was further investigated[276].The anti-inflammatory effect of several somatostatin analogues has also been described[277]. Several octapeptide analogues of somatostatin with N-terminal modifications (D-Nal, D-3(2-naphthyl)alanine; D-Pal, D-3-(3-pyridyl)alanine; D-Qal, D-3-(3-quinolyl)alanine; D-Cl-Phe, D-3-(4-chlorophenyl)alanine, and D-Cl₂-Phe, D-3-(3,4-dichlorophenyl)alanine) have been synthesized and tested for inhibitory effect on GH release[278]. The best results were found in case of the 3-pyridylalanine containing analogue which was 84 times more potent than RC-160.

4.15 Tachykinin (Substance P and Neurokinin) Analogues. – Two analogues of scyliorhinin II, a tachykinin family peptide modified at position 16 were synthesized and characterized by NMR spectroscopy and their binding to the NK₃

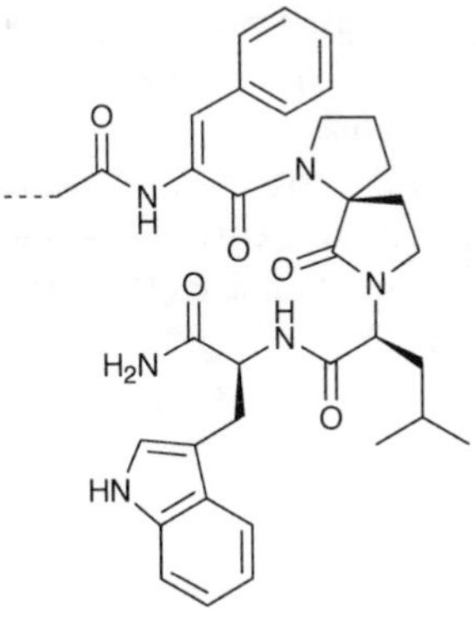

Figure 9

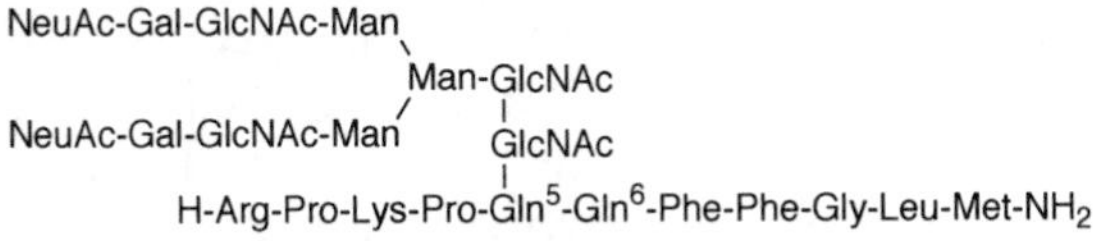

NeuAc-Gal-GlcNAc-Man
    \
      Man-GlcNAc
    / |
NeuAc-Gal-GlcNAc-Man   GlcNAc
                        |
H-Arg-Pro-Lys-Pro-Gln⁵-Gln⁶-Phe-Phe-Gly-Leu-Met-NH₂
(111)

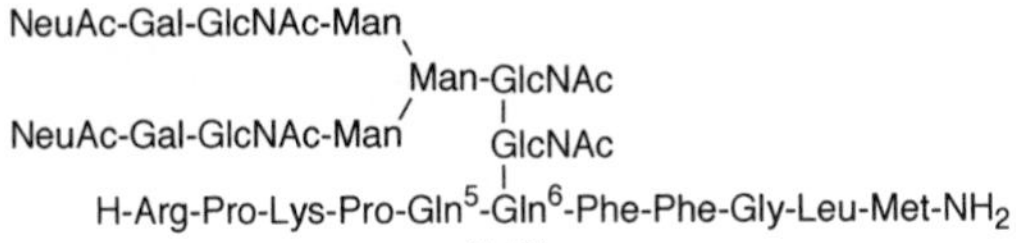

NeuAc-Gal-GlcNAc-Man
    \
      Man-GlcNAc
    / |
NeuAc-Gal-GlcNAc-Man   GlcNAc
                        |
H-Arg-Pro-Lys-Pro-Gln⁵-Gln⁶-Phe-Phe-Gly-Leu-Met-NH₂
(112)

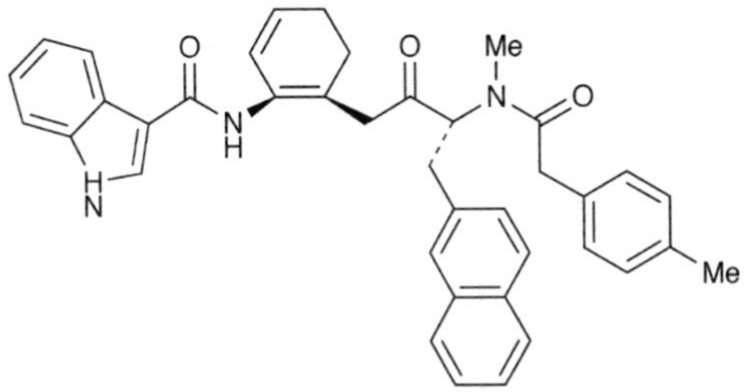

(113) (MEN 11467)

receptor[279]. A spirolactam containing substance P receptor ligand was designed and prepared[280]. The structure of the incorporated constrained molecule is shown in **Figure 9**. Photoreactive amino acid (p-benzoylphenylalanine)-containing substance P analogues havew been prepared and used for the characterisation of neurokinin-1 receptor[281]. The 4-10 fragment of neurokinin A (H-Asp-Ser-Phe-Val-Gly-Leu-Met-NH₂) was used to investigate the importance of natural residues and their chirality for affinity and efficacy at the NK₂ receptor[282]. Octa- to undecapeptide analogues of substance P with the same C-terminal hexapeptide as those of Tyr⁸-substance P, in which amino acids in positions 1-4 were replaced with each of 19 common amino acids have been prepared and investigated for their sialogogic activity[283].

A bicyclic glycopeptide NK_2 receptor antagonist and its by-products were synthesized, and characterised[284] by NMR spectroscopy. The appearance of the observed by-products was mainly caused by isomerization of the sugar moiety in the process of ring opening. Molecular models for the interaction of substance P with its G-protein-coupled receptor, the neurokinin-1 receptor have been developed[285]. Glycosylated analogues of substance P (111 and 112) were prepared by a chemo-enzymatic synthesis[286]. The pharmacological properties of MEN 11467 (113), a selective and orally-effective peptide-mimetic NK_1 receptor antagonist have been investigated[287].

4.16 Vasopressin and Oxytocin Analogues. – *4.16.1   Oxytocin Peptide and Nonpeptide Analogues.* Three conformationally constrained analogues of oxytocin containing an ethylene-bridged dipeptide unit (114) at positions 2 and 3 and two other peptides with N-Me-D-Phe residues were synthesized and tested for vasopressor and uterotonic activities *in vitro*[288]. The structures of these analogues are as follows: Xxx-DPhe-DPhe-Gln-Asn-Cys-Pro-Leu-Gly-NH$_2$, where Xxx: Cys, Mpa or Mcp. The pharmacological tests revealed that these analogues exhibited no pressor or antipressor activity, while two of them displayed weak, but very selective anti-uterotonic activity.

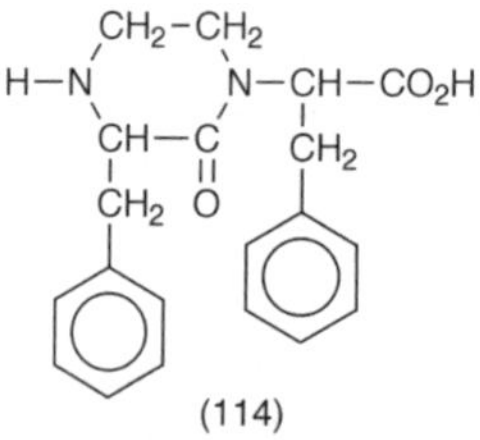

(114)

A set of oxytocin analogues modified on the aromatic ring of amino acids in position two, [Cys-D-Phe(m-OMe)-Ile-Gln-Asn-Cys-Pro-Leu-Gly-NH$_2$, Cys-L-Phe(m-Me)-Ile-Gln-Asn-Cys-Pro-Leu-Gly-NH$_2$,   Cys-L-Phe(m-OMe)-Ile-Gln-Asn-Cys-Pro-Leu-Gly-NH$_2$,   Cys-L-Tyr(o-Me)-Ile-Gln-Asn-Cys-Pro-Leu-Gly-NH$_2$, and Cys-D-Tyr(o-Me)-Ile-Gln-Asn-Cys-Pro-Leu-Gly-NH$_2$] were prepared and tested[289]. The biological measurements were carried out *in vitro* in the presence of Mg^{2+} or *in vivo* in the absence of Mg^{2+}. These results suggested the theory concerning the importance of Mg^{2+} ion for the antagonistic properties of oxytocin analogues. The role of the 6th and 7th amino acid in oxytocin (Cys-Pro) was investigated[290]. The synthesized analogues were the following: [Pen^6]oxytocin, [Pen^6, 5-t-BuPro^7]oxytocin, [Mpa^1, Pen^6]oxytocin, [Mpa^1, Pen^6, 5-t-BuPro^7]oxytocin, [dPen^1, Pen^6]oxytocin, and [dPen^1, Pen^6, 5-t-BuPro^7]oxytocin.

The synthesis of the partially protected 1-6 fragment of oxytocin (tocinoic acid) was optimised[291]. A solid-phase approach applying combined Boc and Fmoc chemistry on 2-chlorotrityl resin proved to be optimal. Twelve new oxytocin antagonists were designed and synthesized[292] during the search for more selective analogues than Atosiban. The structures of these peptides are as follows: Atosiban (d[D-Tyr(Et)^2,Thr^4]OVT), desGly(NH$_2$),d(CH$_2$)$_5$ [Tyr(Me)^2,

Thr^4]OVT, desGly(NH$_2$),d(CH$_2$)$_5$[Thi^2,Thr^4]OVT, desGly(NH$_2$),d-(CH$_2$)$_5$[D-Thi^2,Thr^4]OVT, d(CH$_2$)$_5$[Tyr(Me)^2,Thr^4,Tyr-(NH$_2$)^9]OVT, d(CH$_2$)$_5$[Thi^2,Thr^4, Tyr-(NH$_2$)^9]OVT, d(CH$_2$)$_5$[D-Thi^2,Thr^4,Tyr-(NH$_2$)^9]OVT, d(CH$_2$)$_5$[Tyr(Me)^2, Thr^4,Eda^9]OVT, d(CH$_2$)$_5$[Thi^2,Thr^4,Eda^9]OVT, d(CH$_2$)$_5$[D-Thi^2,Thr^4,Eda^9] OVT, d(CH$_2$)$_5$[Tyr(Me)^2,Thr^4,Eda^9←Tyr^{10}]OVT, d(CH$_2$)$_5$[Thi^2,Thr^4,Eda^9← Tyr^{10}]OVT, d(CH$_2$)$_5$[D-Thi^2,Thr^4,Eda^9←Tyr^{10}]OVT. Approximately half of the peptides displayed much better selectivity of anti oxytocin activity comparing to the parent peptide.

4.16.2 Vasopressin Analogues. The 4-8 fragment of the arginine vasopressin analog (pGlu4, Cyt6) was investigated for its activity on the central nervous system[293]. This peptide proved to be 100 times more effective than the whole molecule. The results suggested that this shorter fragment of AVP might have an enhancing effect on general cognitive abilities.

4.17 Insulins and Chemokines. – *4.17.1 Insulins.* A nonapeptide fragment of the insulin B chain (H-Arg-Gly-Phe-Phe-Tyr-Thr-Pro-Lys-Ala-OH) was found to inhibit dopamine uptake by rat dopamine transporter[294]. A deletion analogue of insulin (missing the amino acids 28 to 30 from the B chain) has been investigated by X-ray crystallography[295]. According to the investigation, the C terminal part of the B chain plays a crucial role in the polypeptide association. The recently discovered insulin 3 **(Figure 10)** has been chemically synthesized by solid-phase technique using continuous flow Fmoc methodology[296].

A-chain SVATNAVHRCCLTGCTQQDLLGLCPH

B-chain EPPEARAKLCGHHLVRALVRVCGGPRWSPEA

Figure 10

4.17.2 Chemokines. The inhibition of the CXCR4 chemokine receptor can lead to anti-HIV activity and cytotoxicity. A set of shorter analogues of T140, a 14-residue long specific CXCR4 inhibitor was prepared for the investigation of serum stability and cytotoxicity[297]. One of the CC chemokine receptors (CCR3) can be antagonized[298] by phenylalanine derivatives **(Figures 11, 12** and **13)**. These derivatives are considered as anti inflammatory agents. The most active antagonist was a p-nitrophenylalanine derivative **(115)** which had an IC$_{50}$ value of 5nM[299]. Based on their lead structure numerous analogues were designed and synthesized. Some of them proved to be potent and selective CCR3 receptor antagonists. HIV-1 Tat protein plays one of the central roles in AIDS pathogenesis, and a recently synthesized neomycin B-hexaarginine conjugate **(116)** inhibited the Tat actions, partially by competing with Tat for the binding to CXCR4 receptor[300].

4.18 Peptide Toxins. For many years, scorpion venoms have provided the neurobiologists with a rich and varied source of toxins that act as selective ion channel blockers. Recently numerous new peptide toxins were isolated and

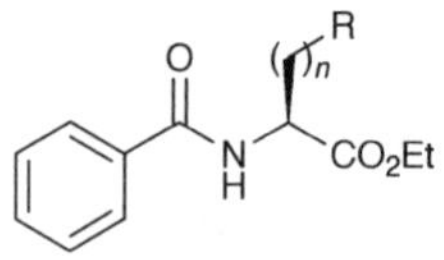

n = 1, R = 3,5-I$_2$-4-OH-Ph, 4-OH-Ph, Ph, H, L-4-NO$_2$-Ph, D-4-OH-Ph, L-3-indolyl
n = 0, R = (±)-Ph
n = 2, R = (±)-Ph

Figure 11

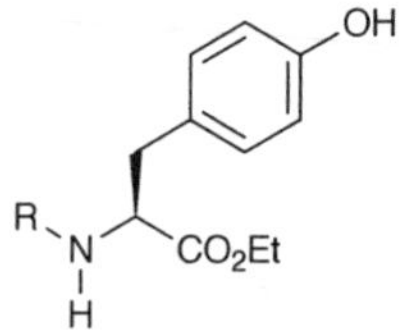

R = COPh, CONHPh, SO$_2$(4-Me)Ph, CO-Z-L-Asn, CO-Z-L-Phe, CO-Z-L-Ser, CO-Z-L-Gln, CO-L-naphthyl,
CO-2-pyridyl, COCH$_2$Ph

Figure 12

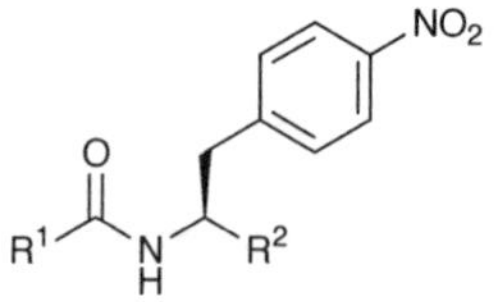

R^1 = naphthyl, R^2 = CO$_2$Et, CO$_2$Me, CO$_2$Pr^i, CO$_2$Bu^t, CO$_2$CH$_2$Ph, (±)COPr, CONHCH$_2$Ph, CONMe$_2$
R^1 = 2,4-Dimethylphenyl, R^2 = CO$_2$Et

Figure 13

identified. A brief summary of them:. tamulustoxin, a novel 35-amino acid peptide (cyclo[Arg-Cys-His-Phe-Val-Val-Cys-Thr-Thr-Asp-Cys-Arg-Arg-Asn-Ser-Pro-Gly-Thr-Tyr-Gly-Glu-Cys-Val-Lys-Lys-Glu-Lys-Gly-Lys-Glu-Cys-Val-Cys-Lys-Ser]) found in the venom of the Indian red scorpion (*Mesobuthus tamulus*) having potassium channel blocker properties[301]; *Centruroides sculpturatus* Ewing peptides that recognize Na$^+$-channels[302]. Altogether 16 different genes were cloned which code 22 different, 63-66 amino acid-long mature peptides; slotoxin (cyclo[Thr-Phe-Ile-Asp-Val-Asp-Cys-Thr-Val-Ser-Lys-Glu-Cys-Trp-Ala-Pro-Cys-Lys-Ala-Ala-Phe-Gly-Val-Asp-Arg-Gly-Lys-Cys-Met-Gly-Lys-Lys-Cys-Lys-Cys-Tyr-Val]) a new peptide blocker of MaxiK channels isolated from *Centruroides noxius* Hoffmann[303]. This peptide belongs to the charybdotoxin sub-family; OsK2, a novel inhibitor of voltage-gated K$^+$ channels from the black scorpion *Orthochirus scrobiculosus*[304]. This peptide is characterized as a 28-residue peptide having six cysteine residues (cyclo[Ala-Cys-Gly-Pro-Gly-Cys-Ser-Gly-Ser-Cys-Arg-Gln-Lys-Gly-Asp-Arg-Ile-Lys-Cys-Ile-Asn-Gly-Ser-Cys-His-Cys-Tyr-Pro]); four K$^+$-toxin-like peptides (BmKK$_1$, BmKK$_2$, BmKK$_3$ and BmKK$_4$) were isolated from the Chinese scorpion *Buthus martensii* Karsch[305]. The upper toxins are 31-mer, three disulfide containing peptides; BmTXKS1, a 31-mer, six cysteine containing peptide was isolated from the same animal[306]; two novel 66-mer α-like-toxins were cloned and characterized from the same Chinese scorpion[307]; a β-toxin-like peptide and two MkTx I homologue 63-mer toxins were cloned and sequenced from scorpion *Buthus martensii*

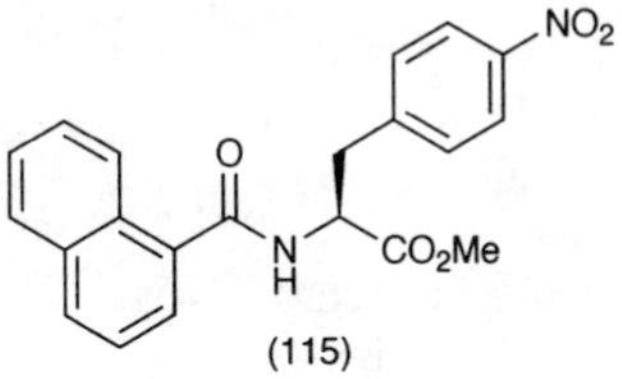

(115)

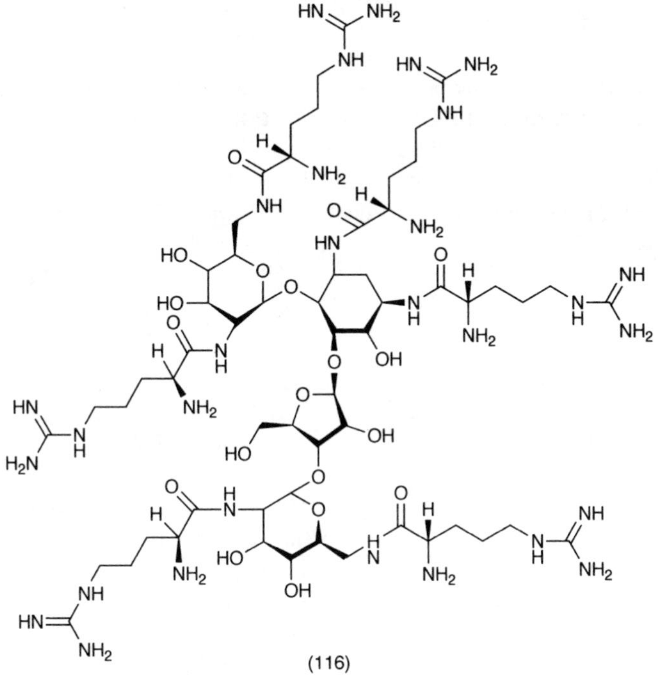

(116)

Karsch[308]; BmTX2, a 37 amino acid- and three disulfide bridge-containing K^+ channel toxin was also cloned[309]; the functional site of bukatoxin, an α-type neurotoxin was identified[310]; a new peptide named Bmk dITAP3 from scorpion *Buthus martensii* Karsch has been identified as possessing dual bioactivity, a depressant neurotoxicity and analgesic effect[311]; maurotoxin, a 34-residue long, four-disulfide bridged scorpion toxin (117) was isolated from the venom of *Scorpio maurus* palmatus and chemically synthesized[312]; the cystine knot motif which is relatively common in small, cysteine-rich toxins was discussed in an article, including their potential use in drug design[115]; the conformation of the P-type cardiotoxin from the cobra snake venom was investigated by NMR spectroscopy[313]; the role of Arg^{13} in μ-conotoxin GIIIA, a peptide toxin isolated from the marine snail *Conus geographus* was investigated[314]; AM-toxin II (118), a host-specific phytotoxin produced by *Alternaria alternata*, the fungus causing leafspot disease of apple trees, has been chemically synthesized[315].

4.19 Miscellaneous. – Chimeric peptides (galanin 1-12 attached to Pro-brady-kinin(2-9)-amide or Pro-spantide-amide) were investigated as galanin receptor

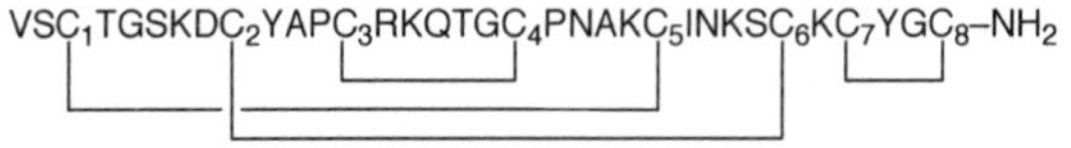

VSC$_1$TGSKDC$_2$YAPC$_3$RKQTGC$_4$PNAKC$_5$INKSC$_6$KC$_7$YGC$_8$–NH$_2$

(117) 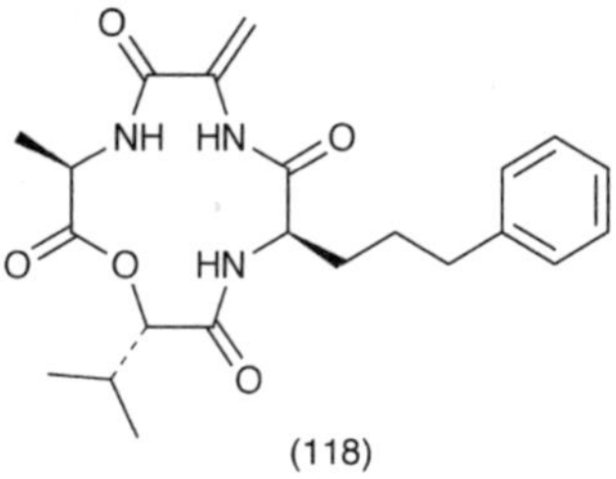   (118)

antagonists on rat exorcine pancreas[316]. An analogue of glucagon-like peptide-1 (7-36)amide containing His-glucitol in position seven was investigated for its glycemic effect and its degradation[317]. Two shorter analogues of CGRP ([Asp^{31}, Pro^{34}, Phe^{35}]CGRP 27-37 and [Asn^{31}, Pro^{34}, Phe^{35}]CGRP 27-37) have been investigated for their modulatory effect[318].

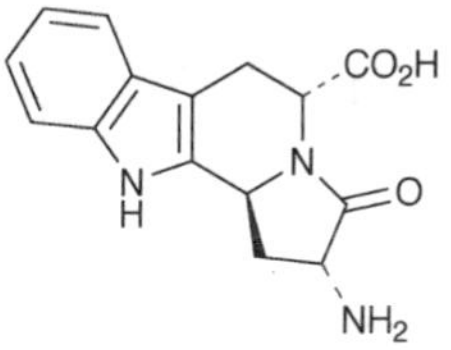

(119) (*S*-IBTM) 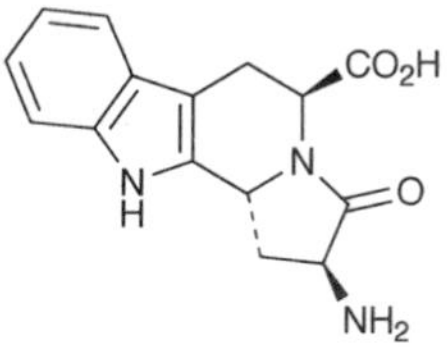   (120) (*R*-IBTM)

Two conformationally constrained PACAP27 analogues have been synthesized and characterized[319]. The incorporated constrained amino acids (119 and 120) were carboline derivatives. The sequence of the resulting peptide analogues were as follows: His-Ser-Asp-Gly-Ile-Phe-Thr-Asp-Ser-[*S*-IBTM^{10,11}] -Arg-Tyr-Arg-Lys-Gln-Met-Ala-Val-Lys-Lys-Tyr-Leu-Ala-Ala-Val-Leu-NH$_2$ and His-Ser-Asp-Gly-Ile-Phe-Thr-Asp-Ser-[*R*-IBTM^{10,11}]-Arg-Tyr-Arg-Lys-Gln-Met-Ala-Val-Lys-Lys-Tyr-Leu-Ala-Ala-Val-Leu-NH$_2$. The conformation and binding affinity of these peptides were investigated. PACAP27 analogues for photoaffinity labelling were prepared and used for the characterization of the PACAP type I receptor[320].

Several fMLF-OMe analogues containing alkyl spacers **(Figure 14)** were synthesized and characterized[321]. A set of TRH analogues (121) has been designed and prepared by modification of the N-terminal pyroglutamic acid residue[322]. Several (*S,S*)-[Pro-Leu]-spirolactam containing short peptides (122) as E2F-1/Cyclin A antagonists have been synthesized and characterized[323].

The synthesis of salmon I calcitonin has been optimized[324], and the total chemical synthesis of the 105 amino acid containing human activin β_A[12-116] described[325]. The ^{99m}Tc-containing complex derivative (123) of the tuftsin receptor binding peptide has been prepared[326]. Eleven new bioactive dahlein peptides were isolated and sequenced from the skin of the Australian aquatic frog *Litoria dahlii*[327], and the α-aminosuberic acid containing deamino-dicarba analogue of eel-calcitonin 1-9 was prepared[328].

The octadecaneuropeptide (ODN, Gln-Ala-Thr-Val-Gly-Asp-Val-Asn-Thr-Asp-Arg-Pro-Gly-Leu-Leu-Asp-Leu-Lys) and its C-terminal octapeptide were

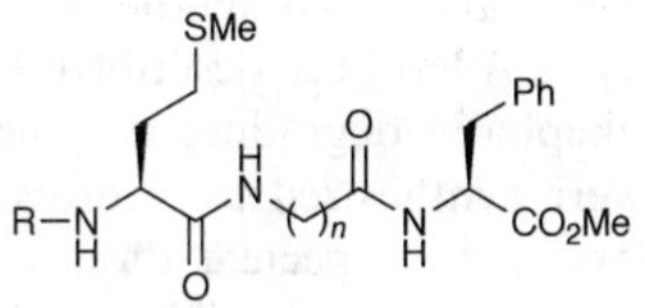

R = Bu^{t}OCO or H–CO,   n = 3, 4, 5   (121)

Figure 14

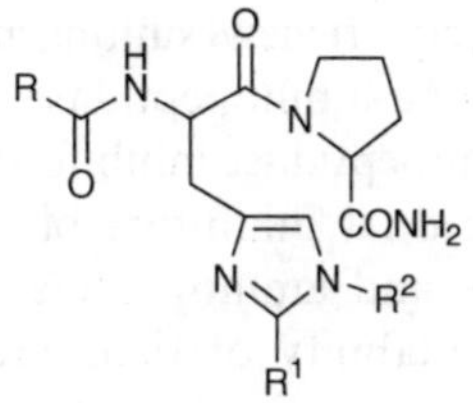

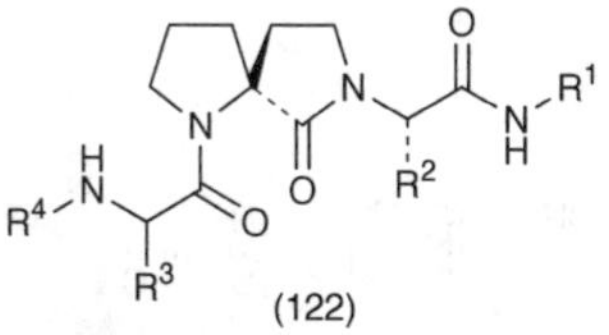

(122) 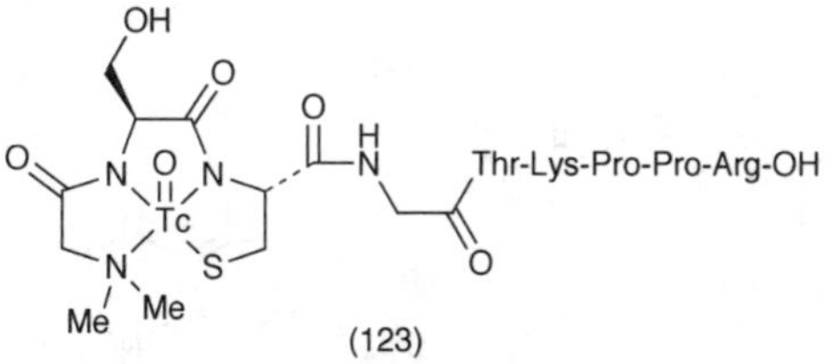   (123)

prepared and used for conformational and biological studies[329]. The cyclic analogues of these peptides proved to be more potent than their parent peptides.

The phosphorylated p21Max protein (1-101) has been synthesized using an expressed peptide as building block[330]. The N-terminal region was synthesized chemically as a phosphorylated thioester and successfully coupled to the expressed part.

5   Enzyme Inhibitors

A number of reviews dealing with enzyme inhibitors were published in 2001. The developments in the field of transmembrane proteases as disease markers and targets for therapy have been discussed[331]. More specialized reviews on the discovery and the clinical development of HIV-protease inhibitors[332, 333], on the strategies for the inhibition of serine proteases[334] and on recent strategies in the development of new human cytomegalovirus inhibitors[335] have been published.

5.1   Aminopeptidase and Deformylase Inhibitors. – A radiolabeled inhibitor of aminopeptidase A was synthesized in order to investigate the localization of the enzyme in the central nervous system and peripheral organs[336]. This compound, [3(*R*)-amino-2(*S*)-sulfhydryl-5-sulfonate]-pentanoyl-(*S*)-3-[^{125}I]-iodotyrosyl-(*S*)-aspartic acid **(124)**, exhibited a K$_i$ value of 4.8 nM for aminopeptidase A and had a specific activity of about 2 000 Ci/mmol at the end of the synthesis. Peptides containing 3*R*-amino-2*S*-hydroxy heptanoic acid (AHHpA) residue may inhibit methionine aminopeptidase-1 with IC$_{50}$ values in the low micromolar range[337]. Methionine aminopeptidase (MetAP), a divalent cobalt metalloprotease essential to the processing of proteins, cleaves the N-terminal methionine from the growing polypeptide chains. Two types of MetAP are known: MetAP-1 and MetAP-2. Prokaryotes have only MetAP-1 and inhibitors of this enzyme could have potential antibacterial activity. (MetAP-2 inhibitors currently show potential use as anti-angiogenesis agents in cancer treatment).

A very concise review summarizes the most effective and most promising inhibitors of Leu-aminopeptidase[338]. A facile and rapid synthesis of naturally occuring aminopeptidase inhibitor tyramycin A has been described[339]. Prodrugs of phosphinic dual inhibitors of the enkephalin degrading enzymes, neutral endopeptidase and aminopeptidase-N were synthesized to improve the poor central bioavailability of their precursors[340]. As expected, these compounds induced long-lasting ($\sim$2h) antinociceptive responses in the hot plate test in mice.

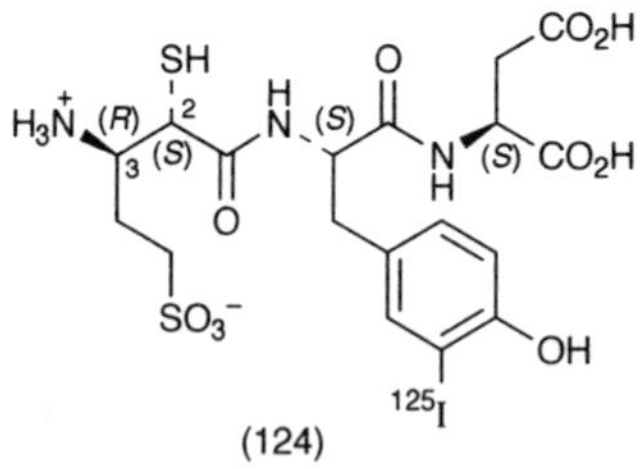

(124)

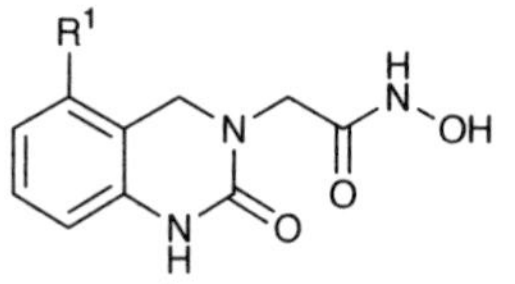

(125) R^1 = H, F, Cl, Br, CF_3

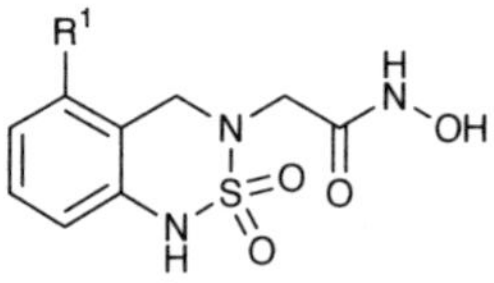

(126) R^1 = H, F, Cl, Br, CF_3

Deformylation of the polypeptide chain by the metalloenzyme deformylase after the ribosomal protein biosynthesis is a crucial step in bacteria. Very recently a human peptide-deformylase was identified which is presumed to be involved in deformylation in the mitochondrion. By screening a library of metalloenzyme inhibitors, an N-formyl-hydroxylamine derivative was identified as a potent inhibitor of *E. coli* peptide deformylase (PDF)[341]. Potent, selective and structurally new inhibitors **(125, 126)** were synthesized and evaluated[342]. Although these compounds showed good selectivity for PDF, their antibacterial acitivity was weak. Screening of a compound collection using *Staphylococcus aureus* PDF afforded a very potent inhibitor of hydroxamic acid structure with an IC_{50} in the low nanomolar range[343]. This compound did not exhibit antibacterial activity.

5.2   Calpain Inhibitors. – Calpains, calcium-activated cysteine proteases widely distributed in mammalian cells, exist in two major forms: calpain I and calpain II. Calpain I is believed to be activated during a biochemical cascade that leads to a delayed degeneration of neurons following ischemia. Potent, membrane permeable calpain I inhibitors are pharmacological targets for putative treatment of neurological diseases. A series of potent dipeptide and tripeptide α-ketohydroxamic esters was prepared as inhibitors of recombinant human calpain I[344]. A dipeptide derivative, Z-Leu-Phe hydroxamate **(127)** displayed the greatest potency against calpain I (IC_{50} = 6 nM), while two tripeptide derivatives, both possessing the Z-Leu-Leu-Phe sequence, were the most potent (IC_{50} = 0.2 µM) in a whole-cell MOLT-4 assay. In another series of experiments, 16 derivatives of the 3,4-dihydro-1,2-benzothiazine-2-carboxylate-1,1-dioxide were synthesized[345] **(Figure 15)** and the effect of 2-, 6- and 7-benzothiazine substituents on the inhibitory effect was examined. The potency of these calpain I inhibitors is particularly dependent upon the 2-substituent, with methyl and ethyl being generally more potent than hydrogen, isopropyl, isobutyl or benzyl. The potency of the best inhibitors in this series (IC_{50} = 5-7 nM) compares favourably with that of the classical Z-dipeptide aldehyde inhibitors bearing L-Leu or L-Val residues. Some negative results: the diketopiperazine calpain inhibitor isolated form *Streptomyces griseus*, *cis*-L,L-3,6-bis-(4-hydroxybenzyl)-1,4-dimethylpiperazine-2,5-dione **(128)** was synthesized and showed no inhibitory activity in an assay with the porcine erythrocyte calpain I[346]. Analogues with a similar structure **(129, 130)** containing N-methylphenylalanine were also unable to inhibit calpain I in the same assay.

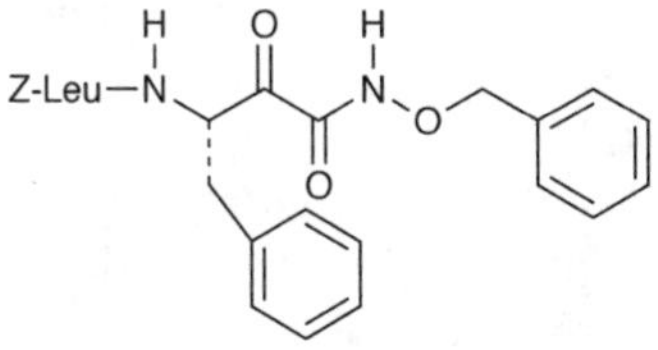

(127)

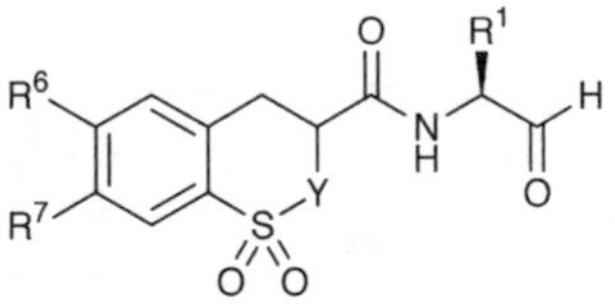

R¹ = Buⁱ, Bn or (CH₂)₄NHSO₂Ph
R⁶ = OMe, H, F, Cl, morpholin-4-yl, OCH₂CH₂O
R⁷ = OMe, H, Cl

Figure 15

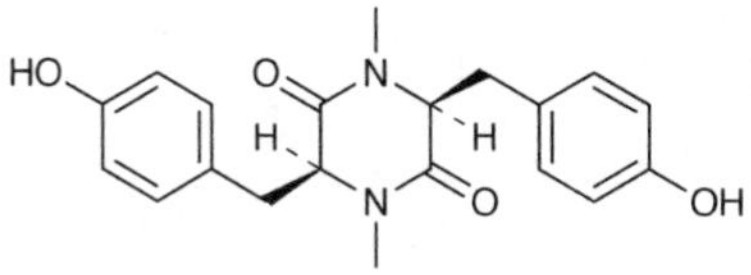

(128) 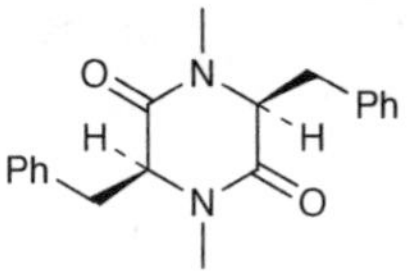   (129) 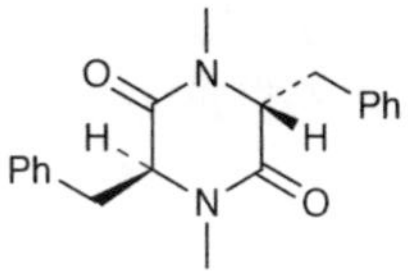   (130)

5.3   Carboxypeptidase Inhibitors. – Carboxypeptidase A (CPA), that cleaves the C-terminal amino acid residue having a hydrophobic side chain, has been most intensively studied, and serves as a prototypic Zn-peptidase. Phosphonopeptides (phosphonate analogues of peptides in which an amide linkage has been replaced with a phosphonate ester or phosphonamide) and thiophosphono-peptides have been prepared as CPA-inhibitors[347]. There have been many examples of phosphonopeptides that strongly inhibit metalloproteases, however, the effectiveness of thiophosphonopeptides as inhibitors of the CPA is demonstrated for the first time. A new inhibitor design strategy was used for CPA, and N-(2-chloroethyl)-N-methylphenylalanine proved to be a potent inhibitor[348] **(131)**. It was hypothesized that the chloroethylamino moiety of the CPA-bound inhibitor undergoes an intramolecular S_N2 reaction and generates a reactive aziridinium ion which reacts with the carboxylate of Glu-270 of the enzyme, leading to covalent modification of the carboxyl groups (irreversible inhibition). The (R)-enantiomer of **(131)** was more potent than the (S)-isomer. β-Lactone-bearing phenylalanine derivatives (*e.g.* (3S,1'S)-3-(1'-Carboxy-2'-phenyl) ethylamino-2-oxetanone) were designed, synthesized, and evaluated also as inhibitors for CPA[349]. This compound inactivates the enzyme irreversibly while the 3R,1'S diastereomer proved to be a weak competitive inhibitor for CPA. From the same research group, a novel class of inhibitor for CPA, N-(hydroxyaminocarbonyl) phenylalanine **(132)** was designed, synthesized and evaluated in enzyme assay[350], exploiting the metal chelating property of hydroxyurea.

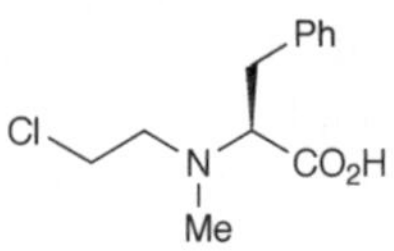

(131) 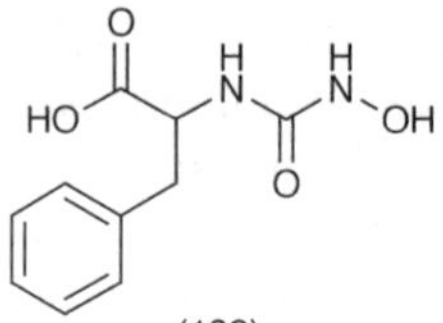   (132) 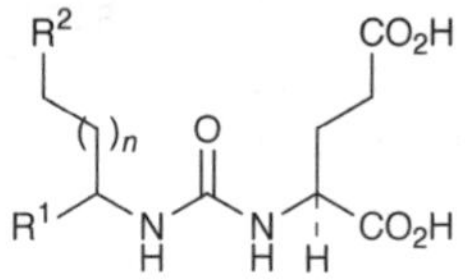   (133)

The designed inhibitor was readily prepared from phenylalanine benzyl ester in two steps. The racemic mixture of **(132)** inhibits CPA in a competitive fashion with the K_i value of 2.09 µM, the D-enantiomer of **(132)** is 3-fold more potent (K_i = 1.54 µM) than its antipode.

N-Acetyl-L-aspartyl-L-glutamate, the peptide neurotransmitter ligand at metabotropic glutamate receptors and a mixed agonist/antagonist at the NMDA receptor, is hydrolysed by the neuropeptidase glutamate carboxypeptidase II. This dipeptidase enzyme is a metalloprotease and can be inhibited by unsymmetrical ureas[351] of general structure **133**. Several of the novel compounds act in IC_{50} of 6-30 nM range and provide significant (up to 69%) neuroprotection by blocking NMDA toxicity *in vitro*.

2-Benzyl-3,4-iminobutanoic acid has been evaluated as a novel class of carboxypeptidase A inhibitor[352].

5.4   Caspase Inhibitors. – The family of caspases, a highly conserved family of cysteine proteases, can be subdivided into 3 groups: a) caspases involved in inflammation (caspase 1, 4, 5 and 13); b) initiator caspases (6, and 8 to 10); and c) effector caspases (2, 3 and 7).

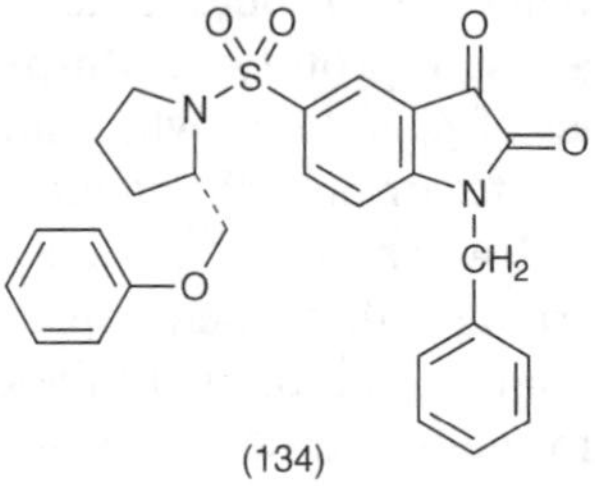

(134) 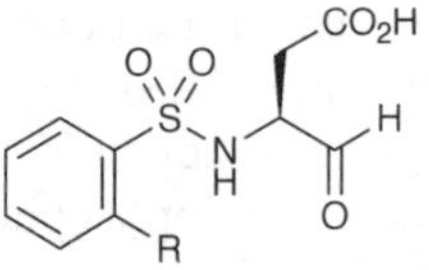   (135)

5-Dialkylaminosulfonylisatins have been identified as potent, nonpeptide inhibitors of caspases 3 and 7[353]. The most active compound within this series **(134)** inhibited both enzymes in the 2-6 nM range and exhibited approximately 1000-fold selectivity for caspases 3 and 7 versus five other caspases (1, 2, 4, 6 and 8). These apoptosis inhibitors are active in three cell-based models. In another laboratory a series of compounds was designed and prepared as inhibitors of the interleukin-1β converting enzyme (ICE), also known as caspase-1[354]. Inhibitors have a diphenyl ether sulfonamide structure **(135)**. The structure-based design and X-ray crystallographic analysis of a new class of active site caspase-1 inhibitors, which contain a biphenyl sulfonamide-aspartic acid aldehyde scaffold, have been reported from the same laboratory[355]. An X-ray crystal structure of a diphenyl ether sulfonamide derivative bound to the active site of caspase-1 shows that the catalytic residue (Cys285) forms a standard covalent bond with the aspartic acid aldehyde of the inhibitor, having a K_i value of 0.62 μM. Structure-based design of a combinatorial library was carried out in order to identify non-peptidic thiomethylketone inhibitors of caspase 3 and 8, and this strategy has proved useful for identifying caspase inhibitors[356]. N-Nitrosoaniline derivatives as NO donors form a new class of caspase-3 inhibitors: NO inhibits the activity of the enzyme by S-nitrosylation[357].

5.5 Cathepsin and Other Cysteine Protease Inhibitors. – Cathepsin C, a dipeptidyl-peptidase, removes dipeptides sequentially from the unsubstituted N-termini of polypeptide substrates with broad specificity. Three dehydrotetrapeptides (*e.g.* **136**, Gly-ΔPhe-Gly-Phe-pNA) were prepared and tested as affectors of cathepsin C[358]. These compounds appeared to be substrates of the enzyme, thus the replacement of an amino acid in a short peptide by the corresponding dehydroamino acid does not prevent cathepsin C in recognizing the dehydropeptide as its substrate.

Using solid phase peptide synthesis, a library of cyclic alkoxyketones **(137)** was prepared as potent inhibitors of cathepsin K[359]. The best compounds show K_i values of some ten nM. In another laboratory, peptidomimetic aminomethyl ketones **(138)** have been identified[360]. Structure-activity relationships were established and certain analogues were characterized with IC_{50} values in the range of 200-500 μM.

The synthesis, *in vitro* activities and pharmacokinetics of a series of azepanone-based inhibitors of cathepsin K are described[361]. These compounds show improved configurational stability. Compound **(139)** is a potent, selective inhibitor of human cathepsin K ($K_i = 0.16$ nM), and pharmacokinetic studies in the rat, show **(139)** to be 42% orally bioavailable. From small-molecule X-ray crystallographic analysis and molecular modelling studies it is concluded that the introduction of a conformational constraint has served the (dual) purpose of increasing inhibitor potency by locking in a bioactive conformation as well as locking out available conformations which may serve as substrates for enzyme systems that limit oral bioavailability.

Two series of cyclic ketones have been designed and identified as cathepsin K inhibitors[362]. One of the inhibitors, a 3-aminotetrahydrofuran-4-one analogue was cocrystallized with the cathepsin K. The inhibitor occupies the unprimed side of the enzyme active site. Compounds containing a 1-cyanopyrrolidinyl ring were identified as potent and reversible inhibitors of cathepsin K and L[363].

Dipeptidyl nitriles have been identified as potent and selective inhibitors of cathepsin B. Compound **(140)** has an IC_{50} of 7 nM of the enzyme, with excellent selectivity over other Cys-cathepsins[364]. Cathepsin B plays a role in apoptotic neuronal cell death: inhibition of cathepsin B contributes to the neuroprotective properties of caspase inhibitor Tyr-Val-Ala-Asp-chloromethyl ketone[365].

Six new peptidyl diazomethylketones have been synthesized as cysteine protease inhibitors structurally based upon the inhibitory centers of cystatins[366]. Inhibitory activities of these compounds against papain and bovine cathepsin B were tested. Dipeptide analogues containing an oxirane (epoxide) ring instead of

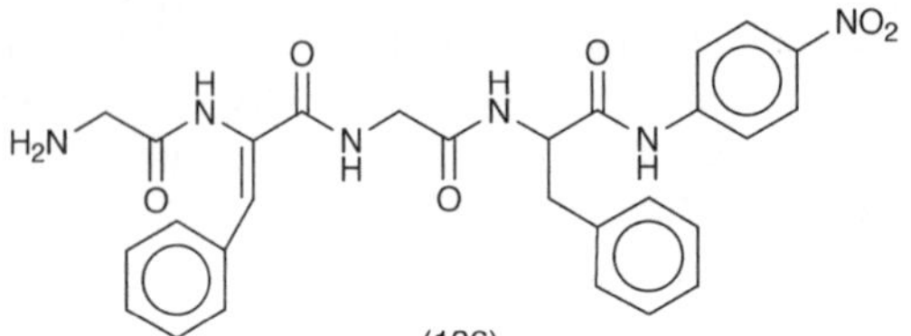

(136)

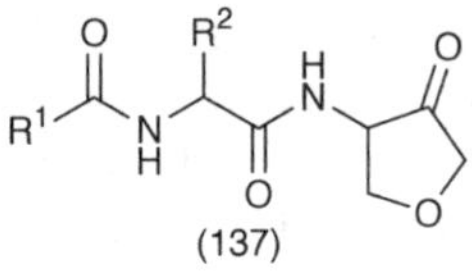

(137)

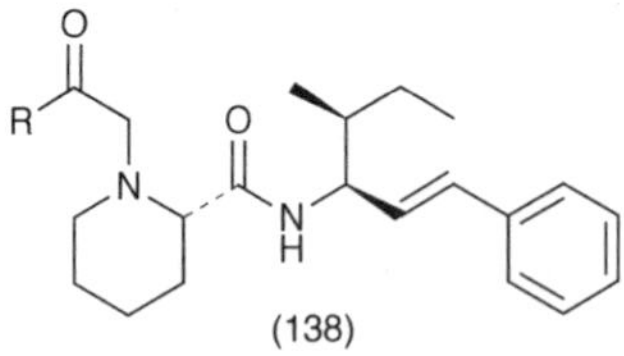

(138)

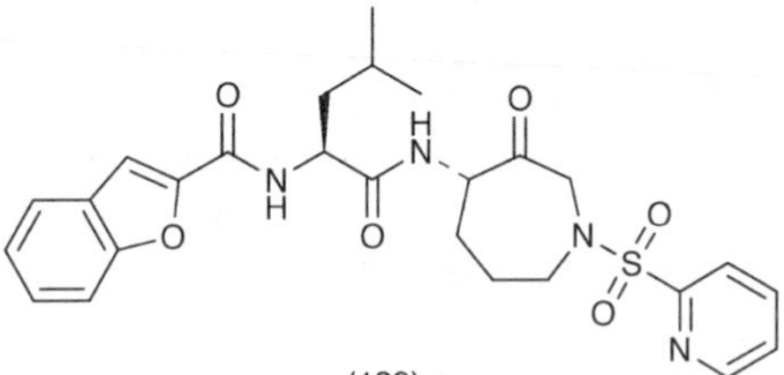

(139)

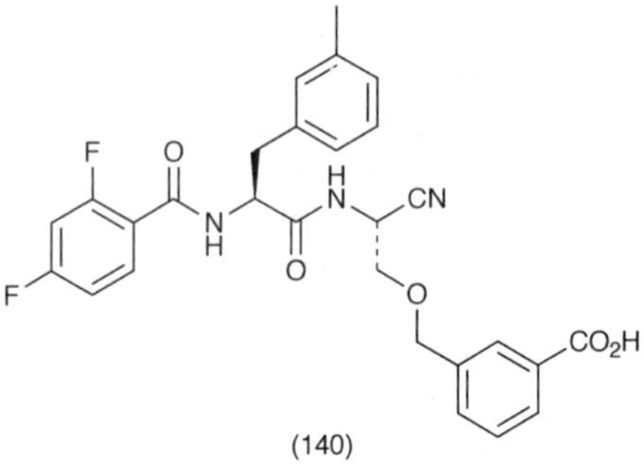

(140)

the peptidic bond were also prepared, some of them proved to be time-dependent reversible inhibitors of cysteine protease but with poor inhibition potency[367].

5.6   Cytomegalovirus and Rhinovirus 3C Protease Inhibitors. – A peptidomimetic library for inhibiting human cytomegalovirus (HCMV) protease was designed based on the interactions between the protease active site and its peptidomimetic inhibitors[368]. The library was synthesized in the liquid phase by the Ugi four-component condensation reaction from four kinds of building blocks: carboxylic acids, amino, oxo compounds and isocyanides and then oxidation performed. No biological data were given.

Selective nonpeptidic inhibitors of herpes simplex virus type 1 and HCMV protease have been synthesized[369]. For treatment of the common cold, the human rhinovirus (HRV) protease inhibitor drug candidate AG7088 was developed[370] **(141)**. An efficient synthetic route to a key intermediate for the preparation of AG7088 has been developed employing a key asymmetric dianionic cyanomethylation of N-Boc-L-glutamic acid dimethyl ester. This methodology enables the preparation of AG7088 in kilogram quantities with an overall yield of 30%.

Depsipeptide-based inhibitors with an ester replacement of an amide bond were designed and synthesized against the HRV 3C protease[371]. These inhibitors

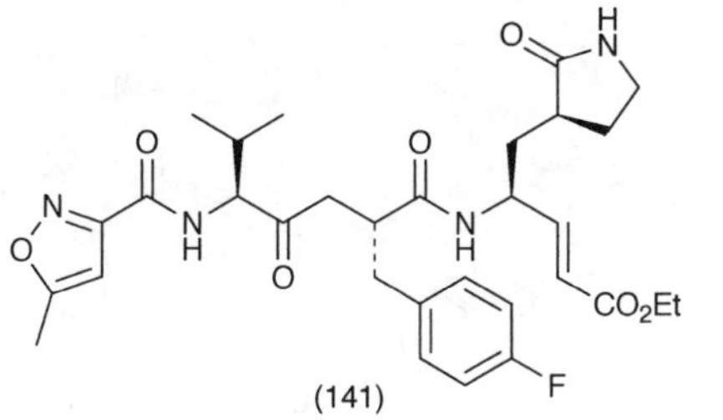

(141) 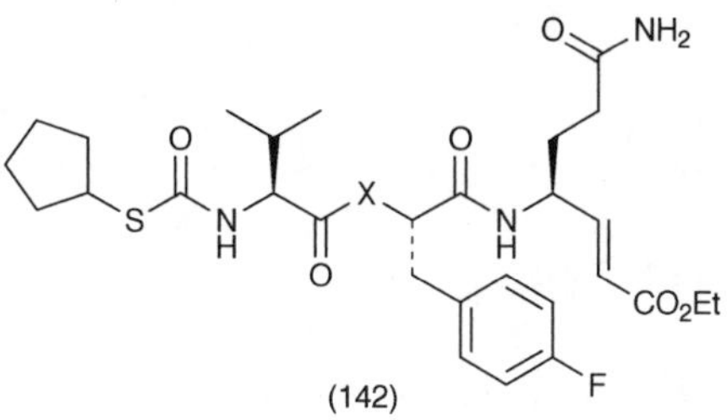   (142)

(*e.g.* **142**) act irreversibly on the enzyme but may not be ideal therapeutic candidates due to their lack of *in vitro* stability.

5.7 Converting Enzymes and Their Inhibitors. – *5.7.1 ACE and Related Enzyme Inhibitors.* Novel tetra-, penta- and hexapeptides as angiotensin converting enzyme (ACE) inhibitors are given in the bulky patent literature, which has not been reviewed here. A new software package, 'Pseudo Atomic Receptor Model' (PARM) has been developed for drug design in situations in which the 3D structure of the target receptor/enzyme is not available[372]. PARM was used for known ACE inhibitors, with favourable cross-validation statistics. A new thioesterase enzyme, isolated from a strain of *Alcaligenes sp.* ISH108, chemoselectively hydrolyses thiol esters and can be applied for the preparation of the ACE-inhibitor captopril[373]. An antihypertensive nonapeptide, with a possible ACE-inhibitor activity was purified from trypsin hydrolysate of hog bone collagen[374]. The nonapeptide contains Ile, His, Ser, Gly, Ala, Pro, Tyr, Leu and Asp and has an IC_{50} value of about 2.6 x 10^{-2} mM. Similarly to these results, two ACE-inhibitory peptides were purified and isolated from thermolysin digest of porcine skeletal muscle protein myosin[375]. The sequences of these inhibitory peptides, named myopentapeptides A and B, are Met-Asn-Pro-Pro-Cys and Ile-Thr-Thr-Asn-Pro and were found in the myosin heavy chain. IC_{50} for ACE-inhibition of these pentapeptides were 945 and 549 µM, respectively. Six other tripeptides of similar sequence (Met-Asn-Pro, Asn-Pro-Pro, Pro-Pro-Lys, Ile-Thr-Thr, Thr-Thr-Asn and Thr-Asn-Pro) showed ACE-inhibitory activity.

5.7.2 Endothelin Converting Enzyme. Endothelin converting enzyme-1(ECE-1), a zinc metalloprotease, catalyses the post-translational conversion of big endothelin-1 to endothelin-1 (ET-1), one of the most potent vasoconstrictive peptides. Potent inhibitors of ET-1 biosynthesis is thus envisaged as an attractive potential therapeutic approach for the treatment of diseases linked with elevated ET-1 levels. Through directed screening of metalloprotease inhibitors, CGS 30084 **(143)** has been identified as a potent ECE-1 inhibitor *in vitro* (IC_{50} = 77 nM). This compound served as a lead compound and after various modifications carried out at the thiol-end and at the biphenyl-portion of CGS 30084, a series of new inhibitors has been synthesized[376]. The thioacetate methyl ester prodrug derivative of compound **(144)** was found to be an orally active and a potent inhibitor of ECE-1 activity in rats.

5.8 Elastase Inhibitors. – The current status and perspectives of elastase in-

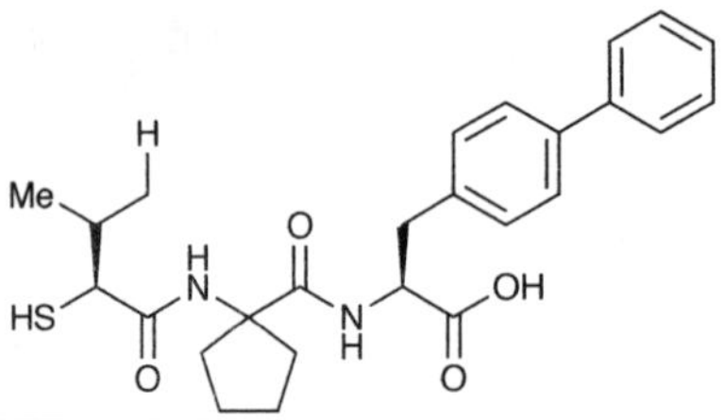

(143) (CGS 30084)

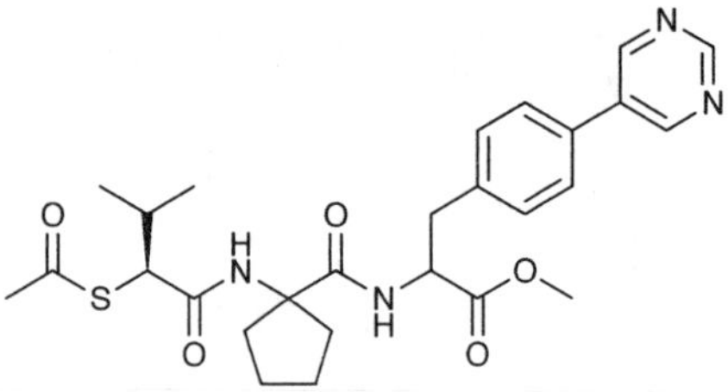

(144)

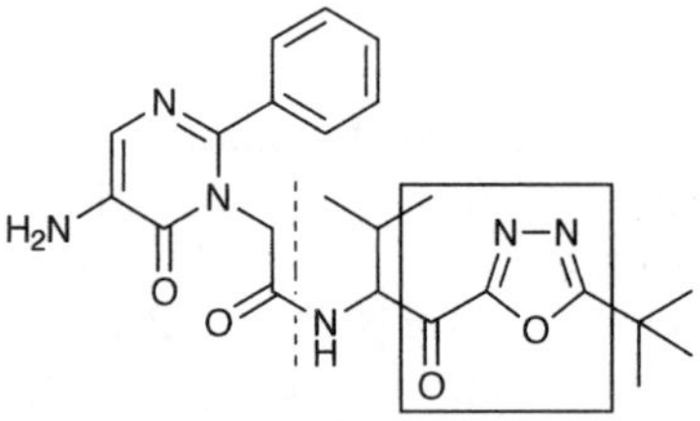

(145) (ONO-6818)

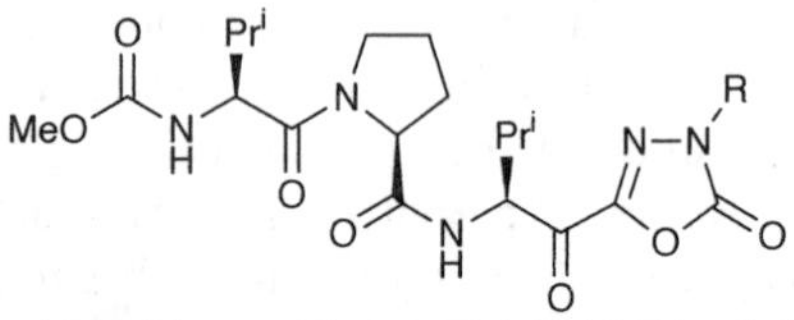

R = Me, Pr^i, Bu, Bu^i, cyclopropylmethyl, iso-pentyl, neopentyl, cyclobutylmethyl, cyclopropylethyl, cyclohexyl, cyclohexylmethyl, benzyl, 3-furanylmethyl, 2-furanylmethyl, phenethyl, 5-benzodioxolanemethyl, $CH_2C_6H_4F$-4, CH_2CH_2OMe, CH_2OEt

Figure 16

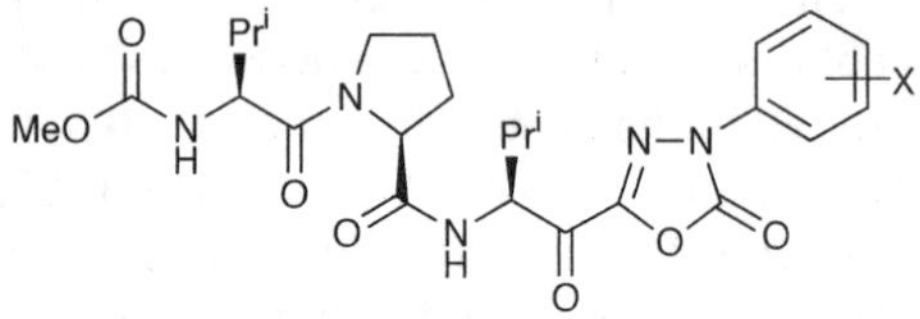

X = H, 2-Me, 3-Me, 4-Me, 2-OMe, 4-OMe, 2-CF$_3$, 4-CF$_3$, 4-F, 4-Cl, 4-CN, 4-OCF$_3$, 4-OEt, 4-OCH$_2$Ph, 4-OH, 4-NMe$_2$

Figure 17

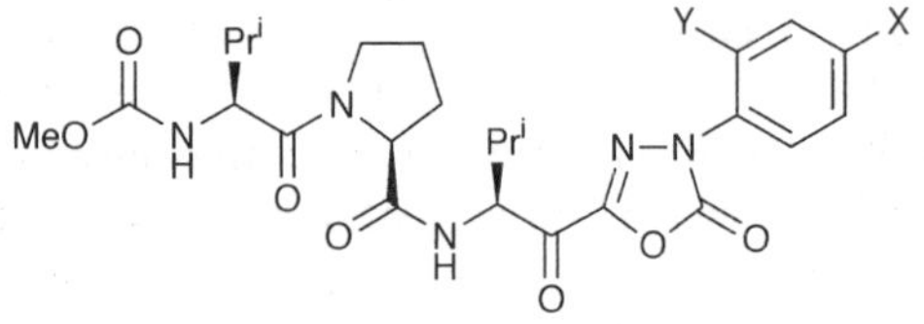

X = OMe, Y = Me; X = Y = OMe; X = F, Y = Me; X = F, Y = OMe; X = Y = F; X = Y = Me

Figure 18

hibitors have been reviewed[377]. Human neutrophil elastase (HNE) is involved in a series of diseases, and despite intensive research work orally active inhibitors with clinical potential are very rare. The low molecular weight HNE inhibitor ONO-6818 **(145)** served as lead compound for identifying new orally active inhibitors[378]. Peptidic derivatives showing more potent inhibitory activity than nonpeptidic inhibitors **(Figures 16, 17** and **18)** have also been discovered. Among these, N-aryl derivatives showed oral activity, and this activity is in good correlation with the metabolic stability. Macrolide antibiotics like erythromycin and flurythromycin show inhibition of HNE[379].

Human leukocyte elastase (HLE) is a serine protease involved in several degenerative lung and tissue diseases. A series of β-lactam derivatives were synthesized and tested to detect the structure-activity relationship for inhibition of HLE[380]. The most potent IC_{50} values were obtained with neutral hydrophobic 7α-methoxy cephalosporanic acid derivatives. Tryptophanyl-9-fluorenylmethyl ester **(146)** and N-benzhydryl piperazine **(147)** derivatives of 7α-methoxy cephalosporanic acid represent two long (beyond 24h) acting novel HLE-inhibitors. New HLE inhibitors, 6-acylamino-2-[(ethylsulfonyl)oxy]-1H-isoindol-1,3diones have also been reported [381].

MeO. and structures (146) and (147)

(146)   (147)

5.9 Farnesyltransferase Inhibitors. – Because of the involvement of farnesylated Ras-proteins in oncogenesis, the inhibition of the protein-modifying enzyme farnesyltransferase is considered as a major emerging strategy in cancer therapy. A novel class of peptidomimetic farnesyltransferase inhibitors based on the benzophenone scaffold is described[382]. The best compounds like **148** have an IC_{50} value in the 70-80 nM range. 4'-Methyl, 4'-chloro, 4'-bromo, and 4'-nitrophenylacetic acid as substituents at the 2-amino group of the benzophenone core structure yield active farnesyltransferase inhibitors; using diphenylacetic acid in this position further improves activity. Flex X docking of **148** confirms the good fit of the molecule into the peptide binding site of farnesyltransferase. The syntheses of new farnesyl diphosphate analogues **(149** and **150)** containing photoactive benzophenone groups has been described[383]. These new analogues can be enzymatically incorporated into Ras-based peptide substrates allowing the preparation of molecules with photoactive isoprenoids that may serve as valuable probes for the study of prenylation function. Trisubstituted cyclopropane-derived pseudopeptides were also designed and synthesized[384]. These peptidomimetics like **(151)** were found to be competitive inhibitors of Ras farnesyltransferase with IC_{50} of 320 nM, but less potent than the tetrapeptide parent C Abu FM. A novel class of highly potent, selective, and nonpeptidic inhibitors, of Ras-farnesyltransferase, based on **(152),** has been found[385].

The binding of farnesyltransferase inhibitors to the enzyme has been studied with computational chemistry[386]. Employing flexible docking of several nonthiol farnesyltransferase inhibitors known from the literature as well as some new model compounds and performing also GRID searches, two regions in the

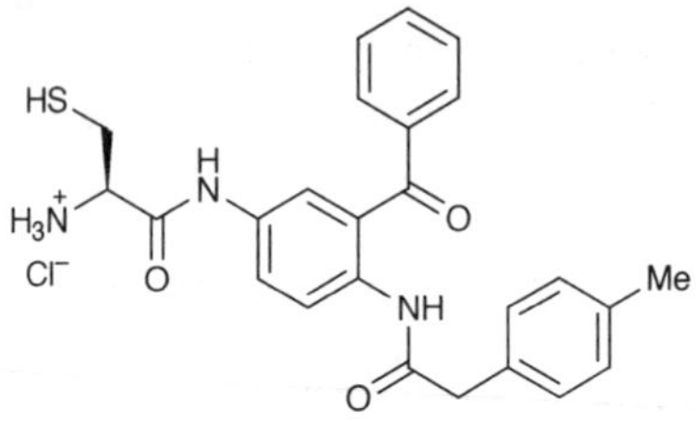

(148)

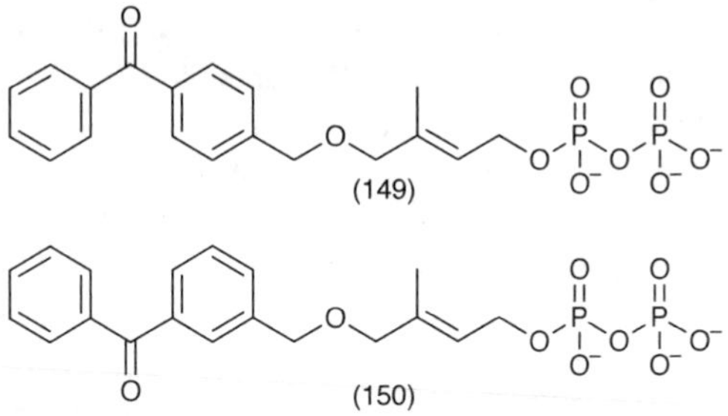

(149)

(150)

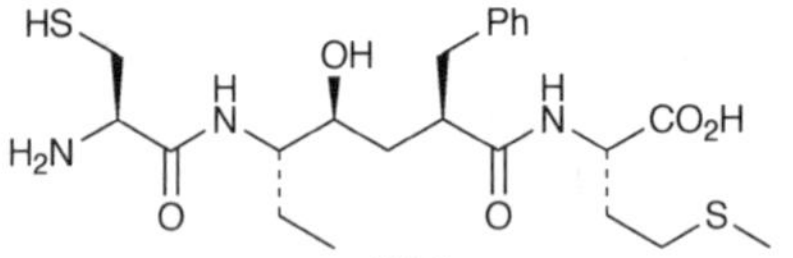

(151)

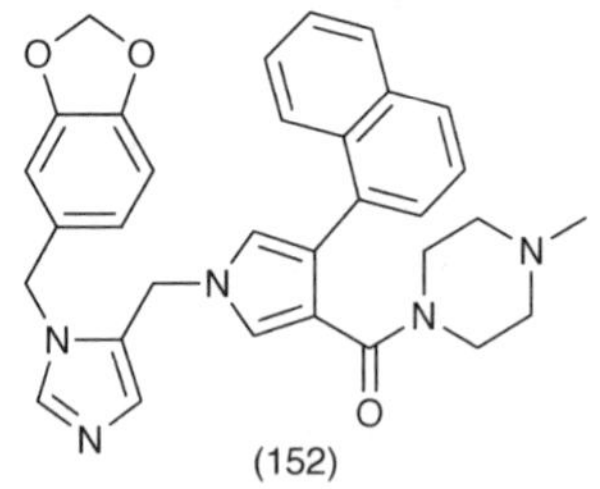

(152)

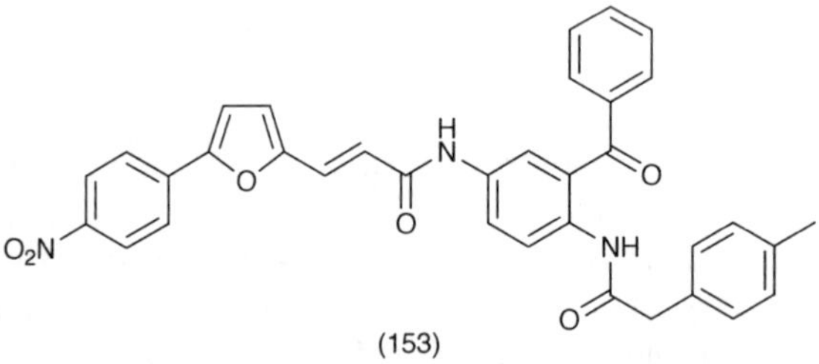

(153)

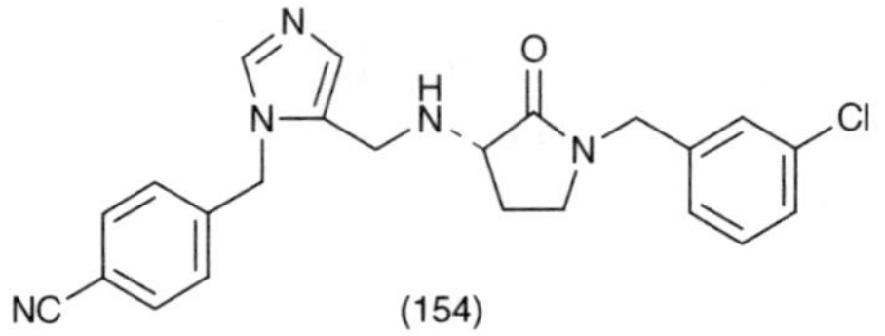

(154)

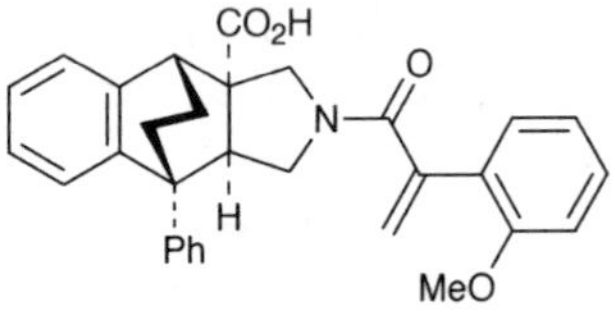

(155) (RPR 115,135)

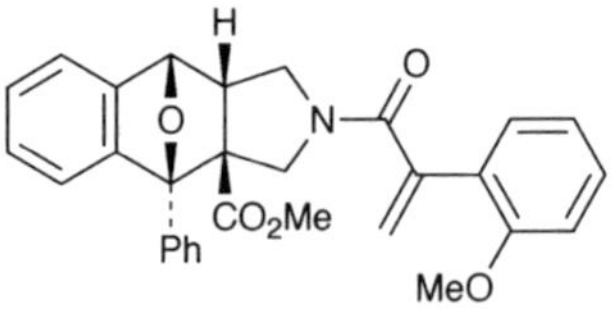

(156) (RPR 225,370)

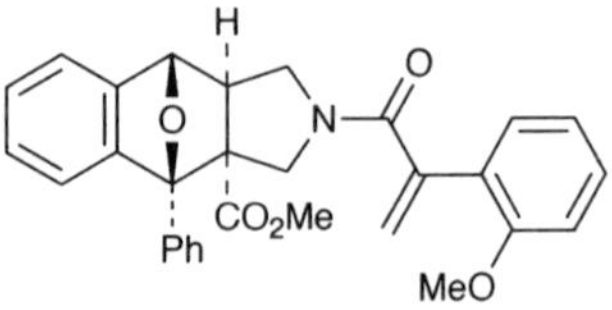

(157) (RPR 222,490)

enzyme's active site were identified as aryl binding regions. On the basis of these results, new inhibitors were designed and synthesized. Compound (153) is a non-thiol inhibitor with an IC_{50} of 35 nM. The synthesis, structure-activity relationships, and biological properties of a novel series of imidazole-containing inhibitors of farnesyltransferase have also been described[387]. Systematic modifications has also provided the compound (154), a non-peptide enzyme inhibitor with excellent bioavailability and with an IC_{50} of 1.9 nM. The farnesyltransferase inhibitor, RPR 115135 (155) and two of its analogues with oxa-bridges (156 and 157) have been synthesized by different routes[388]. The parent compound has low enzyme-inhibition profile with an elevated cellular potency as well as moderate *in vivo* activity. A conformationally restricted inhibitor of farnesyltransferase, 3,8-diazabicyclo[3.2.1]octan-2-one was synthesized[389] and this chemistry was used for producing a conformationally constrained enzyme inhibitor, which aided the elucidation of the enzyme-bound conformation.

5.10 HIV-Protease Inhibitors. – The currently available chemotherapeutic agents for the treatment of AIDS include nucleotide reverse transcriptase inhibitors (AZT, ddI, ddC, 3TC, d4T, *etc.*), nonnucleotide reverse transcriptase inhibitors (*e.g.* nevirapine) and protease inhibitors (*e.g.* ritonavir, indinavir, amprenavir). The recent trend toward early and aggressive intervention with combination chemotherapy requires new and new enzyme inhibitors with low toxicity.

HIV-1 encodes an aspartic acid protease (HIV-1 PR). This enzyme has been an attractive target for the design of inhibitors for effective antiviral therapy. Dimerization of HIV-1 PR is one of the essential events to attain the main structure, which is an enzymatically active C_2-symmetric homodimer. Two new non-peptide inhibitors were designed and synthesized that effectively inhibit dimerization, by a dissociative and also through an active-site directed mechanism[390]. The compounds contain a β-strand mimetic part as shown in (158).

Two stereoselective routes to a series of diastereomeric inhibitors of HIV-protease, monofluorinated analogues of the known Merck inhibitor indinavir, have been described[391]. The most potent inhibitor (159) has about the same inhibitory activity as indinavir (K_i = 2.0 nM).

Derivatized carbohydrates as C_2-symmetric HIV-1 protease inhibitors have been previously described. A series of fluoro substituted P1/P1′ analogues have been synthesized and evaluated for antiviral activity[392]. The potency of the analogues characterized with the general structure (160) was moderate, with K_i values ranging from 1 to 7 nM. A compound with a very similar structure (161) was prepared by a straightforward synthesis involving a thiol nucleophilic ring opening of a diepoxide[393]. Compound (161) was found to be a potent inhibitor of HIV-1 PR, showing good antiviral activity in a cell-based assay. The same research group have described the synthesis of novel potent, diol-based HIV-1 PR inhibitors *via* intermolecular pinacol homocoupling of (2S)-2-benzyloxymethyl-4-phenylbutanal[394]. Compound (162) was found to be a potent inhibitor of HIV-1 PR. The stereochemistry of the central diol of (162) was determined from the X-ray crystallographic structure of its complex with the enzyme. An enantioconvergent synthesis of new pyrrolidin-3-ol and pyrrolidin-3-one peptide conjugates was used to prepare new HIV-1 PR inhibitors[395]. The compounds are mimicking the Phe-Pro dipeptide. Di- and tripeptide analogues containing α-ketoamide as a new core structure and incorporating allophenyl norstatine (Apns) [(2S, 3S)-3-amino-2-hydroxy-4-phenylbutyric acid] as a

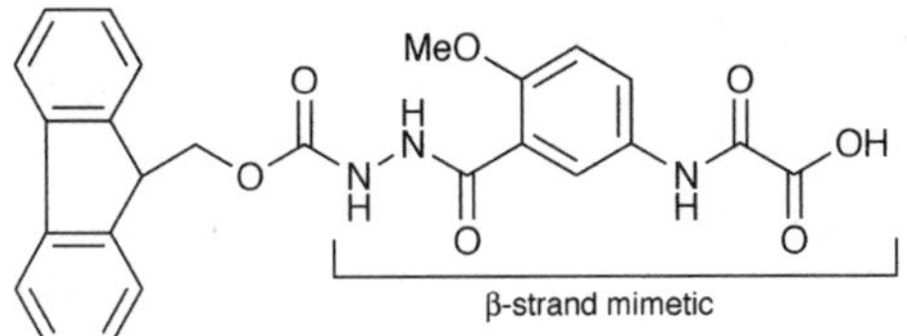

β-strand mimetic

(158)

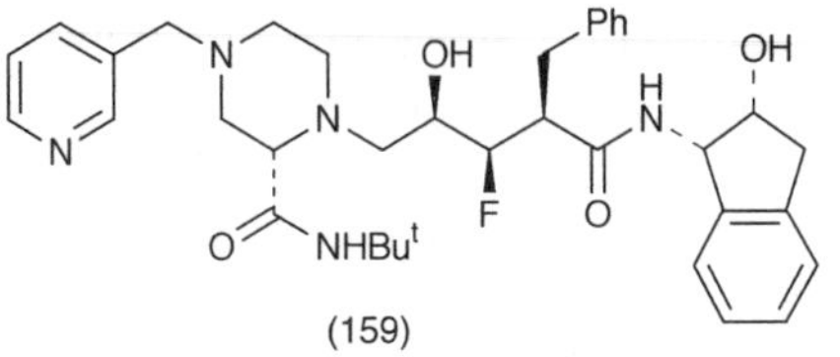

(159)

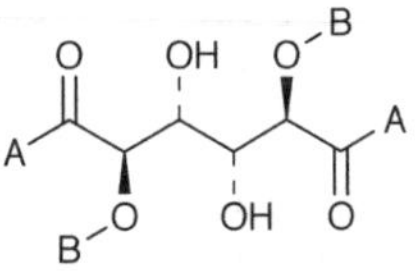

(160) General structure
(A = amine, B = benzyl)

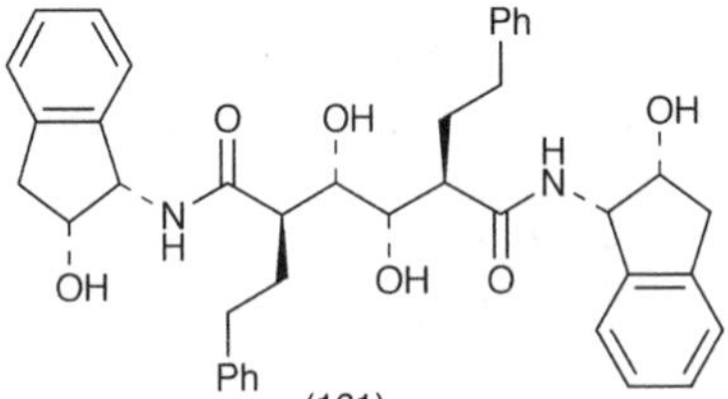

(161)

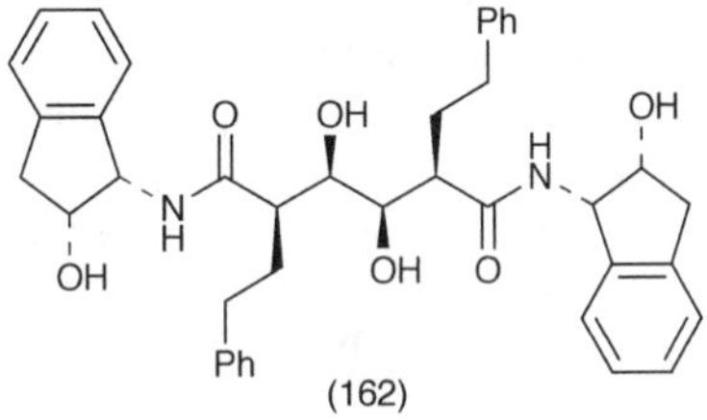

(162)

transition state mimic, were designed and synthesized for obtaining a novel structural type of HIV-1 PR inhibitors[396]. A preliminary evaluation of the inhibitory activity of the synthesized derivatives was determined as the percentage of enzyme inhibition at 5 μM and 50 nM levels. The α-ketoamides displayed a significantly enhanced potency relative to their parent isosteres as inhibitors of HIV-1 PR. Another series of allophenyl norstatine containing HIV-1 PR inhibitors were studied and proved also to be potent against the feline immunodeficiency virus protease (FIV PR), which has been shown to be a useful model for drug – resistant HIV PRs[397]. The most effective compound (VLE776, **163**) showed an IC_{50} value of 8 nM for HIV-PR and 48 nM for FIV-PR. Anomalous tetrapeptides called Mer-N5075A, α- and β-MAPI produced from a species of *Streptomyces* have a potent HIV-1 PR inhibitory effect[398].

The first total synthesis of tetrapeptide inhibitors have been described, achieved simply by a route connecting two dipeptides. The synthetic method is applicable for synthesis of Mer-N5075A analogues such as GE20372A and B (**164** and **165**). The inhibitory activity of all the compounds (IC_{50}) is in the 10 to 100 μM range. Two series of peptidomimetics containing an N-hydroxyamino core structure have been prepared[399] and tested for inhibitory activity against HIV-1 PR. In the N-hydroxy Phe derivatives, Fmoc-Phe-Ψ[CO-N(OH)]-Phe-Pro-NHtBu was the best inhibitor of the series (IC_{50} = 144 nM) showing satisfactory inhibition of HIV replication in cell culture (ED_{50} = 98 nM) and remarkable stability against cell culture and plasma enzymes.

Cyclic tripeptides (**166, 167**) were synthesized in a ring-closing metathesis reaction using Grubb's catalyst as mimics for the constrained conformation of

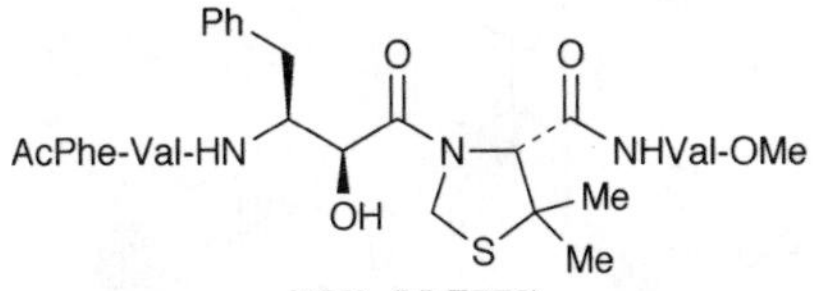

(163) (VLE776)

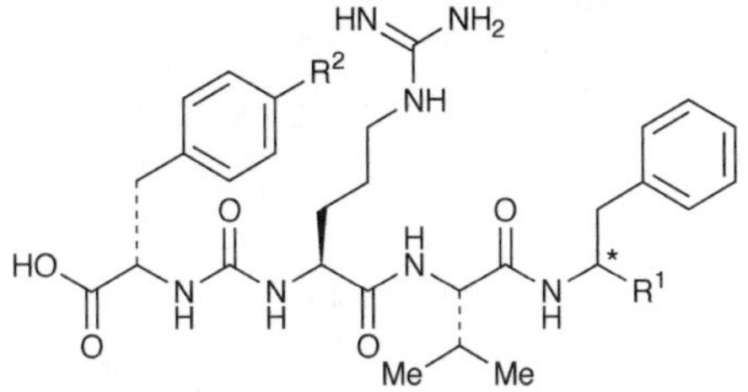

(164) (GE20372A) R^1 = CHO, R^2 = OH, config. at * is *S*
(165) (GE20372B) R^1 = CHO, R^2 = OH, config. at * is *R*

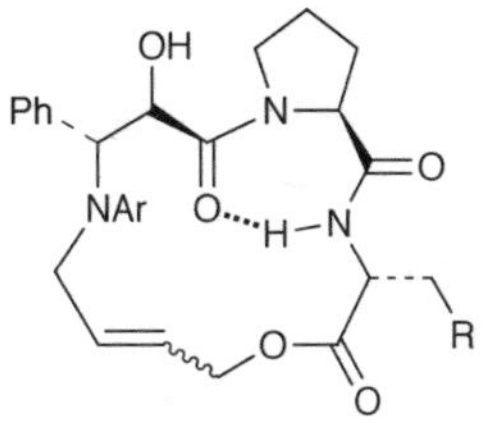

(166) R = CH(Me)$_2$
(167) R = Ph

structural analogues of HIV-1 PR inhibitors[400]. Fully N- and O-sulfated homooligomers from octamer to nonadecamer of Tyr have been obtained as their sodium salts (NaO$_3$S-[Tyr(SO$_3$Na)]$_n$-ONa, n = 8-19] from the reaction mixtures of Tyr with SO$_3$-triethylamine and SO$_3$-pyridine complexes[401,402]. These compounds have a definite polyanionic structure and their anti-HIV activity increased along with the increase of the chain length up to dodecamer.

Based on the 'double-drug' strategy, a potent prodrug-type anti-HIV agent KNI-1039 **(168)** was developed in which an excellent HIV-1 PR inhibitor KNI-727 was coupled through a glutaryl-glycine linker to the nucleoside reverse transcriptase inhibitor AZT[403,404]. MNB 161 exhibited extremely potent anti-HIV activity compared to that of individual components. Compound **(168)** is stable in culture medium but generates the active compounds KNI-727 and AZT in cell homogenate. New water-soluble prodrugs have been synthesized as HIV-1 PR inhibitors. As self-cleavable spacers, succinamide or glutaramide were employed, releasing the parent drug spontaneously *via* intramolecular cyclization-elimination reaction through imide formation[405, 406]. As an example, KNI-727 **(169)**, a water insoluble HIV-1 PR inhibitor was coupled through a spacer to a solubilizing moiety resulting in a water-soluble prodrug **(170)**.

On the basis of substrate transition-state mimic concept of HIV-protease, a series of small-sized dipeptide inhibitors containing a hydrophilic carboxyl group were designed and synthesized[407]. These dipeptide inhibitors such as KNI-357 **(171)** showed good HIV-1 PR inhibitory activity, but their anti-HIV activity was poor, probably due to their inadequate cell membrane permeability caused by the presence of hydrophilic groups.

The HIV-1 PR inhibitor Lopinavir is metabolised rapidly by an oxidative pathway. Three of the major metabolites have been identified and synthesized, their antiviral (HIV) activities also determined[408]. Ritanovir (ABT-538), approved in 1996, is a potent and effective peptidomimetic inhibitor of HIV-1 PR with high oral bioavailability. A new synthesis of the diaminoalcohol core of

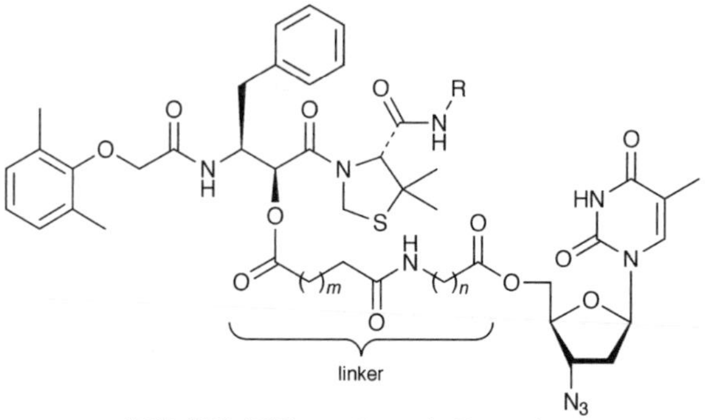

(168) (KNI-1039) *m* = 2, *n* = 1, R = *tert*-butyl

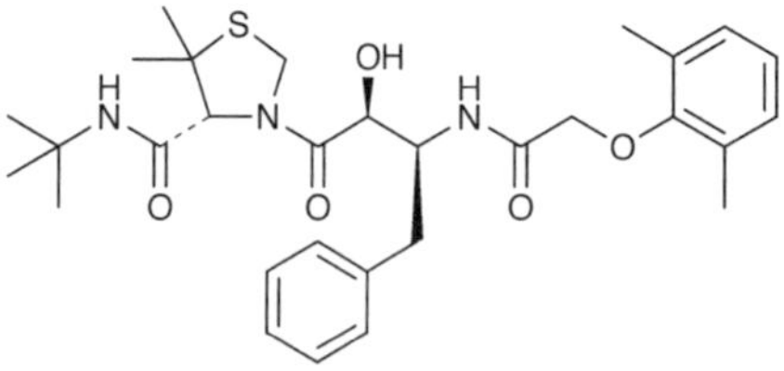

(169) (KNI-727)

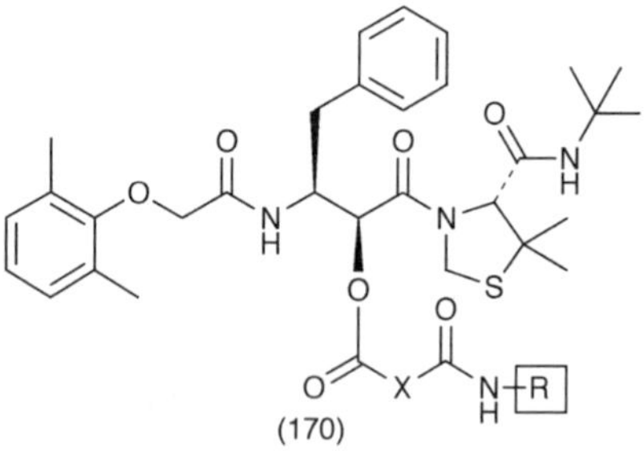

(170)

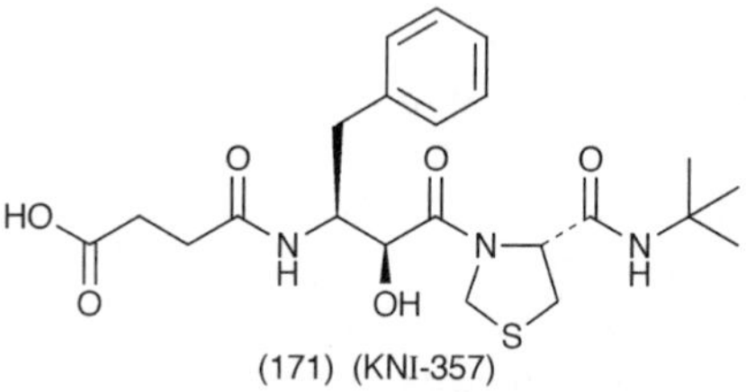

(171) (KNI-357)

Ritanovir was performed, based on regioselective reduction of amino acid-derived epoxyalcohols[409].

The commonly used HIV-1 protease assays rely on measurements of the effect of inhibitors on the hydrolysis rate of synthetic peptides. Recently, an assay based on surface plasmon resonance was introduced. Two new competition assay methods with improved efficiency were developed using either biotin labelling or direct immobilization of the inhibitor to the biosensor surface matrix[410].

A trifluoromethyl analogue of Asp-protease inhibitor pepstatin was synthesized containing γ-trifluoromethyl-γ-amino-β-hydroxy butyric acid units instead of statine units[411]. The new analogue was tested as inhibitor of HIV-1 PR, but the compound did not show any inhibition up to 150 μM concentration.

5.11 Matrix Metalloproteinase Inhibitors. – The matrix metalloproteinase (MMP) family of endopeptidases represent a group of tightly regulated metal-loproteases mediating turnover of the extracellular matrix proteins proteo-glycan, collagen and gelatin. MMPs are involved in rheumatoid arthritis, osteoarthritis, certain cancers, and are implicated in metastasis.

A series of compounds with a free carboxyl group were prepared from N^{α}-substituted 2,3-diaminopropionic acid and were tested for efficacy as MMP inhibitors[412]. Computer aided molecular modelling was performed by 3D structure modelling softwares (Insight II/Discover). Some of the carboxylate compounds, such as carbamate and sulfonamide derivatives proved to be effective MMP-1 inhibitors with IC_{50} values of the order of 10^{-6} M. On the basis of the structure of known MMP inhibitors, an N-hydroxyformamide-class of inhibitors were designed and synthesized, which have a potent broad-spectrum activity towards MMP and TNF-α converting enzyme (TACE)[413]. Compound **(172)** proved to be a good inhibitor of MMP-1 (K_i = 20 nM) and possesses good oral and intravenous pharmacokinetics in the rat and the dog. Design of potentially selective and constrained MMP inhibitors was performed by using the fully automated docking programs AutoDock and DOCK[414]. A comparative study of the two programs in closely approximating the X-ray crystal structures of ten selected MMP inhibitors has shown, that the AutoDock program is highly reliable, efficient and predictive for a set of inhibitors. 3D Quantitative structure-activity relationship (QSAR) models have been obtained using comparative molecular field analysis (CoMFA) for a novel series of piperazine-based MMP-inhibitors[415]. The crystal structure of stromelysine (MMP-3) was used to identify regions of the enzyme and inhibitors where steric and electrostatic effects correlate strongly with biological activity. A complementarity was found between the inhibitors' substituent conformations and the structural characteristics of the MMP-3 S1-S2' binding pockets.

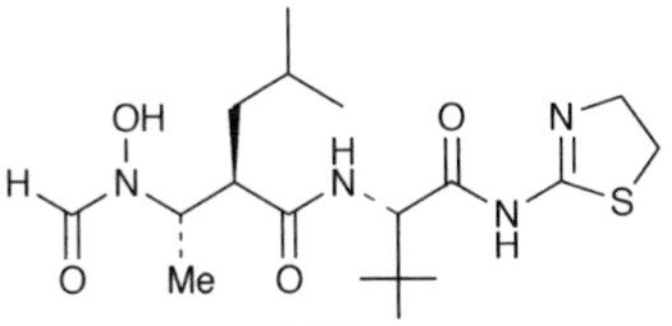

(172) 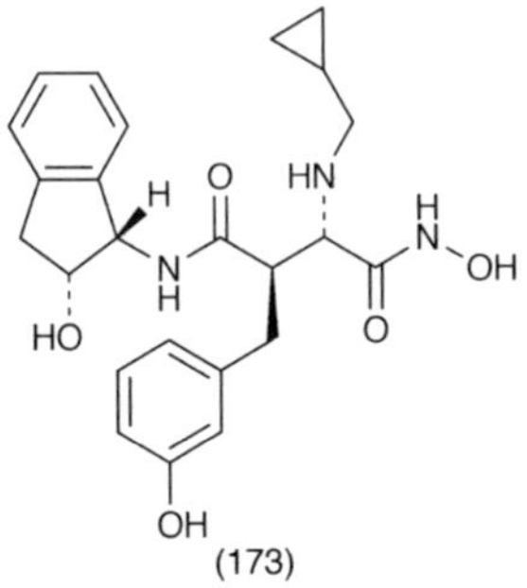   (173)

Potent, selective and orally bioavailable inhibitors have been designed and synthesized against aggrecanase, an enzyme which plays a pivotal role in the catabolism of aggrecan in human arthritic disease[416]. Selective aggrecanase inhibitors may prevent the progression of joint destruction. A pharmacophore model of the P1' site, specific for aggrecanase, was defined using the specificity studies of the MMPs and the similar biological activity of aggrecanase and MMP-8. A *cis*-(1*S*)(2*R*)-amino-2-indanol scaffold was incorporated as tyrosine mimic at P2' position, optimisation resulting in compound **(173)**, a potent and selective inhibitor of aggrecanase (K_{50} = 12 nM).

The rational design of MMP inhibitors has been facilitated through the examination of X-ray crystal structures of enzyme-inhibitor complexes. A new generation of heterocyclic nonpeptide MMP inhibitors derived from a 6H-1,3,4-thiadiazine scaffold have been described[417]. Screening of 6-methyl-1,3,4-thiadiazine compounds resulted in a lead, which on further optimisation of the new compounds, produced the selective inhibitor (2R)-N-[5-(4-bromophenyl)-6H-1,3,4-thiadiazin-2-yl]-2-[(phenylsulfonyl)amino]propanamide **(174)** with high affinity for MMP-9 (K_i = 40 nM). X-ray crystallographic data obtained for CDMMP-8 cocrystallized with N-allyl-5-(4-chlorophenyl)-6H-1,3,4-thiadiazin-2-amine hydrobromide gave detailed information on binding interactions for designing thiadiazine-based MMP inhibitors.

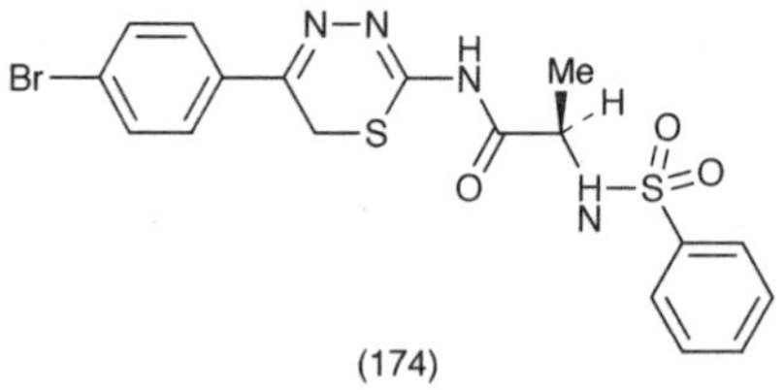

(174)

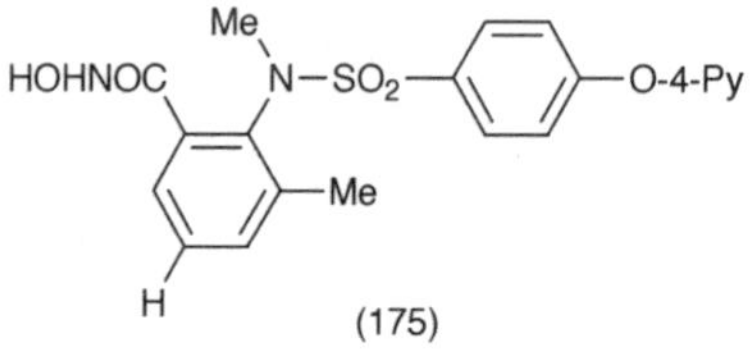

(175) 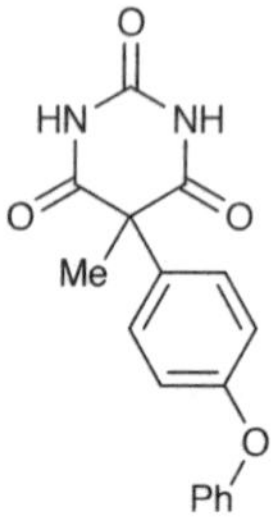   (176)

A novel series of anthranilic acid-based MMP inhibitors of MMP-1, MMP-9, MMP-13 and TACE was prepared and evaluated[418]. Among the new compounds (substituted anthranilate hydroxamic acids) a potent and orally active MMP-13 inhibitor **(175)** was found (IC_{50} = 4 nM). Sulfonamide based hydroxamic acids were found to be potent and selective MMP inhibitors[419]. Design of this new series of compounds was based on a scaffold containing a central ring of six-membered hexahydropyrimidines or seven-membered [1,4] diazepines. An X-ray structure of a stromelysin-inhibitor complex confirmed the predicted arrangement of the inhibitor in the enzyme active site.

A new class of MMP inhibitors, 2,4,6-pyrimidine triones (*e.g.* **176**) has been identified simply by screening a collection of compounds against stromelysin[420]. An X-ray crystal structure of one representative compound bound to the catalytic domain of stromelysin shows that the compounds are bound in an orientation that places the hydrophobic moiety in the deep S1' pocket. The N3 nitrogen atom in the pyrimidine trione ring approaches the active-site zinc most closely, at 2.17Å, and completes a tetrahedral arrangement of nitrogen ligands about the zinc. The pyrimidine triones mimic substrates in forming H-bounds to key residues in the active site.

Macrocyclic biphenyl ether hydroxamic acid inhibitors of collagenase 1 and gelatinases A and B were prepared by intramolecular O-arylation of phenols with phenylboronic acids[421]. The macrocyclization reaction proceeds under mild conditions with copper acetate catalysis.

5.12 NO-synthase Inhibitors. – A family of 3 isoform enzymes, collectively known as nitric oxide synthases (NOS), catalyses the oxidation of a terminal guanidium-N of Arg resulting in NO-biosynthesis. Selective inhibition of the isoforms of NOS could be beneficial in the treatment of certain disease states arising from the overproduction of nitric oxide by NOS. Therefore NOS is an important drug target and isoform-selective inhibitors are needed to block the uncontrolled production of NO in disease states. X-ray crystallographic structures of inducible NOS (iNOS) and endothelial NOS (eNOS) oxygenase domains with both inhibitors and substrates have been previously published and provided more detailed active site information. Now the crystal structure of the catalytic heme domain of eNOS complexed with a small molecule inhibitor 3-bromo-7-nitroindazole (7-NIBr) at 1.65 Å resolution has been described[422]. Glu-363 of eNOS was known to be critical for NOS function prior to the crystal structures which show that Glu-363 forms specific H-bonding interactions with L-Arg. In the 7-NIBr-enzyme complex, eNOS shows an altered conformation, in which the key Glu-363 residue swings out toward one of the heme propionate groups. New putative NOS inhibitors containing ω-nitro-Arg were synthesized. The tripeptide Arg(NO$_2$)-Lys-Arg(NO$_2$) was prepared for inhibiton of iNOS[423] and proved to be more potent than the reference standard L-NNA. In another experiment new dipeptide amides containing Arg(NO$_2$) and D-2,4-diaminobutyric acid (Dbu) were synthesized and evaluated[424]. They are all modest inhibitors of nNOS and only poorly inhibit eNOS and iNOS. L-Canavanine, a unique amino acid from leguminous plants, represents an L-Arg analogue in wich the δ-CH$_2$ group is replaced by oxygen and inhibits NO-production. The new derivatives of L-canavanine (**177, 178**) were prepared by total chemical synthesis and evaluated as inhibitors of iNOS[425]. The structure-activity analysis suggests that the presence of an electron withdrawing group (here oxygen) in the side chain of Arg-derived NOS-inhibitors produces dramatic effects in the interaction with the enzyme which must be considered during future inhibitor design.

The potency and selectivity of a series of 2-iminohexahydroazepines have been examined as inhibitors of the 3 human NOS isoforms[426]. Potencies (IC$_{50}$'s) for the inhibitors are in the low micromolar range. Some peptidomimetic analogues of Arg(NO$_2$)-dipeptide amides containing reduced amide bonds proved to be potent and selective inhibitors of nNOS[427]. The deletion of the CO group from the amide bond either preserves or improves the potency for nNOS. The most potent nNOS inhibitor compound (**179**) [(4S)-N-(4-amino-5-[aminoethyl]aminopentyl)-N-nitroguanidine] has a K$_i$ value of 120 nM and shows a very high selectivity (greater than 2500-fold) over eNOS and 320-fold selectivity over iNOS. The reduced amide bond is an excellent surrogate of the amide bond and it may facilitate the design of new potent and selective inhibitors for nNOS. 7-Methoxyindazole and related substituted indazoles represent a new class of inhibitors of nNOS[428], the indazole nucleus containing an electron-donating

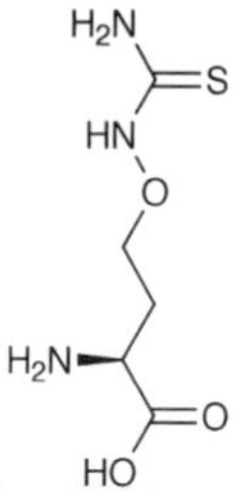

(177) 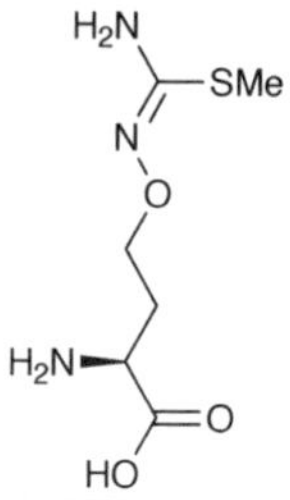   (178) 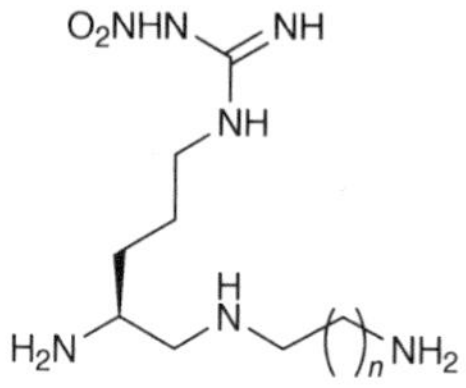   (179) $n = 1$

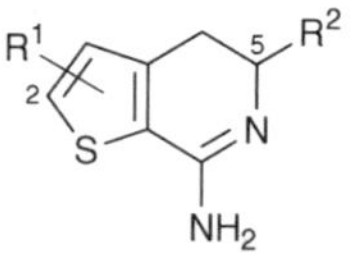

(180) 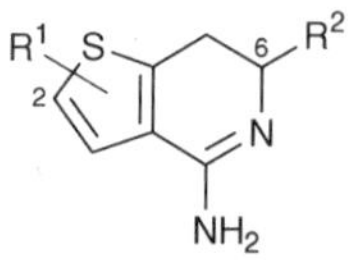   (181)

substituent. Although 7-methoxyindazole is less potent than 7-nitroindazol, the bioassays show that the nitro-substitution is not indispensable to the biological activity of the indazole ring.

5-substituted 7-amino-4,5-tetrahydrothieno [2,3-c] pyridines **(180)** and 6-substituted 4-amino-6,7-dihydrothieno[3,2-c]pyridines **(181)** are described as exceptionally potent inhibitors of iNOS and nNOS[429]. Selectivity and potency could be modulated by variation of the 5- or 6-substitutent. IC_{50} values of the best inhibitors are in the 50-100 nM range.

5.13 Proteasome Inhibitors. – The 26S proteasome, a large cytosolic protease complex, is implicated in many biological processes, including degradation of most cytosolic proteins in mammalian cells. The precise role of the individual enzyme activities in cytosolic proteolysis has remained unclear, however, it is known that proteosomes play a central role in the generation of antigenic peptides presented by MHC I molecules. Proteasome inhibitors (lactacystin, epoxomicin, as well as peptide vinyl sulfones) have proven essential to the study of proteasome activity in living cells. Most proteasome inhibitors are based on short oligopeptidic sequences. Important goals are the development of inhibitors that are subunit-specific and cell-permeable.

Mammalian proteasomes possess three distinct catalytically active species, β1 ('peptidyl-Glu-peptide hydrolysing'), β2 ('tryptic like') and β5 ('chymotryptic like') individual subunits. Extended size peptide-based inhibitors have been synthesized and evaluated[430]. A set of highly potent proteasome inhibitors that target all individual active subunits were found. Modification of the most active compound adamantane-acetyl (6-aminohexanoyl)₃-(Leu)₃-vinyl-(methyl)sulfone (AdaAhx₃L₃VS), itself capable of proteasome inhibition in living cells, afforded a new set of radio- and affinity labels. The extended size peptide inhibitors are more potent in living cells than their shorter peptide vinyl sulfone counterparts.

The natural products TMC-95 A-D **(Figure 19)** are potent proteasome inhibi-

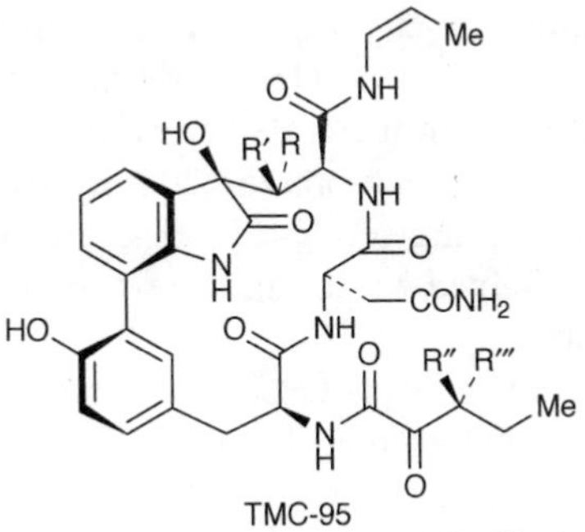

TMC-95

A(1) R = OH, R' = R''' = H, R'' = Me
B(2) R = OH, R' = R'' = H, R'' = Me
C(3) R = R''' = H, R' = OH, R'' = Me
D(4) R = R'' = H, R' = OH, R''' = Me

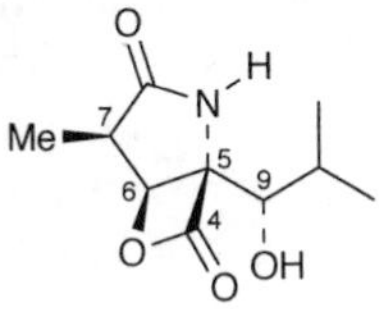

(182) Omuralide

Figure 19

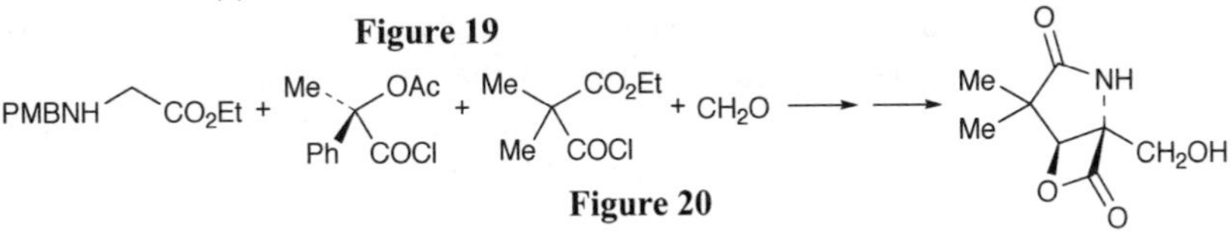

Figure 20

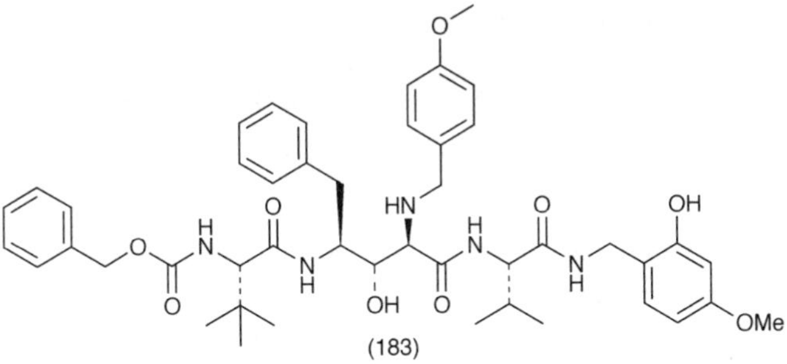

(183)

tors isolated from the fermentation broth of *Apiospora montagnei*. These natural products are unique cyclic peptides containing L-Tyr, L-Asp, an L-Trp derived oxindole, 1-propenylamine, and 3-methyl-2-oxopentanoic acid units. Synthesis of the functionalised macrocyclic core of TMC-95 compounds is a difficult task and several publications are dealing with this problem. One solution is to couple 7-iodoisatin and a Tyr-derived arylboronic acid *via* a Pd(OAc)$_2$-catalysed Suzuki-coupling reaction to the biaryl moiety of TMC-95[431]. Potassium fluoride serves as a base in this reaction. The construction of the biaryl moiety of TMC-95 can be achieved also *via* a Pd-catalysed Stille cross-coupling reaction[432]. The fully functionalised macrocyclic core of TMC-95 was synthesized using as key steps, aldol condensation, Suzuki cross-coupling reaction and macrolactamization to the macrocycle[433]. In other experiments only the protected version of the upper ('northern') part of TMC-95A was synthesized with full stereochemical control[434].

Another natural product for selective and potent inhibition of proteasome function is the β-lactone of lactaxystein, omuralide **(182)**. Four building blocks shown in **Figure 20** can be combined to a simplified omuralide analogue in nine steps, with the use of (*R*)-atrolactic acid as a recoverable chiral controller[435].

Identification and *in vitro* characterization of a series of 2-aminobenzylstatine derivatives that inhibit non-covalently the chymotrypsin-like activity of the 20S proteasome have been described[436]. Compound (183), originally synthesized to target HIV-proteinase, was identified as non-covalently bound proteasome inhibitor by high throughput screening[437]; it inhibits the 20S proteasome with an IC value in the micromolar range. Using an X-ray structure of the human proteasome in complex with (183) helped to design a new inhibitor with one order of magnitude in inhibitory potency.

The structure of tyropeptins A and B, new proteasome inhibitors produced by *Kitasatospora sp.* MK993-dF₂, were determined by analysis of various NMR experiments. The structures of tyropeptins A and B were found to be isovaleryl-L-Tyr-L-Val-DL-tyrosinal and n-butyryl-L-Tyr-L-Leu-DL-tyrosinal, respectively[438]. Four related compounds called phepropeptins A, B, C and D were also isolated as new proteasome inhibitors produced by *Streptomyces sp.*[439]. Phepropeptin-B is a cyclopeptide with the structure cyclo (-L-Leu-D-Phe-L-Pro-L-Phe-D-Leu-L-Val-). The crystal structure of the 20 S proteasome core particle (CP) complexed with the natural non-covalent inhibitor TMC-95A has been described[440]. The inhibitor is bound to the main-chain atoms of the protein.

5.14 Protein Phosphatase Inhibitors. – Three Cdc25 dual specificity phosphatases exist in humans: Cdc25A, Cdc25B and Cdc25C. These enzymes play central roles in coordinating cellular signalling processes and cell proliferation. Cdc25A and B have oncogenic properties and are overexpressed in many human tumors. A new family of potent inhibitors of Cdc25 was discovered after experimental examination of the '1990 compound National Institute of Cancer Institute Diversity Set' and computational selection from 140 000 compound[441]. Eight out of these compounds had an *in vitro* mean inhibitory concentration of < 1μM, the most potent was (184), 6-chloro-7-(2-morpholin-4-ylethylamino)quinoline-5,8-dione (NSC 663284). The first total synthesis of coscinosulfate (185), a metabolite isolated form sea sponge with selective inhibitory effect on Cdc25 has been described[442].

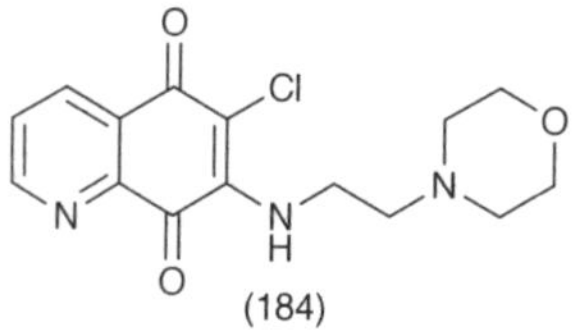

(184)

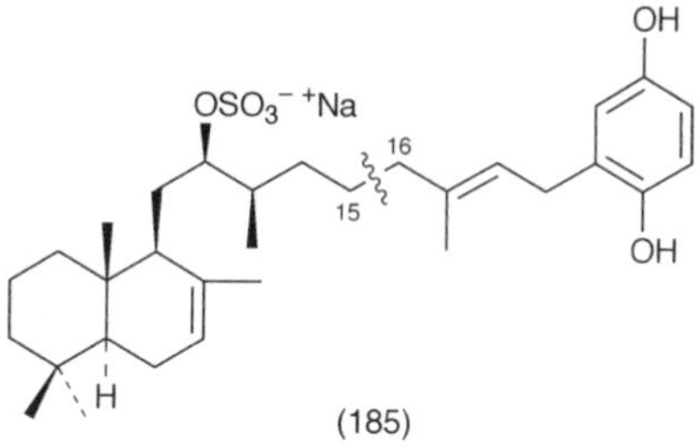

(185)

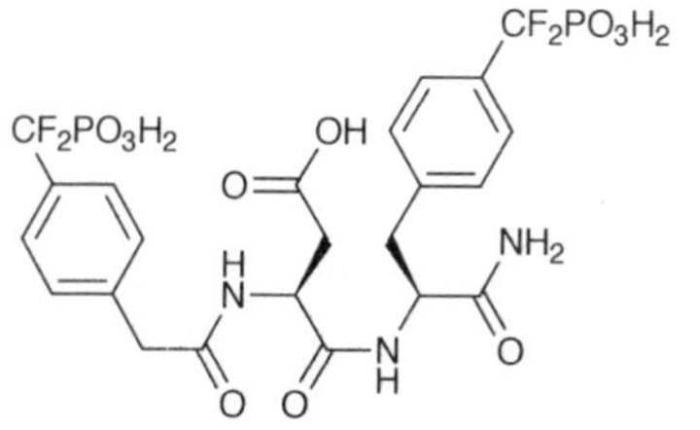

(186)

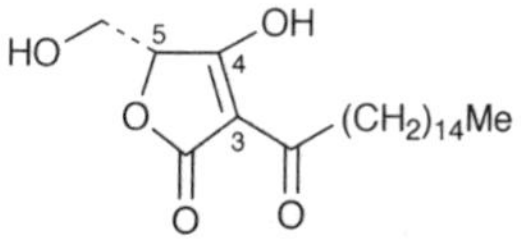

(187)

Protein-tyrosine phophatases (PTPase) form another large family of enzymes that serve as key regulatory components in signal transduction pathways. A novel affinity-based high-throughput assay procedure is described that can be used for PTPase-inhibitor screening[443], on the basis of a combinatorial library. The high throughput/combinatorial library screening protocols furnished a very potent (K_i = 2.4 nM) and selective PTP1B inhibitor **(186)**. Utilization of a peptide in the lead (Ac-Asp-Ala-Asp-Glu-Xxx-Leu-amide) resulted in the discovery of a novel PTP1B binding motif, 6-carboxy-1-naphthoic acid[444]. A series of novel tetronic acid derivatives were synthesized and evaluated as inhibitors of the PTPs and dual-specificity protein phosphatases (VHR)[445]. The best compounds have an IC_{50} value of 4.0 to 10 μM against VHR. The molecular mechanism of VHR inhibition by a known natural product, RK-682 **(187)** has been investigated[446]. Inhibition was competitive and two molecules of RK-682 were required to inhibit one molecule of VHR. A novel dimeric derivative of RK-682 was designed and synthesized, showing increased inhibition of VHR.

Protein phosphatase-1 (PP1) plays a key role in dephosphorylation of very important biological processes. The crystal structure of the molecular complex of the tumor promoter okadaic acid (a C_{38} polyether fatty acid) bound to PP1 to a resolution of 1.9 Å has been described[447]. The inhibitor-bound enzyme shows very little conformational change when compared with two other PP1 structures, except in the inhibitor-sensitive β12-β13 loop region. Nodularin and microcysteins are complex natural cyclic isopeptidic hepatotoxins that serve as subnanomolar inhibitors of the eukaryotic PP1 and PP2A. New PP-inhibitors were designed and prepared, based upon the structures of nodularin and microcystin[448]. The structure-activity correlation of nodularin-based PP1 inhibitors and the role of the 3-methyl and 3-diene groups in the Adda [(2*S*, 3*S*, 4*E*, 6*E*, 8*S*, 9*S*)-3-amino-9-methoxy-2,6,8-trimethyl-10-phenyldeca-4,6-dienoic acid] residue was studied[449].

Two new PP inhibitors, oscillamide B and C **(188, 189)**, were isolated from the cyanobacteria *Oscillatoria agardhii* and *O. pubescens*. The structures of the inhibitors were elucidated by analysis of HR-FAB MS, 1D and 2D NMR spectra, and chemical degradation[450]. An interesting review summarizes the inhibition of phosphatase 1 and 2A enzymes as new targets for rational cancer drug design [451].

5.15 Renin and Other Aspartyl Proteinase Inhibitors. – The future of renin inhibitors and the development programs of renin inhibitors have been highlighted[452]. At the moment it appears that almost all the development programs have been closed, probably owing to the remarkable success of the competitor angiotensin II antagonists. A convergent synthesis of the highly potent human renin inhibitor SPP-100 **(190)**, selected for clinical investigation, was performed using a nitrone intermediate[453]. Non-peptidomimetic renin inhibitors of the piperidine type represent a novel structural class of compounds potentially free of the drawbacks seen with peptidomimetic compounds so far[454]. Synthetic optimisation in two structural series focusing on improvement of potency and metabolic stability, has led to the identification of two candidate compounds

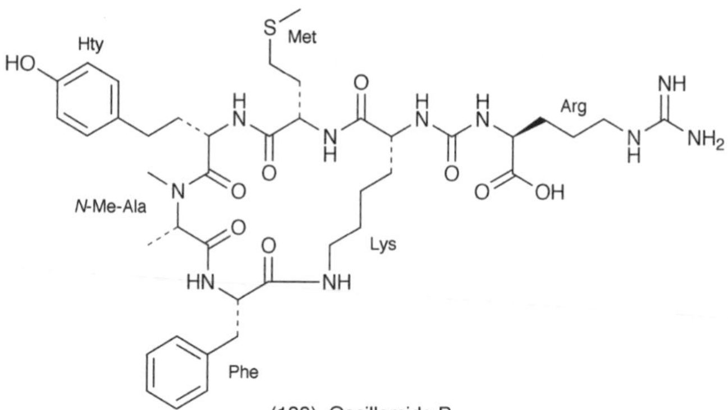

(188) Oscillamide B

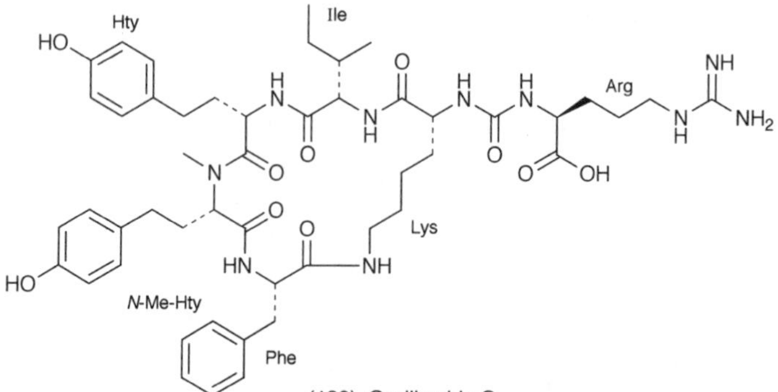

(189) Oscillamide C

(191) and **(192)**. *In vivo* experiments suggest that treatment of chronic renal failure patients with a renin inhibitor might result in a significant improvement of the disease status.

3-Alkoxy-4-arylpiperidine inhibitors of aspartyl proteinases have been designed by use of a structure generating program but only after the enzyme active site conformation was modified in a mechanistically related fashion[455]. New enantioselective syntheses of 3-alkoxy-4-arylpiperidine analogs were described. Cross-linked polystyrene has been used as polymer matrix for simulating aspartyl proteases: three salicylate residues were coupled in close proximity to the polystyrene matrix[456]. The immobilized carboxyl groups can act as the active site of Asp-proteases and effectively hydrolysed albumin into many small fragments. The artificial protease manifested optimum activity at pH 3 the same as Asp-proteases.

Using the known 1.8 Å X-ray crystal structure of pepstatin bound to *Rhizopus* pepsin as the starting point, Asp-protease inhibitors were designed with the structure-generating program GrowMol[457]. The program generated over 20 000 structures. Some of them were selected and synthesized (*e.g.* **193**). Cocrystallization of **(193)** with *Rhizopus* pepsin resulted in a conformation of inhibitor closely related to that predicted by GrowMol. Unsymmetrical ureas form a new class of Asp-protease inhibitors[458]. The design, synthesis, and enzyme inhibition

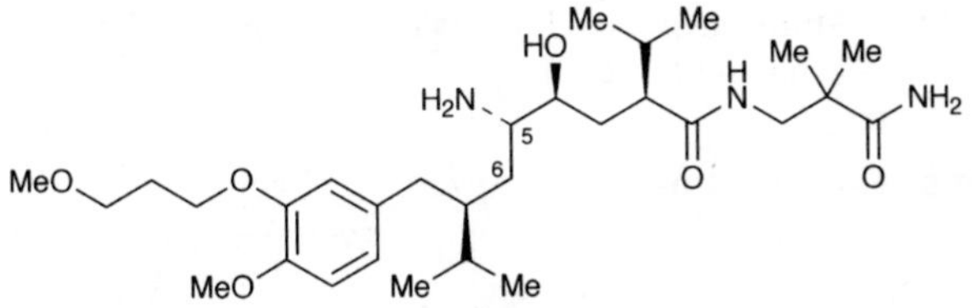

(190)

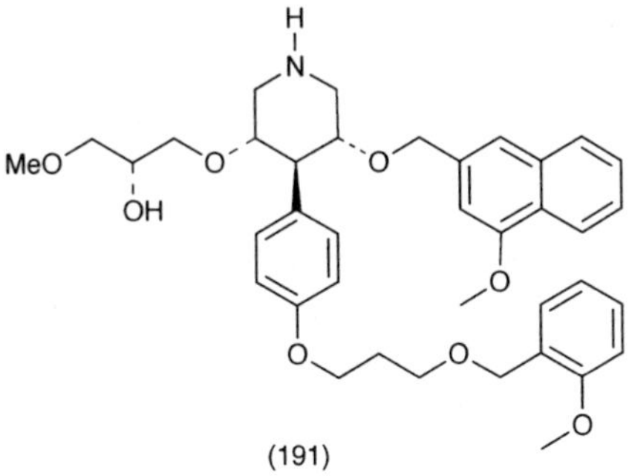

(191)

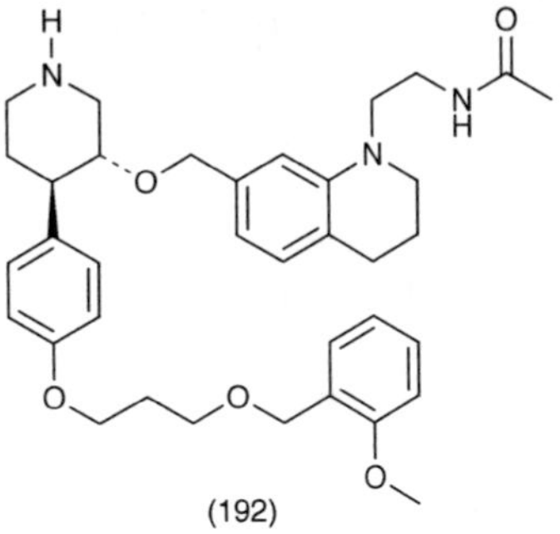

(192)

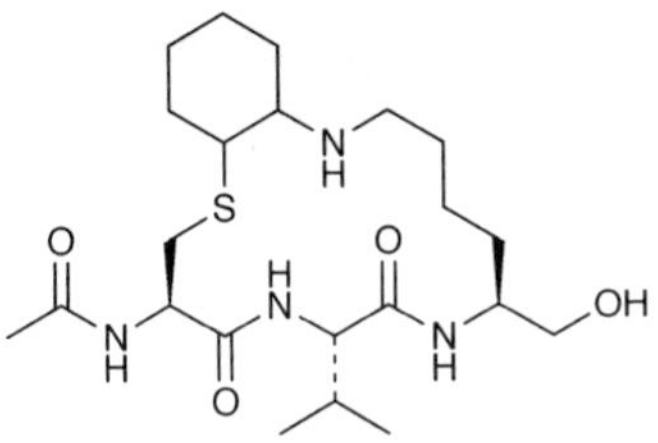

(193)

of these compounds was described. Design was performed using the *de novo* generation of novel structures by GrowMol. The best inhibitors possess a K_i value of 0.3-1 nM for porcine pepsin. 2-Substituted statins as one of the most common transition-state isosteres utilized in Asp-protease inhibitors have been synthesized in a stereocontrolled manner[459]. Peptides containing 2-substituted statins inhibit porcine pepsin with nanomolar IC_{50} values.

Generation of Aβ-peptides in Alzheimer's disease (AD) occurs by proteolysis of amyloid precursor protein (APP) with β- and γ-secretase. Inhibitors against these two enzymes are target compounds of drug development for treating AD at the early stage. Several peptide inhibitors have been developed containing a hydroxyethylene residue against memapsin 2 (β-secretase), the most potent inhibitor possesses a K_i of 3.1 x 10^{-10} M^{460}. Peptide aldehyde MG132, which inhibit proteases and proteasomes, prevented β-secretase cleavage[461] of APP, probably due to a block in APP maturation. Potent β-secretase inhibitors with hydrophobic structure were designed and synthesized[462].

Aspartyl protease pepstatin binds to the presenilins that may play a pivotal role in the pathology of Alzheimer's disease[463]. A novel γ-secretase inhibitor of a very hydrophobic structure has been synthesized and evaluated[464]. A novel γ-secretase assay based on detection of the C-terminal fragment-gamma (CTF-gamma) of 57-59 amino acid residues has been described[465]. Known γ-secretase inhibitors reduced the yield of CTF-gamma.

5.16 Thrombin and Factor-Xa Inhibitors. – Thrombosis is a major cause of mortality in the industrialized world, therefore the control of blood coagulation has become a major target for new therapeutic agents. A set of computer-aided molecular design strategies was reported for developing novel ligands for the study of thrombin inhibitor study[466]. It includes docking simulation by DOCK 3.5 and use of the FRAGMENT + + program. Bicyclic piperazinone-based selective thrombin inhibitors of general structure **(194)** were prepared and evaluated *in vitro* and in *vivo*[467]. These inhibitors, having in common, an electrophilic basic trans-cyclohexylamine P1 residue, diplayed high thrombin affinity, high selectivity against trypsin and good *in vivo* efficacy in the rat arterial thrombosis model. The lead argatobran **(195)** is a known potent competitive inhibitor of thrombin with an IC_{50} of 9 nM and has clinical potential in maintenance anticoagulation therapy[468]. Argatobran was synthesized in seven steps from 4-methylpiperidine. The condensation of (±)*trans*-benzyl 4-methyl-pipecolic acid ester with N^{α}-Boc-N^{ω}-nitro-L-arginine led to two diastereomers. One of them is the precursor of argatobran.

Seven non-natural analogues of Arg and Lys have been substituted in an established Arg-based thrombin inhibitor **(196)**. Four of the new compounds exhibited significant thrombin inhibition with a K_i between 0.53-3.95 μM and were subsequently tested for selectivity against trypsin. The best compound **(197)** gave a selectivity ratio of 962 (trypsin/thrombin), improving upon the parent compound[469].

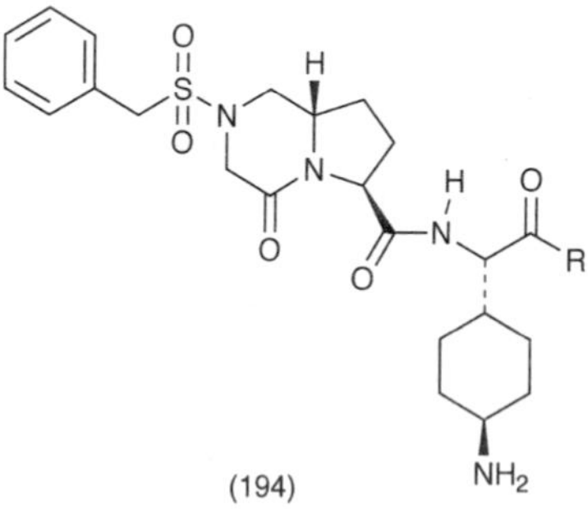

(194)

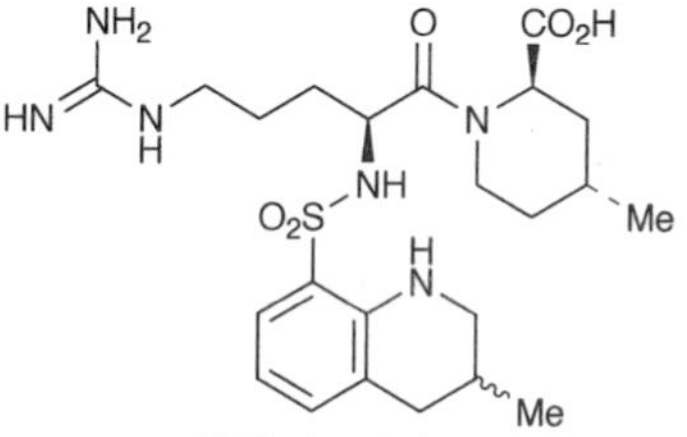

(195) Argatroban

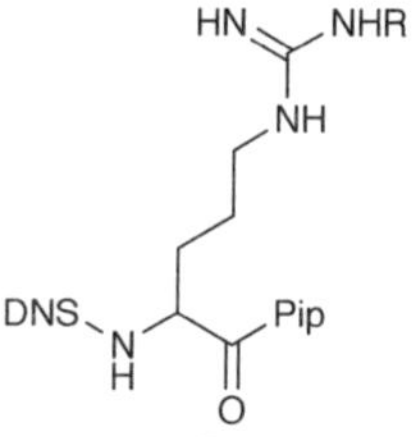

(196) R = H
(197) R = Me

An azapeptide scaffold was incorporated into the central part of the classical tripeptide D-Phe-Pro-Arg inhibitor structure, thus eliminating one stereogenic center from the molecule[470]. A series of compounds **(198)** has been designed and synthesized to optimise occupancy of the S2 pocket of thrombin. Increased hydrophobicity at P2 provides an enhanced fit into this active site S2 pocket. The solution structure and conformational analysis of the inhibitors were also reported.

A simple and economical method was developed for the preparation of biotin derivatives of the thrombin specific inhibitor D-Phe-Pro-Arg-CH$_2$Cl[471]. This chloromethyl ketone has ideal properties for specific labeling of the catalytic sites of thrombin and other Ser-proteases. Specific, weakly basic thrombin inhibitors were designed and synthesized[472] incorporating sulfonyl dicyandiamide moieties in their structure. The best compound, tosyl-D-Phe-Pro-sulfanilyl-dicyandiamide showed an inhibition constant of 9 nM against thrombin.

Factor Xa (FXa) plays a critical role in the coagulation cascade and in the regulation of intravascular thrombus development. FXa has become a target for inhibition as a strategy for the invention of novel antithrombotic agents. A new series of potent and selective inhibitors of this enzyme was designed and synthesized[473]. Inhibitors have a phenylglycine containing benzamidine carboxamide structure **(199)**. The best compounds showed an inhibition constant of 13-15 nM against FXa. A review summarizes the FXa inhibitory drug development process from a lead containing isoxazoline ring up to DPC 423 **(200)**, a drug selected for clinical evaluation[474, 475]. In another review the discovery process of the FXa-inhibitor 2K807834 has been summarized[476].

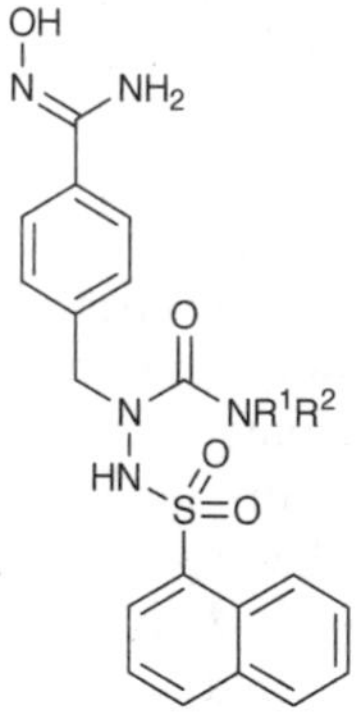

(198) 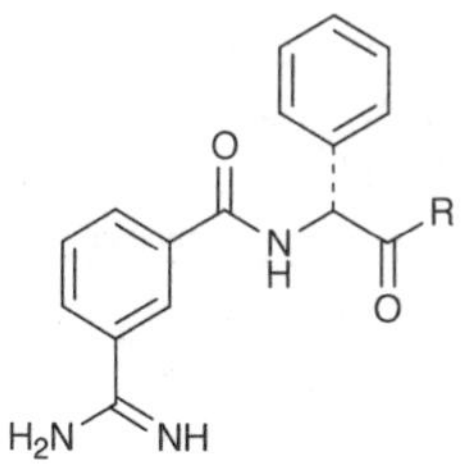   (199) 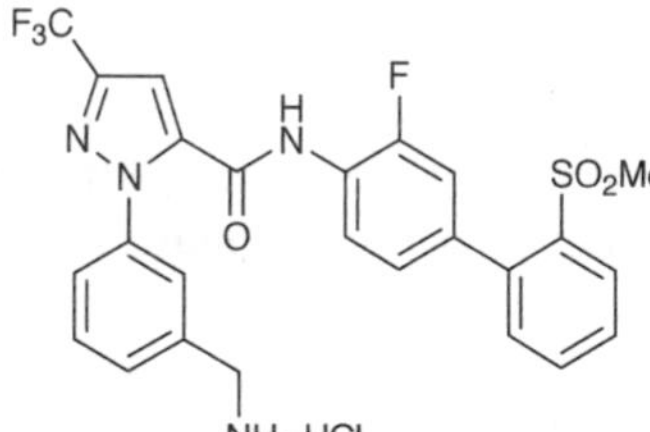   (200) DPC 423

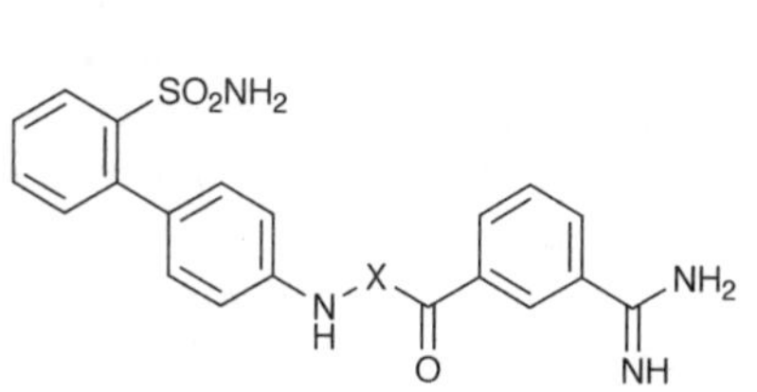

(201) 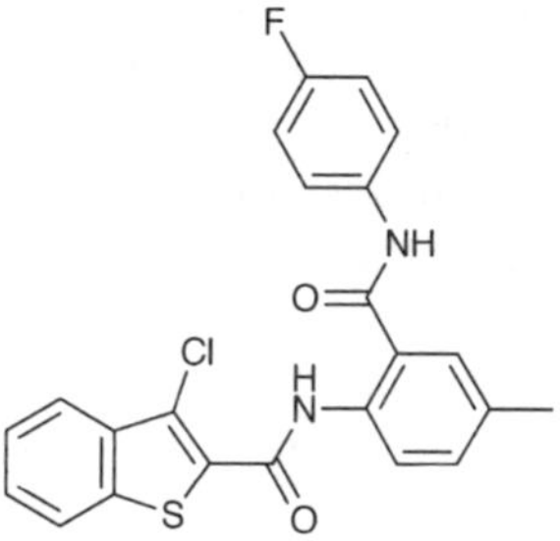   (202)

A series of potent and selective FXa inhibitors, based on the general formula
(201), was synthesized using various readily available amino acids as central
templates[477]. The most potent compound displays an IC_{50} of 3 nM. 1-Arylsul-
fonyl-3-piperazinone derivatives such as compound M55113 (4-[6-chloro-2-
naphthalenyl)sulfonyl]-1-[[1-(4-pyridinyl)-4-piperidinyl]methyl]piperazinone)
proved to be potent FXa inhibitors (IC_{50} = 0.06 μM) with high selectivity for
FXa over trypsin and thrombin[478]. A neutral inhibitor of FXa was identified *via* a
high-throughput screen of a commercial library[479]. The initial lead **(202)** showed
reversible and competitive inhibition kinetics for FXa and possessed a high
degree of selectivity versus other related Ser-proteases. A series of analogues was
synthesized and evaluated.

5.17 Trypsin and Other Serine Protease Inhibitors. – Isothermal titration
calorimetry and protein crystallography have been used for binding studies of
low molecular mass ligands towards trypsin and thrombin[480]. New synthetic
peptide analogues bearing a C-terminal basic α-keto-β-aldehyde moiety have
been prepared as novel inhibitors of the trypsin-like Ser-proteases[481]. Three
compounds (Ac-Leu-Leu-Arg-COCHO, Ac-Arg-Glu-Arg-COCHO and Boc-
Val-Leu-Cys-COCHO) were evaluated kinetically against trypsin and 3 other
enzymes (tryptase, plasmin and thrombin). Results show that α-keto-β-al-
dehydes are potent inhibitors, with similar potency to comparable peptide
aldehydes and provide a useful new template for the development of new
therapeutic agents. The synthesis of a 14-amino acid long cyclopeptide called
sunflower trypsin inhibitior (SFT-1) has been reported[482]. The cyclopeptide has a
bicyclic structure with Cys^3 and Cys^{11} bound with a disulfide bridge. SFT-1 is a
very potent inhibitor, with a K_i of 0.92 nM of the recently identified epithelial
type-II transmembrane serine protease, termed 'matripase'.

The design, synthesis, photoisomerism and biological testing of two peptide-
based photoswitchable inhibitors of α-chymotrypsin have been described[483].
Complex chromatographic methods were used for isolation of a 7.5 kDa peptide
from *Vicia faba* (broad beans), which proved to be a trypsin-chymotrypsin
inhibitor of the Bowman-Birk-type[484]. The peptide possesses antifungal activity.

Peptidyl phosphonate diphenyl ester type irreversible inhibitors for serine
proteases have been synthesized from amide compounds, triphenylphosphite
and phenylacetoaldehyde[485]. New 4-amino-isothiazolidinone oxide derivatives,
designed as bacterial serine protease inhibitors, were prepared using a
stereoselective pathway[486]. Crystal structures of the Pseudomonas serine-car-
boxyl proteinase (PSCP) complexed with a number of inhibitors have been
solved and refined at high-to atomic-level resolution[487]. Inhibitors form covalent
bonds to the active site Ser287 through their aldehyde moieties, while their side
chains occupy subsites S1-S4 of the enzyme. Amidine-type inhibitors, highly
selective for the S1 sites of Ser190 trypsin-like Ser-proteases were selected as drug
targets for urokinase and trypsin[488]. Recently Kunitz-type serine protease inhibitors were isolated from Bauhinia seeds, and more recently the interaction of
human plasma kallikrein with fluorogenic and non-fluorogenic peptides based
on Bauhinia inhibitors' reactive site has been described[489]. A novel series of active

inhibitors against hepatitis C virus (HCV)-NS3 serine protease have been designed, based on the structure of known Zn^{2+} dependent bisbenzymidazole-type inhibitor APC-6336[490].

Therapeutic strategies based on Ser-protease inhibition might be useful in preventing neuronal cell death. It was shown that inhibitors of trypsin-like Ser-proteases (*e.g.* N-tosyl-L-Lysine chloromethylketone, TPCK) prevent DNA-damage induced neuronal death by acting upstream of the mitochondrial checkpoint and of p53 induction[491]. The inhibitors do not act at the level of caspases but prevent apoptosis. Other experiments show dramatic contrast to these results: chymotrypsin-like serine proteases (CSP) may serve an endogeneous neuroprotective role possibly by modulating necrotic cell death[492]. In the latter experiments the inhibitor TPCK was cytotoxic in all cell cultures tested, in 10 to 100 μM concentration.

5.18 tRNA Synthetase Inhibitors. – The increasing emergence of pathogenic bacteria resistant to conventional antibiotics requires the development of novel agents against bacterial targets such as tRNA synthetases. The structure of the natural product SB-219383, a potent and selective inhibitor of bacterial tyrosyl tRNA synthetase (YRS) was the basis for the design of novel YRS-inhibitors[493, 494]. Novel pyranosyl analogues of SB-219383 have been synthesized and highly potent stereoselective and bacterioselective inhibitors of YRS have been identified.

Methionyl- and isoleucyl- tRNA are inhibited with ester and hydroxamate analogues of methionyl and isoleucyl adenylate[495]. The ester analogue **(203)** was found to be a potent inhibitor of E. coli Met-tRNA synthetase. Following this series of compounds, another series of methionyl and isoleucyl phenolic analogues containing bioisosteric linkers mimicking ribose have been investigated[496]. Isoleucyl isovanilloid containing ribose bioisostere **(204)** was found to be a potent inhibitor of Ile-tRNA synthase. A molecular modelling study demonstrated that in **(204)**, isovanillate and hydroxamate served as proper surrogates for adenine and ribose in isoleucyl adenylate.

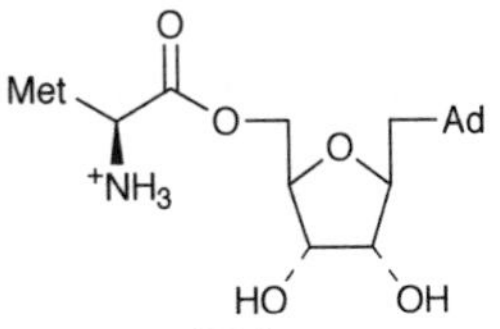

(203) 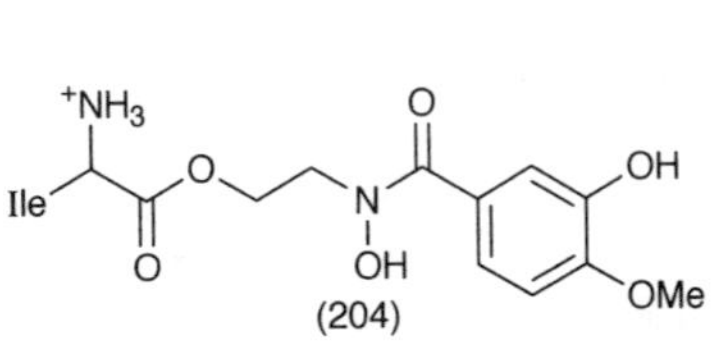   (204)

5.19 Miscellaneous. – Topoisomerase II is a ubiquitous nuclear enzyme adjusting and stabilizing DNA during replication, transcription, and recombination. The synthesis and biological activity of amino acid functionalized β-carboline derivatives, which are structurally related to the established topoisomerase inhibitors azatoxin and tryprostatins, were reported[497]. New *seco*-cyclothialidine derivatives containing a dioxazine moiety (*e.g.* **205** RO-61-6653) were synthesized through a new concise pathway[498]. Despite promising *in vitro* antibacterial activity, only poor activity was found in the neurine infection model.

Chymase possesses a wide variety of actions, including promotion of angiotensin II production and histamine release from mast cells. Human chymase also has exoproteolytic activity. However, due to a lack of effective inhibitors the pathophysiological role of chymase has not yet been fully elucidated. A series of non-peptidic 5-amino-6-oxo-1,6-dihydropyrimidine-containing trifluoromethyl ketones has been designed as putative chymase inhibitors[499]. The most potent compound has a K_i of 50 nM. From the same laboratory the synthesis and structure-activity relationship studies of nonpeptidic difluoromethyl ketones as novel inhibitors of human chymase have been described[500], the most potent compound had a K_i of 2.26 nM. The total synthesis of the natural human chymase inhibitor methyllinderone has been achieved in only four steps[501]. 2-*sec*-Amino-4H-3,1-benzoxazine-4-ones have been evaluated as acyl-enzyme inhibitors of human recombinant chymase[502].

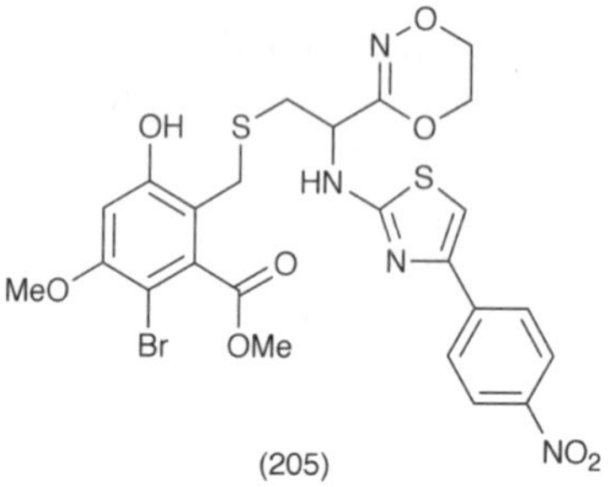

(205) 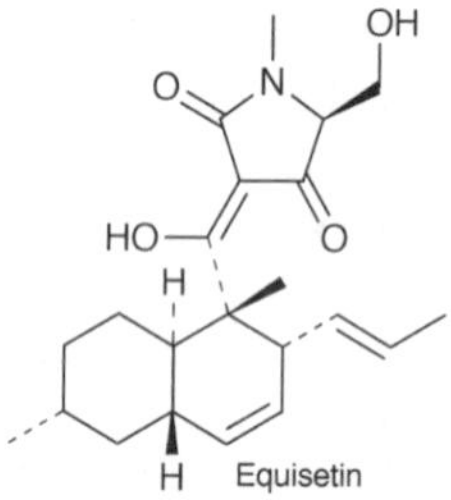   (206) Equisetin

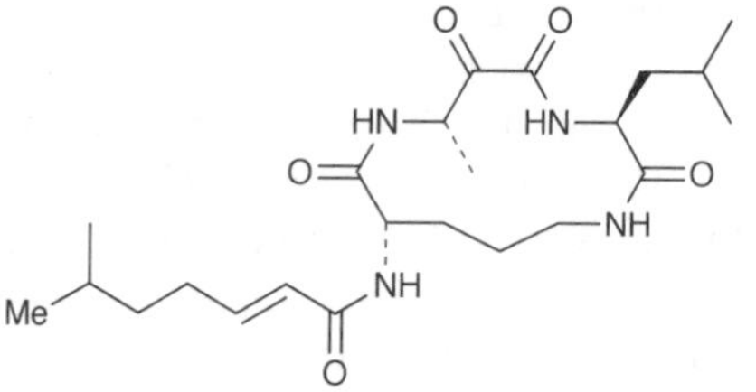

(207) (eurystatin A)

Solid phase synthesis of peptidomimetic inhibitors for the hepatitis C virus NS3-protease has been described[503, 490]. A short stereoselective synthesis was reported for the fusarium toxin equisetin (206), a potent inhibitor of HIV-1 integrase[504].

The results of irreversible inhibition of guinea pig liver transglutaminase by a series of 24 novel dipeptides containing either an α, β-unsaturated amide or an epoxide functional group have been reported[505]. Phosphinic acid pseudopeptide analogues to glutamyl-γ-glutamate have been described, the synthesis and coupling to pteroyl azides leading to potent inhibitors of folyl poly-γ-glutamate synthetase[506].

Ribonucleotide reductase (RR) catalyses the reduction of ribonucleotides to 2'-deoxyribonucleotides, the rate limiting step of DNA biosynthesis. Mammalian RR is a chemotherapeutic target, so a structural model of the enzyme was constructed from the structure of *E.coli* RR enzyme[507].

Telomerase inhibitors are expected to be new candidates as therapeutic agents for cancer. Solid phase synthesis was applied to construct a library of inhibitors having aromatic phosphate, long alkyl chain and tryptophan components for the identification of telomerase inhibitors[508]. From this library a D, D-ditryptophan derivative has been identified as a new potent inhibitor with IC_{50} values of 0.3 μM for telomerase.

A concise total synthesis of the prolyl – endopeptidase inhibitor eurystatin A (**207**) was achieved *via* a novel Passerini reaction – deprotection – acyl migration strategy[509].

Several Cys-dipeptides have been synthesized and shown to be reversible competitive inhibitors of the *Bacillus cereus* metallo-β-lactamase. The thiol anion of the inhibitor displaces hydroxide ion from the active site zinc(II). D, D-Peptides bind to the enzyme better than other diastereoisomers, which is compatible with the predicted stereochemistry of the active site[510].

6 Phage Library Leads

The combinatorial approach offered by phage display has proved to be powerful in obtaining novel variants of canonical inhibitors of Ser-proteinases that show new binding patterns. This strategy was used to search for variants of basic pancreatic trypsin inhibitor (BPTI) that would be strong inhibitors of two enzymes: bovine α-chymotrypsin and porcine pancreatic elastase[511]. A representative library of 3.2×10^4 BPTI variants, randomised at P1, P1′, P2′ and P3′ positions of the proteinase binding loop, was displayed on the surface of phage M13. After four to five rounds of the selection of the target proteinase, consensus sequences of the inhibitor binding loop were obtained. At the final stage a new potent inhibitor of the target enzymes was found with association constants up to $1.9 \times 10^{-9}/_M$ and $3.7 \times 10^{-10}/_M$ for elastase and chymotrypsin, respectively. Thus, the inhibitory properties of BPTI were improved by 7×10^6-fold towards elastase and 420-fold towards chymotrypsin.

Random peptide phage display libraries have been employed widely to identify protein-protein interactions, using as targets either purified proteins, intact cells, or organs. Thus neuroblastoma tumor cell binding peptides were identified through a random peptide phage display[512]. A phage display peptide library was used for targeting plasmodium ligands on mosquito salivary glands and midgut[513]. Highly structured peptide antagonists of the interaction of insulin-like growth factor 1 (IGF-1) and IGF binding protein 1 (IGFB-1) have recently been discovered by phage display of peptide libraries. Detailed analysis of this turn-helix peptide necessary for IGFBP-1 binding proved the importance of hydrophobic patch on one face of the helix[514]. Ala-scanning substitutions confirmed that the hydrophobic residues are necessary for binding. A semi-synthetic library of intrinsically stable antibody fragments derived from a single-framework scaffold was composed[515]. The library allows the isolation of new stable binding specificites. Epitope mapping for four monoclonal antibodies against human plasminogen activator inhibitor type-1 has been performed[516]. A DNA-binding peptide with the amino acid sequence of SVSVGMKPSPRP was selected from a random peptide phage display library[517], and peptides possessing aldolase activity were selected from a phage library using 1,3-diketones designed for the covalent selection of an enamine-based reaction mechanism[518]. A cyclic peptide with the KCHFEECLAY sequence was selected as an α-glucosidase inhibitor by working with phage display[519]. The cyclodecapeptide competitively inhibited *Aspergillus niger* glucoamylase (K_i = 0.2 nM) and rat intestinal α-glucosidase. An attempt has been made for phage-display identification of linear peptides that have high binding specificity to tumor cells[520]. Peptides isolated to date do not bind with high retention to sites *in vivo*. An improved selection procedure for the screening of phage display peptide libraries has been described[521]. Combined use of polystyrene beads and a stepwise decrease in the pH of the elution buffer in the final round of biopanning resulted in the elimination of non-binding clones and an increase in the efficiency in isolating high affinity binding clones. A new method called 'mRNA display', an *in vitro* selection technique to identify peptide aptamers to a protein target, has been reported[522]. mRNA display allows for the preparation of polypeptide libraries with greater complexity than is possible with phage display.

7   Protein-Protein Interaction Inhibitors

7.1   SH2 and SH3 Domain Ligands. – Within the last few years, therapeutic intervention into signal transduction cascades have attracted the interest of pharmaceutical research. A review summarizes the SH2 domain-targeted drug design approach, including new peptidomimetic and non-peptide SH2 domain antagonists for targeting signal transduction[523]. The Scr family of Tyr-kinases play an important role in regulating intracellular signalling. Detailed analysis of Src SH2 binding by peptides containing a novel tricarbonyl-modified phosphotyrosine (pTyr) moiety has been described[524]. Selective inhibition of the SH2 domain of the enzyme by novel thiol-targeting inhibitors was performed. Peptidyl as well as nonpeptidyl inhibitors possessing a 4-α,β-diketoester-modified pTyr mimic, exhibited micromolar affinity to Src SH2. SH2 competitive binding assays serve to identify and characterize SH2 ligands. Two conceptually different biochemical methods were designed to discover Src-SH2 inhibitors: scintillation proximity assay (SPA) and surface plasmon resonance (SPR) have been described[525].

Grb2 SH2 domains, which have been associated with breast cancer, are among the important targets for development of antiproliferative agents. N-Terminal carboxyl- and tetrazole-containing amides such as **(208)** were designed as adjuvants to Grb2 SH2 domain ligand binding[526]. Macrocyclization of a linear Grb2 SH2 domain antagonist resulted in a conformationally constrained compound which showed a 100-fold enhancement to its linear counterpart in extracellular Grb2 SH2 domain binding assay[527]. A diphosphonomethyl phenylalanine-containing Src SH2 inhibitor **(209)** was synthesized and when evaluated has an IC_{50} of 0.35 μM[528].

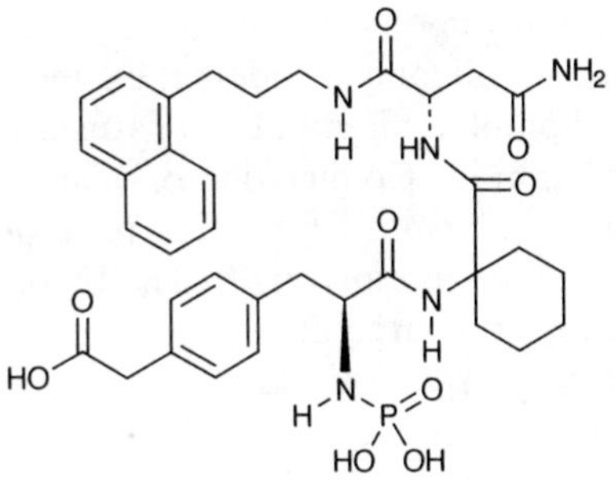

(208) 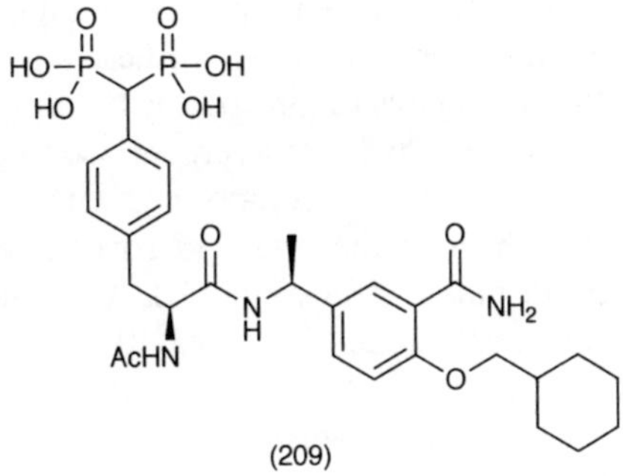   (209)

8   Advances in Formulation/Delivery Technology

Lipoamino acid glycoconjugates have been synthesized for peptide and drug delivery[529]. Lipoamino acid and liposacharide derivatives of the tumor-selective somatostatin analogue TT-232 were also synthesized to modify the physicochemical properties of the parent peptide[530]. Several new lipophilic α-aminophosphonates including macrocyclic (on calyx[4]arene platform) and chiral forms were synthesized by the Kabachnik-Fields reaction[531]. Transport measurement showed that these compounds exhibited remarkable efficiency and selectivity of the transport of hydrophilic substances such as amino acids and amino alcohols across lipophilic membranes. Calix[4]-arene derivatives proved to be new carriers for the membrane transport of α-hydroxy and α-amino acids[532]. The ability of lipopeptides to passively cross the cell membrane opens new opportunities for the intracellular delivery of bioactive peptides. A model peptide, the pseudosubstrate sequence of protein kinease C-ζ(PKC-ζ) was linked to the pentapeptide vector Gly-Arg-Gly-Arg-Lys(Pam)-NH2 through a thiazolidine linkage, and the conjugate was translocated into the cells[533]. Novel analogues of the cell-penetrating peptides transportan and penetratine have been synthesized[534]. Fluorescence studies show that penetratin and transportan do not enter the cell by related mechanisms. The synthesis and cell translocation of a new peptide derived from the third α-helix homeo-domain of Antennapedia has been described[535]. Short oligomers of Arg function as efficient molecular transporter of drugs, so a scalable solution-phase synthesis of octaarginine (Arg)8 has been developed[536] for large-scale synthesis of this peptide vector. Transport characteristics of peptides and peptidomimetics have been studied using N-methylated[537] and hydroxy-ethylamine bioisostere peptidomimetics[538] as substrates for the oligopeptide transporter and P-glycoprotein in the intestinal mucosa. Peptide molecular transporters have been designed, synthesized and evaluated[539]. In the basic domain of TAT (TAT 49-57, RKKRRQRRR) the cationic residues play a principal role in cellular uptake. The guanidinium groups of TAT play a greater role in facilitating cellular uptake than either charge or backbone structure. Based on these results, a class of polyguanidine peptide derivatives were designed and synthesized and a new peptoid transporter has been developed that is superior to TAT 49-57, protease resistant and easily and economically prepared. Cellular uptake and tumor-targeting of oligonuc-leotides was reached by synthesizing somatostatin-oligonucleotide conjugates[540], the conjugates retain specific binding with nanomolar affinities to somatostatin receptors which are overexpressed in various tumors. Conjugation of different drugs and model compounds with L-glutamic acid can improve their penetra-tion through the brain microvessel endothelial cell monolayers[541]. The blood-brain barrier transport of L-Tyr conjugates, by the large neutral amino acid transport carrier, has been studied[542]. Graft polymers were designed and syn-thesized for gene delivery, with the pH-sensitive cationic polymer containing N-acetyl-poly-L-histidine and poly-L-lysine[543]. Model prodrugs have been ap-plied to the study of the human intestinal di/tripeptide carrier, hPep T1[544]. Carbohydrate-based templates were synthesized for drug delivery and prepara-tion of synthetic vaccines[545]. A new macromolecular prodrug, denoted T-0128, containing the anti-cancer agent camptothecin (CPT) bound to dextran through a Gly-Gly-Gly- or similar tripeptide linker was prepared giving a molecular mass of 130 kDa. The triglycine linker exploits lysosomal cathepsin B to liberate CPT slowly and steadily, resulting in improved therapeutic efficacy[546]. Conjuga-tion of camptothecin to poly-(L-Glu) with a molecular mass between 37 and 50 kDa, display enhanced efficacy compared with nonconjugated camptothecins[547]. The design, synthesis, characterization and self-assembling properties of a new class of amphiphilic peptides, constructed from a bifunctional polar core at-tached to totally hydrophobic arms, have been described[548]. Longer peptides form a continuous channel-like structure. The channel is totally hydrophobic in the interior and can selectively encapsulate lipophilic stubstances.

References

1.   J. S. Davies in *Specialist Periodical Reports, Amino Acids, Peptides and Proteins*, 2001, **33**, 135.
2   C. A. Selects on *Amino Acids, Peptides and Proteins*, published by the American Chemical Society and Chemical Abstract Service, Colombus, Ohio, 2001.
3   *Web of Science Service* on http://www.ncbi.nml.nih.gov/entrez/query.fcgi
4   *Web of Science Service* on http://isinet.com/isi/products/citation/wos
5   H. Dyker, J. Scherkenbeck, D. Gondol, A. Goehrt, A. Harder, *J. Org. Chem.*, 2001, **66**, 3760.
6   D. Cabaret, M. G. Gonzalez, M. Wakselman, S. A. Adediran, R. F. Pratt, *European J. Organic Chemistry*, 2001, **1**, 141.
7   A. Cheguillaume, A. Salaunen, S. Sinbandhit, M. Potel, P. Gall, M. Baudy-Floc'h, P. Le Grel, *J. Org. Chem.*, 2001, **66**, 4923.
8   A. Plant, F. Stieber, J. Scherkenbeck, P. Losel, H. Dyker, *Organic Letters*, 2001, **3**, 3427.
9   S. Lucarini, C. Tomasini, *J. Org. Chem.*, 2001, **66**, 727.
10. Y. K. Ramtohul, N. I. Martin, L. Silkin, M. N. G. James, J. C. Vederas, *Chemical Communications (Cambridge, UK)*, 2001, 2740.
11. E. Lodyga-Chruscinska, G. Micera, D. Sanna, J. Olczak, J. Zabrocki, *Polyhedron*, 2001, **20**, 1915.

12. B. C. H. May, A. D. Abell, *Tetrahedron Lett.*, 2001, **42**, 5641.
13. A. S. Kende, H.-Q. Dong, A. W. Mazur, F. H. Ebetino, *Tetrahedron Lett.*, 2001, **42**, 6015.
14. A. Otaka, H. Watanabe, A. Yukimasa, S. Oishi, H. Tamamura, N. Fujii, *Tetrahedron Lett.*, 2001, **42**, 5443.
15. A. Volonterio, P. Bravo, M. Zanda, *Tetrahedron Lett.*, 2001, **42**, 3141.
16. M. Demarcus, M. L. Ganadu, G. M. Mura, A. Porcheddu, L. Quaranta, G, Reginato, M. Taddei, *J. Org. Chem.*, 2001, **66**, 697.
17. R. Gunther, H. J. Hofmann, *J. Am. Chem. Soc.*, 2001, **123**, 247.
18. T. T. Tran, H. Treutlein, A. W. Burgess, *J. Comput. Chem.*, 2001, **22**, 1010.
19. T. T. Tran, H. Treutlein, A. W. Burgess, *J. Comput. Chem.*, 2001, **22**, 1026.
20. S. F. Barker, D. Angus, C. Taillefumier, M. R. Probert, D. J. Watkin, M. P. Watterson, T. D. W. Claridge, N. L. Hungerford, G. W. J. Fleet, *Tetrahedron Lett.*, 2001, **42**, 4247.
21. T. D. W. Claridge, J. M. Goodman, A. Moreno, D. Augus, S. F. Barker, C. Taillefumier, M. P. Watterson, G. W. J. Fleet, *Tetrahedron Lett.*, 2001, **42**, 4251.
22. G. Fridkin, T. Gilon, A. Loyter, C. Gilon, *J. Pept. Res.*, 2001, **58**, 36.
23. J.-M. Hah, L. J. Roman, P. Martasek, R. B. Silverman, *J. Med. Chem.*, 2001, **44**, 2667.
24. M. C. Hillier, J. P. Davidson, S. F. Martin, *J. Org. Chem.*, 2001, **66**, 1657.
25. L. Demange, C. Dugave, *Tetrahedron Lett.*, 2001, **42**, 6295.
26. V. J. Hruby, W. Qiu, T. Okayama, V. A. Soloshonok, *Methods in Enzymology*, 2001, **343**, 91.
27. A. Golebiowski, S. R. Klopfenstein, D. E. Portlock, *Curr. Opin. Drug Discovery Dev.*, 2001, **4**, 428.
28. J. W. Lane, R. L. Halcomb, *Tetrahedron*, 2001, **57**, 6531.
29. A. J. Moreno-Vargas, J. G. Fernández-Bolaños, J. Fuentes, I. Robina, *Tetrahedron Lett.*, 2001, **42**, 1283.
30. C. Alvarez-Ibarra, A. G. Csaky, C. Gomez de la Oliva, E. Rodriguez, *Tetrahedron Lett.*, 2001, **42**, 2129.
31. L. Saniere, M. Schmitt, N. Pellegrini, J. J. Bourguignon, *Heterocycles*, 2001, **55**, 671.
32. A. J. Souers, J. A. Ellman, *Tetrahedron*, 2001, **57**, 7431.
33. K. Burgess, *Accounts of Chemical Research*, 2001, **34**, 826.
34. M. MacDonald, J. Aube, *Current Organic Chemistry*, 2001, **5**, 417.
35. E. Alonzo, F. Lopez-Ortiz, C. del Pozo, E. Peralta, A. Marcias, J. Gonzales, *J. Org. Chem.*, 2001, **66**, 6333.
36. M. Eguchi, M. S. Lee, M. Stasiak, M. Kahn, *Tetrahedron Lett.*, 2001, **42**, 1237.
37. W. D. Kohn, L. Zhang, *Tetrahedron Lett.*, 2001, **42**, 4453.
38. W. M. De Borggraeve, F. J. R. Rombouts, E. V. Van der Eycken, S. M. Toppet, G. J. Hoornaert, *Tetrahedron Lett.*, 2001, **42**, 5693.
39. S. Ro, H.-J. Lee, I.-A. Ahn, D. K. Shin, K.-B. Lee, C. J. Yoon, Y.-S. Choi, *Bioorg. Med. Chem.*, 2001, **9**, 1837.
40. H.-J. Lee, K.-H. Choi, I.-A. Ahn, S. Ro, H. G. Jang, Y.-S. Choi, K.-B. Lee, *J. Mol. Structure*, 2001, **569**, 43.
41. C. Hemmerlin, M. T. Cung, G. Boussard, *Tetrahedron Lett.*, 2001, **42**, 5009.
42. W. Wang, C. Y. Xiong, V. J. Hruby, *Tetrahedron Lett.*, 2001, **42**, 3159.
43. E. Bourguet, J. L. Baneres, J. P. Girard, J. Parello, J. P. Vidal, X. Lusinchi, J. P. Declercq, *Organic Letters*, 2001, **3**, 3067.
44. T. K. Chakraborty, A. Ghosh, R. Nagaraj, A. Ravi Sankar, A. C. Kunwar, *Tetrahedron*, 2001, **57**, 9169.

45.   F. M. Cordero, S. Valenza, F. Machetti, A. Brandi, *Chem. Commun.*, 2001, **17**, 1590.
46.   R. C. F. Jones, J. Dickson, *J. Pept. Sci*, 2001, **7**, 220.
47.   W. Qui, X. Y. Gu, V. A. Soloshonok, M. D. Carducci, V. J. Hruby, *Tetrahedron Lett.*, 2001, **42**, 145.
48.   L. Belsivi, L. Colombo, M. Colombo, M. Di Giacomo, L. Manzoni, B. Vodopivec, C. Scolastico, *Tetrahedron*, 2001, **57**, 6463.
49.   D. Scarpi, E. G. Occhiato, A. Trabocchi, R. J. Leatherbarrow, A. B. E. Brauer, M. Nievo, A. Guarna, *Bioorg. Med. Chem.*, 2001, **9**, 1625.
50.   A. Sidduri, J. P. Lou, R. Campbell, K. Rowan, J. W. Jefferson, *Tetrahedron Lett.*, 2001, **42**, 8757.
51.   V. A. Soloshonok, X. Tang, V. J. Hruby, *Tetrahedron*, 2001, **57**, 6375.
52.   S. Q. Duan, K. D. Moeller, *Tetrahedron*, 2001, **57**, 6407.
53.   M. del Carmen Teran Moldes, G. Costantino, M. Marinozzi, R. Pellicciari, *Farmaco*, 2001, **56**, 609.
54.   P. D. Bailey, N. Bannister, M. Bernad, S. Blanchard, A. N. Boa, *J. Chem. Soc., Perkin Trans. 1*, 2001, 3245.
55.   Z. Feng, W. D. Lubell, *J. Org. Chem.*, 2001, **66**, 1181.
56.   F. Gosselin, D. Tourwe, M. Ceusters, T. Meert, L. Heylen, M. Jurzak, W. D. Lubell, *J. Pept. Res.*, 2001, **57**, 337.
57.   M. A. Estiarte, A. Diez, M. Rubiralta, R. F. W. Jackson, *Tetrahedron*, 2001, **57**, 157.
58.   S. Maricic, A. Ritzen, U. Berg, T. Frejd, *Tetrahedron*, 2001, **57**, 6523.
59.   M. Crisma, A. Moretto, C. Toniolo, K. Kaczmarek, J. Zabrocki, *Macromolecules*, 2001, **34**, 5048.
60.   X. Zhang, W. Jiang, A. C. Schmitt, *Tetrahedron Lett.*, 2001, **42**, 4943.
61.   J. Sakowski, M. Boehm, I. Sattler, H.-M. Dahse, M. Schlitzer, *J. Med. Chem.*, 2001, **44**, 2886.
62.   E. Lattmann, D. C. Billington, D. R. Poyner, S. B. Howitt, M. Offel, *Pharm. Pharmacol. Lett.*, 2001, **11**, 18.
63.   H.-C. Zhang, D. F. McComsey, K. B. White, M. F. Addo, P. Andrade-Gordon, C. K. Derian, D. Oksenberg, B. E. Maryanoff, *Bioorg. Med. Chem. Lett.*, 2001, **11**, 2105.
64.   D. Xiao, M. D. Vera, B. Liang, M. M. Joullie, *J. Org. Chem.*, 2001, **66**, 2734.
65.   J. E. Tarver, Jr., A. J. Pfizenmayer, M. M. Joullie, *J. Org. Chem.*, 2001, **66**, 7575.
66.   A. Romanelli, I. Garella, V. Menchise, R. Iacovino, M. Saviano, D. Montesarchio, C. Didierjean, P. Di Lello, F. Rossi, E. Benedetti, *J. Pept. Sci.*, 2001, **7**, 15.
67.   G. T. Bourne, S. W. Golding, R. P. McGeary, W. D. F. Meutermans, A. Jones, G. R. Marshall, P. F. Alewood, M. L. Smythe, *J. Org. Chem.*, 2001, **66**, 7706.
68.   M. A. Walker, J. Timothy, *Tetrahedron Lett.*, 2001, **42**, 5801.
69.   L. M. A. McNamara, M. J. I. Andrews, F. Mitzel, G. Siligardi, A. B. Tabor, *J. Org. Chem.*, 2001, **66**, 4585.
70.   C. J. Creighton, A. B. Reitz, *Organic Letters*, 2001, **3**, 893.
71.   L. Somogyi, G. Haberhauer, J. Rebek, Jr., *Tetrahedron*, 2001, **57**, 1699.
72.   Z. Xia, C. D. Smith, *J. Org. Chem.*, 2001, **66**, 3459.
73.   B. McKeever, G. Pattenden, *Tetrahedron Lett.*, 2001, **42**, 2573.
74.   C.-T. Cheng, V. Lo, J. Chen, W.-C. Chen, C.-Y. Lin, H.-C. Lin, C.-H. Yang, L. Sheh, *Bioorg. Med. Chem.*, 2001, **9**, 1493.
75.   E. Horikawa, M. Kodaka, Y. Nakahara, H. Okuno, K. Nakamura, *Tetrahedron Lett.*, 2001, **42**, 8337.
76.   S. Gazal, G. Gellerman, E. Glukhov, C. Gilon, *J. Pept. Res.*, 2001, **58**, 527.
77.   L. Polevaya, T. Mavromoustakos, P. Zoumboulakis, S. G. Grdadolnik, P.

Roumelioti, N. Giatas, I. Mutule, T. Keivish, D. V. Vlahakos, E. K. Iliodromitis, D. T. Kremastinos, J. Matsoukas, *Bioorg. Med. Chem.*, 2001, **9**, 1639.

78. K. Alexopoulos, D. Panagiotopoulos, T. Mavromoustakos, P. Fatseas, M. C. Paredes-Carbajal, D. Mascher, S. Mihailescu, J. Matsoukas, *J. Med. Chem.*, 2001, **44**, 328.

79. L. Belsivi, A. Bernardi, A. Checchia, L. Manzoni, D. Potenza, C. Scolastico, M. Castorina, A. Cupelli, G. Giannini, P. Carminati, C. Pisano, *Organic. Lett.*, 2001, **3**, 1001.

80. I. Zeltser, O. Ben-Aziz, I. Schefler, K. Bhargaya, M. Altstein, C. Gilon, *J. Pept. Res.*, 2001, **58**, 275.

81. E. S. Monteagudo, F. Calvani, F. Catrambone, C. I. Fincham, A. Madami, S. Meini, R. Terracciano, *J. Pept. Sci*, 2001, **7**, 270.

82. K. Tickler, C. J. Barrow and J. D. Wade, *J. Pept. Sci.* 2001, **7**, 488.

83. M. R. Nilsson, L. L. Nguyen and D. P. Raleigh, *Anal. Biochem.* 2001, **288**, 76.

84. L. Balaspiri, U. Langel, *J. Pept. Sci.* 2001, **7**, 58.

85. L. Fulop, B. Penke, M. Zarandi, *J. Pept. Sci.* 2001, **7**, 397.

86. P. M. Gorman, A. Chakrabartty, *Biopolymers.* 2001, **60**, 381.

87. K. Watanabe, T. Segawa, K. Nakamura, M. Kodaka, T. Konakahara, H. Okuno, *J. Pept. Res.* 2001, **58**, 342.

88. J. J. Kremer, D. J. Sklansky, R. M. Murphy, *Biochemistry.* 2001, **40**, 8563.

89. H. Mihara, Y. Takahashi, *Trans. Mater. Res. Soc. Jpn.* 2001, **26**, 473.

90. J. Kallijarvi, M. Haltia, M. H. Baumann, *Biochemistry.* 2001, **40**, 10032.

91. M. M. Pallitto, R. M. Murphy, *Biophys. J.* 2001, **81**, 1805.

92. M. Kawahara, M. Kato, Y. Kuroda, *Brain Res. Bull.* 2001, **55**, 211.

93. F. Festy, L. Lins, G. Peranzi, J. N. Octave, R. Brasseur, A. Thomas, *Biochim. Biophys. Acta.* 2001, **154**, 356.

94. D. J. Gordon, K. L. Sciarretta, S. C. Meredith, *Biochemistry.* 2001, **40**, 8237.

95. G. Talaga, *Mini Rev. Med. Chem.* 2001, **1**, 175.

96. L Dumery, F. Bourdel, Y. Soussan, A. Fialkowsky, S. Viale, P. Nicholas, M. Reboud-Ravaux, *Pathol. Biol.* 2001, **49**, 72.

97. T. L Lowe, A. Strzelec, L. L. Kiessling, R. M. Murphy, *Biochemistry.* 2001, **40**, 7882.

98. D. S. Kim, J. Y. Kim, *Bioorg. Med. Chem. Lett.* 2001, **11**, 2541.

99. T. Kortvelyesi, G. Kiss, R.F. Murphy, B. Penke, S. Lovas, *THEOCHEM* 2001, **545**, 215.

100. F. Muehlhauser, U. Liebl, S. Kuehl, S. Walter, T. Bertsch, K. Fassbender, *Clin. Chem. Lab. Med.* 2001, **39**, 313.

101. J. J. Kremer, D. J. Sklansky, R. M. Murphy, *Biochemistry*, 2001, **40**, 8563.

102. S. S. Wang, D. L. Ryme, T. A. Good, *J. Biol. Chem.* 2001, **276**, 42027.

103. J. W. Allen, B. A. Eldadah, X. Huang, S. M. Knoblach, A. I. Faden, *J. Neurosci. Res.*, 2001, **65**, 45-53.

104. L. Botta, P. Valli, A. Asti, P. Perin, G. Zucca, M. Racchi, S. Govoni, A Pascale, *Neuroreport*, 2001, **12.** 2493.

105. Y. Shen, J. Yang, W. Wei L. Liu, S. Xu, *Zhongguo Yaolixue Tongbao*, 2001, **17**, 26.

106. Y. Li, *Zhongguo Yaolixue Yu Dulixue Zazhi*, 2001, **15**, 311.

107. C. Janus, A. L. Phinney, M. A. Chisti, D. Westaway, *Curr Neurol Neurosci Rep*, 2001, **1**, 451.

108. Z. P. Zhuang, M. P. Kung, C. Hou, K. Plossl, D. Skovronsky, T. L. Gur, J. Q. Trojanowski, V. M. Lee, H. F. Kung, *Nucl. Med. Biol.*, 2001, **28**, 887.

109. K. Ghosh, G. Bilcer, C. Harword, R. Kawahama, D. Shin, K. A. Hussain, L. Hong, J. A. Loy, C. Nguyen, G. Koelsch, J. Ermolieff, J. Tang, *J. Med. Chem.*, 2001, 44,

3195.

110. L. Zhang, L. Song, G. Terracina, Y. Liu, B. Pramanik, E. Parker, *Biochemistry*, 2001, **40**, 5049.

111. K. Sennvik, E. Benedikz, J. Fastbom, E. Sundstrom, B. Winblad, M. Ankarcrona, *J. Neurosci. Res.*, 2001, **63**, 429.

112. W. van't Hof, E. C. Veerman, E. J. Helmerhorst, A. V. Amerongen, *Biol. Chem.*, 2001, 382, 597.

113. Y. Shai, Z. Oren, *Peptides*, 2001, **22**, 1629.

114. A. Peschel, L. V. Collins, *Peptides*, 2001, **22**, 1651.

115. D. J. Craik, N. L. Daly, C. Waine, *Toxicon*, 2001, **39**, 43.

116. Z. Yuan, J. Trias, R. J. White, *Drug Discovery Today*, 2001, **6**, 954.

117. J. M. Clements, R. P. Beckett, A. Brown, G. Catlin, M. Lobell, S. Palan, W. Thomas, M. Whittaker, S. Wood, S. Salama, P. J. Baker, H. F. Rodgers, V. Barynin, D. W. Rice, M. G. Hunter, *Antimicrob. Agents Chemother.*, 2001, **45**, 563.

118. C. M. Apfel, S. Evers, C. Hubschwerlen, W. Pirson, M. G. P. Page, W. Keck, *Antimicrob. Agents Chemother.*, 2001, **45**, 1053.

119. C. M. Apfel, H. Locher, S. Evers, B. Takács, C. Hubschwerlen, W. Pirson, M. G. P. Page, W. Keck, *Antimicrob. Agents Chemother.*, 2001, **45**, 1058.

120. S. Fernandez-Lopez, H. S. Kim, E. C. Choi, M. Delgado, J. R. Granja, A. Khasanov, K. Kraehenbuehl, G. Long, D. A. Weinberger, K. M. Wilcoxen, M. R. Ghadiri, *Nature*, 2001, **412**, 452.

121. C. T. Miller, R. Weragoda, E. Izbicka, B. L. Iverson, *Bioorg. Med. Chem.*, 2001, **9**, 2015.

122. K. A. Martemyanov, V. A. Shirokov, O. V. Kurnasov, A. T. Gudkov, A. S. Spirin, *Protein Expression and Purification*, 2001, **21**, 456.

123. M. Trabi, H. J. Schirra, D. J. Craik, *Biochemistry*, 2001, **40**, 4211.

124. H. Yamauchi, *Insect Biochem. Mol. Biol.*, 2001, **32**, 75.

125. T. Unger, Z. Oren, Y. Shai, *Biochemistry*, 2001, **40**, 6388.

126. M. Dathem H. Nikolenko, J. Meyer, M. Beyermann, M. Bienert, *FEBS Letters*, 2001, **501**, 146.

127. J. M. Bland, A. J. DeLucca, T. J. Jacks, C. B. Vigo, *Mol. Cell. Biochem.*, 2001, **218**, 105.

128. G. Kragol, S. Lovas, G. Varadi, B. A. Condie, R. Hoffmann, L. Otvos, Jr., *Biochemistry*, 2001, **40**, 3016.

129. K. C. Nicolaou, S. Y. Cho, R. Hughes, N. Winssinger, C. Smethurst, H. Labischinski, R. Endermann, *Chemistry*, 2001, **7**, 3798.

130. K. C. Nicolaou, S. Y. Cho, R. Hughes, N. Winssinger, C. Smethurst, H. Labischinski, R. Endermann, *Chemistry*, 2001, **7**, 3824.

131. S. Y. Shin, S. H. Lee, S. T. Yang, E. J. Park, D. G. Lee, M. K. Lee, S. H. Eom, W. K. Song, Y. Kim, K. S. Hahm, J. I. Kim, *J. Pept. Res.*, 2001, **58**, 504.

132. Y. Tokunaga, T. Niidome, T. Hatakeyama, H. Aoyagi, *J. Pept. Sci.*, 2001, **7**, 297.

133. S. Y. Shin, E. J. Park, S.-T. Yang, H. J. Jung, S. H. Eom, W. K. Song, Y. Kim, K.-S. Hahm, J. I. Kim, *Biochem. Biophys. Res. Comm.*, 2001, **285**, 1046.

134. B. E. Haug, J. S. Svendsen, *J. Pept. Sci.*, 2001, **7**, 190.

135. B. E. Haug, M. L. Skar, J. S. Svendsen, *J. Pept. Sci.*, 2001, **7**, 425.

136. H. Tsubery, I. Ofek, S. Cohen, M. Fridkin, *Peptides*, 2001, **22**, 1675.

137. H. Hujakka, J. Ratilainen, T. Korjamo, H. Lankinen, P. Kuusela, H. Santa, R. Laatikainen, A. Närvänen, *Bioorg. Med. Chem.*, 2001, **9**, 1601.

138. D. Wade, T. Bergman, J. Silberring, H. Lankinen, *Protein Peptide Letters*, 2001, **8**, 443.

139. D. Avrahami, Z. Oren, Y. Shai, *Biochemistry*, 2001, **40**, 12591.
140. A. Pellegrini, C. Dettling, U. Thomas, P. Hunziker, *Biochimica et Biophysica Acta*, 2001, **1526**, 131.
141. C. Liepke, H.-D. Zucht, W.-G. Forssmann, L. Ständker, *J. Chromatography B*, 2001, **752**, 369.
142. M. Malkoski, S. G. Dashper, N. M. O'Brien-Simpson, G. H. Talbo, M. Macris, K. J. Cross, E. C. Reynolds, *Antimicrob. Agents Chemother.*, 2001, **45**, 2309.
143. P. Mak, D. Chmiel, G. J. Gacek, *Acta Biochimica Polonia*, 2001, **48**, 1191.
144. J. Orivel, V. Redeker, J.-P. Le Caer, F. Krier, A.-M. Revol-Junelles, A. Longeon, A. Chaffotte, A. Dejean, J. Rossier, *J. Biol. Chem.*, 2001, **276**, 17823.
145. I. H. Lee, Y. S. Lee, C. H. Kim, C. R. Kim, T. Hong, L. Menzel, L. M. Boo, J. Pohl, M. A. Sherman, A. J. Waring, R. I. Lehrer, *Biochim. Biophys. Acta*, 2001, **1527**, 141.
146. M. Ferchichi, M. Fathallah, P. Mansuelle, H. Rochat, J.-M. Sabatier, M. Manai, K. Mabrouk, *Biochem. Biophys. Res. Commun.*, 2001, **289**, 13.
147. M. F. Ali, A. M. Soto, F. C. Knoop, J. M. Conlon, *Biochim. Biophys. Acta*, 2001, **1550**, 81.
148. L. Olson III, A. M. Soto, F. C. Knoop, J. M. Conlon, *Biochem. Biophys. Res. Commun.*, 2001, **288**, 1001.
149. I. H. Lee, C. Zhao, T. Nguyen, L. Menzel, A. J. Waring, M. A. Sherman, R. I. Lehrer, *J. Pept. Res.*, 2001, **58**, 445.
150. R. Froidevaux, F. Krier, N. Nedjar-Arroume, D. Vercaigne-Marko, E. Kosciarz, C. Ruckebusch, P. Dhulster, D. Guillochon, *FEBS Letters*, 2001, **491**, 159.
151. H. Mozsolits, H.-J. Wirth, J. Werkmeister, M.-I. Aguilar, *Biocim. Biophys. Acta*, 2001, **1512**, 64.
152. E. J. Prenner, R. N. A. H. Lewis, M. Jelokhani-Niaraki, R. S. Hodges, R. N. McElhaney, *Biocim. Biophys. Acta*, 2001, **1510**, 83.
153. J. S. Tkacz, B. DiDomanico, *Current Opinion in Microbiology*, 2001, **4**, 540.
154. J. Andra, O. Berninghausen, M. Lippe, *Med. Microbiol. Immunol. (Berlin)*, 2001, **189**, 169.
155. S. Y. Hong, T. G. Park, K.-H. Lee, *Peptides*, 2001, **22**, 1669.
156. A. Fujie, H. Muramatsu, S. Yoshimura, M. Hashimoto, N. Shigematsu, S. Takase, *J. Antibiot.*, 2001, **54**, 588.
157. D. Barrett, A. Tanaka, A. Fujie, N. Shigematsu, M. Hashimoto, S. Hashimoto, *Tetrahedron Lett.*, 2001, **42**, 703.
158. D. Barrett, A. Tanaka, K. Harada, H. Ohki, E. Watabe, K. Maki, F. Ikeda, *Bioorg. Med. Chem. Lett.*, 2001, **11**, 479.
159. F. Formaggio, C. Peggion, M. Crisma, C. Toniolo, *J. Chem. Soc. Perkin Trans. 2*, 2001, 1372.
160. X. Sun, D. J. Zeckner, W. L. Current, R. Boyer, C. McMillian, N. Yumibe, S.-H. Chen, *Bioorg. Med. Chem. Lett.*, 2001, **11**, 1875.
161. X. Sun, D. J. Zeckner, Y.-Z. Zhang, R. K. Sachs, W. L. Current, M. Rodriguez, S.-H. Chen, *Bioorg. Med. Chem. Lett.*, 2001, **11**, 1881.
162. G.-H. Gao, W. Liu, J.-X. Dai, J.-F. Wang, Z. Hu, Y. Zhang, D.-C. Wang, *Int. J. Biol. Macromol.*, 2001, **29**, 251.
163. M. Jelokhani-Niaraki, E. J. Prenner, L. H. Kondejewski, C. M. Kay, R. N. McElhaney, R. S. Hodges, *J. Pept. Res.*, 2001, **58**, 293.
164. H. Wang, T. B. Ng, *Biochem. Biophys. Res. Commun.*, 2001, **288**, 765.
165. D. G. Lee, D.-H. Kim, Y. Park, H. K. Kim, H. N. Kim, Y. K. Shin, C. H. Choi, K.-S. Hahm, *Biochem. Biophys. Res. Commun.*, 2001, **282**, 570.
166. X. Y. Ye, T. B. Ng, *J. Protein Chem.*, 2001, **20**, 353.

167. V. G. Bashkatova, V. B. Koshelev, O. E. Fadyukova, A. A. Alexeev, A. F. Vanin, K. S. Rayevsky, I. P. Ashmarin, D. M. Armstrong, *Brain Research*, 2001, **894**, 145.

168. L. Polevaya, T. Mavromoustakos, P. Zoumboulakis, S. G. Grdadolnik, P. Roumelioti, N. Giatas, I. Mutule, T. Keivish, D. V. Vlahakos, E. K. Iliodromitis, D. Th. Kremastinos, J. Matsoukas, *Bioorg. Med. Chem.*, 2001, **9**, 1639.

169. S. Lindman, G. Lindeberg, A. Gogoll, F. Nyberg, A. Karlén, A. Hallberg, *Bioorg. Med. Chem.*, 2001, **9**, 763.

170. E. A. Kramár, D. L. Armstrong, S. Ikeda, M. J. Wayner, J. W. Harding, J. W. Wright, *Brain Research*, 2001, **897**, 114.

171. M. J. Warburton, F. Bernardini, *FEBS Letters*, 2001, **500**, 145.

172. J. G. Darker, S. J. Brough, J. Heath, D. Smart, *J. Pept. Sci.*, 2001, **7**, 598.

173. C. MacKinnon, C. Waters, D. Jodrell, C. Haslett, T. Sethi, *J. Biol. Chem.*, 2001, **276**, 28083.

174. S. A. Mantey, D. H. Coy, T. K. Pradhan, H. Igarashi, I. M. Rizo, L. Shen, W. Hou, S. J. Hocart, R. T. Jensen, *J. Biol. Chem.*, 2001, **276**, 9219.

175. H. C. Weber, J. Walters, J. Leyton, M. Casibang, S. Purdom, R. T. Jensen, D. H. Coy, C. Ellis, G. Clark, T. W. Moody, *Eur. J. Pharmacol.*, 2001, **412**, 13.

176. J. F. Valliant, R. W. Riddoch, D. W. Hughes, D. G. Roe, T. K. Fauconnier, J. R. Thornback, *Inorganica Chimica Acta*, 2001, **325**, 155.

177. I. Chatzistamou, A. V. Schally, K. Szepeshazi, K. Groot, F. Hebert, J. M. Arencibia, *Cancer Letters*, 2001, **171**, 37.

178. T. Abiko, Y. Kimura, *Current Pharmaceutical Biotechnology*, 2001, **2**, 201.

179. E. S. Monteagudo, F. Calvani, F. Catrambone, C. I. Fincham, A. Madami, S. Meini, R. Terracciano, *J. Pept. Sci.*, 2001, **7**, 270.

180. C. Giragossian, E. Nardi, C. Savery, M. Pellegrini, S. Meini, C. A. Maggi, A. M. Papini, D. F. Mierke, *Biopolymers*, 2001, **58**, 511.

181. P. Rovero, M. Pellegrini, A. Di Fenza, S. Meini, L. Quartara, C. A. Maggi, F. Formaggio, C. Toniolo, D. F. Mierke, *J. Med. Chem.*, 2001, **44**, 274.

182. Derdowska, A. Prahl, K. Neubert, B. Hartrodt, A. Kania, D. Dobrowolski, S. Melhem, H. I. Trzeciak, T. Wierzba, B. Lammek, *J. Pept. Res.*, 2001, **57**, 11.

183. A. Hasan, M. Warnock, S. Srikanth, A. H. Schmaier, *Thrombosis Research*, 2001, **104**, 451.

184. T. A. Morinelli, G. P. Meier, K. L. Schey, H. S. Margolius, *Peptides*, 2001, **22**, 2169.

185. L. Biondi, F. Filira, M. Gobbo, B. Scolaro, R. Rocchi, R. Galeazzi, M. Orena, A. Zeegers, T. Piek, *J. Pept. Sci.*, 2001, **7**, 626.

186. K. Kitagawa, C. Aida, H. Fujiwara, T. Yagami, S. Futaki, M. Kogire, J. Ida, K. Inoue, *J. Org. Chem.*, 2001, **66**, 1.

187. T. Abiko, S. Nakatsubo, *Protein and Peptide Lett.*, 2001, **8**, 153.

188. S. De Luca, D. Tesauro, P. Di Lello, R. Fattorusso, M. Saviano, C. Pedone, G. Morelli, *J. Pept. Sci.*, 2001, **7**, 386.

189. J. M. Bartolome-Nebreda, M. T. Garcia-Lopez, R. Gonzalez-Muniz, E. Cenarruzabeitia, M. Latorre, J. Del Rio, R. Herranz, *J. Med. Chem.*, 2001, **44**, 4196.

190. M. V. Silva Elipe, M. A. Bednarek, Y. D. Gao, *Biopolymers*, 2001, **59**, 489.

191. M. Roumi, E. Kwong, R. Deghenghi, V. Locatelli, S. Marleau, P. Du Souich, R. Béliveau, H. Ong, *Peptides*, 2001, **22**, 1129.

192. M. R. Melis, S. Succu, M. S. Spano, R. Deghenghi, A. Argiolas, *Neuropharmacology*, 2001, **41**, 254.

193. C. L. Edward, W. J. Hoekstra, M. F. Addo, P. Andrade-Gordon, B. P. Damiano, J. A. Kauffman, J. A. Mitchell, B. E. Maryanoff, *Bioorg. Med. Chem. Lett.*, 2001, **11**, 2619.

194.   H. U. Stilz, W. Guba, B. Jablonka, M. Just, O. Klinger, W. Konig, V. Wehner, G. Zoller, *J. Med. Chem.*, 2001, **44**, 1158.

195.   S. Kitamura, H. Fukushi, T. Miyawaki, M. Kawamura, Z. Terashita, H. Sugihara, T. Naka, *Chem. Pharm. Bull. (Tokyo)*, 2001, **49**, 258.

196.   A. Manaka, M. Sato, M. Aoki, M. Tanaka, T. Ikeda, Y. Toda, Y. Yamane, S. Nakaike, *Bioorg. Med. Chem. Lett.*, 2001, **11**, 1031.

197.   M. S. Smyth, J. Rose, M. M. Mehrotra, J. Heath, G. Ruhter, T. Schotten, J. Seroogy, D. Volkots, A. Pandey, R. M. Scarborough, *Bioorg. Med. Chem. Lett.*, 2001, **11**, 1289.

198.   A. Pandey, J. Seroogy, D. Volkots, M. S. Smyth, J. Rose, M. M. Mehrotra, J. Heath, G. Ruther, T. Schotten, R. M. Scarborough, *Bioorg. Med. Chem. Lett.*, 2001, **11**, 1293.

199.   T. M. Sielecki, J. Liu, S. A. Mousa, A. L. Racanelli, E. A. Hausner, R. R. Wexler, R. E. Olson, *Bioorg. Med. Chem. Lett.*, 2001, **11**, 2201.

200.   S. Kitamura, H. Fukushi, T. Miyawaki, M. Kawamura, N. Konishi, Z. Terashita, T. Naka, *J. Med. Chem.*, 2001, **44**, 2438.

201.   J. Jiang, W. Wang, D. C. Sane, B. Wang, *Bioorg. Chem.*, 2001, **29**, 357.

202.   K. Hojo, Y. Susuki, M. Maeda, I. Okazaki, M. Nomizu, H. Kamada, Y. Yamamoto, S. Nakagawa, T. Mayumi, K. Kawasaki, *Bioorg. Med. Chem. Lett.*, 2001, **11**, 1429.

203.   M. Matsuda, S.-I. Nishimura, *J. Med. Chem.*, 2001, **44**, 715.

204.   D. Ma, C. Xia, *Org. Lett.*, 2001, **3**, 2583.

205.   A. Peyman, K.-H. Scheunemann, D. W. Will, J. Knolle, V. Wehner, G. Breipohl, H. U. Stilz, D. Carniato, J.-M. Ruxer, J.-F. Gourvest, M. Auberval, B. Doucet, R. Baron, M. Gaillard, T. R. Gadek, S. Bodary, *Bioorg. Med. Chem. Lett.*, 2001, **11**, 2011.

206.   M. Royo, W. Van Den Nest, M. del Fresno, A. Frieden, D. Yahalom, M. Rosenblatt, M. Chorev, F. Albericio, *Tetrahedron Lett.*, 2001, **42**, 7387.

207.   C. Gibson, G. A. G. Sulyok, D. Hahn, S. L. Goodman, G. Holzemann, H. Kessler, *Angew. Chem. Int. Ed. Engl.*, 2001, **40**, 165.

208.   G. A. G. Sulyok, C. Gibson, S. L. Goodman, G. Holzemann, M. Wiesner, H. Kessler, *J. Med. Chem.*, 2001, **44**, 1938.

209.   S. Nagashima, S. Akamatsu, E. Kawaminami, S. Kawazoe, T. Ogami, Y. Matsumoto, M. Okada, K. I. Suzuki, S. I. Tsukamoto, *Chem. Pharm. Bull. (Tokyo)*, 2001, **49**, 1420.

210.   D. Boturyn, P. Dumy, *Tetrahedron Lett.*, 2001, **42**, 2787.

211.   M. Kawaguchi, R. Hosotani, S. Ohishi, N. Fujii, S. S. Tulachan, M. Koizumi, E. Toyoda, T. Masui, S. Nakajima, S. Tsuji, J. Ida, K. Fujimoto, M. Wada, R. Doi, M. Imamura, *Biochem. Biophys. Res. Commun.*, 2001, **288**, 711.

212.   N. Moitessier, S. Dufour, F. Chretien, J. P. Thiery, B. Maigret, Y. Chapleur, *Bioorg. Med. Chem.*, 2001, **9**, 511.

213.   D. L. Boger, J. Goldberg, S. Silletti, T. Kessler, D. A. Cheresh, *J. Am. Chem. Soc.*, 2001, **123**, 1280.

214.   L. Schiebler, D. F. Mierke, G. Bitan, M. Rosenblatt, M. Chorev, *Biochemistry*, 2001, **40**, 15117.

215.   A. J. Duplantier, G. E. Beckius, R. J. Chambers, L. S. Chupak, T. H. Jenkinson, A. S. Klein, K. G. Kraus, E. M. Kudlacz, M. W. McKechney, M. Pettersson, C. A. Whitney, A. J. Milici, *Bioorg. Med. Chem. Lett.*, 2001, **11**, 2593.

216.   S. Wattanasin, B. Weidmann, D. Roche, S. Myers, A. Xing, Q. Guo, M. Sabio, P. von Matt, R. Hugo, S. Maida, P. Lake, M. Weetall, *Bioorg. Med. Chem. Lett.*, 2001, **11**, 2955.

217. W. K. Hagmann, P. L. Durette, T. Lanza, N. J. Kevin, S. E. de Laszlo, I. E. Kopka, D. Young, P. A. Magriotis, B. Ling, L. S. Lin, G. Yang, T. Kamenecka, L. L. Chang, J. Wilson, M. MacCoss, S. G. Mills, G. Van Riper, E. McCauley, L. A. Egger, U. Kidambi, K. Lyons, S. Vincent, R. Stearns, A. Colletti, J. Teffera, S. Tong, J. Fenyk-Melody, K. Owens, D. Levorse, P. Kim, J. A. Schmidt, R. A. Mumford, *Bioorg. Med. Chem. Lett.*, 2001, **11**, 2709.

218. G. Müller, M. Albers, R. Fischer, G. Heßler, T. E. Lehmann, H. Okigami, M. Tajimi, K. Bacon, T. Rölle, *Bioorg. Med. Chem. Lett.*, 2001, **11**, 3019.

219. J. Boer, D. Gottschling, A. Schuster, M. Semmrich, B. Holzmann, H. Kessler, *J. Med. Chem.*, 2001, **44**, 2586.

220. J. Boer, D. Gottschling, A. Schuster, B. Holzmann, H. Kessler, *Angew. Chem. Int. Ed. Engl.*, 2001, **40**, 3870.

221. D. Gottschling, J. Boer, A. Schuster, B. Holzmann, H. Kessler, *Bioorg. Med. Chem. Lett.*, 2001, **11**, 2997.

222. P. M. Green, S. B. Ludbrook, D. D. Miller, C. M. T. Horgan, S. T. Barry, *FEBS Letters*, 2001, **503**, 75.

223. T. Beckers, M. Bernd, B. Kutscher, R. Kühne, S. Hoffmann, T. Reissmann, *Biochem. Biophys. Res. Commun.*, 2001, **289**, 653.

224. S. Rahimipour, I. Bilkis, V. Peron, G. Gescheidt, F. Barbosa, Y. Mazur, Y. Koch, L. Weiner, M. Fridkin, *Photochemistry and Photobiology*, 2001, **74**, 226.

225. S. Rahimipour, N. Ben-Aroya, M. Fridkin, Y. Koch, *J. Med. Chem.*, 2001, **44**, 3645.

226. I. Chatzistamou, A. V. Schally, K. Szepeshazi, K. Groot, F. Hebert, J. M. Arencibia, *Cancer Letters*, 2001, **171**, 37.

227. B. H. Woo, J. W. Kostanski, S. Gebrekidan, B. A. Dani, B. C. Thanoo, P. P. DeLuca, *J. Controlled Release*, 2001, **75**, 307.

228. G. F. Walker, R. Ledger, I. G. Tucker, *Int. J. Pharmaceutics*, 2001, **216**, 77.

229. O. Linhart, S. D. Mims, B. Gomelsky, A. E. Hiott, W. L. Shelton, J. Cosson, M. Rodina, D. Gela, *Aquatic Living Resources*, 2001, **13**, 455.

230. S.-K. Lim, S.-Z. Li, C.-H. Lee, C.-J. Yoon, J.-H. Baik, W. Lee, *Chemistry & Biology*, 2001, **8**, 857.

231. C. R. Nakaie, S. R. Barbosa, R. F. F. Vieira, R. M. Fernandez, E. M. Cilli, A. M. L. Castrucci, M. A. Visconti, A. S. Ito, M. T. Lamy-Freund, *FEBS Letters*, 2001, **497**, 103.

232. J. Chluba, D. L. de Souza, B. Frisch, F. Schuber, *Biochim. Biophys. Acta*, 2001, **1510**, 198.

233. R. Muceniec, I. Mutule, F. Mutulis, P. Prusis, M. Szardenings, J. E. S. Wikberg, *Biochim. Biophys. Acta*, 2001, **1544**, 278.

234. M. A. Bednare, T. MacNeil, R. N. Kalyani, R. Tang, L. H. T. Van der Ploe, D. H. Weinberg, *J. Med. Chem.*, 2001, **44**, 3665.

235. A. Hosmalin, M. Andrieu, E. Loing, J.-F. Desoutter, D. Hanau, H. Gras-Masse, A. Dautry-Varsat, J.-G. Guillet, *Immunology Letters*, 2001, **79**, 97.

236. D. H. Zimmerman, J. P. Lloyd, D. Heisey, M. D. Winship, M. Siwek, E. Talor, P. S. Sarin, *Vaccine*, 2001, **19**, 4750.

237. J. Sidney, S. Southwood, D. L. Mann, M. A. Fernandez-Vina, M. J. Newman, A. Sette, *Human Immunology*, 2001, **62**, 1200.

238. G. Kragol, L. Otvos, Jr., J.Q. Feng, W. Gerhard, J. D. Wade, *Bioorg. Med. Chem. Lett.*, 2001, **11**, 1417.

239. F. Peri, L. Cipolla, M. Rescigno, B. La Ferla, F. Nicotra, *Bioconjug. Chem.*, 2001, **12**, 325.

240. A. Balasubramaniam, V. C. Dhawan, D. E. Mullins, W. T. Chance, S. Sheriff, M.

Guzzi, M. Prabhakaran, E. M. Parker, *J. Med. Chem.*, 2001, **44**, 1479.

241. M. Langer, R. La Bella, E. Garcia-Garayoa, A. G. Beck-Sickinger, *Bioconjugate Chem.*, 2001, **12**, 1028.
242. M. Langer, F. Kratz, B. Rothen-Rutishauser, H. Wunderli-Allenspach, A. G. Beck-Sickinger, *J. Med. Chem.*, 2001, 44, 1341.
243. M. M. Berglund, I. Lundell, H. Eriksson, R. Söll, A. G. Beck-Sickinger, D. Larhammar, *Peptides*, 2001, **22**, 351.
244. Y. Sasaki, M. Hirabuki, A. Ambo, H.Ouchi, Y. Yamamoto, *Bioorg. Med. Chem. Lett.*, 2001, **11**, 327.
245. Y. Lu, G. Weltrowska, C. Lemieux, N. N. Chung, P. W. Schiller, *Bioorg. Med. Chem. Lett.*, 2001, **11**, 323.
246. Z. Kříž, P. H. J. Carlsen, J. Koča, *J. Mol. Structure: THEOCHEM*, 2001, **540**, 231.
247. J. Malicka, M. Groth, C. Czaplewski, J. Karolczak, A. Liwo, W. Wiczk, *Biopolymers*, 2001, **59**, 180.
248. F. Gosselin, D. Tourwe, M. Ceusters, T. Meert, L. Heylen, M. Jurzak, W. D. Lubell, *J. Pept. Res.*, 2001, **57**, 337.
249. S. Gupta, S. Pasha, Y. K. Gupta, D. K. Bhardwaj, *Brain Research Bulletin*, 2001, **55**, 51.
250. M. Yamazaki, T. Suzuki, M. Narita, A. W. Lipkowski, *Life Sciences*, 2001, **69**, 1023.
251. D. Page, A. Naismith, R. Schmidt, M. Coupal, M. Labarre, M. Gosselin, D. Bellemare, K. Paysa, W. Brown, *J. Med. Chem.*, 2001, **44**, 2387.
252. T. Lovekamp, P. S. Cooper, J. Hardison, S. D. Bryant, R. Guerrini, G. Balboni, S. Salvadori, L. H. Lazarus, *Brain Research*, 2001, **902**, 131.
253. M. Keller, C. Boissard, L. Patiny, N. N. Chung, C. Lemieux, M. Mutter, P. W. Schiller, *J. Med. Chem.*, 2001, **44**, 3896.
254. C. L. Neilan, T. M.-D. Nguyen, P. W. Schiller, G. W. Pasternak, *Eur. J. Pharm.*, 2001, **419**, 15.
255. S. Liao, M. Shenderovich, K. E. Kover, Z. Zhang, K. Hosohata, P. Davis, F. Porreca, H. I. Yamamura, V. J. Hruby, *J. Pept. Res.*, 2001, **57**, 257.
256. S. E. Schullery, D. W. Rodgers, S. Tripathy, D. E. Jayamaha, M. D. Sanvordekar, K. Renganathan, C. Mousigian, D. L. Heyl, *Bioorg. Med. Chem.*, 2001, **9**, 2633.
257. J. Kotlińska, P. Suder, A. Sciubisz, A. Łęgowska, J. Eilmes, K. Rolka, J. Silberring, *Eur. J. Pharmacol.*, 2001, **419**, 33.
258. R. Guerrini, G. Caló, R. Bigoni, D. Rizzi, D. Regoli, S. Salvadori, *J. Pept. Res.*, 2001, **57**, 215.
259. R. Guerrini, G. Caló, R. Bigoni, D. Rizzi, A. Rizzi, M. Zucchini, K. Varani, E. Hashiba, D. G. Lambert, G. Toth, P. A. Borea, S. Salvadori, D. Regoli, *J. Med. Chem.*, 2001, **44**, 3956.
260. A. Ambo, N. Hamazaki, Y. Yamada, E. Nakata, Y. Sasaki, *J. Med. Chem.*, 2001, **44**, 4015.
261. M. Xu, V. K. Kontinen, P. Panula, E. Kalso, *Peptides*, 2001, **22**, 33.
262. H. Mazarguil, C. Gouardères, J.-A. M. Tafani, D. Marcus, M. Kotani, C. Mollereau, M. Roumy, J.-M. Zajac, *Eur. J. Pharmacol.* 2001, **428**, 29.
263. Mauborgne, S. Bourgoin, H. Poliénor, M. Roumy, G. Simonnet, J.-M. Zajac, F. Cesselin, *Eur. J. Pharmacol.*, 2001, **430**, 273.
264. Pertovaara, U. Keski-Vakkuri, J. Kalmari, H. Wei, P. Panula, *Neuroscience*, 2001, **105**, 457.
265. H. Wei, P. Panula, A. Pertovaara, *Brain Research*, 2001, **900**, 234.
266. E. Falb, Y. Salitra, T. Yechezkel, M. Bracha, P. Litman, R. Olender, R. Rosenfeld, H. Senderowitz, S. Jiang, M. Goodman, *Bioorg. Med. Chem.*, 2001, **9**, 3255.

267. B. A. Hay, B. M. Cole, F. DiCapua, G. W. Kirk, M. C. Murray, R. A. Nardone, D. J. Pelletier, A. P. Ricketts, A. S. Robertson, T. W. Siegel, *Bioorg. Med. Chem. Lett.*, 2001, **11**, 2731.

268. Janecka, M. Zubrzycka, T. Janecki, *J. Pept. Res.*, 2001, **58**, 91.

269. J. K. Amartey, R. S. Parhar, I. Al-Jammaz, *Nucler Medicine and Biology*, 2001, **28**, 225.

270. K. Gademann, T. Kimmerlin, D. Hoyer, D. Seebach, *J. Med. Chem.*, 2001, **44**, 2460.

271. S. K. Jiang, S. Gazal, G. Gelerman, O. Ziv, O. Karpov, P. Litman, M. Bracha, M. Afargan, C. Gilon, M. Goodman, *J. Pept. Sci.*, 2001, **7**, 521.

272. W. G. Rajeswaran, S. J. Hocart, W. A. Murphy, J. E. Taylor, D. H. Coy, *J. Med. Chem.*, 2001, **44**, 1305.

273. P. Bernhardt, L. Kolby, V. Johanson, S. A. Benjegard, O. Nilsson, H. Ahlman, E. Forssell-Aronsson, *Nuclear Medicine and Biology*, 2001, **28**, 67.

274. S. Pervez, A. Mushtaq, M. Arif, *Applied Radiation and Isotopes*, 2001, **55**, 647.

275. H. Kiaris, A. V. Schally, A. Nagy, K. Szepeshazi, F. Hebert, G. Halmos, *Eur. J. Cancer*, 2001, **37**, 620.

276. T. Vantus, G. Keri, Z. Krivickiene, M. Valius, A. Stetak, S. Keppens, P. Csermely, P. I. Bauer, G. Bokonyi, W. Declercq, P. Vandenabeele, W. Merlevede, J. R. Vandenheede, *Cellular Signalling*, 2001, **13**, 717.

277. Z. Helyes, E. Pinter, J. Nemeth, G. Keri, M. Than, G. Oroszi, A. Horvath, J. Szolcsanyi, *British J. Pharmacol.*, 2001, **134**, 1571.

278. Janecka, M. Zubrzycka, *Polish J. Chem.*, 2001, **75**, 1877.

279. S. Rodziewicz-Motowid, A. Lesner, C. Czaplewski, A. Liwo, K. Rolka, R. Patacchini, L. Quartara, *J. Pept. Res.*, 2001, **58**, 159.

280. R. Millet, L. Goossens, K. Bertrand-Caumont, J. F. Goossens, R. Houssin, J. P. Hénichart, *J. Pharmacy and Pharmacology*, 2001, **53**, 929.

281. A. Bremer, S. E. Leeman, N. D. Boyd, *J. Biol. Chem.*, 2001, **276**, 22857.

282. F. J. Warner, P. Mack, A. Comis, R. C. Miller, E. Burcher, *Biochem. Pharm.*, 2001, **61**, 55.

283. K. Higa, C. Gao, W. Motokawa, K. Abe, *Archives of Oral Biology*, 2001, **46**, 313.

284. H. Weisshoff, T. Nagel, A. Hansicke, A. Zschunke, C. Mugge, *FEBS Letters*, 2001, **491**, 299.

285. M. Pellegrini, A. A. Bremer, A. L. Ulfers, N. D. Boyd, D. F. Mierke, *J. Biol. Chem.*, 2001, **276**, 22862.

286. K. Haneda, T. Inazu, M. Mizuno, R. Iguchi, H. Tanabe, K. Fujimori, K. Yamamoto, H. Kumagai, K. Tsumori, E. Munekata, *Biochim. Biophys. Acta*, 2001, **1526**, 242.

287. R. Cirillo, M. Astolfi, B. Conte, G. Lopez, M. Parlani, G. Sacco, R. Terracciano, C. I. Fincham, S. Sisto, S. Evangelista, C. A. Maggi, S. Manzini, *Neuropeptides*, 2001, **35**, 137.

288. E. Lempicka, I. Derdowska, B. Jastrzebska, P. Kuncarova, J. Slaninova, B. Lammek, *J. Pept. Res.*, 2001, **57**, 162.

289. J. Slaninova, L. Maletinska, J. Vondrasek, Z. Prochazka, *J. Pept. Sci.*, 2001, **7**, 413.

290. L. Belec, L. Maletinska, J. Slaninova, W. D. Lubell, *J. Pept. Res.*, 2001, **58**, 263.

291. J. Hlavacek, U. Ragnarsson, *J. Pept. Sci.*, 2001, **7**, 349.

292. M. Manning, S. Stoev, L. L. Cheng, N. C. Wo, W. Y. Chan, *J. Pept. Sci.*, 2001, **7**, 449.

293. A. Dietrich, J. T. Taylor, C. E. Passmore, *Brain Research*, 2001, **919**, 41.

294. Z. Liu, Y. Wang, W. Zhao, J. Ding, Z. Mei, L. Guo, D. Cui, J. Fei, *Neuropharmacology*, 2001, **41**, 464.

295. J. Ye, W. Chang, D, Liang, *Biochim. Biophys. Acta*, 2001, **1547**, 18.

296. K. J. Smith, J. D. Wade, A. A. Claasz, L. Otvos, C. Temelcos, Y. Kubota, J. M. Hutson, G. W. Tregear, R. A. Bathgate, *J. Pept. Sci.*, 2001, **7**, 495.
297. H. Tamamura, A. Omagari, K. Hiramatsu, K. Gotoh, T. Kanamoto, Y. Xu, E. Kodama, M. Matsuoka, T. Hattori, N. Yamamoto, H. Nakashima, A. Otaka, N. Fujii, *Bioorg. Med. Chem. Lett.*, 2001, **11**, 1897.
298. D. Dhanak, L. T. Christmann, M. G. Darcy, A. J. Jurewicz, R. M. Keenan, J. Lee, H. M. Sarau, K. L. Widdowson, J. R. White, *Bioorg. Med. Chem. Lett.*, 2001, **11**, 1441.
299. D. Dhanak, L. T. Christmann, M. G. Darcy, R. M. Keenan, S. D. Knight, J. Lee, L. H. Ridgers, H. M. Sarau, D. H. Shah, J. R. White, L. L. Zhang, *Bioorg. Med. Chem. Lett.*, 2001, **11**, 1445.
300. A. Litovchick, A. Lapidot, M. Eisenstein, A. Kalinkovich, G. Borkow, *Biochemistry*, 2001, **40**, 15612.
301. P. N. Strong, G. S. Clark, A. Armugam, F. A. De-Allie, J. S. Joseph, V. Yemul, J. M. Deshpande, R. Kamat, S. V. Gadre, P. Gopalakrishnakone, R.M. Kini, D.G. Owen, K. Jeyaseelan, *Archives of Biochemistry and Biophysics*, 2001, **385**, 138.
302. M. Corona, N. A. Valdez-Cruz, E. Merino, M. Zurita and L. D. Possani, *Toxicon*, 2001, **39**, 1893.
303. J. Garcia-Valdes, F. Z. Zamudio, L. Toro, L. D. Possan, *FEBS Letters*, 2001, **505**, 369.
304. E. E. Dudina, Y. V. Korolkova, N. E. Bocharova, S. G. Koshelev, T. A. Egorov, I. Huys, J. Tytgat, E. V. Grishin, Biochem. Biophys. Res. Commun., 2001, **286**, 841.
305. X.-C. Zeng, F. Peng, F. Luo, S.-Y. Zhu, H. Liu, W.-X. Li, *Biochimie*, 2001, **83**, 883.
306. S. Y. Zhu, W. X. Li, X. C. Zeng, *Toxicon*, 2001, **39**, 1291.
307. J.-G. Ye, J. Chen, X.-P. Zuo, Y.-H. Ji, *Toxicon*, 2001, **39**, 1191.
308. X.-C. Zeng, W.-X. Li, S.-Y. Zhu, F. Peng, Z.i-H. Zhu, H. Liu, X, Mao, *Toxicon*, 2001, **39**, 225.
309. X.-C. Zeng, Z.-H. Zhu, W.-X. Li, S.-Y. Zhu, F. Peng, X. Mao, H. Liu, *Toxicon*, 2001, **39**, 407.
310. K. N. Srinivasan, S. Nirthanan, T. Sasaki, K. Sato, B. Cheng, M. C. E. Gwee, R. M. Kini, P. Gopalakrishnakone, *FEBS Letters*, 2001, **494**, 145.
311. R. J. Guan, C. G. Wang, M. Wang, D. C. Wang, *Biochim. Biophys. Acta*, 2001, **1549**, 9.
312. E. Carlier, Z. Fajloun, P. Mansuelle, M. Fathallah, A. Mosbah, R. Oughideni, G. Sandoz, E. Di Luccio, S. Geib, I. Regaya, J. Brocard, H. Rochat, H. Darbon, C, Devaux, J. M. Sabatier, M. de Waard, *FEBS Letters*, 2001, **489**, 202.
313. P. V. Dubovskii, D. V. Dementieva, E. V. Bocharov, Y. N. Utkin, A. S. Arseniev, *J. Mol. Biol.*, 2001, **305**, 137.
314. M. Nakamura, Y. Niwa, Y. Ishida, T. Kohno, K. Sato, Y. Oba, H. Nakamura, *FEBS Letters*, 2001, **503**, 107.
315. E. Horikawa, M. Kodaka, Y.Nakahara, H. Okuno, K. Nakamura, *Tetrahedron Lett.*, 2001, **42**, 8337.
316. Kisfalvi, Jr., G. Rácz, A. Bálint, M. Máté, A. Oláh, T. Zelles, E. S. Vizi, G. Varga, *Journal of Physiology-Paris*, 2001, **95**, 385.
317. F. P. M. O'Harte, M. H. Mooney, C. M. N. Kelly, A. M. McKillop, P. R. Flatt, *Regulatory Peptides*, 2001, **96**, 95.
318. D.-D. Malis, B. Rist, K. Nicoucar, A. G. Beck-Sickinger, D. R. Morel, J.-S. Lacroix, *Regulatory Peptides*, 2001, **101**, 101.
319. R. González-Muñiz, M. Martín-Martínez, C. Granata, E. de Oliveira, C. M. Santiveri, C. González, D. Frechilla, R. Herranz, M. T. García-López, J. Del Río, M. A. Jiménez,D. Andreu, *Bioorg. Med. Chem.*, 2001, **9**, 3173.

320. Y.-J. Cao, G. Gimpl, *Biochim. Biophys. Acta*, 2001, **1548**, 139.
321. G. P. Zecchini, E. Morera, M. Nalli, M. P. Paradisi, G. Lucente, S. Spisani, *Il Farmaco*, 2001, **56**, 851.
322. R. Jain, J. Singh, J. H. Perlman, M. C. Gershengorn, *Bioorg. Med. Chem.*, 2001, **10**, 189.
323. S. K. Sharma, T. M. Ramsey, Y. N. P. Chen, W. C. Chen, M. S. Martin, K. Clune, M. Sabio, K. W. Bair, *Bioorg. Med. Chem. Lett.*, 2001, **11**, 2449.
324. D. Gatos, C. Tzavara, *J. Pept. Res.*, 2001, **57**, 168.
325. H. H. Keah, N. Allen, R. Clay, R. I. Boysen, T. Warner, M. T. W. Hearn, *Biopolymers*, 2001, **60**, 279.
326. E. Wong, S. Bennett, B. Lawrence, T. Fauconnier, L. F. L. Lu, R. A. Bell, J. R. Thornback, D. Eshima, *Inorg. Chem.*, 2001, **40**, 5695.
327. K. L. Wegener, C. S. Brinkworth, J. H. Bowie, J. C. Wallace, M. J. Tyler, *Rapid Communications In Mass Spectrometry*, 2001, **15**, 1726.
328. V. Gut, V. Cerovsky, M. Zertova, E. Korblova, P. Malon, H. Stocker, E. Wunsch,*Amino Acids*, 2001, **21**, 255.
329. J. Leprince, H. Oulyadi, D. Vaudry, O. Masmoudi, P. Gandolfo, C. Patte, J. Costentin, J. L. Fauchere, D. Davoust, H. Vaudry, M. C. Tonon, *Eur. J. Biochem.*, 2001, **268**, 6045.
330. T. Kawakami, K. Hasegawa, K. Teruya, K. Akaji, M. Horiuchi, F. Inagaki, Y. Kurihara, S. Uesugi, S. Aimoto, *J. Pept. Sci.*, 2001, **7**, 474.
331. C. Antczak, I. De Meester, B. Banvois, *J. Biol. Regul. Homeost. Agents*, 2001, **15**, 130.
332. S. Ren, E. J. Lien, *Prog. Drug Res.* 2001, Spec No: 1-34. Development of HIV protease inhibitors: a survey.
333. J. R. Huff, J. Kahn, *Adv. Protein Chem.*, 2001, **56**, 213.
334. B. Walker, J. F. Lyans, *Cell. Mol. Life Sci.*, 2001, **58**, 596.
335. A. Martinez, A. Castro, C. Gil, C. Perez, *Med. Res. Rev.*, 2001, **21**, 227.
336. C. David-Basei, L. Bischoff, M.-C. Fournie-Zaluski, B. P. Roques, *J. Labelled Compd. Radiopharm.*, 2001, **44**, 89.
337. S. J. Keding, D. H. Rich, *Pept. New Millennium, Proc. Am. Pept. Symp.* 16 th. 2000, 431.
338. J. Grembecka, P. Kafarski, *Mini Rev. Med. Chem.*, 2001, **1**, 133.
339. S. Mangaleswaran, N. P. Argade, *J. Org. Chem.*, 2001, **66**, 5259.
340. H. Chen, F. Noble, B. P. Roques, M.-C. Fourne-Zaluski, *J. Med. Chem.*, 2001, **44**, 3523.
341. L. M. Pratt, R. P. Beckett, S. J. Davies, S. B. Launchbury, A. Miller, Z. M. Spavold, R. S. Todd, M. Whittaker, *Bioorg. Med. Chem. Lett.*, 2001, **11**, 2585.
342. C. Apfel, D. W Banner, D. Dietz, C. Hubschwerlen, H. Locher, F. Marlin, R. Masciadri, W. Pirson, H. Stadler, *J. Med. Chem.*, 2001, **44**, 1847.
343. A. Thorarensen, M. R. Jr Deibel, D. C. Rohrer, A. F. Vosters, A. W. Yem, D. V. Marshall, J. C. Lynn, M. J. Bohanon, P. K. Tomich, G. E. Zurenko, M. T. Sweeney, R. M. Jensen, J. W. Nielsen, E. P. Seest, L. A. Dolak, *Bioorg. Med. Chem. Lett.*, 2001, **11**, 1355.
344. K. A. Josef, F. W. Kauer, R. Bihovsky, *Bioorg. Med. Chem. Lett.*, 2001, **11**, 2615.
345. G. J. Wells, M. Tao, K. A. Josef, R. Bihovsky, *J. Med. Chem.*, 2001. **44** 3488.
346. I. O. Donkor and M. L. Sanders, *Bioorg. Med. Chem. Lett.*, 2001, **11**, 2647.
347. H. Fan, Y. Zhao, L. Byer, R. P. Hammer, *Pept. New Millennium, Proc. Am. Pept. Symp.*, 16th. 2000, 91-93. Edited by G. B. Fields, J. P. Tam, G. Barany. Kluwer Academic Publishers, Dordrecht, Neth.

348. J. D. Park, K. J. Lee, D. H. Kim, *Bioorg. Med. Chem.*, 2001, **9**, 237.
349. M. Lee, D. H Kim, *Bull. Korean Chem. Soc.*, 2001, **22**, 1236.
350. S. J. Chung, D. H. Kim, *Bioorg. Med. Chem.*, 2001, **9**, 185.
351. A.P. Kozikowski, F. Nan, P. Conti, J. Zhang, E. Ramadan, T. Bzdega, B. Wroblewska, J. H. Neale, S. Pshenichkin, J. T. Wroblewski, *J. Med. Chem.*, 2001, **44**, 298.
352. J. I. Park, D. H. Kim, *Bioorg. Med. Chem. Lett.*, 2001, **11**, 2967.
353. D. Lee, S. A. Long, J. H. Murray, J.L. Adams, M. E. Nuttall, D. P. Nadeau, K. Kikly, J. D. Winkler, C. M. Sung, M. D. Ryan, M. A. Levy, P. M. Keller, W. E. Jr. DeWolf, *J. Med. Chem.*, 2001, **44**, 2015.
354. A. B. Shahripour, M. S. Plummer, E. A. Lunney, T. K. Sawyer, C. J. Stankovic, M. K. Connolly, J. R. Rubin, N. P. Walker, K. D. Brady, H. J. Allen, R. V. Talanian, W.W. Wong, C. Humblet, *Bioorg. Med. Chem. Lett.*, 2001, **11**, 2779.
355. B. Shahripour, M. S. Plummer, E. A. Lunney, H.P. Albrecht, S. J. Hays, C. R. Kostlan, T. K. Sawyer, N. P. C. Walker, K. D. Brady, H. J. Allen, R. W.Talanian, W. W. Wong, C. Humblet, *Bioorg. Med. Chem.*, 2001, **7**, 533.
356. M. S. Head, M. D. Ryan, D. Lee, Y. Feng, C. A. Janson, N. O. Concha, P. M. Keller, W. E. Jr. DeWolf, *J. Comput. Aided Mol. Des.*, 2001, **15**, 1105.
357. Z. Gou, M. Xian, W. Zhang, A. McGill, P. G. Wang, *Bioorg. Med. Chem.*, 2001, **9**, 99.
358. M. Makowski, M. Pawelczak, R. Latajka, K. Nowak, P. Kafarski, *J. Pept. Sci.*, 2001, **7**, 141.
359. A. E. Fenwick, B. Garnier, A. D. Gribble, R. J. Ife, A. D. Rawlings, J. Witherington, *Bioorg. Med. Chem. Lett.*, 2001, **11**, 195.
360. R. A. Smith, A. Bhargava, C. Browe, J. Chen, J. Dumas, H. Hatoum-Mokdad, R. Romero, *Bioorg. Med. Chem. Lett.*, 2001, **11**, 2951.
361. R. W. Marquis, Y. Ru, S. M. LoCastro, J. Zeng, D. S. Yamashita, H.-J. Oh, K. F. Erhard, L. D. Davis, T. A. Tomaszek, D. Tew, K, Salyers, J Proksch, K. Ward, B. Smith, M. Levy, M. D. Cummings, C. R. Haltiwanger, G. Trescher, B. Wang, M. E. Hemling, C. J. Quinn, H-Y Cheng, F. Lin, W. W. Smith, C. A. Janson, B. Zhao, M. S. McQueney, K. D'Alessio, C-P. Lee, A. Marzulli, R. A. Dodds, S. Blake, S.-M. Hwang, I. E. James, C. J. Gress, B. R. Bradley, M. W. Lark, M. Gowen, D. F. Veber, *J. Med. Chem.*, 2001, **44**, 1380.
362. R. W. Marquis, Y. Ru, J. Zeng, R. E. Trout, S. M. LoCastro, A. D. Gribble, J. Witherington, A. E. Fenwick, B. Garnier, T. Tomaszek, D. Tew, M. E. Hemling, C. J. Quinn, W. W. Smith, B. Zhao, M. S. McQueney, C. A. Janson, K. D'Alessio, D. F. Weber, *J. Med. Chem.*, 2001, **44**, 725.
363. J. P. Falgueyret, R. M. Oballa, O. Okamoto, G. Wesolowski, Y. Aubim, R. M. Rydzewski, P. Prasit, D. Riendeau, S. B. Rodan, M. D. Percival, *J. Med. Chem.*, 2001, **44**, 94.
364. P. D. Greenspan, K. L. Clark, R. A. Tommasi, S. D. Cowen, L. W. McQuire, D. L. Farley, J. H. van Duzer, R. L. Goldberg, H. Zhou, Z. Du, J. J. Fitt, D. E. Coppa, Z. Fang, W. Macchia, L. Zhu, M. P. Capparelli, R. Goldstein, A. M. Wigg, J. R. Doughty, R.S. Bohacek, A. K. Knap, *J. Med. Chem.*, 2001, **44**, 4524.
365. J. Gray, M. M. Haran, K. Schneider, S. Vesce, A. M. Ray, D. Owen, I. R. White, P. Cutler, J. B. Davis, *J. Biol.Chem.*, 2001, **276**, 32750.
366. F. Kasprzykowski, R. Kasprzykowska, P. Kania, K. Plucinska, A. Grubb, M. Abrahamson, C. Schalen, *Pol. J. Chem.*, 2001, **75**, 831.
367. M. Demarcus, M. L. Ganadu, G. M. Mura, A. Porcheddu, L. Quaranta, G. Reginato, M. Taddei, *J. Org. Chem.*, 2001, **66**, 697.
368. P. Xu, W. Lin, *Zhongguo Yaowu Huaxue Zazhi*, 2001, **11**, 32.

369. M. Matsumoto, S. Misawa, N. Chiba, H. Takaku, H. Hayashi, *Biol. Pharm. Bull.*,
      2001, **24**, 236.
370. Q. Tian, N. K. Nayyar, S. Babu, L. Chen, J. Tao, S. Lee, A Tibbetts, T. Moran, J.
      Liou, M. Guo, T. P. Kennedy, *Tetrahedron Lett.*, 2001, **42**, 6807.
371. S. E. Webber, J. T. Marakovits, P. S. Dragovich, T. J. Prins, R. Zhou, S. A.
      Fuhrman, A. K. Patick, D. A. Matthews, C. A. Lee, B. Srinivasan, T. Moran, C. E.
      Ford, M. A. Brothers, J. E. V. Harr, J. W. Meador, R. A. Ferre, S. T. Worland,
      Bioorg. Med. Chem. Lett., 2001, **11**, 2683.
372. J. Pei, J. Zhou, G. Xie, H. Chen, X.He, *J. Mol. Graph. Model.*, 2001, **19**, 448.
373. I. Kumar, R. S. Jolly, *Org. Lett.*, 2001, **3**, 283.
374. X. Gneg, Y. Li, Y. Shang, G. Li, S. Wang, X. Sun, W. Ren, *Xi'an Yike Daxue
      Xuebao*, 2001, **22**, 418.
375. K. Arihara, Y. Nakashima, T. Mukai, S. Ishikawa, M. Itoh, *Meat Sci.*, 2001, **57**, 319.
376. F. Firooznia, C. Gude, K. Chan, C. A. Fink, Y. Qiao, Y. Satoh, N. Marcopoulos, P.
      Savage, M. E. Beil, C. W. Bruseo, A. J. Trapani, A. Y Jeng, *Bioorg. Med. Chem. Lett.*,
      2001, **11**, 375.
377. M. Reboud-Ravaux, *J. Soc. Biol.*, 2001, **195**, 143.
378. K. Ohmoto, M. Okuma, T. Yamamoto, H. Kijima, T. Sekioka, K. Kitagawa, S.
      Yamamoto, K. Tanaka, K. Kawabata, A. Sakata, H. Imawaka, H. Nakai, M. Toda,
      Bioorg. Med. Chem., 2001, **9**, 1307.
379. M. Gurrini, A. Lupi, S. Viglio, F. Pamparana, G. Getta, P. Iadarola, M. Powers.
      Luisetti, *Am. J. Respir. Cell Mol. Biol.*, 2001, **25**, 492.
380. K. P. Koteva, A. M. Cantin, W. A. Neugebauer, E. Escher, *Can. J. Chem.*, 2001, **79**,
      377.
381. L. M. Vagnoni, M. Gronostaj, J. E. Kerrigan, *Bioorg. Med. Chem.*, 2001, **9**, 637.
382. J. Sakowski, M. Boehm, I. Sattler, H.-M. Dahse, M. Schlitzer, *J. Med. Chem.*, 2001,
383. T. C. Turek, I. Gaon, M. D. Distefano, C. L. Strickland, *J. Org. Chem.*, 2001, **66**,
      3253.
384. M. C. Hillier, J. P. Davidson and S. F. Martin, *J. Org. Chem.*, 2001, **66**, 1657.
385. H. Lee, J. Lee, S. Lee, Y. Shin, W. Jung, J. H. Kim, K. Park, K. Kim, H. S. Cho, S.
      Ro, S. Lee, S. W. Jeong, T. Choi, H. H. Chung, J. S. Koh, *Bioorg. Med. Chem. Lett.*,
      2001, **11**, 3069.
386. M. Bohm, A. Mitsch, P. Wissner, I. Sattler, M. Schlitzer, *J. Med. Chem.*, 2001, **44**,
      3117.
387. I. M. Bell, S. N. Gallicchio, M. Abrams, D. C. Beshore, C. A. Buser, J. C. Culberson,
      J. Davide, M. Ellis-Hutchings, C. Fernandes, J. B. Gibbs, S. L. Graham, G. D.
      Hartman, D. C. Heimbrook, C. F. Homnick, J. R. Huff, K. Kassahun, K. S. Koblan,
      N. E. Kohl, R. B. Lobell, J. J. Jr. Lynch, P. A. Miller, C. A. Omer, A. D. Rodrigues, E.
      S. Walsh, T. M. Williams, *J. Med. Chem.*, 2001, **44**, 2933.
388. C. Martin, P. Mailliet, J. Maddaluno, *J. Org. Chem.*, 2001, **66**, 3797.
389. C. J. Dinsmore, J. M. Bergman, M. J. Bogusky, J. C. Culberson, K. A. Hamilton, S.
      L. Graham, *Org. Lett.*, 2001, **3**, 865.
390. M. Song, S. Rajesh, Y. Hayashi, Y. Kiso, *Bioorg. Med. Chem. Lett.*, 2001, **11**, 2465.
391. E. G. Myers, J. K. Barbay, B. Zhong, *J. Amer. Chem. Soc.*, 2001, **123**, 7207.
392. D. Pyring, J. Lindberg, A. Rosenquist, G. Zuccarello, I. Kvarnstroem, H. Zhang, L.
      Vrang, T. Unge, B. Classon, A. Hallberg and B. Samuelsson, *J. Med. Chem.*, 2001,
      44, 3083.
393. A. Muhlman, B. Classon, A. Hallberg, B. Samuelsson, *J. Med. Chem.*, 2001, **44**,
      3402.
394. A. Muhlman, A. Lindberg, B. Classon, A. Hallberg, B. Samuelsson, *J. Med. Chem.*,

2001, **44**, 3407.

395. J. Courcambeck, F. Bihel, C. De Michelis, G. Quelever, J. L. Kraus, *J. Chem. Soc. Perkin. Trans. 1*, 2001, **12**, 1421.
396. M. M. Sheha, N. M. Mahfouz, H. Y. Hassan, A. F. Youssef, T. Mimoto, Y. Kiso, *Eur. J. Med. Chem.*, 2000, **35**, 887.
397. V.-D. Le, C. C. Mak, Y.-C. Lin, J. H. Elder, C.-H. Wong, *Bioorg. Med. Chem.*, 2001, **9**, 1185.
398. Y. Konda, Y. Takahashi, S. Arima, N. Sato, K. Takeda, K. Dobashi, M. Baba, Y. Harigaya, *Tetrahedron*, 2001, **57**, 4311.
399. M. Marastoni, M. Bazzaro, S. Salvadori, F. Bortolotti, R. Tomatis, *Bioorg. Med. Chem.*, 2001, **9**, 939.
400. E. N. Prabhakaran, V. Rajesh, S. Dubey, J. Igbal, *Tetrahedron Lett.*, 2001, **42**, 339.
401. M. Ueki, S. Watanabe, Y. Ishii, O. Okunaka, K. Uchino, T. Saitoh, K. Higashi, H. Nakashima, N. Yamamoto, H. Ogawara, *Bioorg. Med. Chem.*, 2001, **9**, 487.
402. M. Ueki, S. Watanabe, Y. Ishii, O. Okunaka, K. Uchino, T. Saitoh, K. Higashi, H. Nakashima, N. Yamamoto, H. Ogawara, *Bioorg. Med. Chem.*, 2001, **9**, 477.
403. Y. Kiso, H. Matsumoto, T. Hamawaki, H. Ota, T. Kimura, T. Goto, K. Sano, Y. Hayashi, *Pept. Sci.*, 2001, **37**, 213.
404. H. Matsumoto, T. Kimura, T. Hamawaki, A. Kumagai, T. Goto, K. Sano, Y. Hayashi, Y. Kiso, *Bioorg. Med. Chem.*, 2001, **9**, 1589.
405. H. Matsumoto, Y. Sohma, D. Shuto, T. Hamawaki, T. Kimura, Y. Hayashi, Y. Kiso, *Pept. Sci.*, 2001, **37**, 237.
406. H. Matsumoto, Y. Sohma, T. Kimura, Y. Hayashi, Y. Kiso, *Bioorg. Med. Chem. Lett.*, 2001, **11**, 605.
407. H. Matsumoto, T. Matsuda, S. Nakata, T. Mitoguchi, T. Kimura, Y. Hayashi, Y. Kiso, *Bioorg. Med. Chem.*, 2001, **9**, 417.
408. H. L. Sham, D. A. Betebenner, T. Herrin, G. Kumar, A. Saldivar, S. Vasavanonda, A. Molla, D. J. Kempf, J. J. Plattner, D. W. Norbeck, *Bioorg. Med. Chem. Lett.*, 2001, **11**, 1351.
409. F. Bernedetti, S. Norbedo, *Chem. Commun (Cambridge UK)*., 2001, 203.
410. M. Alterman, H. Sjobom, P. Safsten, P.-O. Markgren, U. H. Danielson, M. Hamalainen, S. Lofas, J. Hulten, B. Classon, B. Samuelsson, A. Hallberg, *Eur. J. Pharm. Sci.*, 2001, **13**, 203.
411. C. Pesenti, A. Arnone, S. Bellosta, P. Bravo, M. Canavesi, E. Corradi, M. Frigerio, S. V. Meille, M. Monetti, W. Panzeri, F. Viani, R. Venturini, M. Zanda, *Tetrahedron*, 2001, **57**, 6511.
412. T. Fujisawa, S.-I. Katakura, S. Odake, Y. Morita, J. Yasuda, I. Yasumatsu, T. Morikawa, *Chem. Pharm. Bull.*, 2001, **49**, 1272.
413. D. L. Musso, M. W. Andersen, R. C. Andrews, R. Austin, E. J. Beaudet, J. D. Becherer, D. G. Bubacz, D. M. Bickett, J. H. Chan, J. G. Conway, D. J. Cowan, M. D. Gaul, K. C. Glennon, K. M. Hedeen, M. H. Lambert, M. A. Leesnitzer, D. L. McDougald, J. L. Mitchell, M. L. Moss, M. H. Rabinowitz, M. C. Rizzolio, L. T. Schaller, J. B. Stanford, T. Tippin, J. R. Warner, L. G. Whitesell, R. W. Wiethe, *Bioorg. Med. Chem. Lett.*, 2001, **11**, 2147.
414. S. Hanessian, N. Moitessier, E. Therrien, *J. Comput. Aided Mol. Des.*, 2001, **15**, 873.
415. E. Ambrose Armin, W. J. Welsh, *J. Med. Chem.*, 2001, **44**, 3849.
416. W. Yao, Z. R. Wasserman, M. Chao, G. Reddy, E. Shi, R. Q. Liu, M. B. Qian, M. D. Ribadeneira, D. Christ, R. R. Wexler, C. P. Decicco, *J. Med. Chem.*, 2001, **44**, 3347.
417. J. Schroder, A. Henke, H. Wenzel, H. Brandstetter, H. G. Stammler, A. Stammler, W. D. Pfeiffer, H. Tschesche, *J. Med. Chem.*, 2001, **44**, 3231.

418. J. I. Levin, J. Chen, M. Du, M. Hogan, S. Kincaid, F. C. Nelson, A. M Venkatesan, T. Wehr, A. Zask, J. DiJoseph, L. M. Killar, S. Skala, A. Sung, M. Sharr, C. Roth, G. Jin, G. Cowling, K. M. Mohler, R. A. Black, C. J. March, J. S. Skotnicki, *Bioorg. Med. Chem. Lett.*, 2001, **11**, 2189.

419. S. Pikul, K. M. Dunham, N. G. Almstead, B. De, M. G. Natchus, Y. O. Taiwo, L. E. Williams, B. A. Hynd, L. C. Hsieh, M. J. Janusz, F. Gu, G. E. Mieling, *Bioorg. Med. Chem. Lett.*, 2001, **11**, 1009.

420. P. Dunten, U. Kammlott, R. Crowther, W. Levin, L. H. Foley, P. Wang, R. Palermo, *Protein Sci.*, 2001, **10**, 923.

421. C. P. Decicco, Y. Song, D. A. Evans, *Org. Lett.*, 2001, **3**, 1029

422. Cs. Raman, H. Li, P. Martasek, G. Southan, B. S. Masters, T. L. Poulos, *Biochemistry*, 2001, **40**, 13448.

423. J. Yang, S. Wu, K. Zhou, S. Peng, S. Meng, Z. Wu, Y. Ruan, F. Dong, *Zhongguo Yaowu Huaxue Zazhi*, 2001, **11**, 13.

424. H. Huang, P. Martasek, L. J. Roman, R. B. Silverman, *J. Enzyme Inhib.*, 2001, **16**, 233.

425. X.. Li, R. N. Atkinson, S. Bruce King, *Tetrahedron*, 2001, **57**, 6557.

426. A. E. Moorman, S. Metz, M. V. Toth, W. M. Moore, G. Jerome, C. Kornmeier, P. Manning, D. W. Jr. Hansen, B. S. Pitzele, R. K. Webber, *Bioorg. Med. Chem. Lett.*, 2001, **11**, 2651.

427. J. M. Hah, L. J. Roman, P. Martasek, R. B Silverman, *J. Med. Chem.*, 2001, **44**, 2667.

428. P. Schumann, V. Collot, Y. Hommet, W. Gsell, F. Dauphin, J. Sopkova, E. T. MacKenzie, D. Duval, M. Boulouard, S. Rault, *Bioorg. Med. Chem. Lett.*, 2001, **11**, 1153.

429. H. Beaton, N. Boughton-Smith, P. Hamley, A. Ghelani, D. J. Nicholls, A. C. Thinker, A. V. Wallace, *Bioorg. Med. Chem. Lett.*, 2001, **11**, 1027.

430. B. M. Kessler, D. Tortorella, M. Altun, A. F Kisselev, E. Fiebiger, B. G. Hekking, H. L. Ploegh, H. S. Overkleeft, *Chem. Biol.*, 2001, **8**, 913.

431. D. Ma, Q. Wu, *Tetrahedron Lett.*, 2001, **42**, 5279.

432. B. K. Albrecht, R. N. Williams, *Tetrahedron Lett.*, 2001, **42**, 2755.

433. S. Lin, S. J. Danishefsky, *Angew. Chem. Int. End.*, 2001, **40**, 1967.

434. M. Inoue, H. Furuyama, H. Sakazaki, M. Hirama, *Org. Lett.*, 2001, **3**, 2863.

435. S. N. Crane, E. J. Corey, *Org. Lett.*, 2001, **3**, 1395.

436. C. Garcia-Echeverria, P. Imbach, D. France, P. Frust, M. Lang, A. M. Noorani, D. Scholz, J. Zimmermann, P. Furet, *Bioorg. Med. Chem. Lett.*, 2001, **11**, 1317.

437. P. Furet, P. Imbach, P. Furst, M. Lang, M. Noorani, J. Zimmermann, C. Garcia-Echeverria, *Bioorg. Med. Chem. Lett.*, 2001, **11**, 1321.

438. I. Momose, R. Sekizawa, S. Hirosawa, D. Ikeda, H. Naganawa, H. Iinuma, T. Takeuchi, *J. Antibiot.*, 2001, **54**, 1004.

439. R. Sekizawa, I. Momose, N. Kinoshita, H. Naganawa, M. Hamada, Y. Muraoka, H. Iinuma, T. Takeuchi, *J. Antibiot.*, 2001, **54**, 874.

440. M. K. Groll, Y. Koguchi, R. Huber, J. Kohno, *J. Mol. Biol.*, 2001, **311**, 543.

441. J. S. Lazo, D. C. Aslan, E. C. Southwick, K. A. Cooley, A. P. Ducruet, B. Joo, A. Vogt, P. Wipf, *J. Med. Chem.*, 2001, **44**, 4042.

442. N. Poigny, S. Nouri, A. Chiaroni, M. Guyot, M. Samadi, *J. Org. Chem.*, 2001, **66**, 7263.

443. K. Shen, Y. F. Keng, L. Wu, X. L. Gou, D.S. Lawrence, Z. Y. Zhang, *J. Biol. Chem.*, 2001, **276**, 47311.

444. Y. Gao, J. Voigt, H. Zhao, G. C. Pais, X. Zhang, L. Wu, Z. Y. Zhang, T. R. Burke Jr, *J. Med. Chem.*, 2001, **44**, 2869.

445. M. Sodeoka, R. Sampe, S. Kojima, Y. Baba, T. Usui, K. Ueda, H. Osada, *J. Med. Chem.*, 2001, **44**, 3216.

446. T. Usui, S. Kojima, S. Kidokoro, K. Ueda, H. Osada, M. Sodeoka, *Current Opin. Chem. Biol.*, 2001, **8**, 1209.

447. J. T. Maynes, K. S. Bateman, M. M. Cherney, A. K. Das, H. A. Luu, C. F. Holmes, M. N. James, *J. Biol. Chem.*, 2001, **276**, 44078.

448. K. L. Webster, A. B. Maude, M. E. O'Donell, A. P. Mehrotra, D. Gani, *J. Chem. Soc. Perkin Trans 1*, 2001, **14**, 1673.

449. M. E. O'Donell, J. Sanvoisin, D. Gani, *J. Chem. Soc. Perkin Trans 1*, 2001, **14**, 1696.

450. T. Sano, T. Usui, K. Ueda, H. Osada, K. Kaya, *J. Nat. Prod.*, 2001, **64**, 1052.

451. A. McCluskey, S. P. Ackland, E. Gardiner, C. C. Walkom, J. A. Sakoff, *Anticancer Drug Des.*, 2001, **16**, 291.

452. N. D. Fisher, N. K. Hollenberg, *Expert Opin. Investig. Drugs.*, 2001, **10**. 417.

453. A. Dondoni, G. De Lathauwer, D. Perrone, *Tetrahedron Lett.*, 2001, **42**, 4819.

454. H. P. Marki, A. Binggeli, B. Bittner, V. Bohner-Lang, V. Breu, D. Bur, P. H. Coassolo, J. P. Clozel, A. D'Arcy, H. Doebeli, W. Fischli, C. H. Funk, J. Foricher, T. Giller, F. Gruninger, A. Guenzi, R. Guller, T. Hartung, G. Hirth, C.H. Jenny, M. Kansy, U. Klinkhammer, U. Lave, B. Lohri, F. C. Luft, E. M. Mervaala, D. N. Muller, F. Montavon, C. H. Oefner, C. Qiu, A. Reichel, P. Sanwald-Ducray, M. Scalone, M. Schleimer, R. Schmid, H. Stadler, A Treiber, O. Valdenaire, E. Vieira, P. Waldmeimer, R. Wiegand-Chou, M. Wilhelm, W. Wostl, M. Zell, R. Zell, *Farmaco*, 2001, **56**, 21.

455. M. G. Bursavich, C.W. West, D. H. Rich, *Org. Lett.*, 2001, **3**, 2317.

456. S. Oh, W. Chang, J. Suh, *Bioorg. Med. Chem. Lett.*, 2001, **11**, 1469.

457. A. S. Ripka, K. A. Satyshur, R. S. Bohacek, D. H. Rich, *Org. Lett.*, 2001, **3**, 2309.

458. N. A. Dales, R. S. Bohacek, K. A. Satyshur, D. H. Rich, *Org. Lett.*, 2001, **3**, 2313.

459. J. M. Travins, M. G. Bursavich, D. F. Veber, D. H. Rich, *Org. Lett.*, 2001, **3**, 2725.

460. R.T. Turner, G. Koelsch, P. Castanheira, J. Ermolieff, A. K, Ghosh, J. Tang, P. Castenheira, A. Ghosh, *Biochemistry*, 2001, **40**, 10001.

461. M. L Steinhilb, R. S. Turner, J. R. Gaut, *J. Biol. Chem.*, 2001, **276**, 4476.

462. A. K. Ghosh, G. Bilcer, C. Harwood, R. Kawahama, D. Shin, K. A. Hussain, L. Hong, J. A. Loy, C. Nguyen, G. Koelsch, J. Ermolieff, J. Tang, *J. Med. Chem.*, 2001, **44**, 2865.

463. G. Evin, R. A. Sharples, A. Weidemann, F. B. Reinhard, V. Carbone, J. G. Culvenor, R. M. Holsinger, M. F. Sernee, K. Beyreuther, C. L. Masters, *Biochemistry*, 2001, **40**, 8359.

464. D. Beher, J. D. Wrigley, A. Nadin, G. Evin, C. L. Masters, T. Harrison, J. L. Castro, M. S. Shearman, *J. Biol. Chem.*, 2001, **276**, 45394.

465. I. Pinnix, U. Musunuru, H. Tun, A. Sridharan, T. Golde, C. Eckman, C. Ziani-Cherif, L. Onstead, K. Sambamurti, *J. Biol. Chem.*, 2001, **276**, 481.

466. S. Makino, T. Kayahara, K. Tashiro, M. Takahashi, T. Tsuji, M. Shoji, *J. Comput. Aided Mol. Des.*, 2001, **15**, 553.

467. B. Bachand, M. Tarazi, Y. St-Denis, J. J. Edmunds, P. D. Winocour, L. Leblond, M. A. Siddiqui, *Bioorg. Med. Chem. Lett.*, 2001, **11**, 287.

468. J. Cossy, D. Belotti, *Bioorg. Med. Chem. Lett.*, 2001, **11**, 1989.

469. K. P. Kokko, C. E. Arrigoni, T. A. Dix, *Bioorg. Med. Chem. Lett.*, 2001, **11**, 1947.

470. A. Zega, G. Mlinsek, P. Sepic, S. Golic Gradolnik, T. Solmajer, T. B. Tschopp, B. Steiner, D. Kikelj, U. Urleb, *Bioorg. Med. Chem.*, 2001, **9**, 2745.

471. P. J. Anderson, P. E. Bock, *Anal. Biochem.*, 2001, **296**, 254.

472. B, W. Clare, A. Scozzafava, C. T. Supuran, *J. Enz. Inhibition*, 2001, **16**, 1

473. S. D. Jones, J. W. Liebeschuetz, P. J. Morgan, C. W. Murray, A. D. Rimmer, J. M. E. Roscoe, B. Waszkowycz, P. M. Welsh, W. A. Wylie, S. C. Young, H. Martin, J. Mahler, L. Brady, K. Wilkinson, *Bioorg. Med. Chem. Lett.*, 2001, **11**, 733.

474. D. J. Pinto, M. J. Orwat S. Wang, J. M. Fevig, M. L. Quan, E. Amparo, J. Cacciola, K. A. Rossi, R. S. Alexander, A. M. Smallwood, J. M. Luettgen, L. Liang, B. J. Aungst, M. R. Wright, R. M. Knabb, P. C. Wong, R.R. Wexler, P. Y. Lam, *J. Med. Chem.*, 2001, **44**, 566.

475. M. L. Quan, R. R. Wexler, *Curr. Top. Med. Chem.*, 2001, **1**, 1028.

476. D. R. Light, W. J. Guilford, *Curr. Top. Med. Chem.*, 2001, **1**, 121.

477. T. Su, Y. Wu, B. Doughan, Z. J. Jia, J. Woolfrey, B. Huang, P. Wong, G. Park, U. Sinha, R. M.. Scarborough, B.-Y. Zhu, *Bioorg. Med. Chem. Lett.*, 2001, **11**, 2947.

478. H. Nishida, Y. Miyazaki, Y. Ktamuta, M. Ohashi, T. Matsusue, A. Okomoto, Y. Hosaka, S. Ohnishi, H. Mochizuki, *Chem. Pharm. Bull.*, 2001, **49**, 1237.

479. W.D. Shrader, W. B. Young, P. A. Spengeler, J. C. Sangalang, K. Elrod, G. Carr, *Bioorg. Med. Chem. Lett.*, 2001, **11**, 1801.

480. F. Dullweber, M. T. Stubbs, D. Musil, J. Sturzebecher, G. Klebe, *J. Mol. Biol.*, 2001, **313**, 593.

481. J. F. Lynas, S. L. Martin, B. Walker, *J. Pharm. Pharmacol.*, 2001, **53**, 473.

482. Y.-Q- Long, S.-L. Lee, C.-Y. Lin, I. J. Enyedy, S. Wang, P. Li, R. B. Dickson, P. P. Roller, *Bioorg. Med. Chem. Lett.*, 2001, **11**, 2515.

483. A. J. Harvey, A. D. Abell, *Bioorg. Med. Chem. Lett.*, 2001, **11**, 2441.

484. X. Y. Ye, T. B.Ng, P. F. Rao, *Biochem. Biophys. Res. Commun.*, 2001, **289** 91.

485. M. Umezaki, F. Kondo, T. Fujii, T. Yoshimura, S. Ono, *J. Pept. Sci.*, 2001, **37**, 33.

486. Z. Chen, T. P. Demuth, F. C. Wireko, *Bioorg.Med. Chem. Lett.*, 2001, **11**, 2111.

487. A. Wlodawer, M. Li, A. Gustchina, Z. Dauter, K. Uchida, H. Oyama, N. E. Goldfarb, B. M. Dunn, K. Oda, *Biochemistry*, 2001, **40**, 15602.

488. B. A. Katz, P. A. Sprengeler, C. Luong, F. Verner, K. Elrod, M. Kirtley, J. Janc, J. R.. Spencer, J. G. Breitenbucher, H. Hiu, D. McGee, D. Allen, A. Martelli, R. L. Mackman, *Chem. Biol.*, 2001, **8**, 1107.

489. M. L. Oliva, E. M. Santomauro-Vaz, S. A. Andrade, M. A. Juliano, V. J. Pott, M. U. Sampaio, C. A. Sampaio, *Biol. Chem.*, 2001, **382**, 109.

490. K. S. Yeung, N. A. Meanwell, Z. Qiu, D. Hernandez, S. Zhang, F. McPhee, S. Weinheimer, J. M. Clark, J. W. Janc, *Bioorg. Med. Chem. Lett.*, 2001, **11**, 2355.

491. H. J. Rideout, E. Zang, M. Yeasmin, R. Gordon, O. Jabado, D. S. Park, L. Stefanis, *Neuroscience*, 2001, **107**, 339.

492. V. A. Movsesyan, A. G. Yakovlev, L. Fan, A. I. Faden, *Exp. Neurol.*, 2001, **167**, 366.

493. P.Brown, D. S. Eggleston, R. C. Haltiwanger, R. L. Jarvest, L Mensah, P. J. O'Hanlon, A. J. Pope, *Bioorg. Med. Chem. Lett.*, 2001, **11**, 711.

494. R. L. Uarvest, J. M. Berge, P. Brown, D. V. Hamprecht, O. J. McNair, L. Mensah, P. J. O'Hanlon, A. J. Pope, *Bioorg. Med. Chem. Lett.*, 2001, **11**, 715.

495. J. Lee, S. U. Kang, S. Y. Kim, S. E. Kim, M. K. Kang, Y. J. Jo, S. Kim, *Bioorg. Med. Chem. Lett.*, 2001, **11**, 961.

496. J. Lee, S. U. Kang, S. Y. Kim, S. E. Kim, Y. J. Jo, S. Kim, *Bioorg. Med. Chem. Lett.*, 2001, **11**, 965.

497. A. M. Deveau, M. A. Labroli, C. M. Dieckhaus, M. T. Barthen, K. S. Smith, T. L. MacDonald, *Bioorg. Med. Chem. Lett.*, 2001, **11**, 1251.

498. J. Rudolph, H. Theis, R. Hanke, R. Endermann, L. Johannsen, F.-U. Geschke, *J. Med. Chem.*, 2001, **44**, 619.

499. F. Akahoshi, A. Ashimori, T. Yoshimura, T. Imada, M. Nakajima, N. Mitsutomi, S. Kuwahara, T. Ohtsuka, A. Fukaya, M. Miyazaki, N. Nakamura, *Bioorg. Med.*

Chem., 2001, **9**, 301.

500. F. Akahoshi, A. Ashimori, H. Sakashita, T. Yoshimura, M. Eda, T. Imada, M. Nakajima, N. Mitsutomi, S. Kuwahara, T. Ohtsuka, C. Fukaya, M. Miyazaki, N. Nakamura, *J. Med. Chem.*, 2001, **44**, 1297.

501. Y. Aoyama, K. Konoike, A. Kanda, N. Naya, N. Nakajima, *Bioorg. Med. Chem. Lett.*, 2001, **11**. 1695.

502. U. Neumann, N. M. Schechter, M. Gutschov, *Bioorg. Med. Chem.*, 2001, **9**, 947.

503. M-A. Poupart, D. R. Cameron, C. Chabot, E. Ghiro, N. Goudreau, S. Goulet, M. Poirier, Y. S. Tsantrizos, *J. Org. Chem.*, 2001, **66**, 4743.

504. L. T. Burke, D. J. Dixon, S. V. Ley, F. Rodriguez, *Org. Lett.*, 2000, **2**, 3611.

505. C. Marrano, P. de Macedo, J. W. Keillor, *Bioorg. Med. Chem.*, 2001, **9**, 1923

506. N. Valiaeva, D. Bartley, T. Konno, J. K. Coward, *J. Org. Chem.*, 2001, **66**, 5146.

507. B. A. Pender, X. Wu, P. H. Axelsen, B. S. Cooperman, *J. Med. Chem.*, 2001, **44**, 36.

508. S. Sasaki, T. Ehara, M. Rowshon Alam, Y. Fujino, N. Harada, J. Kimura, H. Nakamura, M. Maeda, *Bioorg. Med. Chem. Lett.*, 2001, **11**, 2581.

509. T. D. Owens, G-L. Araldi, R. F. Nutt, J. E. Semple, *Tetrahedron Lett.*, 2001, **42**, 6271.

510. S. Bounaga, M. Galleni, A. Laws, M. I. Page, *Bioorg. Med. Chem.*, 2001, **9**, 503.

511. L. Kiczak, M. Kusztura, K. Koscielska-Kasprzak, M. Dadlez, J. Otlewski, *Biochim. Biophys. Acta*, 2001, **1550,** 153.

512. J. Zhang, H. Spring, M. Schawab, *Cancer Lett.*, 2001, **171** 153.

513. A. K. Ghosh, P. E. Ribolla, and M. Jacobs-Lorena, *Proc. Natl. Acad. Sci. U S A*, *2001*, **98,** 13278.

514. N. J. Skelton Y. M. Chen, N. Dubree, C. Quan, D. Y. Jackson, A. Cochran, K. Zobel, K. Deshayes, M. Baca, M. T. Pisabarro, H. B. Lowman, *Biochemistry, 2001*, **40**, 8487.

515. A. Desiderio, R. Franconi, M. Lopez, M. E. Villani, F. Viti, R. Chiaraluce, V. Consalvi, D. Neri, E. Benvenuto, *J. Mol. Biol.*, 2001, **310**, 603.

516. T. Wind, M. A. Jensen, P. A. Andreasen, *Eur. J. Biochem.*, 2001, **268**, 1095.

517. J. Wolcke, E. Weinhold, *Nucleosides, Nucleotides, Nucleic Acids*, 2001, **20**, 1239.

518. F. Tanaka, C. F. Barbas, *Chem. Commun (Cambridge UK).*, 2001, 769.

519. J. Li, K. Wang, *Shengwu Huaxue Yu Shengwu Wuli Xuebao*, 2001, **33**, 513.

520. S. J. Kennel, S. Mirzadeh, G. B. Hurst, L. J. Foote, T. K. Lankford, K. A. Glowienka, L. L Chappell, J. R. Kelso, S. M Davern, A. Safavy, M. W. Brechbiel, *Nucl. Med. Biol.* 2000, **27**, 815.

521. F. D'Mello, C. R. Howard *J. Immunol. Methods*, 2001, **247** 191.

522. D. S. Wilson, A. D. Keefe, J. W. Szostak., *Proc. Natl. Acad. Sci. U S A*, 2001, **98**, 3750.

523. G. Muller, *Top. Curr. Chem.*, 2001, **211**, 17.

524. R. Sundaramoorthi, C. Siedem, C. B. Vu, D. C. Dalgarno, E. C. Laird, M. C. Botfield, A. B. Combs, S. E. Adams, R. W. Yuan, M. Weigele, S. S. Narula, *Bioorg. Med. Chem. Lett.*, 2001, **11**, 1665.

525. E. Mandine, D. Gofflo, V. Jean-Baptiste, E. Sarubbi, G. Touyer, P. Deprez, D. Lesuisse, *J. Mol. Recognit.*, 2001, **14**, 254.

526. Y. Gao, J. X. Viogt Vu, D. Yang, T. R. Burke Jr, *Bioorg. Med. Chem. Lett.*, 2001, **11**, 1889.

527. T. R. Burke Jr, Z.-Y. Yao, Y. Gao, J. X. Wu, X. Zhu, J. H. Luo, R. Guo, D. Yang, *Bioorg. Med. Chem. Lett.*, 2001, **9**, 1439.

528. S. Bohacek, D. C. Dalgarno, M. Hatada, V. A. Jacobsen, B. A. Lynch, K. J. Macek, T. Merry, C. A. Metcalf III, S. S. Narula, T. K. Sawyer, W. C. Schakespeare, S. M.

Violette, M. Weigele, *J. Med. Chem.*, 2001, **44**, 660.

529. R. A. Falconer, I. Toth, *Pept. New Millennium, Proc. Am. Pept. Symp*, 16th 1999 (Pub. 2000) 777-778. Edited by G. B. Fields, J. P. Tam, G. Barany. Kluwer Academic Publishers, Dordrecht, Neith.

530. J.P. Malkinson, Gy. Keri, P. Artursson, I. Toth, *Pept. New Millennim, Proc. Am. Pept. Symp*, 16 th. 1999, (Pub. 2000) 779-781. Edited by G. B. Fields, J. P. Tam, G. Barany, Kluwer. Academic Publisher, Dordrecht, Neth.

531. I. I. Stoikov, S. A. Repejkov, I. S. Antipin, A. I. Konovalov, *Heteroat. Chem.*, 2000, **11**, 518.

532. I. I. Stoikov, A. A. Khrustalev, I. S. Antipin, A. I Konovalov, *Dokl. Akad. Nauk.*, 2000, **374**, 202.

533. D. Bonnet, K. Thaim, E. Loing O. Melnyk, H. Gras-Masse, *J. Med. Chem.*, 2001, **44**, 468.

534. M. Lindgren, X. Gallet, U. Soomets, M. Haellbrink, E. Brkenhielm, M. Pooga, R. Brasseur, U. Langel, *Bioconjugate Chem.*, 2000, **11**, 619.

535. C. Garcia-Echeverria, L. Jiang, T. M. Ramsey, S. K. Sharma, Y.-N. Chen, *Bioorg. Med. Chem. Lett.*, 2001, **11**, 1363.

536. P. A Wender, T. C. Jessop, K. Pattabiraman, E. T. Pelkey, C. L. VanDeusen, *Org. Lett.*, 2001, **3**, 3229.

537. J. Gao, M. Sudoh, J. Aube, R. T. Borchardt, *J. Pept. Res.*, 2001, **57**, 316.

538. J. Gao, S.L. Winslow, D. Vander Velde, J. Aube, R. T. Borchardt, *J. Pept. Res.*, 2001, **57**, 361.

539. P.A. Wender, D.J. Mitchell, K. Pattabiraman, E.T. Pelkey, L. Steinman, J.B. Rothbard, *Proc. Natl. Acad. Sci. U.S.A,.* 2000, **97**, 13003.

540. W. Mier, R. Eritja, A. Mohammed, U. Haberkorn, M. Eisenhut, *Bioconjugate Chem.*, 2000, **11**, 855.

541. T. Sakaeda, Y. Tada, T. Sugawara, T. Ryu, F. Hirose, T. Yoshikawa, K. Hirano, L. Kupczyk-Subotkowska, J. T. Siahaan, K. L. Audus, V. J. Stella, *J. Drug Targeting*, 2001, **9**, 23.

542. T. Ohnishi, T. Maruyama, S. Higashi, S. Awazu, *J. Drug Targeting*, 2000, **8**, 395.

543. J. M. Benns, J-S. Choi, R. I. Mahato, J-S. Park, S. W. Kim, *Bioconjugate Chem.*, 2000, **11**, 637.

544. C. U. Nielsen, R. Andersen, B. Brodin, S. Frokjaer, B. Steffansen, *J. Controlled Release*, 2001, **73**, 21.

545. R. P. McGeary, I. Jablonkai, I. Toth, *Tetrahedron*, 2001, **57**, 8733.

546. M. Harada, H. Sakakibara, T. Yano, T. Suzuki, S. Okuno, *J. Controlled Release*, 2000, **69**, 399.

547. J. W. Singer, P. De Vries, R. Bhatt, J. Tulinsky, P. Klein, C. Li, L. Milas, R. A. Lewis, S. Wallace, *Ann. N. Y. Acad. Sci.*, 2000, **922**, 136.

548. D. Ranganathan, S. Kurur, A. C. Kunwar, A. V. S. Sarma, M. Vairamani, I. L. Karle *J. Pept. Res*, 2000, **56**, 416.

3
Cyclic, Modified and Conjugated Peptides

BY JOHN S. DAVIES

1 Introduction

The sub-divisions of this Chapter have again worked out to be similar to previous volumes, and in the absence of a detailed index, it is hoped that this continuity will aid researchers wishing to carry out year by year reviewing of a particular sub-topic. Core references for the Chapter were again obtained from the databases of the Chemical Abstracts Service, as compiled in their publication, CA Selects on Amino Acids, Peptides and Proteins (up to issue 12, 2002)[1]. However with the expansion in the availability of journals in electronic format, computer scanning of published titles has become very much easier, and a great aid to this has been the Web of Science databases[2]. Another efficient way of communicating knowledge is the International Conference and the fruits of the 26th European Peptide Symposium at Montpellier, published[3] as a 1050 page tome contains many accounts of direct relevance to this Chapter. But as in the past, conference reports are not reviewed in these Reports. The ever-increasing discoveries of new cyclic structures tend to saturate the mind as to the correlation between names and structures. Therefore a useful glossary of structures and names for the more established structures published within a new monograph[4] is a reference base to be welcomed.

Based on the number of references retrieved for reviewing, there has been increased activity in this subject area this year. Quite significant themes have been the subject of a number of authoritative reviews. A most welcomed update on the synthesis of cyclic peptides covering the 1997-2000 era has been published[5], and the role of enzymes in peptide cyclisation has also been reviewed[6]. A comprehensive compilation[7] of natural products found in marine cyanobacteria reveals that they are a rich source of cyclic structures, while β-turn mimetics have been shown[8] to be a very productive endeavour in the enkephalin, somatostatin area, as well as targeting many receptors associated with a number of diseases. The synthesis of cyclic peptidomimetics often requires special reagents and reactions, and the contributions to this area by Murray Goodman and his group have been reviewed[9]. Designer hybrid cyclopeptides for membrane ion transport and tubular structures has been the subject of a short review[10], mainly concentrating on the author's own work. A 406-reference update[11] of new trends in

peptide coupling reagents is of direct relevance to the continuing demands for
efficient synthetic routes to cyclic peptides.

2   Cyclic Peptides

2.1   General Considerations. – Another addition to the armoury of solid phase
resin linkers for on-resin cyclisation is based on a phenyl hydrazide[12] as summarised in Scheme1. Excellent purity of product has been obtained, although the
yields depend on the steric demands of the amino acid residues. Isolated thioesterase domains have been shown[13] to catalyse the cyclisation of synthetic
peptide thioester substrates. Tyrocidine non-ribosomal peptide synthetase not
only coped well with the cyclisation of the tyrocidine linear precursor peptide (as
its N-acyl thioester), but could also cyclise linear precursors of gramicidin S and

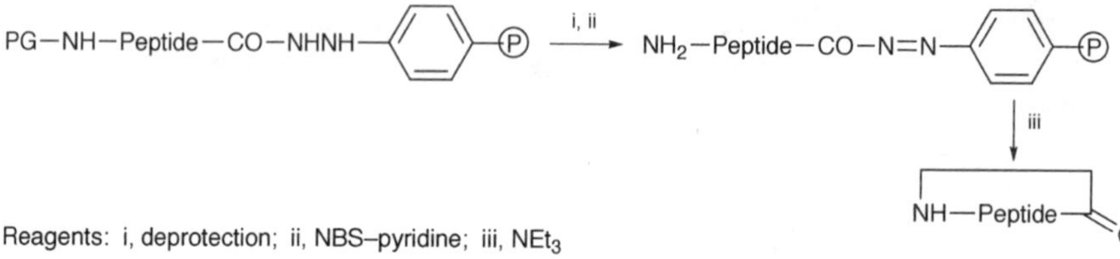

Reagents: i, deprotection; ii, NBS–pyridine; iii, NEt$_3$

Scheme 1

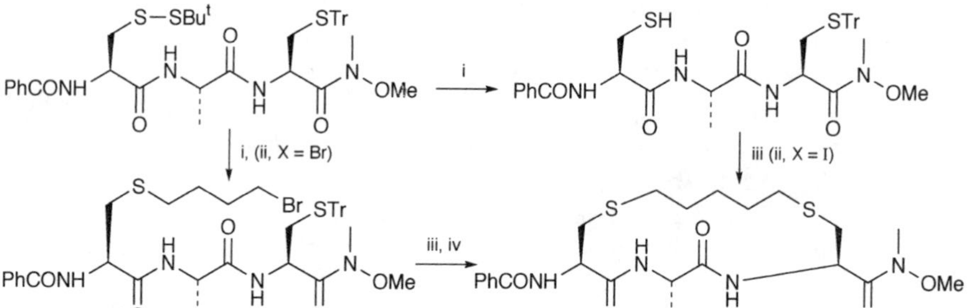

Reagents: i, 4eq. Bu$_3$P, THF/H$_2$O; ii, X–(CH$_2$)$_4$–X, DBU, PhH; iii, 2eq. Et$_3$SiH, 3:1 CH$_2$Cl$_2$:TFA;
iv, DBU, PhH

Scheme 2

a lipopeptide analogue of surfactin A. The Ugi four-component condensation
method has been used as a strategy[14] for the synthesis of orthogonally protected
diaminoglutaric acid containing peptides, to replace cystine in the somatostatin
analogues, Sandostatin and TT-232. Hplc aided the separation of the four
diastereomeric peptides produced via the Ugi strategy. A rapid method[15] for
producing a diverse set of cyclopeptidomimetics based on side chain alkyl linkers
(summarised in Scheme 2), also uses as C-terminus the Weinreb amide, which
will yield aldehydes or ketones, useful as serine protease inhibitors.

The journey from γ,δ-unsaturated amino acid synthesis to the preparation of
cyclic peptides by ring closing metathesis has been reviewed[16], while an example
of this ring closing process has been used[17] to make macrocyclic helical peptides
as exemplified in Scheme 3. Conformational studies showed that the cyclic

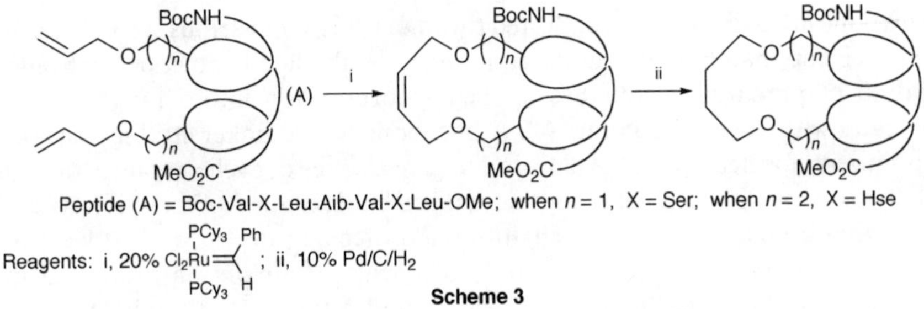
BocNH
MeO2C
(A)
i
ii
Peptide (A) = Boc-Val-X-Leu-Aib-Val-X-Leu-OMe; when n = 1, X = Ser; when n = 2, X = Hse
Reagents: i, 20% Cl2Ru(PCy3)2=C(Ph)H ; ii, 10% Pd/C/H2

Scheme 3

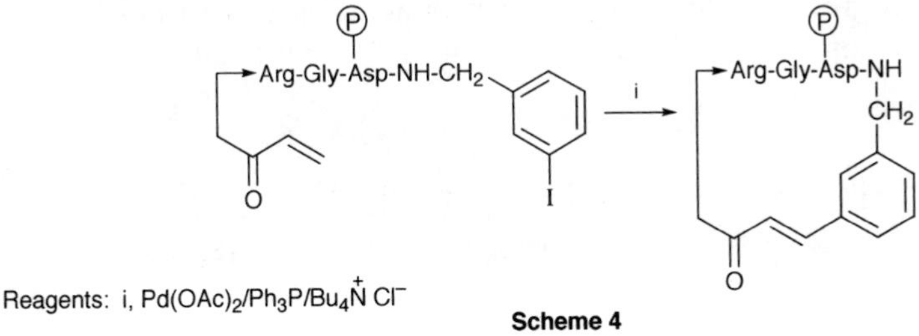
Arg-Gly-Asp-NH-CH2
i
Arg-Gly-Asp-NH
CH2
Reagents: i, Pd(OAc)2/Ph3P/Bu4N+ Cl−

Scheme 4

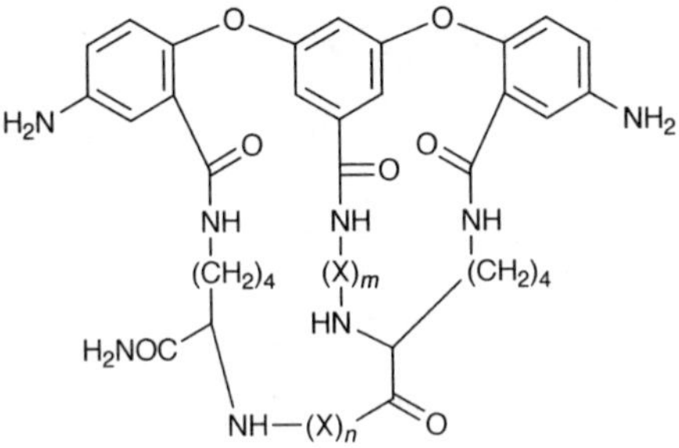
H2N
NH2
NH
(CH2)4
(X)m
HN
H2NOC
NH—(X)n

(1)

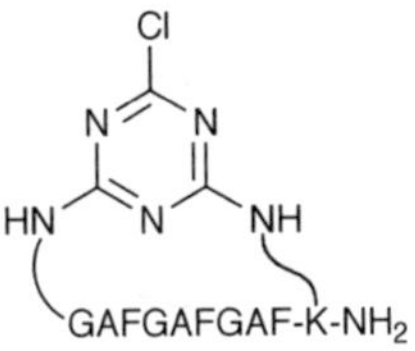
Cl
HN
NH
GAFGAFGAF-K-NH2

(2)

compounds exist as right handed 3_{10} helices up to the 5th residue. The Heck reaction (Scheme 4) has been chosen[18] for the construction of cyclic RGD libraries and uses a side-chain attachment of a chlorotritylchloride resin to the β-carboxyl of the aspartyl residue as anchorage. The S_NAr cyclisation method, well-developed in the solution phase for the formation of biphenyl ether links in the synthesis of compounds related to vancomycin has been adapted[19] to the solid phase as exemplified by the synthesis of (1) in a single cyclisation step. Sequential S_NAr reactions of e.g. 2,4,6-trichloro[1,3,5]-triazine on cellulose membranes have effected[20] the synthesis of cyclic di- to deca-peptides having structures such as (2).

The impetus towards making cyclic peptide libraries has spawned many new approaches to synthesis. An allyl/allyloxycarbonyl strategy[21] for the preparation of side chain-to-side chain cyclic peptides has been used in the construction of a small library with lactam bridges. Removing the Allyl/Alloc groups under specific neutral conditions [$Pd(PPh_3)/PhSiH_3/DCM$] was key to the success of the process. Further advantages over the conventional lactam bridge construction has been claimed[22] for the method of Gilon and co-workers, wherein any two backbone nitrogens can be connected through bridges of various sizes. There is a requirement that building units containing N-alkylated moieties such as (3) should be prepared in advance of their incorporation into a range of cyclic peptides with general structure (4). A backbone amide linker strategy has been optimised[23] for head to tail cyclisation to give a series of cyclic peptides targeted at the somatostatin receptor as summarised in Scheme 5. A new safety-catch linker has been developed[24] which utilises a protected catechol derivative, which allows Boc-type solid phase assembly of linear precursors on-resin, and just before cyclisation, the linker is activated, and after neutralisation of the N-terminal amino group, cyclisation with concomitant cleavage from the resin takes place as summarised in Scheme 6. A 400-member library of cyclic

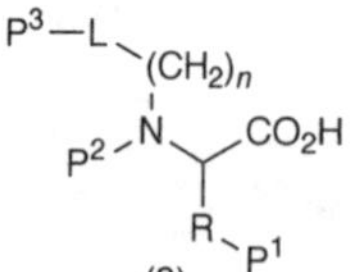

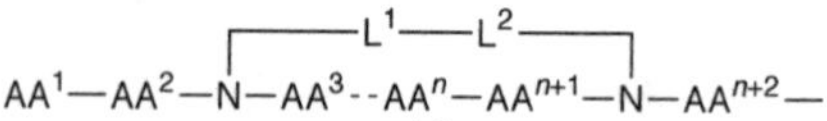

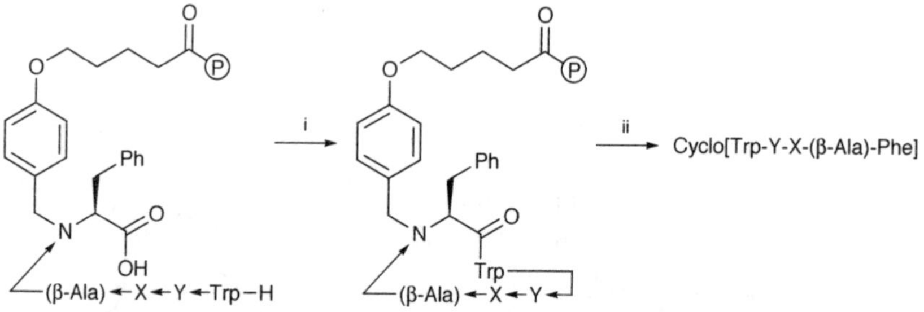

Reagents: i, BOP/DIEA/DMF 61% yield; ii, HF or strong acid + scavengers (1 hr)

Scheme 5

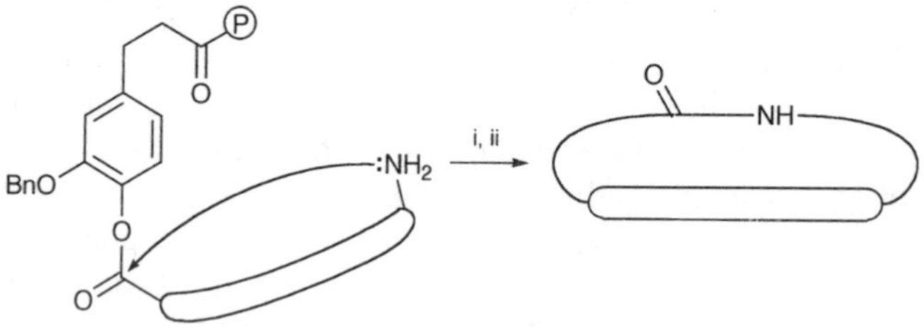

Reagents: i, HF/cresol; ii, 2% DIEA in DMF or 20% piperidine

Scheme 6

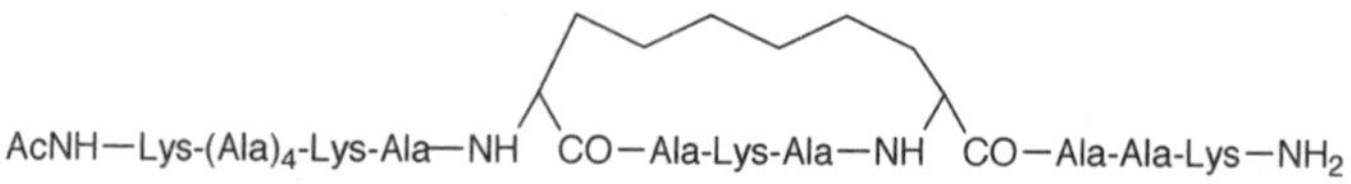

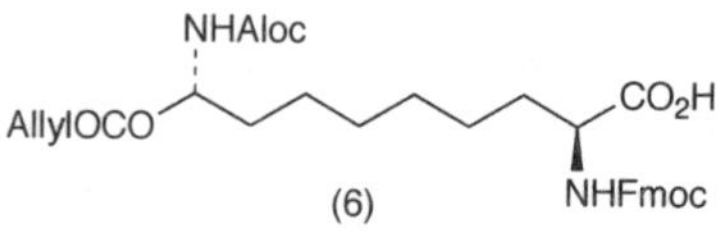

peptides was assembled using this linker technology. A polymethylene bridge has been incorporated[25] into a variant (5) of the AK peptide, by using the orthogonal protection offered by the di-amino acid derivative (6). While the aim for this i to i+4 link was to stabilise helicity, it turned out that the cyclic peptide showed a strong β-turn propensity. An efficient approach[26] to an amphipathic bicyclic peptide involves on-resin intramolecular thioester ligation, followed by an off-resin disulfide formation, while intracellular libraries of small cyclic peptides[27] (ranging from 4 to 9 residues) can be generated using DnaE intein from *Synechoeystis* sp. PCC6803.

2.2 Cyclic Dipeptides (Including Dioxopiperazines). – On many occasions in the past this sub-section has been the preserve of the 2,5 dioxopiperazines, but whilst they still remain the predominant structures, the thrust towards pep-tidomimetics, has generated an interest in other structural variants which can be accommodated by the term cyclic dipeptide. However nature sticks loyally to the 2,5 dioxopiperazine forms, as shown by the presence[28] of lumpidin (7) in the food-spoiling fungus *Penicillium nordicum*. The culture broth of *Streptomyces* sp. (strain GT 051237) has yielded[29] the new dioxopiperazines, maremycins C1 (8),

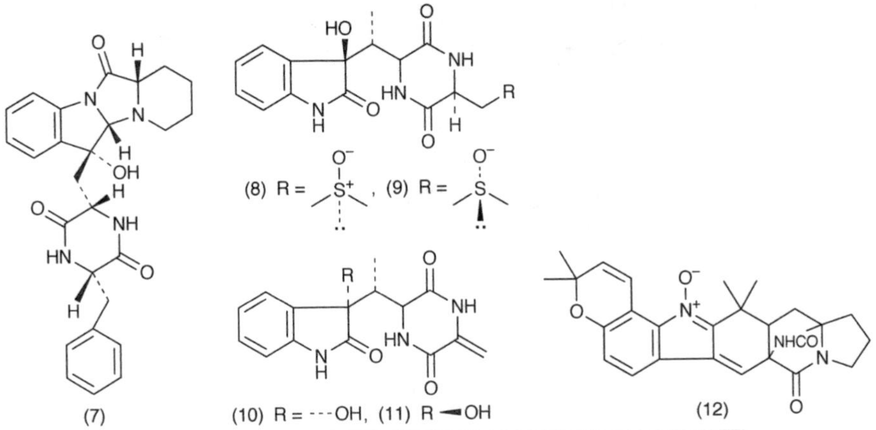

(8) R = ; (9) R =

(7)　(10) R = ---OH, (11) R ←OH　(12)

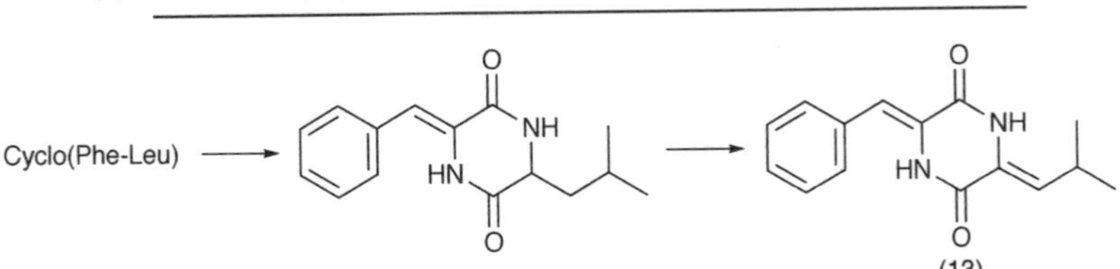

Cyclo(Phe-Leu)

Scheme 7

(13)

C2 (9), D1 (10) and D2 (11), while the broth from *Aspergillus ochraceus* CL41582 has given the antibiotic[30], CJ-17,665 (12). This antibiotic inhibits the growth of multi-drug resistant *S. aureus*, *S. pyogenes* and *Enterococcus faecalis* with MIC values of 12.5, 12.5 and 25 μg/ml. The enzyme, cyclic dipeptide oxidase, which specifically catalyses the dehydrogenation (Scheme 7) of the cyclic dipeptide precursor of the antibiotic, albonourisin (13) has been purified and partially

characterised[31] and found to be able to catalyse a wider variety of dehydrogena-
tions than first thought. Cyclic dipeptides containing hydrophobic amino acids,
have been shown to interact[32] with hepatic cytochrome P450 and the anti-fungal
action of cyclic dipeptides such as cyclo (Phe-Trp) and cyclo (Phe-Phe) has been
rationalised[33] as involving membrane disruption, and mediation/activation of
intracellular signalling.

Enantioselective total syntheses[34] have been achieved for ditryptophenaline
(14) and ent-WIN 64821 (15), using the ready availability of a bisoxindole system
as the starting point. In the syntheses of circumdatin F and sclerotigenin, a
photolabile nitrobenzyl group on the amide nitrogen of the dioxopiperazine[35]
has enabled an aza-Wittig procedure to be used for the intermediate
fumiquinazoline G (16). Fluoroisosteres of gypsetin and brevianamide E have
been prepared[36] via a fluorination/cyclisation process, as exemplified in Scheme 8
for the formation of fluorogypsetin in 77% yield.

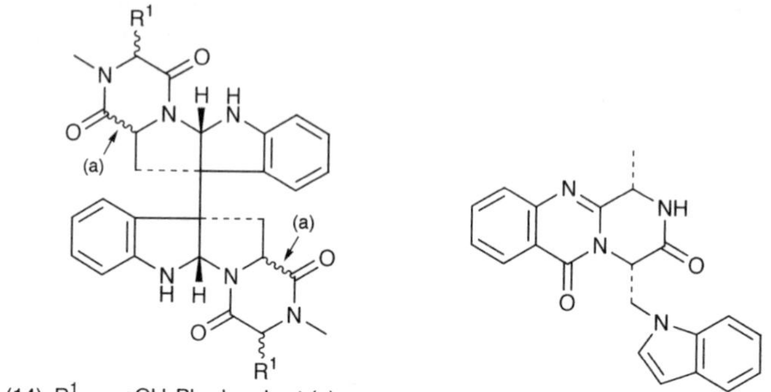

(14) R^1 = —CH$_2$Ph, bonds at (a) —
(15) R^1 = ---CH$_2$Ph, bonds at (a) ----

(16)

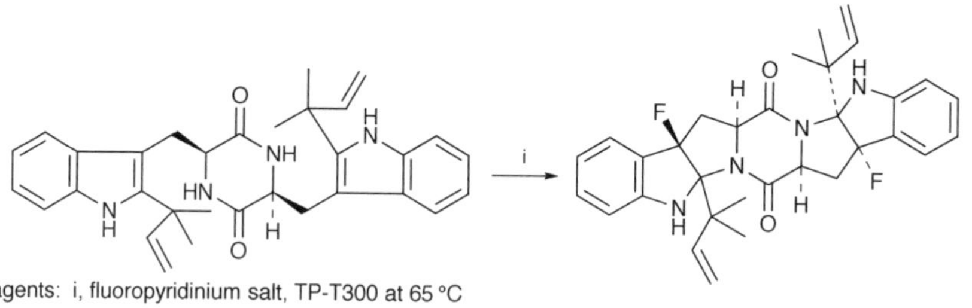

Reagents: i, fluoropyridinium salt, TP-T300 at 65 °C

Scheme 8

The role of dioxopiperazines as enzyme inhibitors have been investigated. But
it was disappointment that awaited the researchers that synthesised[37] (17), believ-
ing it to be an inhibitor of calpain. The crystalline product synthesised, contras-
ted with the active oil originally isolated form *S. griseus*, in that it showed no
calpain inhibitory activity. Better luck was obtained[38] with a series of dioxo-
piperazine-based inhibitors of plasminogen activator inhibitor PAI-1, with (18)
having an *in vitro* IC$_{50}$ of 2 μM. Topoisomerase inhibitors have little or no
antineoplastic activity unless they are topoisomerase poisons as well. It has been

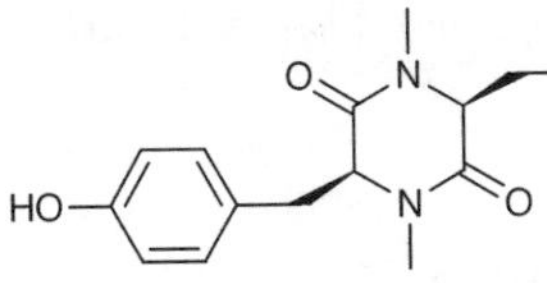

(17)

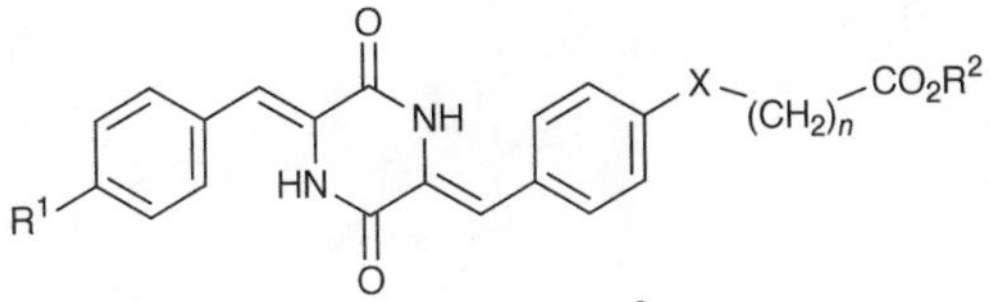

(18) R^1 = NHCO–(2-thiophene), R^2 = H, X = O, $n = 7$

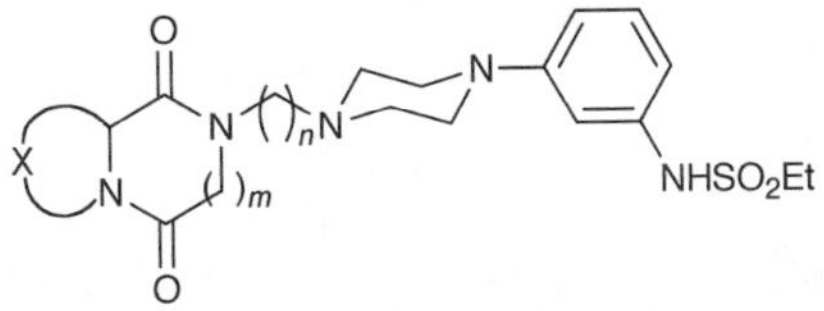

(19) X = $(CH_2)_3$, $m = 0$, $n = 4$

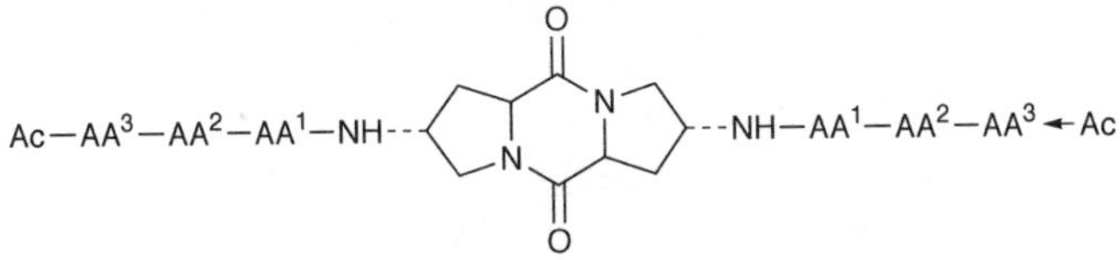

Ac—AA^3—AA^2—AA^1—NH-- [structure] --NH—AA^1—AA^2—AA^3 ← Ac

(20)

H—Phe—Asp—Leu—Trp—N [structure] N ← Trp ← Leu ← Asp ← Phe—NH$_2$

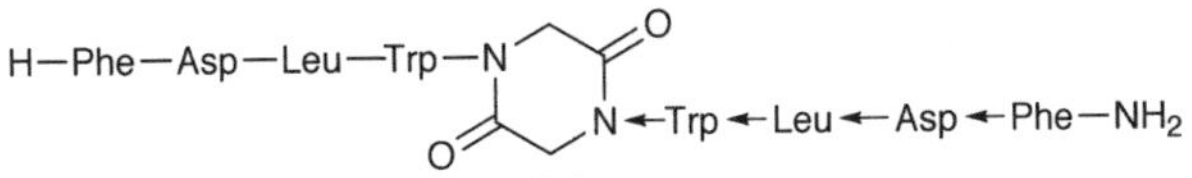

(21)

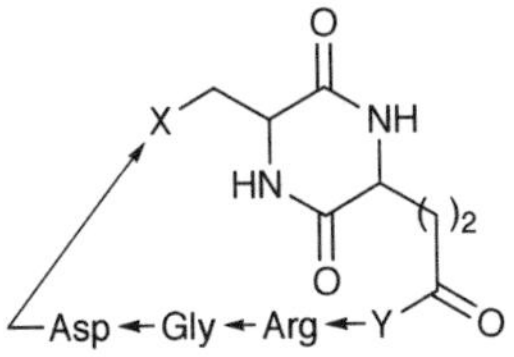

(22)

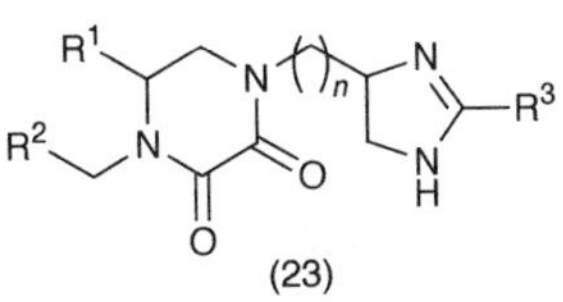

(23)

shown[39] that meso-2,3-bis(2,6-dioxopiperazin-4-yl) butane (ICRF-193) gives significant topoisomerase II poisoning, consistent with β-isozyme selectivity. A series of amide arylpiperazine derivatives have been designed and synthesised[40] to assess their 5-HT1A/α (1) adrenergic receptor affinity. Of the series examined, the best profile was attained by (19), which turns out to be a hydantoin. 'Two-armed' dioxopiperazine structures such as (20), derived from 4-OH-Pro can function[41] as highly selective receptors for small peptides, while the dioxopiperazine ring can also function as a scaffold[42] for constructing dimeric structures, e.g. (21) a N-terminal cholecystokinin tetrapeptide dimer. Dioxopiperazines can be incorporated[43] on-resin for the formation of constrained RGD scaffolds such as (22). The best binding efficiency for the $α_vβ_3$ receptor was obtained for (22, X and Y not being present) having an IC_{50} value of ca. 4μM. Solid phase syntheses[44] of imidazoline-tethered 2,3-dioxopiperazines (23) have been described as part of a broader series of compounds based on cyclic ureas and thioureas.

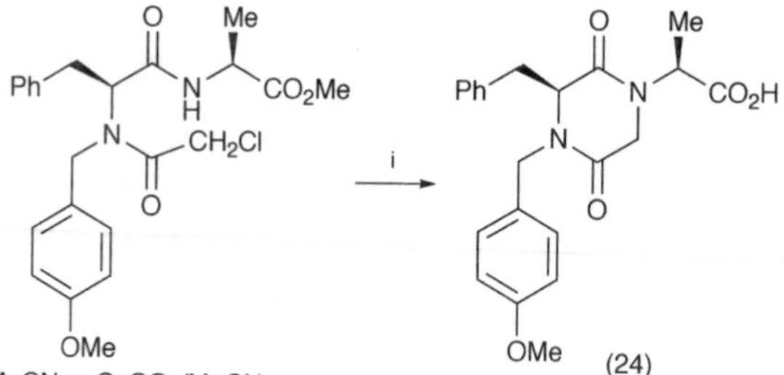

Reagents: i, NaH/MeCN or CsCO₃/MeCN

Scheme 9

Synthetic protocol, which gave a versatile route to 3-unsubstituted 4-alkyl-4-carboxy-2-azetidinones from amino acid derivatives, when applied[45] to dipeptides gave dioxopiperazine (24) according to Scheme 9. Aminoxydioxopiperazine (25) has performed[46] a key role in the stereospecific synthesis of 2(S)-amino-6(R)-hydroxy-1,7-heptanedioic acid dimethyl ester hydrochloride, while cyclo -(Leu-Leu), -(norLeu-nor Leu), -(Val-Val) and -(Phe-Phe) have been converted[47] by POCl₃ to dichlorodihydropyrazines which then condense to form polyethers such as (26). N-Terminal aldehydes have been synthesised[48] on a solid support, and on reductive amination with amino esters led to N-terminal dioxopiperazine peptides such as (27). Anomeric mannopyranose amino acid derivatives such as (28) have been synthesised[49] via N-acylated bicyclic [2.2.2] lactones, while (S)-N, N'-bis(p-methoxybenzyl)-3-isopropylpiperazine-2,5-dione, as a glycine enolate, exhibited[50] high levels of enantiodiscrimination towards racemic 2-bromo-propionate esters. The related chiral auxiliary (29), on 1,4-conjugate addition of organocuprates, affords[51] homochiral α – amino acids in excellent yields.

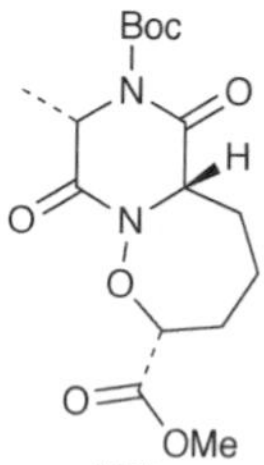

(25)

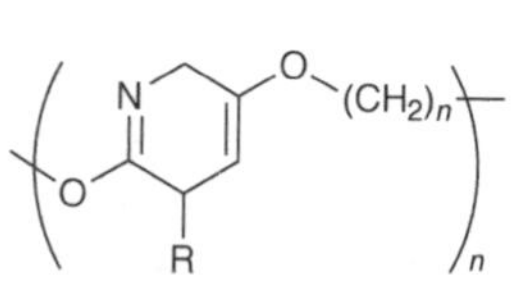

(26) R = CH₂CHMe₂, (CH₂)₃Me, CHMe₂ or CH₂Ph

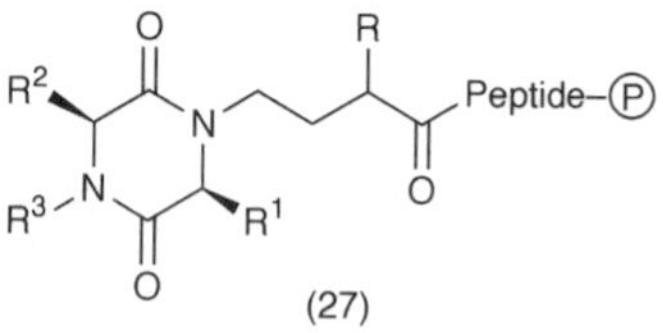

(27)

The term 'cyclised dipeptides' defines work in the peptidomimetic field, which does not necessarily involve cyclisation to the homodetic dioxopiperazine ring. For example Gly-Gly units can be tethered[52] via aminocaproic acid linkers to form macrocycles such as (30). This model secured a β-turn conformation for the dipeptide sequence and was further explored through substitution of benzyl groups in turn into position $R^2 - R^6$. The highest yielding macrocyclisation conditions were DECP(diethylcyano phosphate)/Et₃N/DMF which gave an yield of 26%. Ring closing metathesis of a diallylglycine precursor has pro-

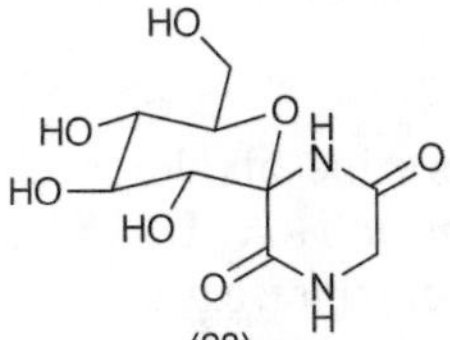

(28) 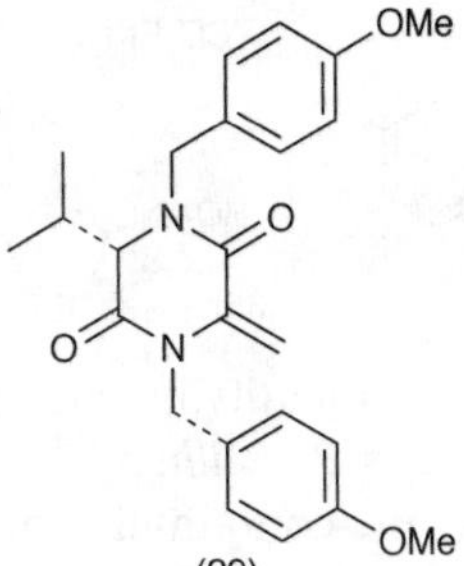 (29) 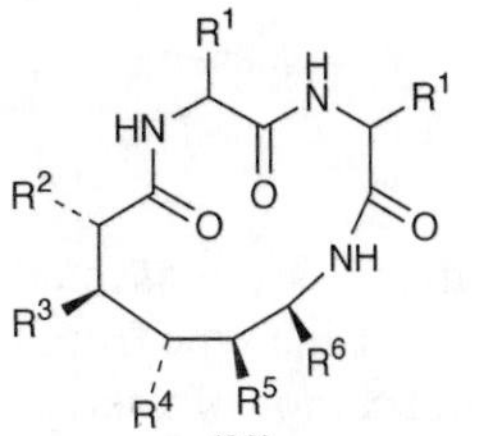 (30)

duced[53] a cyclic derivative, which on deprotection gave cyclic dipeptide (31) which may be useful as a scaffold for more peptidomimetic work. Cyclisation[54] of a linear protected N-Ala-β-amino acid sequence using BOP/DIEA/DMF yielded the 2,5-imidazolidinedione (32), believed to have involved a 7-membered cyclodipeptide ring on its way to the final product. Methodology developed in the solution phase has now been adapted[55] for use in the solid phase synthesis of 1,3-disubstituted 2,6-dioxopiperazines (33), which are of interest in inhibitor and anti-cancer research. The potential of 3-aminopiperidin-2,5-diones, such as (34) has been explored[56] as conformationally – constrained surrogates of Ala-Gly dipeptides.

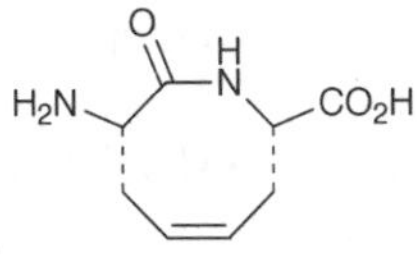

(31) 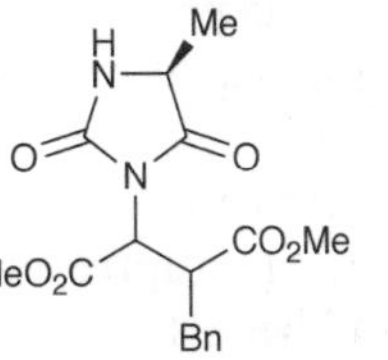 (32) 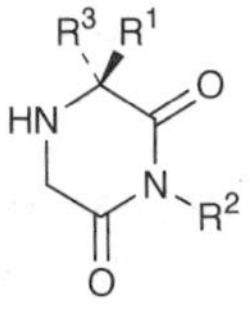 (33) 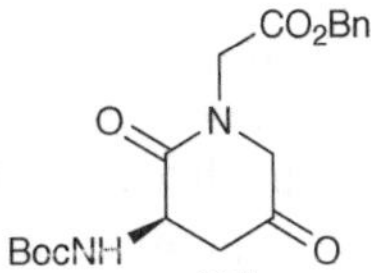 (34)

The possibility of 2,5-diketopiperazine formation from commercially important dipeptide derivatives has been a subject of intense discussion for a number of years. The sweetener aspartame (Asp-Phe-OMe), undergoes different degradation pathways depending on the pH of the solution[57]. The major degradation product found in the pH range 7-10 was the corresponding 2,5-dioxopiperazine, which is also the main product detected by X-ray powder diffractometry[58], if aspartame is heated in powder form to 180°C. The solid state degradation of the pharmaceutical, enalapril (35) as its maleate salt, studied[59] using FT-IR microspectroscopy in the range 120-130°C proceeds via intramolecular cyclisation to its dioxopiperazine. An activation energy of 195 ± 12 kJ was required for the cyclisation. A pre-dissociation *cis-trans* isomerisation of the N-terminal amide bond of protonated triglycyl units has been subjected to detailed quantum-mechanical and RRKM modelling[60], which came to the conclusion that the mass spectral fragmentation process via the dioxopiperazine pathway is kinetically controlled. Glyco-conjugates based on D-glucosyl esters of the carboxy groups of peptides[61] such as Tyr-Pro, Tyr-Pro-Phe or Tyr-Pro-Phe-Val can undergo intramolecular transformations, including the formation of cyclo (Tyr-Pro). Uv absorption spectra and chromatographic techniques have

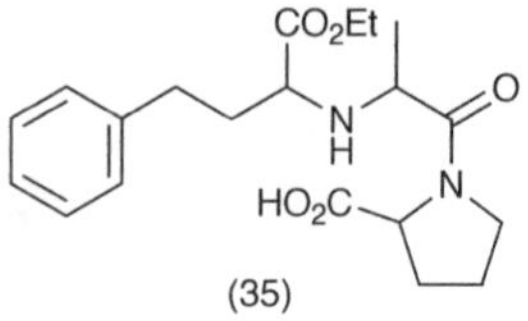

(35)

been assessed[62] as methods for identifying dioxopiperazine alkaloids of the roquefortine type from the fungus *Penicillium chrysogenum* VKM F-1987. Zinc complexation of a series of cysteine-containing dioxopiperazines, e.g. cyclo-(Gly-Cys), -(His-Cys), and -(Cys-Cys) showed[63] that cyclisation of the peptides did not enhance the complex stabilities over and above their linear counterparts. Studies[64] on the gas phase fragmentation ions $[M+H]^+$ of the methyl esters of histidine and histidine-containing di- and tri-peptides have revealed that the side-chain protonated dioxopiperazine unit is thermodynamically favoured over all other isomeric structures. The histidyl residue appears to play a key role in the cleavage of the peptide bonds via neighbouring group participation.

2.3   Cyclotripeptides. – Since homodetic cyclic tripeptides are difficult to make, this year again enlarging the ring has produced examples, which can be reviewed under this heading. Thus replacement of α- by β-amino acid residues has produced[65] useful analogues in the form of cyclo [β^3- HSer(OBn)]$_3$, and cyclo (β^3-Hmet)$_3$. Pentafluorophenyl esters of linear precursors were used in the cyclisation steps, but required trifluoroethanol as solvent due to the inherent insolubility of the β-peptides. When these two analogues were tested, together with previously synthesised analogues, in cancer cell lines, their cytotoxic activity were much lower (LC$_{50}$>100μM) than their antiproliferative activity (GI$_{50}$ 70nM). Cyclotripeptides with benzyl esters and benzyl ethers in the side-chains gave the strongest activity. In the synthesis[66] of building blocks such as cyclic peptide conjugates to 2-N-alkyl-1,2,3,4-tetraisoquinoline, cyclisation with only a triglycyl unit in the peptide portion P in (36) failed but was made possible by increasing the number of β-Ala units. A γ-turn induced cyclisation of tripeptides has been performed[67] via a ring-closing metathesis reaction using Grubb's catalyst [Cl$_2$(Cy$_3$P)$_2$Ru=CH$_2$Ph]. Cyclisation was made easier by the presence of a turn-inducing Pro residue as in the sequence shown in (37).

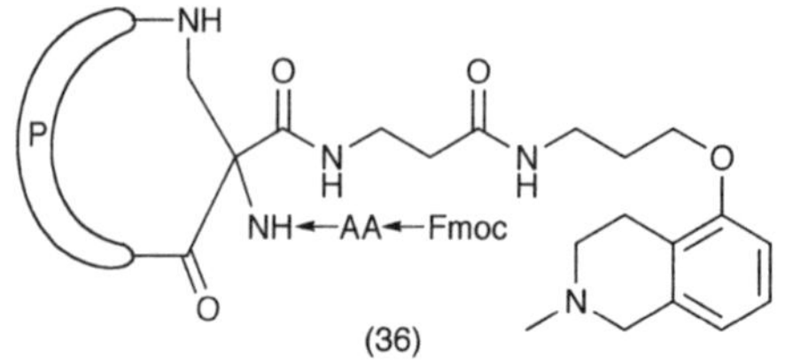

(36) 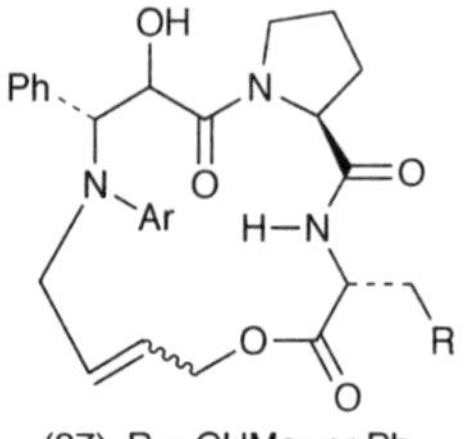   (37) R = CHMe$_2$ or Ph

2.4   Cyclotetrapeptides. – Work on the apicidin family dominate this sub-section. Apicidin (38), was first isolated by the Merck research group, and identified as a histone deacetylase inhibitor with *in vitro* and *in vivo* efficacy against

Plasmodium beghei malaria at less than 10mg/Kg. Two new syntheses of (38) and apicidin A (39) have been reported. Both involve macrocyclisation between the carboxyl group of the pipecolinic acid residue and the amino group of 2-amino-8-oxodecanoic acid. Scheme 10 summarises the synthetic pathways, path A being taken by one research group[68] to form (38) and path B by the other group[69] to form (39). Path A gave an overall yield of 44% over the four steps, while the pentafluorophenyl ester approach for cyclisation via path B was the only successful method, with DPPA, PyBOP and DCC/HOBt proving unsuccessful. The

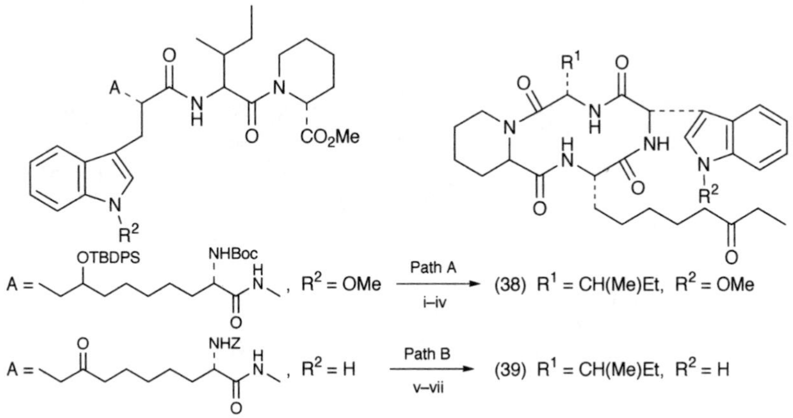

Reagents: i, LiOH; ii, HCl; iii, pentafluorodiphenyl phosphinate (FDPP); iv, TBAF; v, OH⁻; vi, H₂/Pd/C; vii, cyclisation via pentafluorophenyl esters

Scheme 10

biosynthetic machinery of the original producing organism has also been modi-fied[70] to produce apicidins B [with Pro replacing the pipecolinic acid residue in (38)] and C [with Ile residue replaced by Val in (38)]. Apicidins B and C inhibited the binding of [^{3}H] apicidin A to *Eimera tenella* parasite histone deacetylase with IC_{50} values of 10 and 6nM respectively. The structural requirements associated with the 8-oxo-decanoic acid side chain have been explored[71] through the synthesis of analogues with changes in the side chain. Picomolar activity as histone deacetylase inhibitors was achieved in this way and a similar SAR study[72] on Trp residue replacements deduced that the indole region is a key area for enzyme binding, and modifications in this region can give rise to analogues with between 20 and 100-fold parasite selectivity.

The potent inhibitors of histone deacetylase, trichostatin A and trapoxin, inhibit either by blocking catalytic action via chelation of a zinc ion in the active site pocket through a hydroxamic acid group (trichostatin A), or using an epoxyketone (trapoxin) moiety to alkylate the enzyme. A novel hybrid trapoxin analogue (40) has now been synthesised[73] which has a hydroxamic moiety instead of an epoxyketone, but still inhibits histone deacetylase at the low nanomolar concentrations. The change however makes it a reversible inhibition in contrast to the irreversibility of trapoxin inhibition. Cyclic tetrapeptide, cyclo (Tyr-Asp-Pro-Ala) and a pentapeptide analogue cyclo (Tyr-Asp-Pro-Ala-Pro), were amongst analogues[74] synthesised to investigate their insect development inhibitory (oostatic) activity. Linear precursors were assembled on a chlorotrityl chloride resin, cleaved off the resin and cyclised in solution using HOBt/TPTU. Both analogues had only low oostatic activity. Two novel 6-aminohexanoic acid (Ahx) containing cyclic tetrapeptides, cyclo (Lys-Tyr-Lys-Ahx) and cyclo (Lys-Trp-Lys-Ahx) have been synthesised[75] using the pentafluorophenyl esters at the hexanoic acid residue of the linear precursors. The former analogue was devoid of DNA-nicking properties, while the latter possessed DNA-nicking activity on supercoiled wX174 DNA with a rate of 50.7 pM min^{-1}.

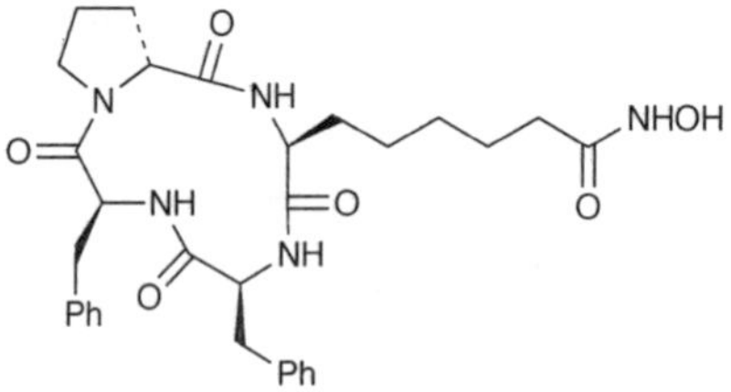

(40)

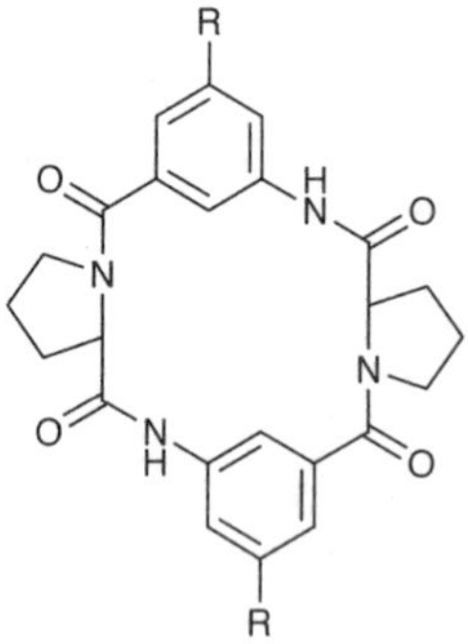

(41) R = H or NHBoc

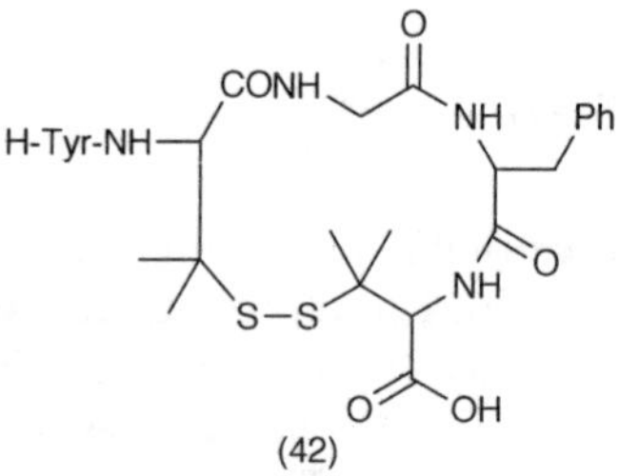

(42)

The cyclic tetrapeptides (41), when surveyed[76] in solution and in the solid phase, showed a *cis* conformation at the Pro amides and followed the *cis-trans-cis-trans* pattern usually found in such peptides. However they had a significantly smaller cation affinity when compared to similar hexapeptides. A theoretical conformational study[77] using a CICADA programme package on the cyclic enkephalin (42) and a smaller ring analogue has been carried out, and the results compared with spectroscopic data. Spectroscopically-derived H-bonding properties and the different conformational behaviour of aromatic rings were confirmed by the calculations. A series of nine lactam-bridged cyclic peptide inhibitors of mammalian ribonucleotide reductase, previously synthesised, have been sequenced[78] using an ion-trap mass spectrometer with an ESI source.

2.5 Cyclopentapeptides. – Although not strictly homodetic in structure, the protein phosphatase inhibitors, oscillamides B (43) and C (44) isolated[79] from *Planktothrix agardhii* and *P. rubescens* still have five residues in the ring. The most effective inhibitor against protein phosphatases PP1 and PP2A was C (44) with IC_{50} values of 0.90 and 1.33µM respectively. The Arg and N-methyl homotyrosine residues seem to be associated with their activity. Stripped-down versions of the protein phosphatase inhibitors nodularin and microcystin, have been synthesised[80] in both solution and solid phase, using strategies that allow further elaboration after macrocyclisation. Analogues (45) and (46) were found to be weak inhibitors of PP1 with IC_{50} values of 2.9 and 2.7mM respectively. The role of the 2-methyl and 3-diene groups in the Adda (2S,3S,4E,6E,8S -3-amino-9-methoxy,2,6,8-trimethyl-10-phenyldeca-4,6-dienoic acid) residue found in nodularins at the equivalent of position 1 in structures (47) and (48), has been explored through the synthesis[81] of (47) and (48) and an analogue with Pro replacing Sar. All three analogues were competitive inhibitors.

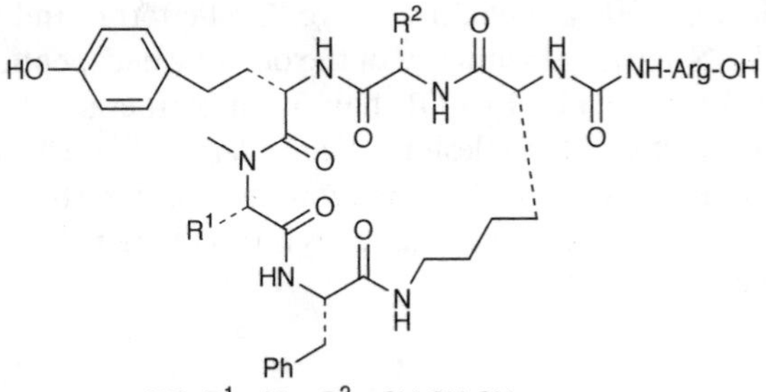

(43) R^1 = Me, R^2 = CH$_2$CH$_2$SMe
(44) R^1 = HOPhCH$_2$CH$_2$, R^2 = CH(Me)Et

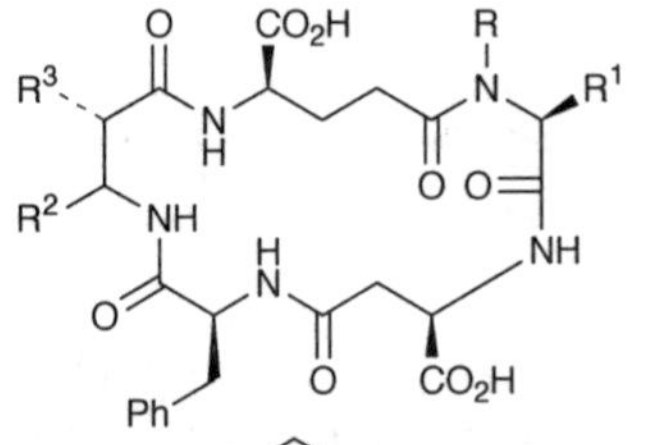

(45) R = Me, R^1 = H, R^2 = Ph..., R^3 = H

(46) R = R^1 = −(CH$_2$CH$_2$CH$_2$)−, R^2 = Ph..., R^3 = H

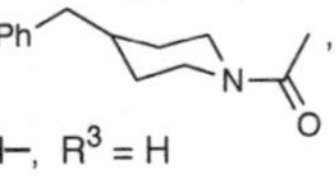

(47) R = Me, R^1 = H, R^2 = PhCH=CH−, R^3 = H
(48) R = Me, R^1 = H, R^2 = PhCH=CH−, R^3 = Me

A novel motilin antagonist has emanated from studies[82] on cyclic N-terminal motilin partial peptides. The best antagonists amongst the structures synthesised were (49, n = 1 or 2). Linear precursor peptides were synthesised on a PAM-resin, with the final cyclisation carried out between the azide of the Tyr residue and the side-chain amino group in the solution phase. In order to further clarify the characteristics of the proteolytic cleavage site of pro-oxytocin, the relationship between the secondary structure and enzyme recognition has been explored[83] using type-II β-turn proline-containing cyclic pentapeptide rings as seen in (50). The cyclisation to form the ring was performed off-resin using HATU/DIEA, and the NMR studies showed a 1→4 H-bonded type II β-turn

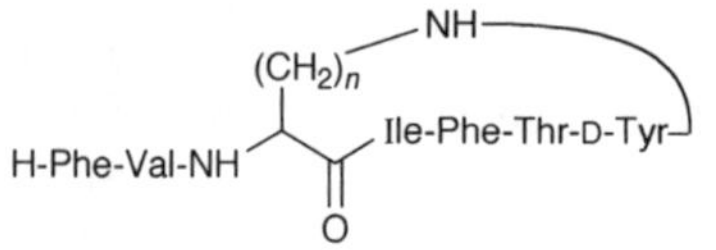

(49)

γ-Abu-Gly-Pro-Gly-Asp-Lys-Arg-Ala-Val-Leu-NH$_2$

(50)

with an H-bond detectable between Gly^{2}CO and Asp^{5}NH. β-Turn antagonists of the human B-2 kinin receptor were amongst the ten analogues synthesised[84] within the structure cyclo (Pro-aa^i-D-Tic-Oic-aa^{i+3}), where Oic is octahydroin-dole-2-carboxylic acid. The best antagonist (pK_i = 6.2) was doubly-charged at positions i and i + 3. It has been proven[85] that the His residue in the melanocortin receptor-binding peptide, cyclo (His-D-Phe-Arg-Trp-Gly) does not participate in binding at the receptor since cyclo(Asn-D-Phe-Arg-Trp-Gly), and cyclic tet-rapeptides lacking His are equally active. Cyclic tetra- and penta-peptides de-rived from the N-terminal sequence of the oostatic decapeptide H-Tyr-Asp-Pro-Ala-Pro-Pro-Pro-Pro-Pro-Pro-OH have been synthesised[86] and assayed in their effect on reproduction of the flesh fly *Neobellieria bullata*. Cyclisation decreased the oostatic activity by one order of magnitude, implying from NMR studies that the space structure of the cyclic analogues is too restricted to adopt the receptor-binding conformation.

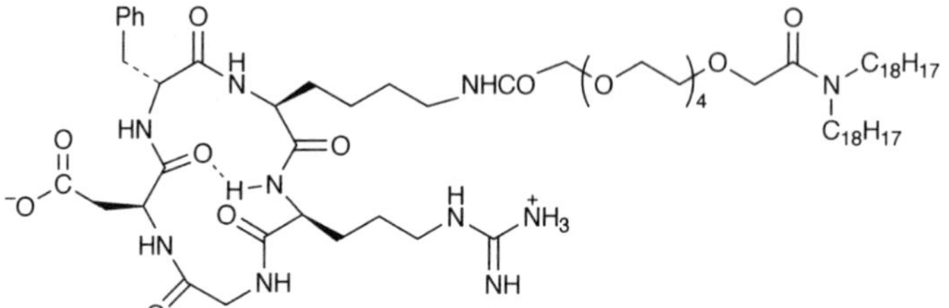

(51)

The well-documented cyclo (Arg-Gly-Asp-D-Phe-Lys) which is selective for $\alpha_v\beta_5$ and $\alpha_v\beta_3$ integrin receptors has been modified[87] to include increased lipophilicity via the modified Lys side chain as in (51). Selective adhesion of endothelial cells to artificial membranes incorporating this analogue has been observed. The Lys side chain has also has been the focus of other derivatisations. Thus regioselective derivatisation of Lys has yielded[88] cyclo [Arg-Gly-Asp-D-Phe-Lys(X)], where X = NH$_2$-O-CH$_2$CO-, OCHCO- or biotin (β-Ala)$_n$- with n = 0 to 5, and in cyclo [Arg-Gly-Asp-D-Tyr-Lys (SAA)], the Lys side chain has been conjugated[89] with SAA, 3-acetamido-2,6-anhydro-3-deoxy-β-O-glycero-D-gulo-heptonic acid and radio-iodinated by the iodo-Gen method. Specific bind-ing of this radio-analogue to $\alpha_v\beta_3$ integrin was shown, and proof was obtained of high tumour uptake due to the favourable bio-kinetics of the analogue. The widely used metal chelator, DTPA (diethylene triamine pentaacetic acid has been attached[90] to the Lys side chain to give cyclo (Arg-Gly-Asp-D-Phe-Lys)DTPA, and [cyclo (Arg-Gly-Asp-D-Phe-Lys)]$_2$ DTPA which have then been used to complex with ^{90}Y. The rate of ^{90}Y chelation to the conjugated vitronectin receptor antagonist 2-(1,4,7,10-tetraaza-4,7,10-tris(carboxymethyl)-1-cyclododecyl)acetyl-Glu-(cyclo[Lys-Arg-Gly-Asp-D-Phe])(cyclo[Lys-Arg-

Gly-Asp-D-Phe]) has been studied[91] while the same cyclic peptide system attached to a hydrazinonicotinamide conjugate[92] has been further chelated with [99mTc], the final complex showing comparable binding affinity for the vitronectin receptor to that of the uncomplexed form.

2.6 Cyclohexapeptides. – Three new antimitotic bicyclic peptides, celogentins A(52), B(53) and C(54), all related to moroidin, have been isolated[93] from the seeds of *Celosia argentea*. All three inhibited the polymerisation of tubulin, with (54) being four times more potent than moroidin in inhibitory activity. Phepropeptin B, a new proteasome inhibitor from *Streptomyces* sp has been found[94] to have the structure, cyclo(Leu-D-Phe-Pro-Phe-D-Leu-Val), while its congeners, phepropeptins A, C, and D differ in two of the six amino acid residues. They inhibit proteasomal chymotrypsin-like activity but not α – chymotrypsin. Neuroprotectins A (55) and B (56) together with the known complestatin have been isolated and characterised[95] from *Streptomyces* sp. Q27107. Their structures possess an oxindolylalanine moiety in place of the Trp present in complestatin.

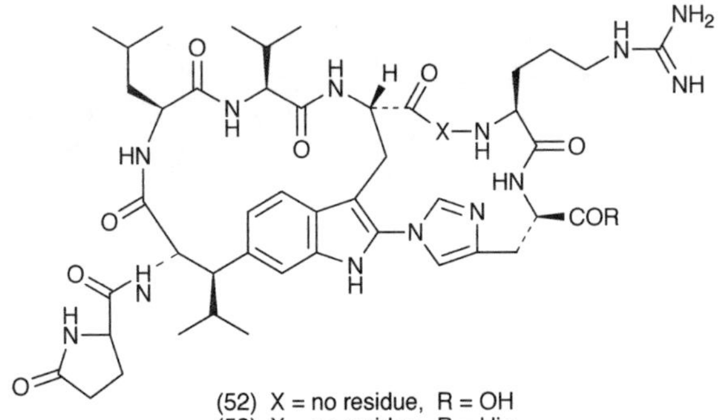

(52) X = no residue, R = OH
(53) X = no residue, R = His
(54) X = Pro, R = OH

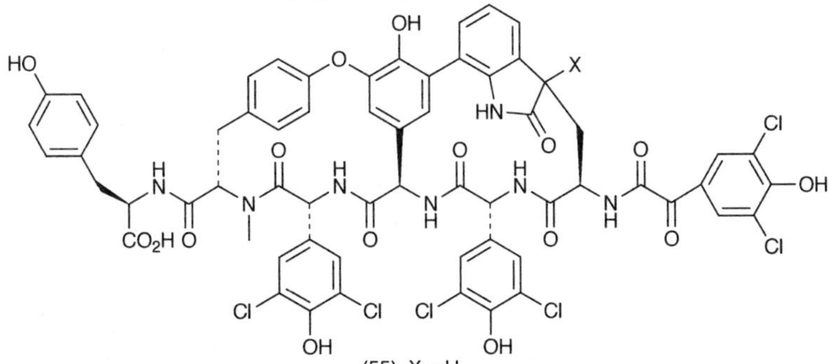

(55) X = H
(56) X = OH

The first synthesis of segetalin A, cyclo (Gly-Val-Pro-Val-Trp-Ala), and analogues has been reported[96]. Linear peptide precursors were assembled on Sasrin resin, and the protected linear precursor cyclised off-resin by activating the Ala carboxyl with DPPA, which gave a 45% yield of cyclised material. PyBrop only gave a 10% yield of cyclic product, while the pentafluorophenyl ester gave no results. D-Amino acid analogues of the anti-tumour bicyclic hexapeptide RA-VII have been produced[97] using selective epimerisation of L-analogues through their conversion into oxazole intermediates. Using the previously known conversion of RA-VII to [Tyr$^3\Psi$[CSNH]Ala^4] RA-VII using Lawesson's reagent, this thioamide was treated with AgBF$_4$ or Hg(OAc)$_2$ to produce isomers(57) and (58), convertible to [D-Ala^4] RA-VII and [D-Ala^2] RA-VII respectively on treatment with BF$_3$ etherate. The analogues showed much weaker cytotoxic activity than the parent molecules. The bridged 14-membered cyclodityrosine residue in this family of compounds RA-I to RA-XVI represents the pharmacophore. A first synthesis of the 14-membered ring of RA-IV has been accomplished[98] via an intramolecular S$_N$Ar reaction to give (59). Horseradish peroxidase/hydrogen peroxide have provided the phenolic oxidation conditions[99] necessary to form dimers such as (60) in low yield. The cyclic peptide (61), cyclo (Phe-Leu-Leu-Arg-εLys-Dap) and a D-Phe analogue, both having the minimal peptide sequence Phe-Leu-Leu-Arg needed to exhibit activity at the thrombin receptor, have been synthesised[100]. Cyclisation was carried out via a C-terminal Leu residue using TBTU/DIPEA, and it turned out that (61) was a 100-fold more active in the rat aorta relaxation assay than its D-analogue. Nepadutant, a NK2 receptor antagonist has been synthesised[101] and fully characterised as bicyclic cyclohexapeptide (62). NMR studies confirmed a βI and βII turn arrangement, with the active dipeptide unit Trp-Phe occupying the i + 1 and i + 2 positions of the βI turn.

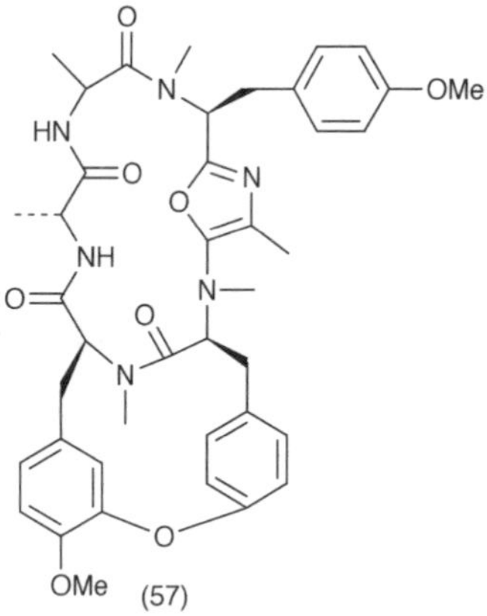

(57) 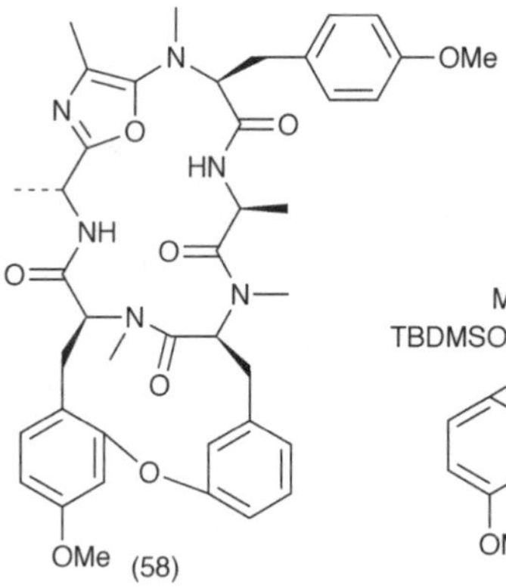   (58) 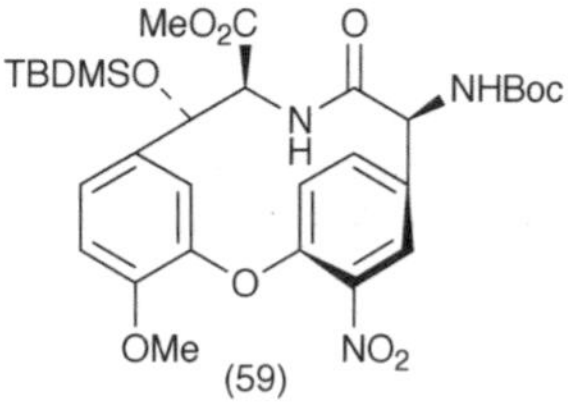   (59)

As interactions between integrin $\alpha_4\beta_7$ and mucosal addressin cell adhesion molecule-1 (MAdCAM-1) play a crucial role in many situations e.g. development of mucosa-associated lymphoid organs, generation of mucosal immune responses, chronic inflammatory bowel disease and type I diabetes, it is not surprising that finding antagonists would be highly prized. A lead antagonist molecule, cyclo (Leu-Asp-Thr-Ala-D-Pro-Ala) has been found[102] in a library of cyclo penta- and hexa-peptides, and this has been followed up with further cyclohexapeptides, cyclo (Leu-Asp-Thr-Ala-D-Pro-Phe), cyclo (Leu-Asp-Thr-Asp-D-Pro-Phe), cyclo (Leu-Asp-Thr-Asp-D-Pro-His), and cyclo (Leu-Asp-Thr-Asp-D-Pro-Tyr) with enhanced activity towards inhibiting $\alpha_4\beta_7$ integrin me-

OH   OH

Val-Ala–NH   NH–Ala-Val
Ala ← Val ← Ala   O   O   Ala ← Val ← Ala
(60)

O   Phe-Leu-Leu-Arg
H_2N NHCO—CH—$(CH_2)_4$—NH
NH$_2$
(61)

Asn(GlcNAc) → Asp-Trp-Phe
Leu—Dpr
(62) Dpr = 2,3-diaminopropanoic acid

diated cell adhesion to MAdCAM-1. It has also been suggested[103] that integrins at the cell surface may be connected with the immunomodulatory activity of HLA-DQ fragments, which has been corroborated by the cyclodimeric peptide cyclo (Arg-Gly-Asp-Arg-Gly-Asp), a known selective $\alpha_v\beta_3$ inhibitor, strongly suppressing both the humoral and cellular immune response.

A library of cyclohexapeptides designed to be selectin antagonists, as summarised (63), has been constructed[104] through attaching the hydroxy group of HOPro to a polymer via a linker based on tetrahydropyrenyl chemistry. Cyclohexapeptides with alternating Pro and 3-aminobenzoic acid units substituted in the manner represented in (64-66), have been shown[105] to bind monosaccharides, with compound (65) showing the greatest affinity. Pro residues alternating with 6-aminopicolinic acid as in (67) have been shown[106] to bind halide and sulfate anions, by holding them via H-bonds in a cavity formed through aggregation of two cyclopeptide molecules. Conformational studies[107] confirm that it is easier to fine-tune the complexing properties of these series of cyclopeptides, if the substituents are in the 4-position on the aromatic ring as in compounds (68-72). With this orientation the compounds lose their anion complexation characteristics and become better cation complexation agents with a range of stability constants from $K_a = 140$ to 10,800 M^{-1} for (68) and (72) respectively, in the presence of n-butyltrimethylammonium picrate. The interaction of cyclic peptides with model peptides covalently bonded to silicon surfaces, has been investigated[108] by ATR-FTIR spectroscopy. Within the series cyclo (Lys-X^1-Lys-X^2-Lys-X^3), with X^1, X^2 and X^3 permutated between Phe(NO$_2$), Glu, Gln, Ala and Leu, cooperative effects, strong bonding between functional groups or no interaction were detected. Based on the success of cyclo (Arg-Lys-Tyr-Pro-Tyr^5-βAla) as a chiral selector in capillary electrophoresis[109] further analogues with various substitutions for Tyr^5 have been assessed. Increased electron-withdrawing power at position 5 improved the separation properties. ε-(3-Hydroxy-1,4-dihydro-2-methyl-4-oxo-1-pyridyl), norleucine and Gly or β-Ala containing cyclohexapeptides have been shown[110] to form 1:1 iron (III) complexes, while alkali and alkaline earth metal picrate complexes with cyclo (Pro-Gly)$_n$ ionophores with n = 3 or 4, facilitate[111] the migration of metal ions across a bulk liquid dichloromethane membrane. Cyclic hexapeptides ranging form cyclo (Ala)$_6$ to cyclo (Gly)$_6$ have been used[112] in the analysis of the transferability of atomic multipoles for amino acids, in modelling macromolecular charge distribution fragments.

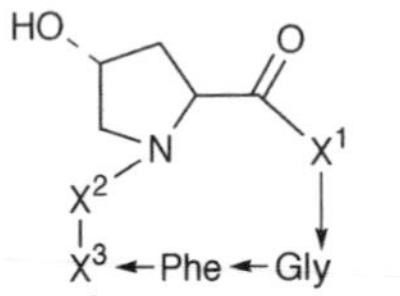

(63) X¹ = Asp, Ser or Lys
X² = Gly(D or L)
X³ = Lys(D or L)

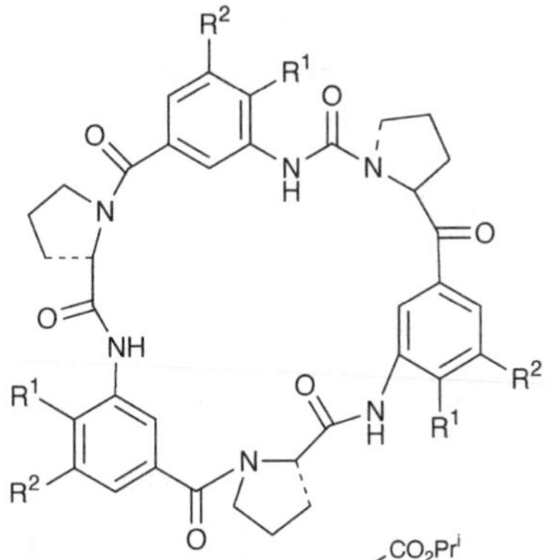

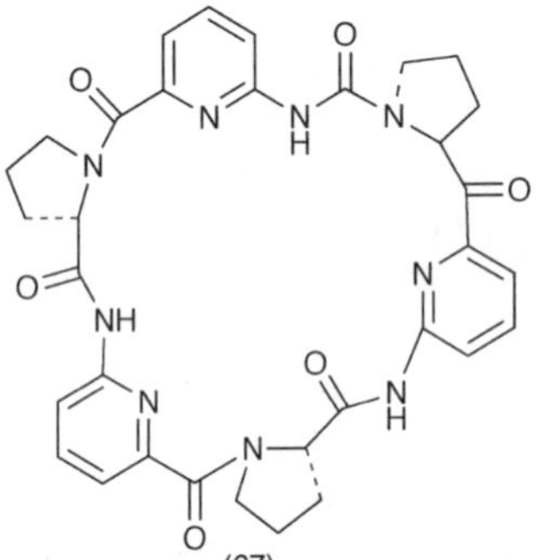

(67)

(64) R¹ = H, R² = CONH...CO₂Prⁱ / CO₂⁻ N(Buⁿ)₄⁺

(65) R¹ = H, R² = CONH...CO₂Prⁱ / CO₂⁻ N(Buⁿ)₄⁺

(66) R¹ = H, R² = CONH...CO₂Prⁱ / CO₂⁻ N(Buⁿ)₄⁺

(68) R¹ = Me, R² = H
(69) R¹ = Cl, R² = H
(70) R¹ = CH₂OMe, R² = H
(71) R¹ = OMe, R² = H
(72) R¹ = CO₂Me, R² = H

2.7 Cycloheptapeptides. – The cyanobacterium *Nostoc* sp ATCC53789 from a lichen collected on the Isle of Arran, Scotland, contains[113] the nostocyclopeptides A1 (73) and A2 (74). An unusual feature is the imino linkage found in the macrocycle, but neither compound showed any significant activity in many biological assays attempted. A Fijian marine sponge *Stylotella aurantium* has yielded[114] wainunuamide, cyclo (Phe-Pro-His-Pro-Pro-Gly-Leu). The His-Pro bond is *trans* while the Pro-Pro bond is *cis*, but it only has weak cytotoxic activity. Mahafacyclin B, cyclo (Gly-Thr-Phe-Phe-Gly-Phe-Phe) from the latex of *Jatropha mahafalensis* has anti-malarial activity (IC₅₀ = 2.2 μM), and has been

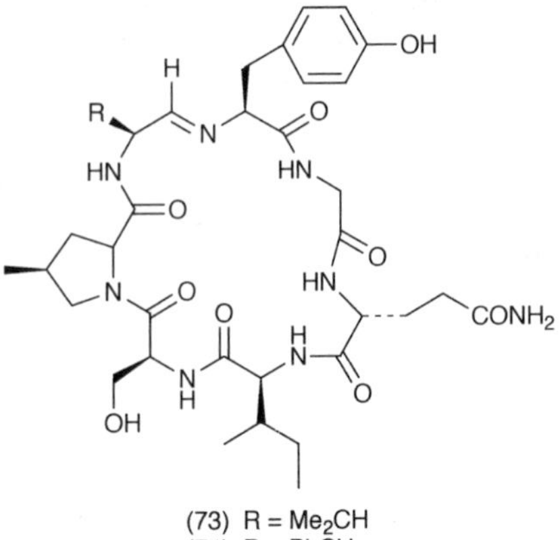

(73) R = Me₂CH
(74) R = PhCH₂

synthesised[115] in the solution phase by cyclisation at the Gly-Thr bond using TBTU under high dilution conditions. A β-bulge conformation, with two β-turns around Phe⁴-Gly and Phe⁷-Gly¹ typical of cycloheptapeptides, has been confirmed by NMR, although it does not contain a Pro residue. When the marine cyclopeptide, hymenamide C, cyclo (Leu-Trp-Pro³-Phe-Gly-Pro⁶-Glu) was synthesised[116] on-resin using the side chain of Glu as a link to the resin (PAC-PEG-PS), the major product was identical to hymenamide C which has a *cis* -Pro³ and a *trans* -Pro⁶ bond geometry, but an isomer with a *cis* -Pro³ and a *cis* -Pro⁶ conformation was present as a minor product. Macrocyclisation was achieved in 27% overall yield using HBTU/HOBt. Hymenamide C caused 52% inhibition of the elastase degranulating release at 10 μM concentration, while its isomer caused 45% inhibition.

An X-ray structural and thermodynamic analysis[117] of the binding between the monoclonal antibody tAb2 and selected peptides which include cyclic peptides (75) and (76) has been reported. Compound (75) was bound in a loop conformation, with its cyclic structure counteracting the exchange of aspartate in the epitope sequence to glutamate. High energy tandem mass spectrometry (MS/MS) has been helpful[118] in the elucidation of the structure of members of the surfactin family from *Bacillus subtilis*, while purification schedules[119] have been reported for the purification of the microcystins.

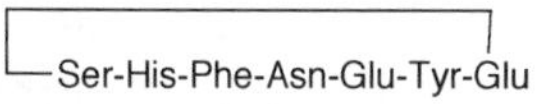

Ser-His-Phe-Asn-Glu-Tyr-Glu

(75)

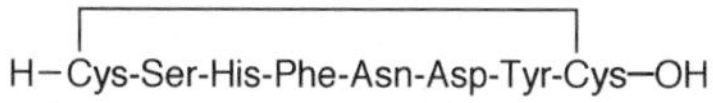

H–Cys-Ser-His-Phe-Asn-Asp-Tyr-Cys–OH

(76)

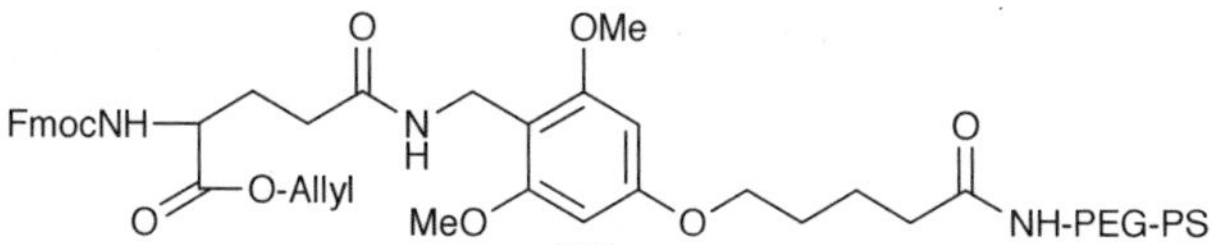

(77)

2.8 Cyclooctapeptides/Cyclononapeptides. – The roots of the Chinese herb, *Brachystemma calycinum* D. don is a source[120] of four new cyclic octapeptides, brachystemins A, cyclo (Pro-Phe-Leu-Ala-Thr-Pro-Ala-Gly), B cyclo (Pro-Ala-Phe-Trp-Asp-Pro-Leu-Gly), C cyclo (Pro-Ile-Gly-Pro-Val-Ala-Ala-Tyr) and D cyclo (Pro-OMet-Trp-Ile-Gly-Ala-Leu-Asp). In a separate study[121] the same herb and a relative *Cucubalus baccifer* have yielded a fifth, brachystemin E cyclo (Pro-Leu-Ile-Gly-Pro-Ile-Trp-Asn). Polycarponins B, cyclo (Gly-Ile-Val-Leu-Val-Gly-Leu-Pro) and C, cyclo (Pro-Thr-Leu-Pro-Pro-Val-Leu-Phe) have been identified[122] in the whole plants of *Polycarpon prostratum* while the seeds of *Goniothalus leiocarpus* have yielded[123] leiocyclocins A cyclo (Leu-Phe-Ser-Ala-Pro-Gln-Ile-Gly) and B cyclo (Ala-Leu-Pro-Pro-Ala-Pro-Trp-Val). A methanol extract of the whole plant, *Schnabelia oligophylla*, has produced[124] a new cyclopeptide, schnabepeptide, cyclo (Ile-Val-Trp-Pro-Val-Pro-Ser-Gly) which shows immunosuppressive activity on T/B lymphocytes. Linseed seeds (*Linum usitatissiman*) have been shown to be a source[125] of cyclolinopeptides CLF-CLI and found to have the structures, cyclo (Pro-Phe-Phe-Trp-Val-Mso-Leu-Mso), cyclo

(Pro-Phe-Phe-Trp-Ile-Mso-Leu-Mso), cyclo (Pro-Phe-Phe-Trp-Ile-Mso-Leu-Met), and cyclo (Pro-Phe-Phe-Trp-Val-Met-Leu-Mso), respectively, where Mso = methionine sulfoxide. In contrast to the previously studied CLB, these four analogues did not show any immunosuppressive activity.

Solid phase methodologies[126] have been shown to be superior to solution phase techniques for the synthesis of phakellistatin 11, cyclo (Ile-Phe-Pro-Gln-Pro-Phe-Pro-Phe). On-resin cyclisation was carried out with PyAOP in 17% overall yield, the resin assembly having started from the PAL resin attachment (77), utilising the side-chain of Glu. With the synthesis[127] of pure stereoisomers of 4-amino-3-hydroxy(and 3-oxo)-1-cyclohexane carboxylic acids, there has been an opportunity to insert them into cyclinopeptide A analogues (78) and (79) as twisted *cis* - amide bond mimetics. Only the *cis*-substituted cyclohexane versions could be synthesised since they are backbone turn inducers in contrast to the *trans*-forms, which induce extended conformations. Because increases in intracellular Ca^{2+} in rat astrocytes, through activation of a metabotropic receptor coupled to phospholipase C, is a property of the C-terminal octapeptide (Arg-Pro-Gly-Leu-Leu-Asp-Leu-Lys) of octadecaneuropeptide, cyclic analogues[128] have been sought to increase activity. Cyclo (Arg-Pro-Gly-Leu-Leu-Asp-Leu-Lys) synthesised by on-resin cyclisation was shown to be three times more potent and 1.4 times more efficacious than its linear analogue in increasing Ca^{2+} in rat astrocytes. A γ-turn over Pro^2-Leu^4 and a type III β-turn over Leu^5-Lys^8 were features seen in its conformation. The N-terminal sequence of deamino-dicarba-eel calcitonin (80) has been synthesised[129] by condensing suitable pentapeptide fragments with Asu (aminosuberic acid) tripeptides in the solution phase, followed by the zipping up of the macrocycle between Ser and Asu using HBTU at high dilution. Strong binding affinities to somatostatin hsst 2 and hsst 5 receptor sub-types (5.2 and 1.2 nM respectively) have been realised[130] through the synthesis of a heterodetic cyclic somatostatin analogue (81). Cyclisation was accomplished through formation of a disulfide bond between Cys^1 and thioethyl N-alkylated Gly^8. In an effort[131] to stabilise the conformation of nonapeptide analogues of the N-terminus of human growth hormone hGH (6-14), side chain cyclisation between Lys^9 and Glu^{13} or betweenGlu^9 and Lys^{13}, was more efficient using DIC/HOBt rather than Castro's reagent.

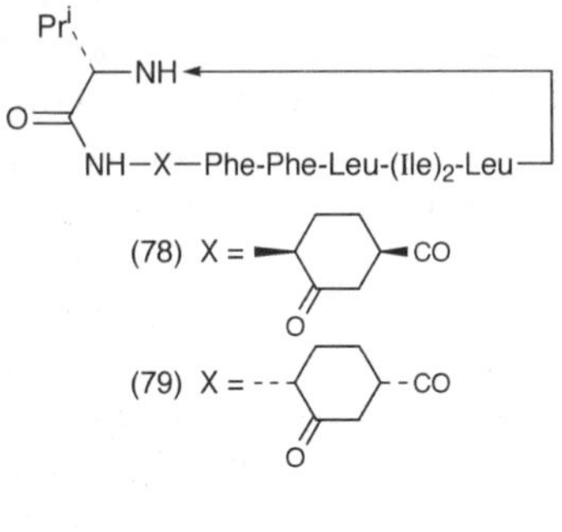

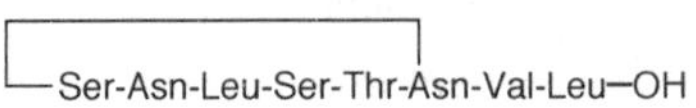

(78) X =

(79) X =

(80)

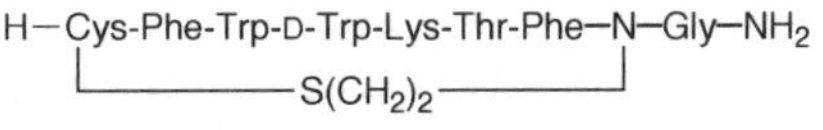

(81)

X-Ray studies[132] have been used to assess aryl-aryl interactions in cyclo (Phe-D-MeAla)₄ as compared to cyclo (Phe-D-MeAla-hPhe-D-MeAla)₂. The

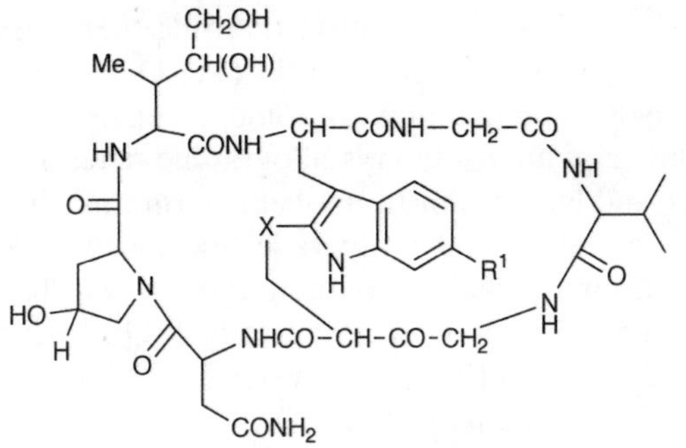

(82) R^1 = OH, X = (R)SO
(83) R^1 = OMe, X = (R)SO
(84) R^1 = OH, X = S
(85) R^1 = OH, X = (S)SO
(86) R^1 = OH, X = SO$_2$

one carbon homologation of Phe in the latter completely altered the packing, probably by increasing mobility of the aromatic residue. The electronic structures and conformational analysis of the toxic cyclopeptides α-amanitin (82) and congeners ((83-86) have been obtained[133] from semiempirical AM1 and *ab initio* parameters. The molecules are shaped like a 'bean' with concave and convex sides, with the indole ring above the carbocyclic moiety. An inner hole or cavity of fixed dimensions is present, but may show slight distortions owing to the changes in oxidation state of sulfur. The conformation of bicyclic nonapeptide BCP2, cyclo (Glu^1,Lys^5) (Glu-Ala-Pro-Gly-Lys-Ala-Pro-Gly-Gly) has been investigated[134] by NMR spectroscopy. The cyclic peptide itself in CD$_3$CN solution exists in at least four conformational forms, but on complexing with Ca^{2+} a new set of resonances appeared, believed to be due to the more symmetric structure of the Ca complex, with four CO's from the main ring pointing towards Ca, and an involvement also from two carbonyls from the bridged section.

2.9   Cyclodecapeptides and Higher Cyclic Peptides. – The structures of streptocidins A-D have been elucidated[135] as cyclo (Val-Orn-Leu-D-Phe-Pro-Leu-X^7-Asn-Gln-X^{10}) with X^7 = D-Trp (for A, B and C) or D-Phe for D and X^{10} = Tyr (in A), Trp (in B and D) or D-Trp (in C). NMR studies on streptocidins C and D showed that they had molecular topology similar to those of tyrocidine A and gramicidin S. Topological similarity to gramicidin S has also been seen[136] in solution for a series of cyclodecapeptides having the structures, cyclo (Pro-Phe-Phe-Ala-Xaa)$_2$ where Xaa can be Glu(OBu^t), Lys (ClZ), Leu or Ala. The conformation of the Ca^{2+} complexes of these peptides changed to saddle shape structures with two β-turns and two *cis* peptide bonds, similar to those observed for antamanide and analogues. Replacement[137] of the Val and Leu residues in gramicidin S (GS) forming the analogues [Asn$^{1,1'}$, Trp$^{3,3'}$] GS, [Asn^1, Asp$^{1'}$, Trp$^{3,3'}$] GS, [Asp$^{1,1'}$, Trp$^{3,3'}$] GS and [Gln$^{1,1'}$, Trp$^{3,3'}$] GS gave an increase in H-bonding potential for binding to carbohydrates. Nuclear Overhauser effects were observed between [Gln$^{1,1'}$, Trp$^{3,3'}$]GS and mannose indicating that the interaction of the peptide with the sugar occurred in the hydrophobic environment created by Trp and Phe residues. Surprisingly, up to this time, there has been no report of a solid state structure of gramicidin S cyclo (Val-Orn-Leu-D-Phe-Pro)₂ itself being determined by X-ray diffraction , although complexes with urea have been studied. To fill this gap, an X-ray crystallographic determination[138] of N, N-bis(trichloroacetyl) and N, N-bis(m-bromobenzoyl) gramicidin S has shown that the conformations are based on antiparallel pleated β-sheets , with pseudo two-fold symmetries, and β-turns over the fragments D-Phe-Pro. Biophysical data[139] accumulated for the gramicidin S analogues, GS 10 cyclo (Val-Lys-Leu-D-Tyr-Pro)₂, GS 12 cyclo (Val-Lys-Leu-Lys-D-Tyr-Pro-Lys-Val-Lys-Leu-Tyr-Pro), GS 14 cyclo (Val-Lys-Leu-Lys-Val-D-Tyr-Pro-Leu-Lys-Val-Lys-Leu-Tyr-Pro) and [D-Lys⁴]-GS 14 have shown that they were mainly monomeric structures in aqueous buffer, but GS 14 tended to aggregate in higher concentration conditions. Good correlation was found between conformational structures, amphipathic and hydrophobic properties and their haemolytic activity, but correlation with antimicrobial activity is more complex. Gramicidin S analogues[140] bearing a pyrenyl and a p-nitrophenyl group at different positions, when subjected to photoinduced electron transfer from an excited pyrenyl group, showed a complex dependence on the number of spacer units, or the edge-to-edge distance, between two aromatic groups. After rationalisation, however, the results correlate with the tunnelling pathway model on both α-helix and β-sheet model peptides.

The prototype[141] for a new generation of regioselectively addressable functionalised templates (RAFT) for use in protein de novo design, has the structure, cyclo [Phe(p-NO₂)-D-Pro-Gly-Phe(p-NO₂)-Ala-Phe(p-NO₂)-D-Pro-Gly-Phe(p-NO₂)-Ala], and the conformation determined by X-ray diffraction, and NMR techniques consists of an antiparallel β-sheet spanned by heterochirally-induced type II' β-turns. Bicyclic decapeptide (87) has been synthesised[142] on a gram scale by solid phase techniques, and incorporates four quasi-orthogonal protecting groups on lysines, to allow site selective assembly of building blocks for combinatorial work in the solution phase.

In a review[143] of bioactive compounds from coral reef invertebrates, reference has been made of the discovery, in 2000, of the cyclic undecapeptides barangamides A-D, which were discussed in this Chapter in Vol 33. It has been reported from China[144] that cyclosporin H can be isolated from cultures *Fusarium solani* sp. No. 4-11. A review[145] to update knowledge on the conformations of cyclosporins determined in the solid state, and their correlation with biological activity has appeared. Using ion trap mass spectrometry, it has been found[146] that the MeBmt¹ residue (88) in protonated cyclosporins, [cyclosporin A is (MeBmt¹-Abu²-Sar³-MeLeu⁴-Val⁵-MeLeu⁶-Ala⁷-D-Ala⁸-MeLeu⁹-MeLeu¹⁰-MeVal¹¹)] undergo N→O acyl shifts, which is a useful initiation for further sequencing by tandem mass spectrometry. The presence of the N-Me group in MeBmt is essential for rearrangement to proceed. A biotinylated derivative of cyclosporin A has been synthesised[147] that retains the capacity of binding to cyclophilin A.

Cyclotides is now a well recognised term to define a group of large macrocyclic peptides, cyclised via disulfide links to form a 'cysteine knot' and isolated from

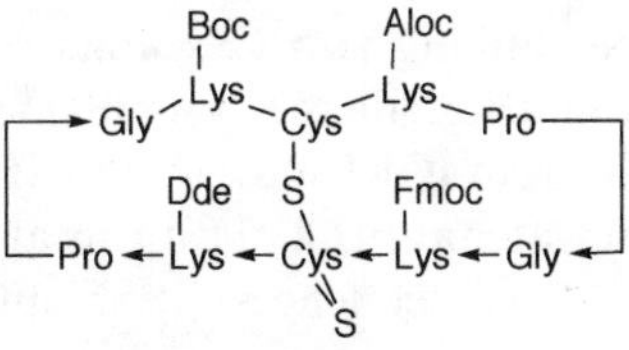

(87)

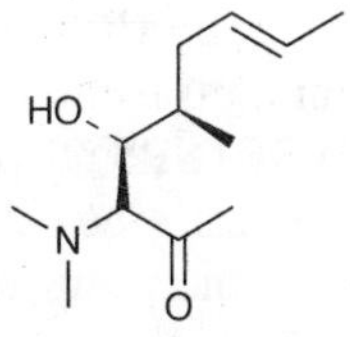

(88)

the Rubiaceae and Violaceae families. A new example containing 30 amino acids has now been identified[148] from *Hybanthus* (Violaceae), named Hypa A, and found to have the structure cyclo (SCVYIPCTITALLGCSCKNKVCYNGIP-CAE), with Cys-Cys bridges occurring at C^2-C^{17}, C^7-C^{22} and C^{15}-C^{28}. Attempts have been made to understand the biosynthesis[149] of the cyclotides from *Oldenlandia affinis* by isolation of the cDNA that encodes the cyclotide kalata E11. Evidence suggests that the cyclotide domains are excised and cyclised from predicted precursor proteins, with the processing sited on the N-terminal side of the strongly conserved Gly-Leu-Pro or Ser-Leu-Pro sequence that flanks both sides of the cyclotide domain. Two novel cyclotridecapeptides, tolybyssidins A (89) and B (90) have been isolated[150] from the cyanobacterium *Tolypothrix byssoidea* (EAWAG 195) and both contain dehydrohomoalanine (Dhha). Both compounds showed quite moderate antifungal activity against *Candida albicans*. In investigations relating to the development of peptide vaccines, the central part (residues 42-61) of the N-terminal parasitic Merozoite Surface Protein1 (Peptide 1513) has been chosen[151] as a candidate to achieve specificity in peptide receptor interactions. Cyclisation was achieved using the backbone cyclisation method built up through aminoethyl glycine linked via diacids to give cyclic analogues (91-94). Analogues (91-93) showed increased specificity in ligand-receptor interactions with a 19kDa receptor on the RBC membrane. A series of cyclic analogues of the inhibitory sequence of the protein tendamistat have been prepared[152] and their porcine pancreatic α-amylase (PPA) inhibitory activity measured. Cyclic analogue (95) provided the strongest inhibitory activity with a K_i value of 0.27 μM, and cd studies showed that the g-hairpin structure had been stabilised in this analogue. NMR data have been accumulated[153] on two cyclic

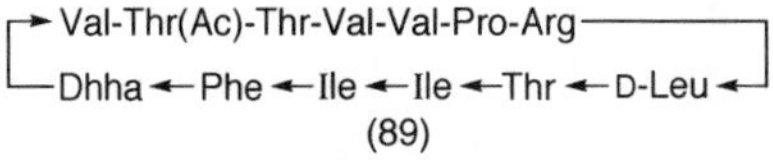

(89)

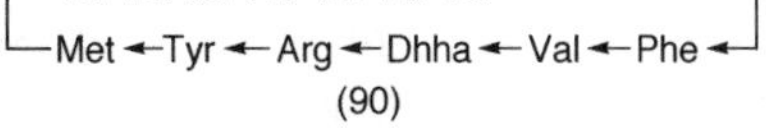

(90)

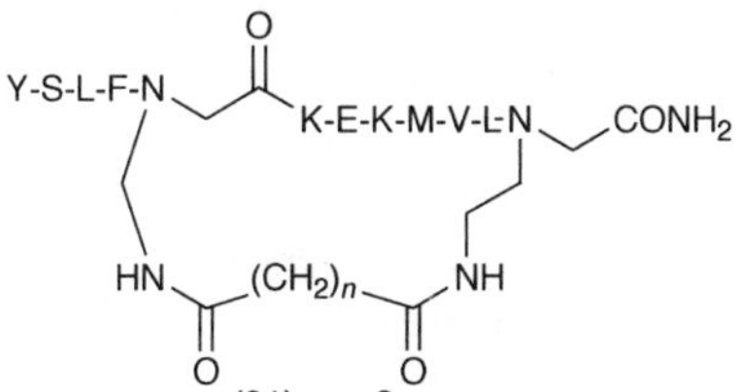

(91) $n = 2$
(92) $n = 3$

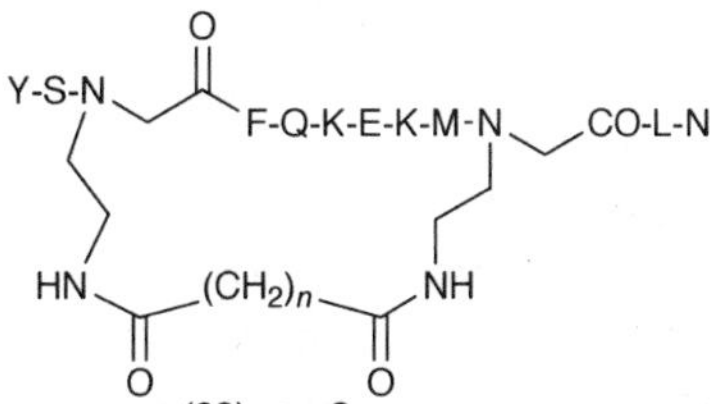

(93) $n = 2$
(94) $n = 4$

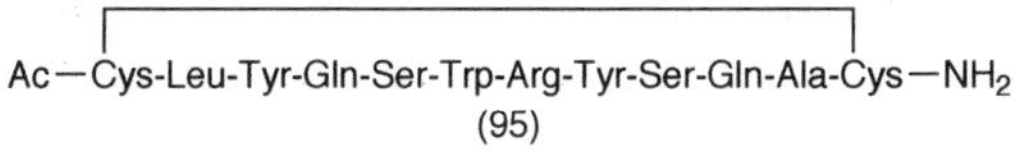

Ac—Cys-Leu-Tyr-Gln-Ser-Trp-Arg-Tyr-Ser-Gln-Ala-Cys—NH₂

(95)

peptides formed by disulfide bridges linking two Cys units at the termini of tandem repeat units from the C-terminal domain of RNA polymerase II. Data on cyclo (CRDYSPTSPSYSRDC) and cyclo (CRDYSPTSPSYSPTSPNYSRDC) were consistent with the existence in water at pH 4 of β-turns at both SPTS and SPSY in the former, and the two SPSTS sequences, SPSY and SPNY in the latter.

Cyclo (Trp-Dap-Leu-D-Ala-Trp-Ser-Val-D-Ala-Trp-Ser-Ile-Gly) where Dap = diaminopropionic acid has shown evidence of creating an artificial ion channel in lipid bilayer membranes[154]. The pH dependence of ion conductance showed that the β-amino group of Dap may play a role in the conductance of the peptide channels.

2.10 Peptides containing Thiazole/Oxazole Rings. – Many synthetic feasts have been derived from studies in this area on biologically active peptides of aquatic origin. Some of these have been reviewed[155] by active participants in the field, who conclude that while their own DPPA coupling agent works well in most circumstances, FDPP (96) will sometimes give a better result. Recent studies on cyclic peptides from marine sponges have been reviewed[156], while oxazole and thiazole peptides are part of the coverage in a recent Report[157].

The marine sponge *Haliclona* sp. has yielded[158] haliclonamides A (97) and (B) (98) which were found to have Fe(III) binding properties that could explain why the sponge's tissue had a significantly higher than average Fe(III) concentration. A second component originally isolated[159] with nostocyclamide from *Nostoc* 31, has been identified as nostocyclamide M (99) which has the D-Val residue of nostocyclamide replaced by D-Met. Yet another member of the patellamide family, prepatellamide A (100), has been identified[160] from the ascidian *Lisso-clinum patella*. As its structure differs from patellamide A, only in the way that Thr is present instead of an oxazole ring, it is suggested that (100) might be a biosynthetic precursor of patellamide A. Although not cyclic in structure, it is worth noting here, that the cyanobacterium *Lyngbya majuscula* produces[161] six dysideaprolines A-F (e.g 101, dysideaproline A) and barbaleucamides A (102) and B (103).

The powerful cytotoxic qualities of diazonamide A (104) and the difficulties of accumulating significant amounts from the natural resource, has initiated world wide efforts on its synthesis. One research group[162] after a demanding 16-step synthesis managed to synthesise structure (104), only to find that there were differences in physical data and biologically activity between the synthesised and natural form. On re-assessment[163] of the data the authors now believe that the structure should be re-assigned to be (105), confirmed by no valine being found in the hydrolysates, and from a re-assessment of X-ray data. Synthesis of the oxygen analogue (106) of (105) does give a product with equivalent activity to the natural diazonamide A. Another variation, named seco diazonamide A (107) was also synthesised by the same research group[164] from its benzofuran diol precursors, but did not have the same mechanism of cellular action as diazonamide A. Taking the original published structure (104) for diazonamide A as their syn-

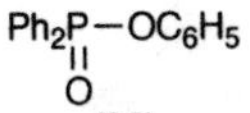

(96)

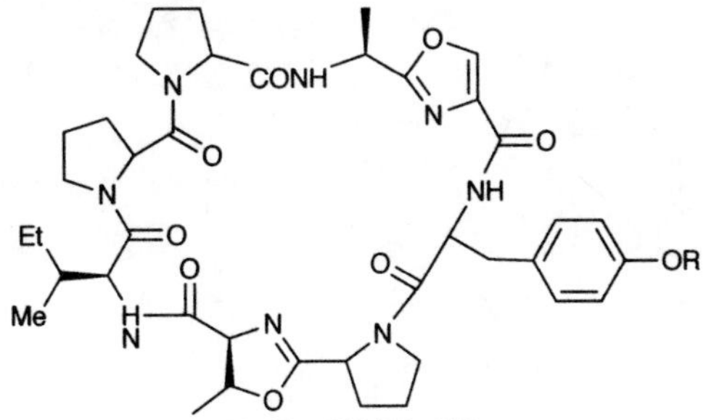

Ph₂P—OC₆H₅ — shown as $Ph_2P-OC_6H_5$ with $\overset{\parallel}{O}$

(96)

(97) R = CH₂CH=CMe₂
(98) R = H

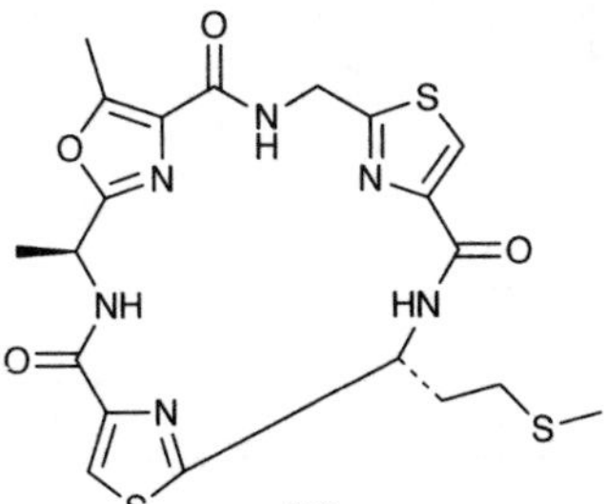

(99)

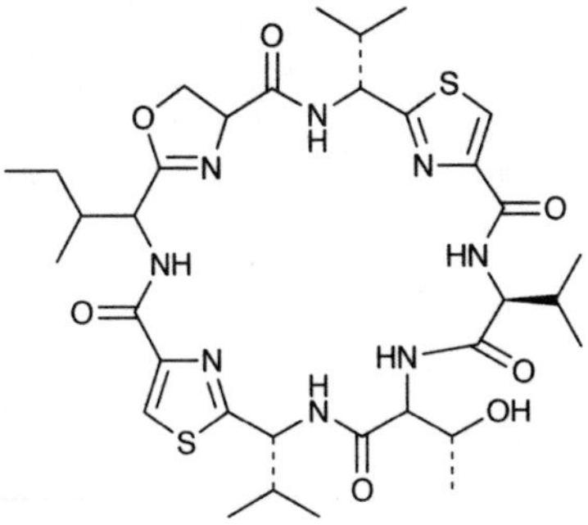

(100)

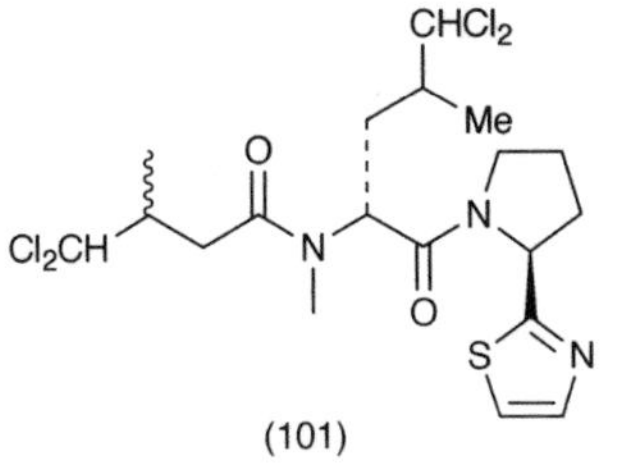

(101)

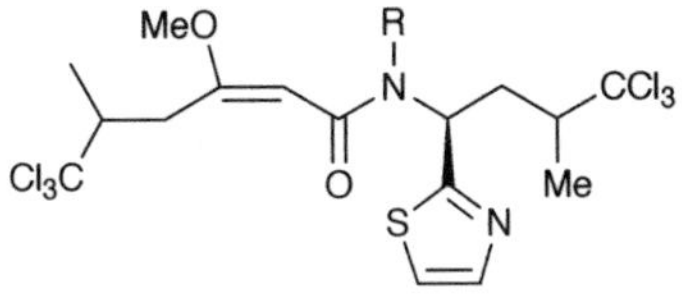

(102) R = H
(103) R = Me

thetic challenge, a number of synthetic studies have aimed at the core structure. Thus a hetero pinacol macrocyclisation cascade reaction[165] using SmI₂, summarised in Scheme 11, has provided the complete aromatic core. Irradiation causing a Photo-Fries Rearrangement was the last step used[166] in the synthesis of the macrocycle (108), while the indole-oxazole fragment (109) was prepared[167] by cyclodehydration with p-toluene sulfonic acid, after a Chan-type rearrangement of a tertiary amide. A Dieckmann-type cyclisation, following on from a Suzuki coupling of fragments has been used[168] to construct macrocyclic ketone (110). Hennoxazole A (111), although not a macrocyclic compound, does contain the bisoxazole unit seen in the diazonamides, which was constructed[169] via an oxidation/cyclodehydration step.

Dendroamide A (112), known for its ability in reversing multiple drug resistance in tumour cells, has been synthesised[170] from three fragments derived from junctions (a), (b) and (c) in the structure, with the final macrocyclisation occurring at (b) in 56% yield using DPPA. Two stable isomers, (*cis, cis* and *trans, trans*

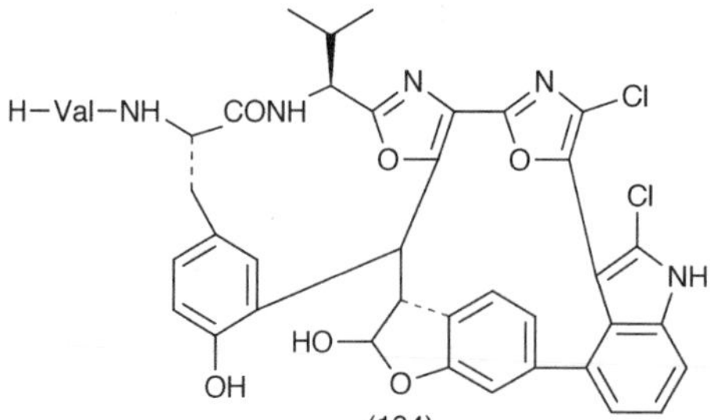

(104)

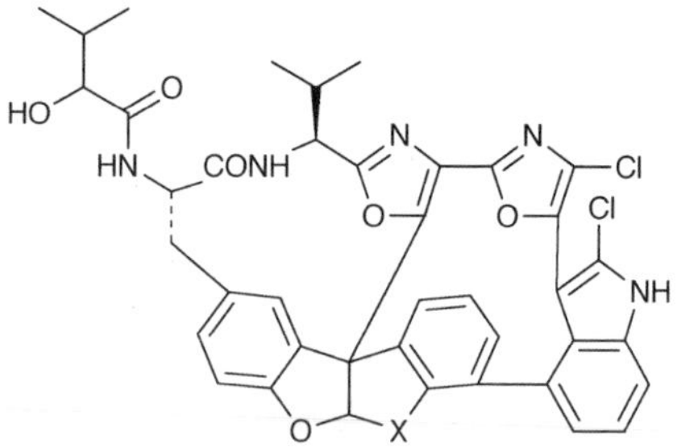

(105) X = NH
(106) X = O

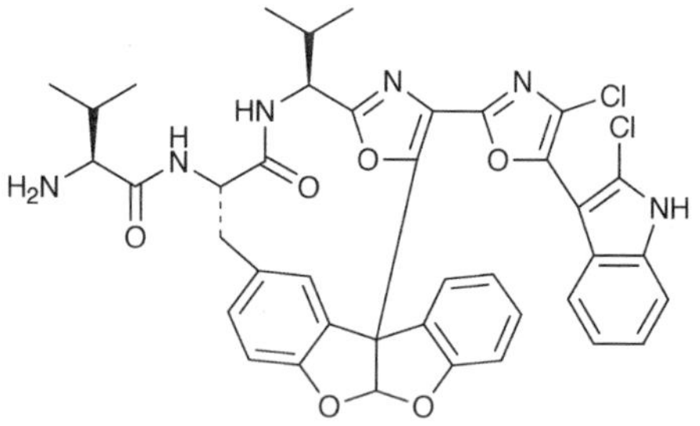

(107)

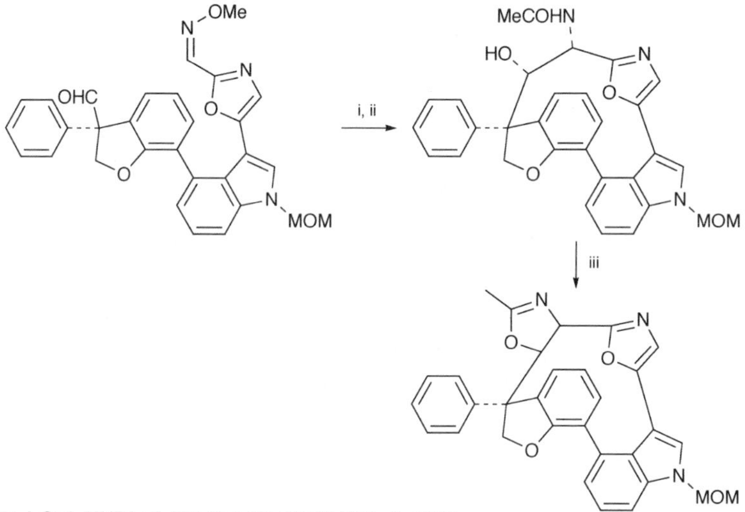

Reagents: i, SmI₂/HMPA; ii, NH₄Cl, AcOH/EDC/HOBt; iii, POCl₃

Scheme 11

around the Pro bonds) of ceratospongamide (113) have been isolated from the red alga *Ceratodictyon spongiosum* and it is the *cis, cis* form (113) that was the first to be synthesised[171] using a (5 + 2) convergent strategy to form a linear precursor which allowed cyclisation by activation of the thiazole carboxyl group. Macrocyclisation with DPPA gave a 31% yield, while FDPP gave 63% with cyclodehydration to form the oxazoline kept to the last step and achieved using

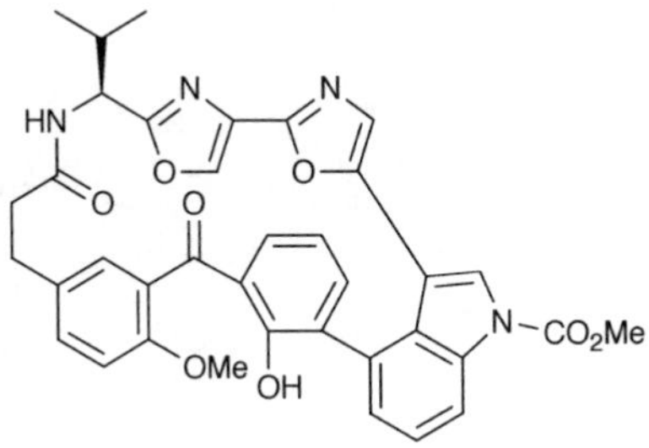

(108)

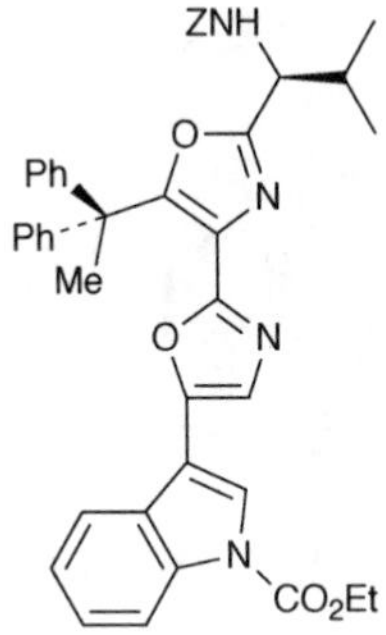

(109)

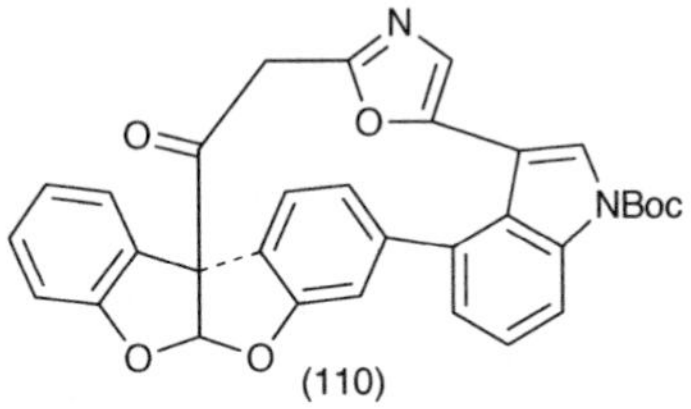

(110)

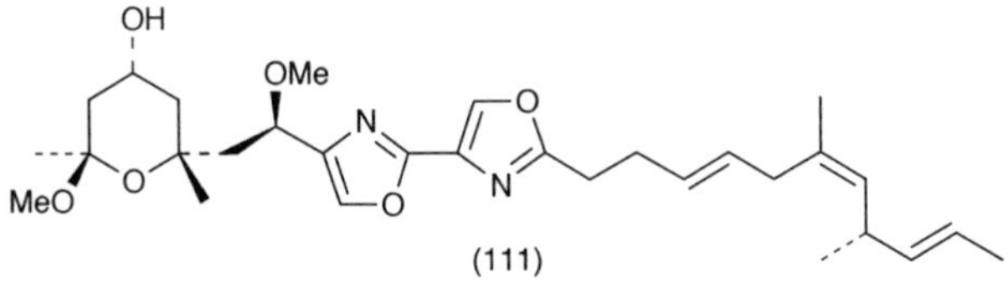

(111)

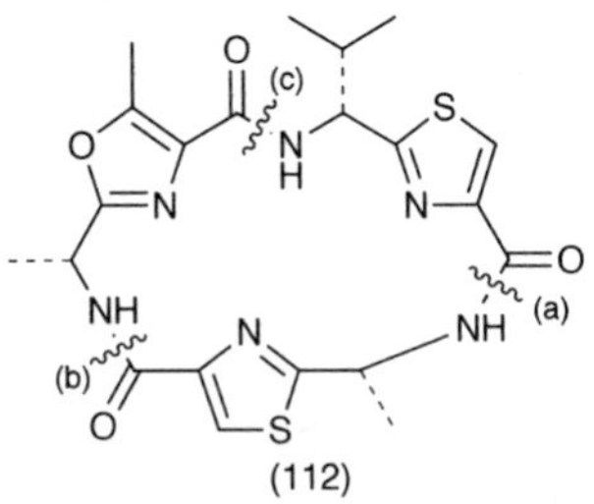

(112)

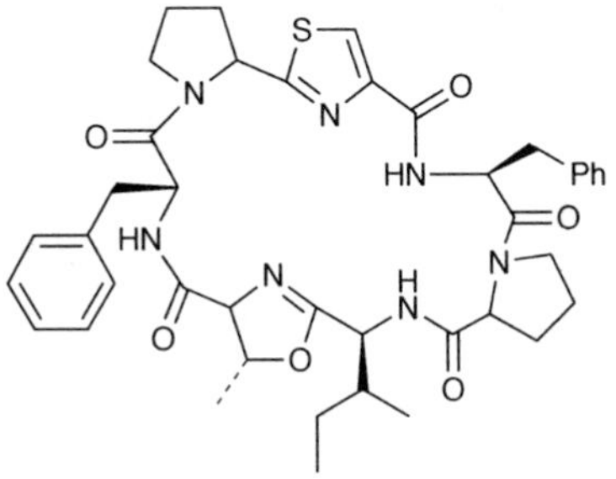

(113)

bis(2-methoxyethyl)aminosulfur trifluoride. Macrocyclisation[172] with amide formation at the thiazole carboxyl group using DPPA (58% yield) was also the strategy used for the total synthesis of lyngbyabellin A (114), the novel peptolide from the marine cyanobacterium *Lyngbya majuscula*. Trunkamide A (115), currently in preclinical trials has been synthesised[173] using solid phase techniques on chlorotrityl resin. Assembly was carried out starting with Pro linked to the resin, with the final two residues being D-PheΨ[CSNH]Ser added as the precursor to the thiazole ring which was obtained via dehydration using diethylaminosulfur trifluoride. A total synthesis of nostacyclamide (116) and analogues has been accomplished successfully[174] using cyclisation with FDPP. The presence of metal ions in the cyclisation mixture greatly influences the proportion of (116) obtained, relative to other cyclic analogues.

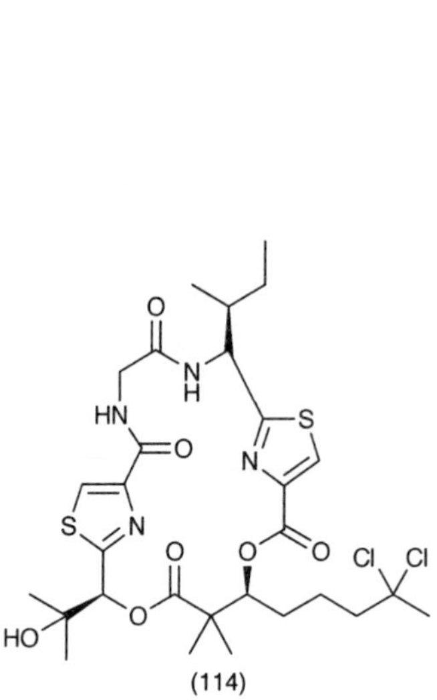

(114)

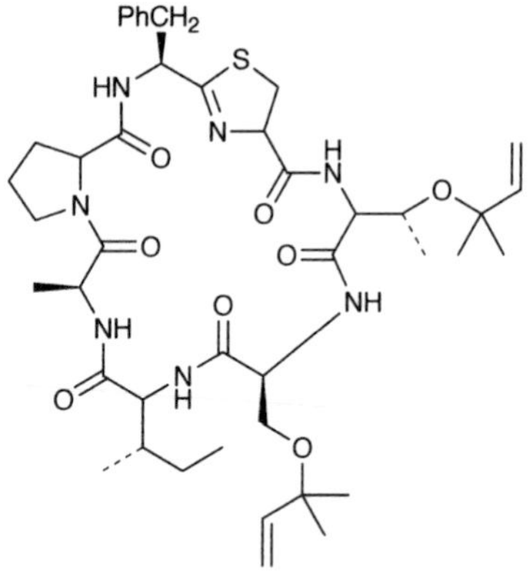

(115)

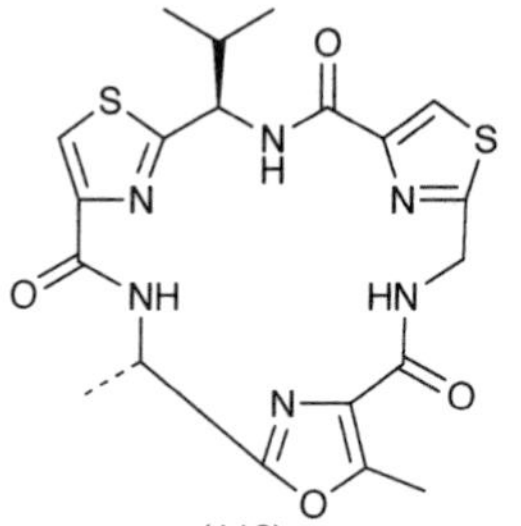

(116)

In order to improve on a previous synthesis of molecular platforms relating to dolastatins and dendroamides, insertion of Me groups into the oxazole or thiazole rings creates more rigidity, and an increased yield at macrocyclisation to (117) is obtained[175] using PyBOP. In a totally different context[176] an oxazole ring has served as an activator of a terminal carboxyl group to enable a macrocyclodepsipeptide ring to form according to Scheme 12. Synthesis[177] of the linear precursor (118) of the antibiotic GE 2270 A has been achieved, with the next challenge being the cyclisation between the thiazole methyl ester position and the amino group (protected by Boc). The tetradehydro segment (119) has been constructed[178] to be a precursor of the main skeleton of the macrocyclic antibiotic, berninamycin B.

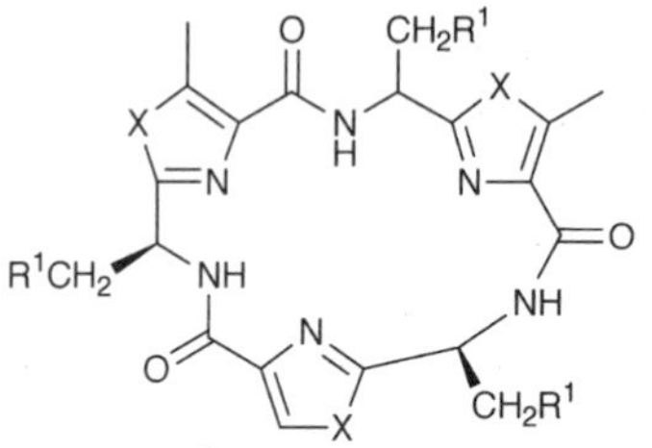

(117) R^1 = CO$_2$Bzl, X = O or S

Two research groups have succeeded in synthesising interesting tubular and cage structures by utilising cyclic thiazole peptides. Thus (120) and (121) have been condensed[179] together in the presence of FDPP/Pr^i $_2$NEt under high dilution to form (122) in 30% yield. Condensation[180] of (121) with (123) using

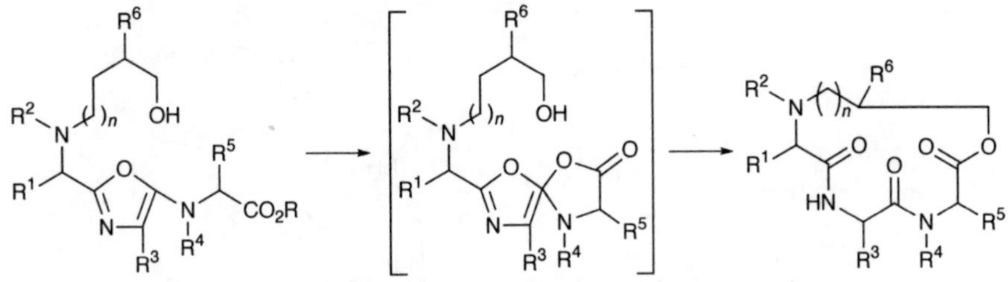

Reagent: i, trifluoroacetic acid *e.g.* $R^1 = C_6H_{13}$, $R^2 = R^5 = R^6 = H$, $R^3 = PhCH_2$, $R^4 = Me$, $n = 4$

Scheme 12

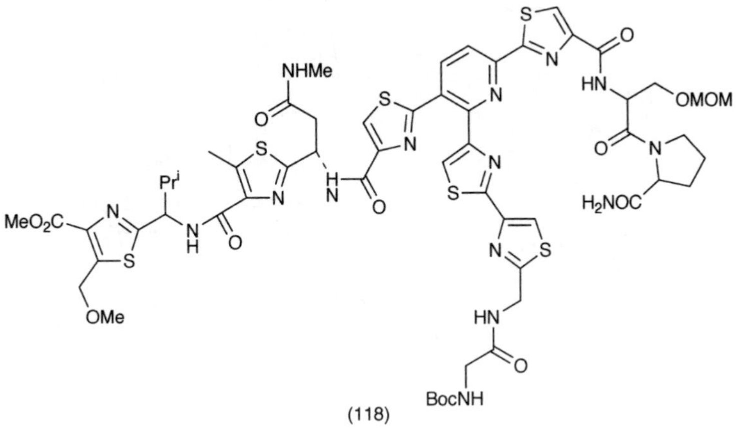

(118)

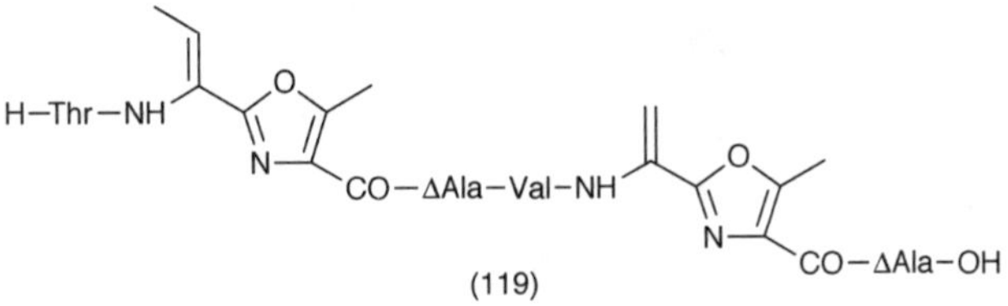

(119)

BOP/DIPEA gave rise to (124) which showed potential for forming host-guest pairs based on the evidence of trapping HMPA and hydroxybenztriazole molecules during synthesis. When the trithiazole cyclopeptide skeleton was condensed with 1,4,7-triazacyclononane a conical cage formed which was capable of interaction with Cu^{2+} ions.

The application of ion-trap-electrospray ionisation mass spectrometry (ESI MS) to the structural analysis of lissoclinamides 9 and 10 has been reviewed[181]. These lissoclinamides[182] and related metabolites[183], patellamides A, and C, and ulithiacyclamide A have been subjected to metal binding studies. It was found that lissoclinamide 10 , patellamide C and ulithiacyclamide A were selective for Cu^{2+} whereas lissoclinamide 9 and patellamide A were less selective. Absolute stereochemistry and solution conformations have been worked out[184] for promothiocins A (125) and B (126). The Ala and Val residues were found to be L-form, and the oxazole and thiazole residues were in the S-form. The dehydroalanine units in the side chain have an effect on activity but they do not seem to be essential. X-ray crystal structures have been obtained[185] for four cyclic octapept-

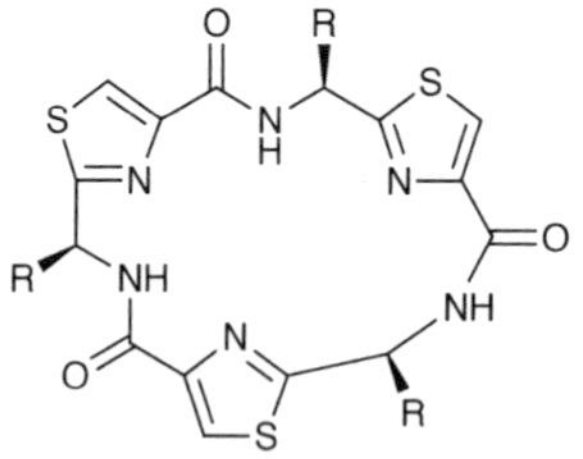

(120) R = (CH₂)₃NH₂
(121) R = (CH₂)₂CO₂H
(123) R = (CH₂)₄NH₂

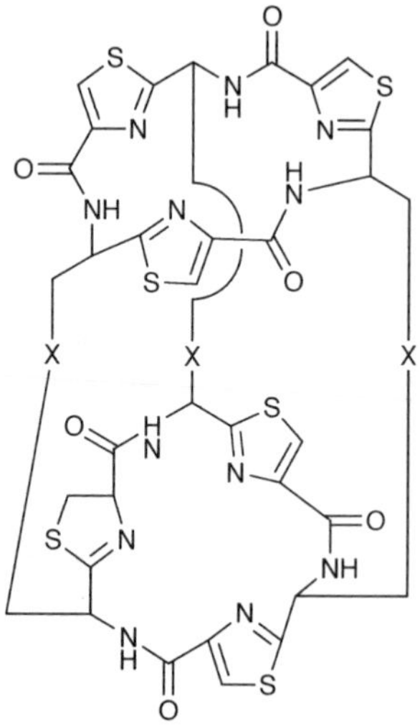

(122) X = (CH₂)₂CONH(CH₂)₃—
(124) X = (CH₂)₂CONH(CH₂)₄—

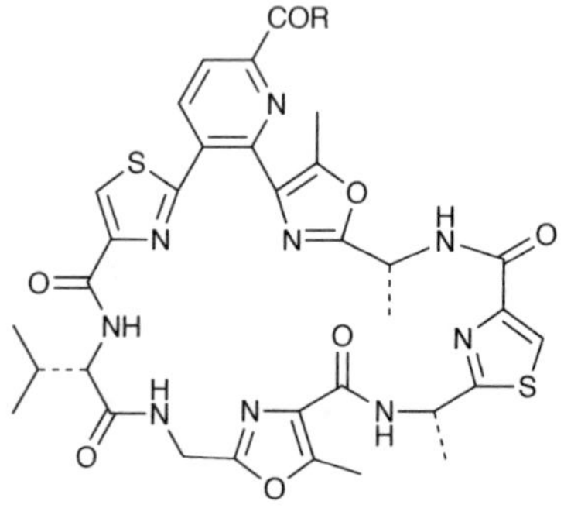

(125) R = NH

(126) R = NH

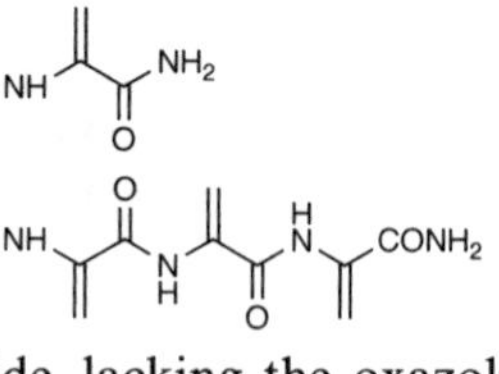

ide analogues of ascidiacyclamide, lacking the oxazoline ring. Thus for cyclo
(Ile-aThr-D-Val-Thz)₂, cyclo (Ala-aThr-D-Val-Thz-Ile-aThr-D-Val-Thz), cyclo
(Val-aThr-D-Val-Thz-Ile-aThr-D-Val-Thz) and cyclo (Ile-aThr-Val-Thz-Ile-
aThr-D-Val-Thz), all four had a folded structure, and modifications at positions
1 and 3 introduced local conformational changes in agreement with the greater
conformational flexibility in these desoxazoline analogues. A dimeric analogue
of desoxazoline ascidiacyclamide has also been synthesised[186], and its structure
was not folded, but possessed the flat conformation of a β-sheet, stabilised by
H-bond interactions between thiazole and carbonyl oxygen atoms.

2.11 Cyclodepsipeptides. – The novel structures found under this category has revealed once more the diversity of Nature, especially the variety found in the marine environment. The Vanuatu marine sponge *Haliclona* sp produces haliclamide (127)[187], which exhibited cytotoxicity in bronchopulmonary, non-small-cell-lung carcinoma cell lines with an IC_{50} value of 4.0 μg/mL. The same sponge also can claim[188] the first reported source of units based on 1,2 oxazetidine-4-methyl-4-carboxylic acid, 3-hydroxy-2, 2, 4-trimethyl-7-methoxy(and 7-hy-droxy)decanoic acid, and N-methyl-δ-hydroxyisoleucine as assembled in the halipeptins A (128) and B (129). Halipeptin A caused about 60% inhibition of edema in mice at a dose of 300μg/Kg. The sponge *Sidonops microspinosa* is a source[189] of microspinosamide (130), which is the first example of a peptide containing a β-hydroxy, *p*-bromophenylalanine residue. Microspinosamide inhibited the cytopathic effect of HIV-1 infection in an XTT-based *in vitro* assay with an EC_{50} value of 0.2μg/mL. Clavariopsins A and B (131) have been extracted and identified[190] from an aquatic hyphomycetes, *Clavariopsis aquatica* and found to be related to the aureobasidins and cyclopeptolides. Apratoxin A from the marine cyanobacterium *Lyngbya majuscula* Harvey ex Gomont has been shown[191] to have the structure (132). Of mixed peptide/polyketide biogenesis, apratoxin A possessed good *in vitro* cytotoxic IC_{50} values (0.36-0.52 nM), but it was only marginally active *in vivo* against a colon tumour and ineffective against a mammary tumour. From the mouth of an Australian river bearing a cyanobacterium species, evidence has been accumulated[192] that the cyanobacterium's main secondary metabolite is the cyclodepsipeptide, georgamide (133).

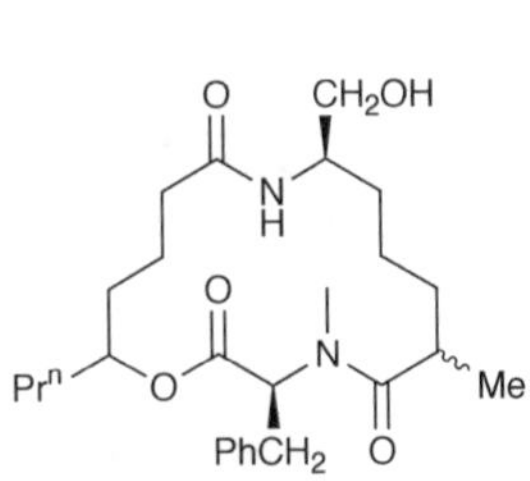

(127)

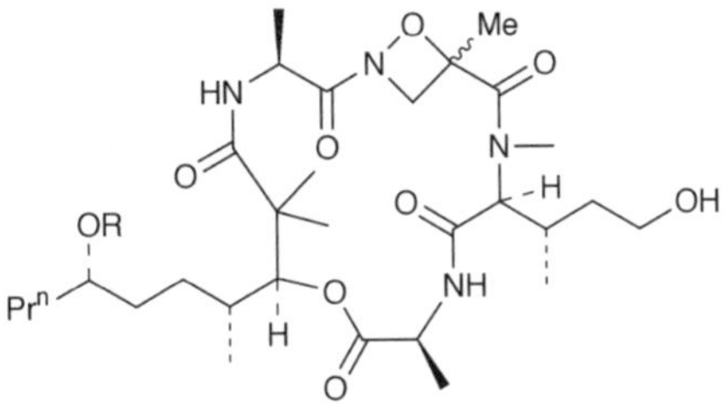

(128) R = Me
(129) R = H

Strong antibacterial activity against MSRA and VanA-type vancomycin-resistant enterococci with MIC's ranging from 0.39-3.13 μg/mL have given notice[193] of the potential of katanosin B (134) and plusbacin A (135) as inhibitors of peptidoglycan biosynthesis via a mechanism different from that of vancomycin. Pseudodestruxins A (136) and B (137) have been isolated[194] from the cultures of the corophilous fungus, *Nigrosabulum globosum*, their structures having been derived from 2D NMR and X-ray crystallography. No great biological activity could be associated with these cyclodepsipeptides. Novel antitumour depsipeptides termed SW-163C (138) and E (139) have been shown[195] to belong to the quinomycin family, while the non-toxic metabolites scyptolin A (140) and B (141), appear[196] to be related to the microcystins, both exhibiting selective inhibition of porcine pancreatic elastase with IC_{50} values of 3.1 μg/mL.

Previously established structures in this field have become major synthetic challenges in many laboratories. Thus linear precursors (142) and (143) were

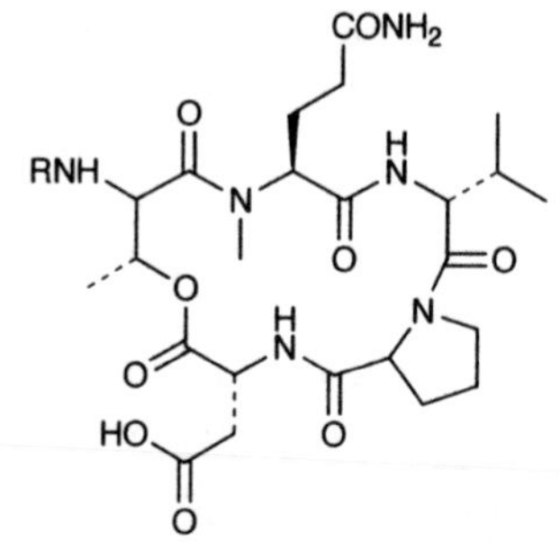

(130) R =

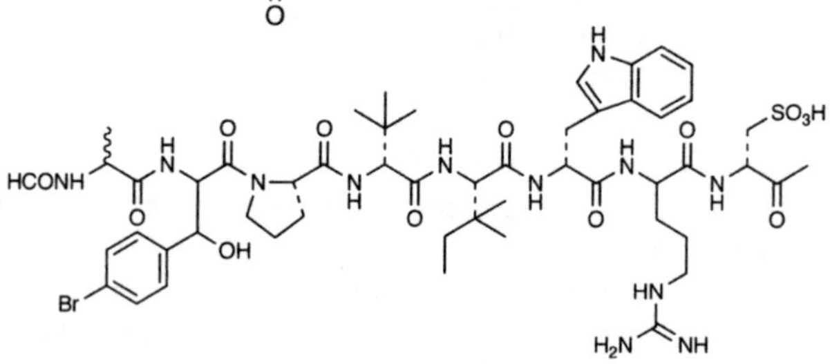

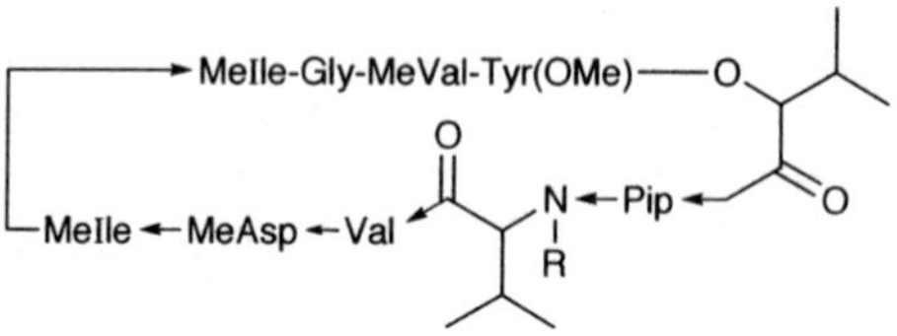

(131) In (A) R = Me, In (B) R = H

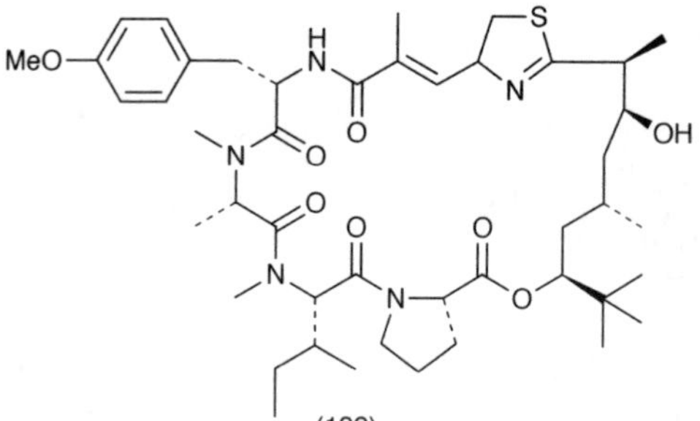

Melle-Gly-MeVal-Tyr(OMe) —— O

Melle ← MeAsp ← Val — N ← Pip
                        |
                        R

(132)

MePhe —— Pro —— O

Thr — O ← Val ← MeVal

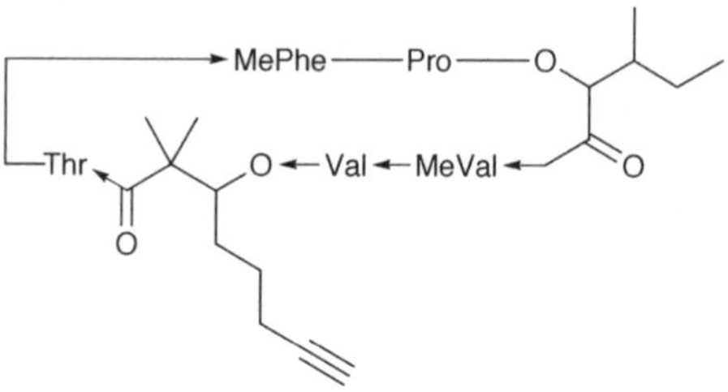

(133)

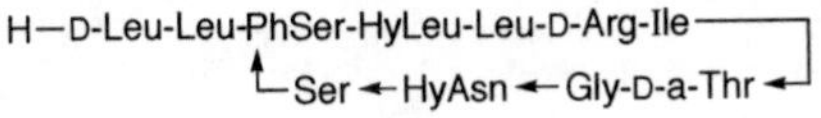

H—D-Leu-Leu-PhSer-HyLeu-Leu-D-Arg-Ile ⌐
└ Ser ← HyAsn ← Gly-D-a-Thr ◄┘

(134)

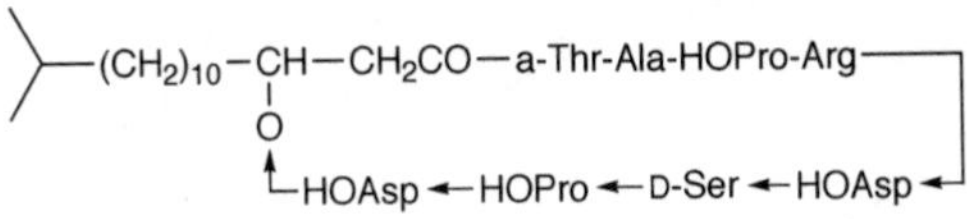

(135)

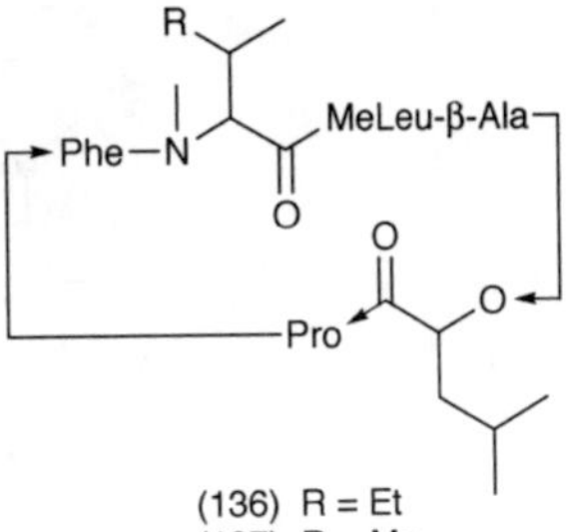

(136) R = Et
(137) R = Me

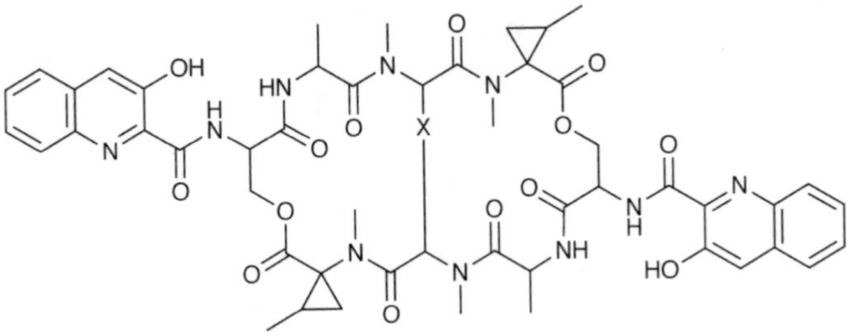

(138) X = —S–S—

(139) X = —S⌐S—

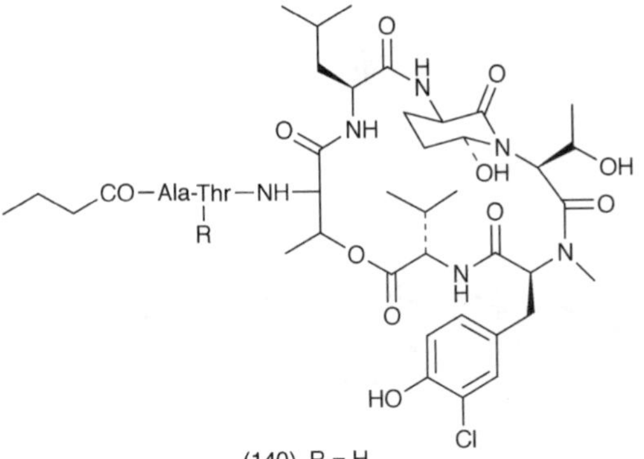

(140) R = H
(141) R = ⌐CO—Ala-

successfully synthesised[197], before macrocyclisation using HATU/DIEA (63% yield) to give the core cyclic structure of tamandarins A and B, which was further elaborated through side chain attachment of (144). It was also established through testing analogues, that the tamandarins mimic didemnin B. A previously used macrocyclodimerisation step, starting with (145) has been successful in synthesising[198] luzopeptin C (146), with the quinaldic acid moieties being added as the final stage. A new solid phase synthesis[199] of AM toxin II (147) has been achieved by cyclisation of deprotected (148) using FDDP/DIEA, followed by

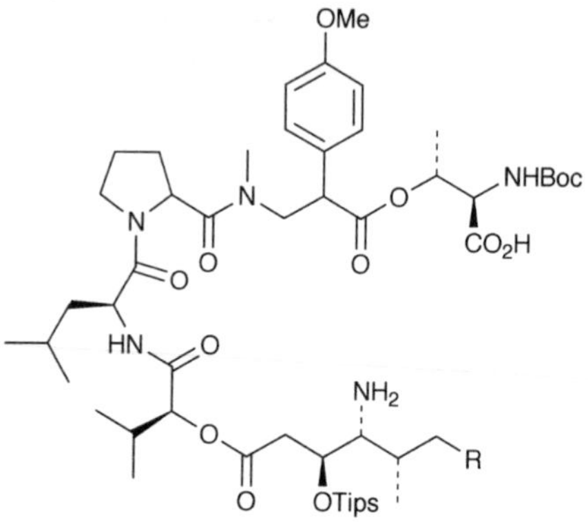

(142) R = Me
(143) R = H

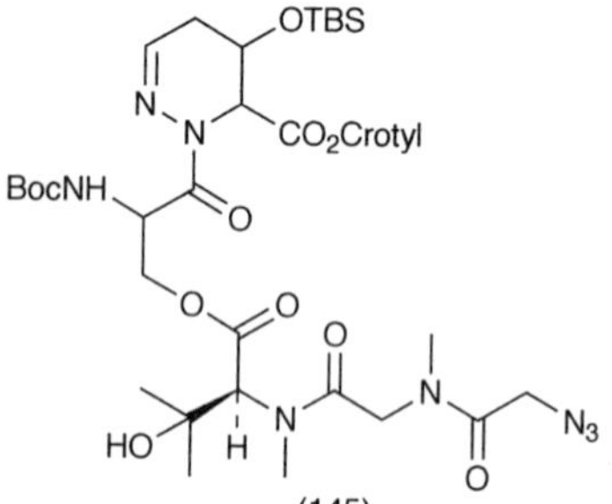

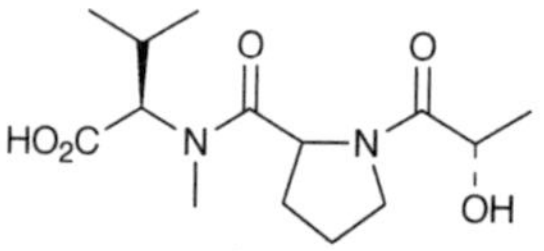

(144)

(145)

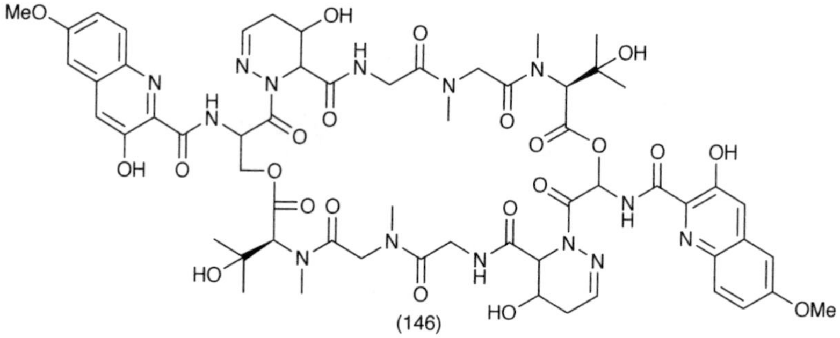

(146)

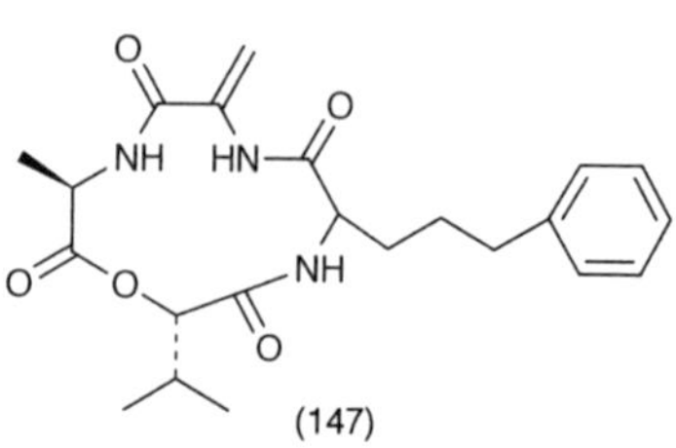

(147)

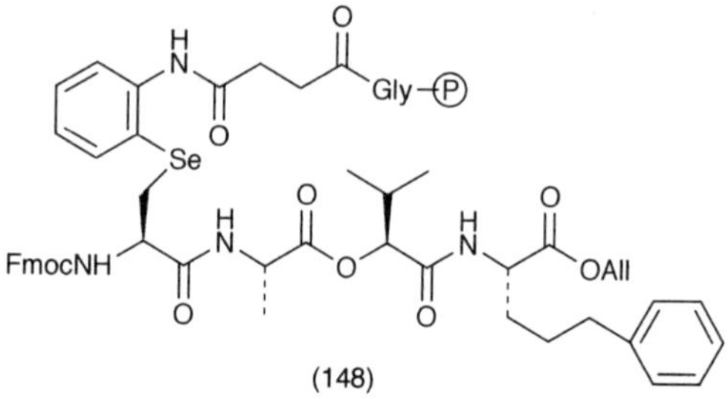

(148)

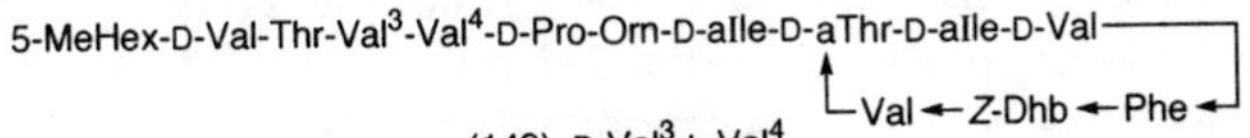

5-MeHex-D-Val-Thr-Val³-Val⁴-D-Pro-Orn-D-alle-D-aThr-D-alle-D-Val────┐
└Val◄─Z-Dhb◄─Phe◄┘

(149) D-Val³-L-Val⁴
(150) L-Val³-D-Val⁴

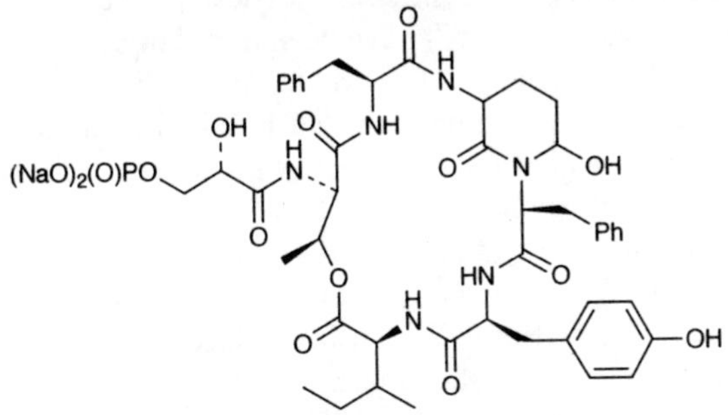

(151)

oxidative cleavage of the selenyl link to generate the dehydro residue. Previous syntheses of kahalalide F, now in clinical trials for treatment of prostate cancer, have aimed at two possible diastereoisomeric structures (149) and (150). A solid phase assembly[200] of a linear precursor, followed by off-resin cyclisation between D-Val and Phe using PyBOP/DIEA confirmed that it was (149) that represented the natural product. Small differences in the NMR data between synthetic and natural samples of micropeptin T-20 (151) have been revealed from its total synthesis[201]. The 3-amino-6-hydroxy-2-piperidone residue, which is critical to the conformation of the molecule, was introduced as the final step through oxidation and cyclisation of a homoserine unit. Completion of the macrocycle took place by activation (with FDPP) of the Phe carboxyl, which linked on to the homoserine amino group in 84% yield. The spectral anomalies were associated with the glyceric acid residue and could be a pH effect.

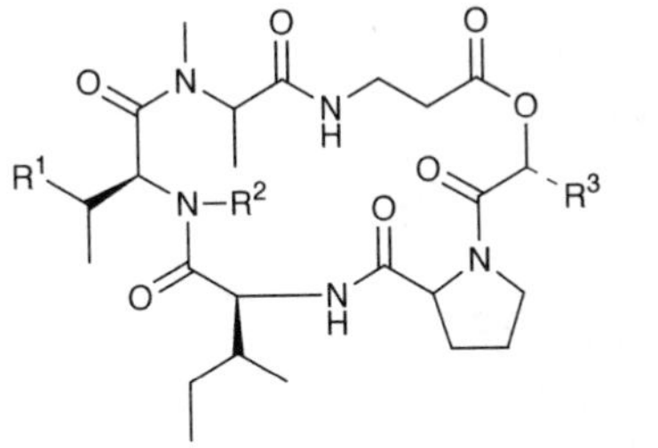

(152) $R^1 = R^2 = Me$, $R^3 = CH_2CH=CH_2$
(153) $R^1 = R^2 = Me$, $R^3 = CH_2CHMe_2$
(154) $R^1 = Et$, $R^2 = Me$, $R^3 = CH_2CHMe_2$
(155) $R^1 = Me$, $R^2 = H$, $R^3 = CH_2CHMe_2$
(156) $R^1 = R^2 = Me$, $R^3 = CH_2C(OH)Me_2$
(157) $R^1 = Et$, $R^2 = Me$, $R^3 = CH_2C(OH)Me_2$

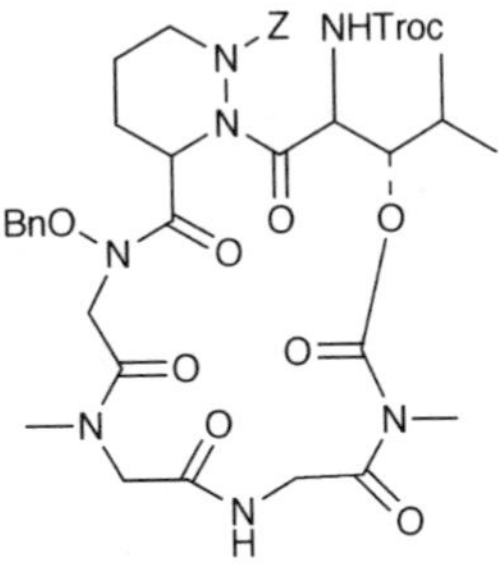

(158)

Destruxin A (152), of interest because of its insecticidal properties, together with analogues having variable allyl groups, have been synthesised[202] using solid phase techniques for the first time. The Pro residue was directly linked to a trityl resin and the linear precursor assembled on-resin including the depside link where the hydroxy acid was incorporated as 2-(3-Fmoc-aminopropionyloxy) pent-4-enoic acid. Macrocyclisation was achieved using PyBOP or PyAOP, but was not possible using DPPA. Syntheses[203] of destruxin B (153), and both natural and synthetic analogues (154-157) have been useful in probing the host selective phytotoxicity of the destruxins. The MeAla-β-Ala linkage was chosen as the cyclisation point, utilising in a novel way a Boc hydrazide protecting group on the MeAla carboxyl, which on activation with nitrous acid did not allow the formation of dioxopiperazines. Yields of 58-65% were achieved in this way, and of the analogues synthesised, (154) proved to be the most phytotoxic. A synthetic strategy[204] based on a [2+2+2] fragment condensation, has been worked out to achieve the assembly of the cyclodepsipeptide core (158) of the antitumour antibiotic verucopeptin. Macrolactamisation between Gly and Sar to form (158) was achieved in 67% yield using 10eq HATU for 48hr.

(159) R^1 = Me, R^2 = heptyl, R^3 = CHMe$_2$ (160)

Another synthesis of hapalosin (159), which is capable of reversing multidrug resistance in tumour cells, has been achieved[205] mainly by solid phase, although macrolactamisation was carried out off-resin. The latter was achieved between the hydroxy acid carboxyl group and the amino group of valine using TBTU/HOBt, and when analogues centred on changes in R^1, R^2 and R^3 were also included, yields ranging from 33-64% were recorded. A review[206] has also recorded the chemical and biological studies carried out on hapalosin. The potent antitumour depsipeptide doliculide (160) has been synthesised[207] using a strategy of separately forming fragments based on the points (a) (b) and (c) in (160). The final step was the cycloamidation at point (a), using BOP/DMAP, achieved in 82% yield. In order to overcome the pH-dependent instability of the depside bond in the antifungal depsipeptide FR901469 (161) an amide analogue (162) has been synthesised[208] via a biotransformation of (161) using the organism, *Actinolanes utahensis* IFO-13244 at pH 7.8, which created a linear derivative which could be re-condensed with fragment (163) after the necessary deprotection of groups. Final macrolactamisation to make (162) occurred in 83% yield using WSC/HOBt. A similar approach[209] has also been used to replace the ornithinyl residue at position 1 in (161) as this residue is implicated in the instability of the depside link. Reduced potency of the cytotoxic depsipeptide arenastatin A (164) in blood has been associated with the metabolic degradation of the depside links, so syntheses of non-depside analogues (165-168) have been reported[210], as well as a rationalised synthesis of the parent (164). The best stability in blood was found

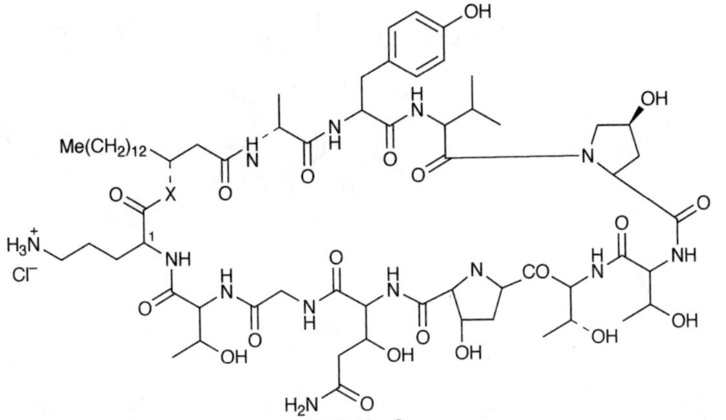

(161) X = O
(162) X = NH

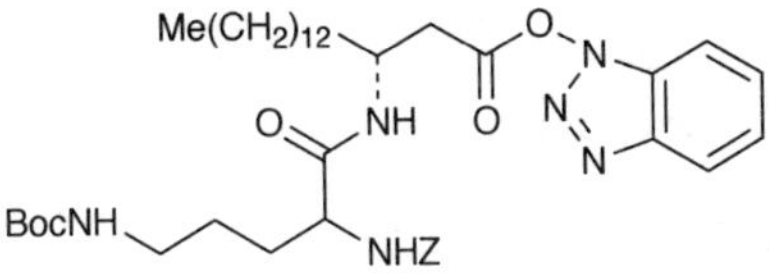

(163)

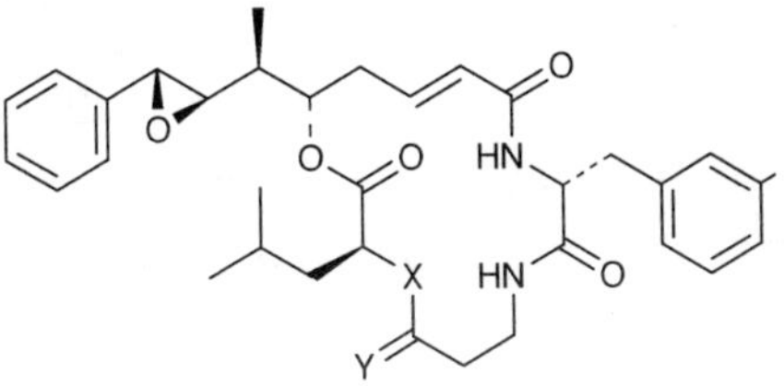

(164) X = Y = O
(165) X = NH, Y = O
(166) X = CH₂, Y = O
(167) X = O, Y = H₂
(168) X = S, Y = O

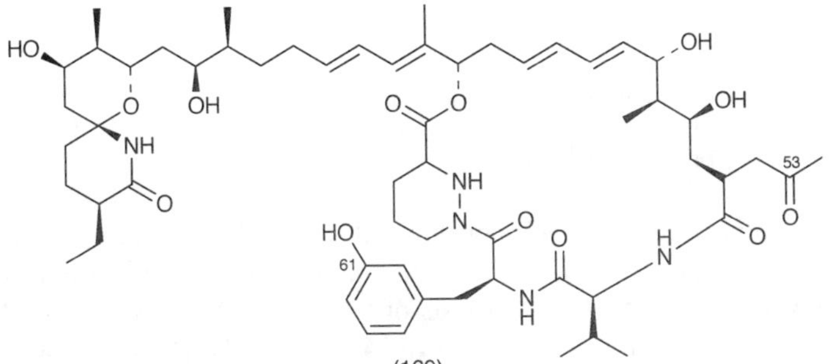

(169)

with the two analogues (166) and (167). In a survey[211] of which functional groups
are required for activity in the immunosuppressant sanglifehrin A (169), chemical
removal of CO at 53 using borohydride and phenylchlorothionocarbonate gave
an analogue with no loss of immunosuppressive ability nor in binding with
cyclophilin. However removal of the phenol group at 61, by first totally hydro-
genating the phenolic ring, did destroy its immunosuppression.

By far the most efficient way for cyclising linear depsipeptides is via the
formation of amide links. This has been borne out by the failure of two lactonisa-
tion attempts from linear precursors in a synthetic study on stevastelin B (170).

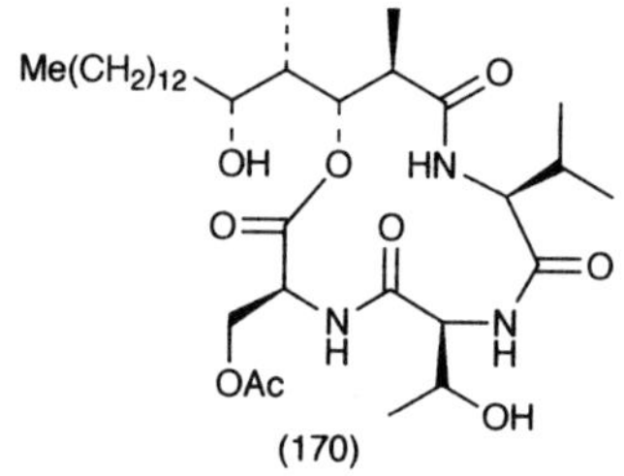

Me(CH$_2$)$_{12}$

OH   O   HN

O

O   O

N   N
H   H

OAc

OH

(170)

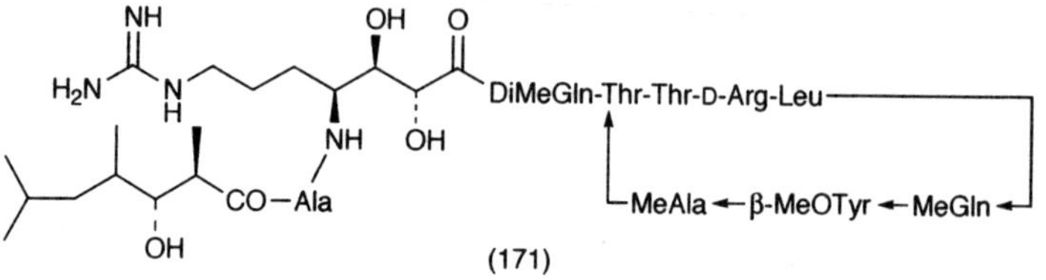

NH

H$_2$N   OH   O

NH   OH

DiMeGln-Thr-Thr-D-Arg-Leu

CO—Ala

MeAla ← β-MeOTyr ← MeGln

OH

(171)

HO$_2$C   OH

O   OH

O   NH   Cl

HO   NH

NH   O

R^{3}NH   NH

--HNCO

OH

O   NH

N   NH   (CH$_2$)$_n$Me
H

CO$_2$H O   NHR^1

NHR^2

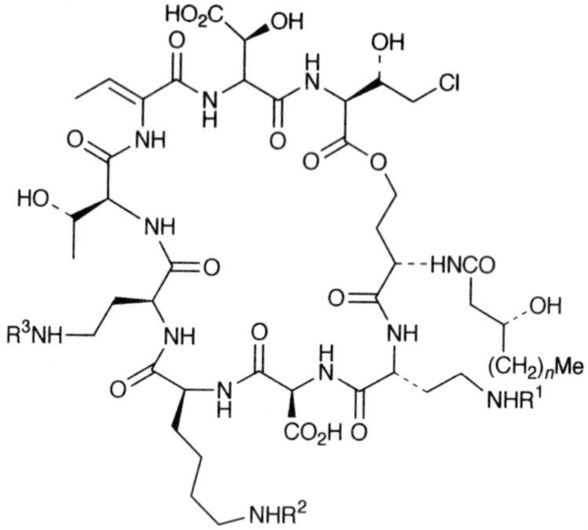

(172) $n = 10$, R^1 = R^2 = H, R^3 = DOXO
(173) $n = 10$, R^1 = R^2 = DOXO, R^3 = H   DOXO =
(174) $n = 12$, R^1 = R^2 = R^3 = DOXO   —CO$_2$

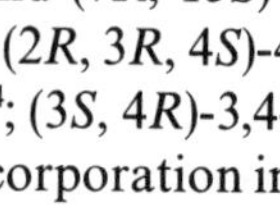

Many naturally occurring cyclodepsipeptides have quite complicated non-peptidic chiral components, which are quite demanding to synthesise. Examples in this review year include: the synthesis of (7S, 15S)- and (7R, 15S)-dolatrienoic acids for incorporation[213] into dolastatin 14; protected (2R, 3R, 4S)-4,7-diamino-2, 3-dihydroxy heptanoic acid for callipeptin A (171)[214]; (3S, 4R)-3,4-dimethyl-S-glutamine and the equivalent pyroglutamic acid for incorporation into papuamides and callipeltins[215, 216]. In studies[217] on the cardiac sodium-calcium exhanger, callipeltin A (171) induces a positive inotropic effect, which has been linked to the inhibition of the Na$^+$/Ca^{2+} exchanger. The potential clinical utility of the pseudomycins is compromised by the undesirable irritation, which occurs at the injection site, so the synthesis[218] of pro-drug versions (172-174) was aimed at overcoming this problem. The analogues retained the original activity and were free of tail vein irritation. By using re-combinant enzyme Vgb from *Staphylococcus aureus* it has been deduced[219] that the deactivation of the streptogramin B antibiotics occurs via an elimination (lyase) mechanism rather than hydrolysis of the depside bond as depicted in the diagrammatical scheme 13.

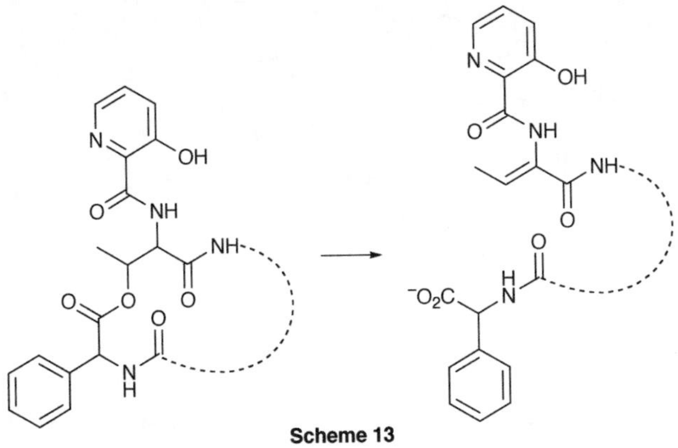

Scheme 13

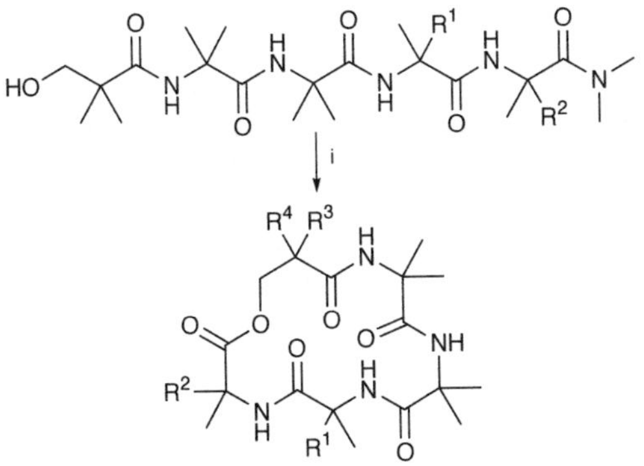

(175) R^1 = Bn, R^2 = R^3 = R^4 = Me
(176) R^1 = R^3 = R^4 = Me, R^2 = Bn
(177) R^1 = R^2 = Me, R^3 = Ph, R^4 = H

Reagent: i, HCl/toluene/100 °C

Scheme 14

Cyclodepsipeptides (175) and (176) were very surprisingly made[220], by just heating the linear amide precursors with HCl in toluene as in scheme 14. No doubt the steric restrictions associated with the α,α,-dialkyl substitution, assembled via the azirine methodology, influence the ease of macrolactonisation. A conformational analysis on the cyclodepsipeptide (177), using X-ray diffraction and other techniques[221] confirmed the presence of two H-bonds, stabilising two β-turns. Dehydrodidemnin (Aplidine) (178) the potent anti-tumoral natural product, currently in Phase II clinical trials, is known to exist as two slowly interconverting conformations (1:1 mixture). NMR spectroscopy[222] has shown that these two forms are due to restricted rotation around the pyruvyl-Pro bond in the side chain. The overall 3D-structures are similar despite the conformational change, with each structure stabilised by a H-bond between the pyruvyl carbonyls (different one for each isomer) and the Thr^6 residue. The characteristic ions produced by electrospray ionisation/ion trap mass spectrometry of ring-

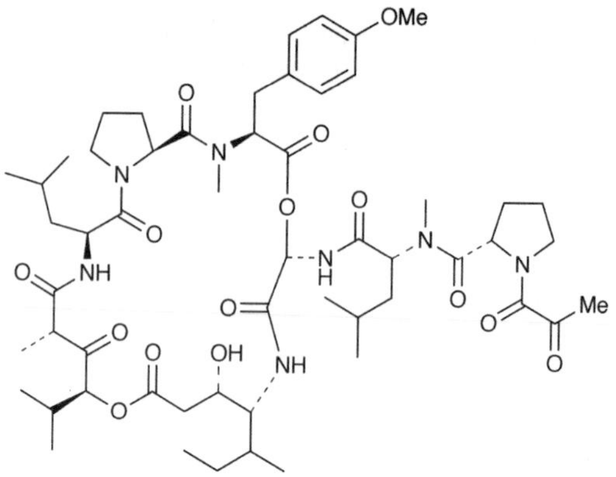

(178)

opened LI-F cyclodepsipeptide antibiotics could be useful in the characterisation of the sequences[223].

3   Modified and Conjugated Peptides

This continually expanding section of these annual reports is dedicated to a discussion of molecules that bear non-peptidic conjugate groups, which often have a significant bearing on the biological activity of the molecules. They demand special attention as they often contain links that can be lost in less than delicate conditions.

3.1   Phosphopeptides. – Since phosphorylation of Ser, Thr and Tyr residues is an extremely important modulator of protein function there has been increasing demand for ways and means of locating these phosphorylated residues within proteins. While full characterisation of phosphoproteins is still a formidable task, techniques allied to mass spectrometry, seem to be emerging as a possible answer to the challenges. But as yet the analysis is far from routine. The methodologies available have been the subject of a review[224] which highlights the combination of techniques necessary to reach an understanding of the sequence and location of phosphorylation. To overcome the complexity of using mass spectrometry with mixtures of peptides, pre-treatment with barium hydroxide has been suggested[225], which produces dehydroamino acid residues that then can be further derivatised with alkane thiols. Alkaline hplc/ electrospray ionisation skimmer collision-induced dissociation mass spectrometry[226] is a rather cumbersome label for a rapid on-line method for identifying phosphorylated peptides in enzyme digests. The pH10.5 conditions enhanced the generation of phophopeptide-specific fragment ions from the phosphorylated residues. Chemical modification and affinity purification prior to mass spectral analysis have been investigated[227] and involve β-elimination/1,2 -ethanedithiol addition followed by reversible biotinylation. Being able to use mass spectrometry in chiral recognition is something of a surprise, but enantio-discrimination of *O*-phosphoserine in bio-logical specimens has been used[228] successfully in quantifying the enantiomeric excess of phosphoserine in proteomic analysis. Key to the technique is the generation of charged trimeric clusters generally referred to as $[M(ref)_2A]^+$ with ref = *S*-(+)-(1-aminoethyl)phosphonic acid or *O*-phospho-L-threonine as typi-fied by the ion $[Na(O$-phospho-L-Thr$)_2 O$-phospho-Ser$]^+$. Some of the complex-ity of the mass spectra of phosphoproteins, can be better clarified if elastase digestion to produce smaller peptide fragments is carried out[229] which can then be followed by tandem mass scans concentrating on the loss of phosphoric acid from specific residues. Scanning of tryptic digests has also been investigated[230] by applying a 4-s IR laser pulse at 7-11W to the digest, and then monitoring the phosphopeptide signature ions formed using ESI-FITCR-MS in the negative mode with infrared multiphoton dissociation.

A multitude of advanced mass spectrometric techniques, without ancillary support have been checked out for possible application. These include, a multi-dimensional ES-MS approach[231], positive and negative ion matrix-assisted laser desorption/ionisation MS[232], ion scanning of the immonium ion of phos-photyrosine using TOF(Time of Flight)-MS in the positive mode[233], quadrupole TOF-MS[234] and MALDI-TOFMS[235], the latter benefiting from a further analysis with nanoelectrospray quadrupole TOF-MS.

Methods for the synthesis of phosphopeptides have been reviewed[236], with the main emphasis on the preparation of phosphoamino acid building blocks. A comparison [237] of the Fmoc protected building block approach to that of post-assembly global phosphorylation (with N,N-diethyl-dialkyl phos-phoramidite) has been made. The longer the peptide, the lower the phosphorylat-ing efficiency, and the stepwise phosphorylation strategy worked best for tyrosyl residues. Ammonium t-Bu H-phosphonate has been used[238] for phosphorylation of Tyr- and Ser-containing peptides derived from an Fmoc synthetic strategy. Good quality monoprotected peptide phosphates were made, without pyrophos-phate formation or β-elimination catalysed by piperidine. N-Phosphodipeptides have been synthesised[239] by coupling N-phosphoamino acids and amino acid methyl esters with dichlorotriphenylphosphorane, conditions and methodology which could be more generally applicable. Studies towards the synthesis of the antibiotic alafosfaline have been recorded[240] in a Journal not accessible to the reviewer, but the abstract indicates that both N- and O-protected alafosfaline have been made starting from 2-alkylphosphonopropanoic acid.

3.2 Glycopeptide Antibiotics. – The challenge to overcome antibiotic resis-tance continues, with the hope that synthesis of many of the complicated structures will lead eventually to improved or re-engineered molecules. A review of the most recent synthetic and mechanistic studies on vancomycin, teicoplanin and ramo-planin by the D.L. Boger group has appeared[241]. Some of the formi-dable difficulties in the synthesis of glycopeptide antibiotics, such as atro-pisomerism, formation of biphenyl and biaryl ether linkages have been high-lighted[242] in the context of the potential of the Suzuki coupling methodology.

In a major contribution to synthetic studies in this area, Evans *et.al* [243] have

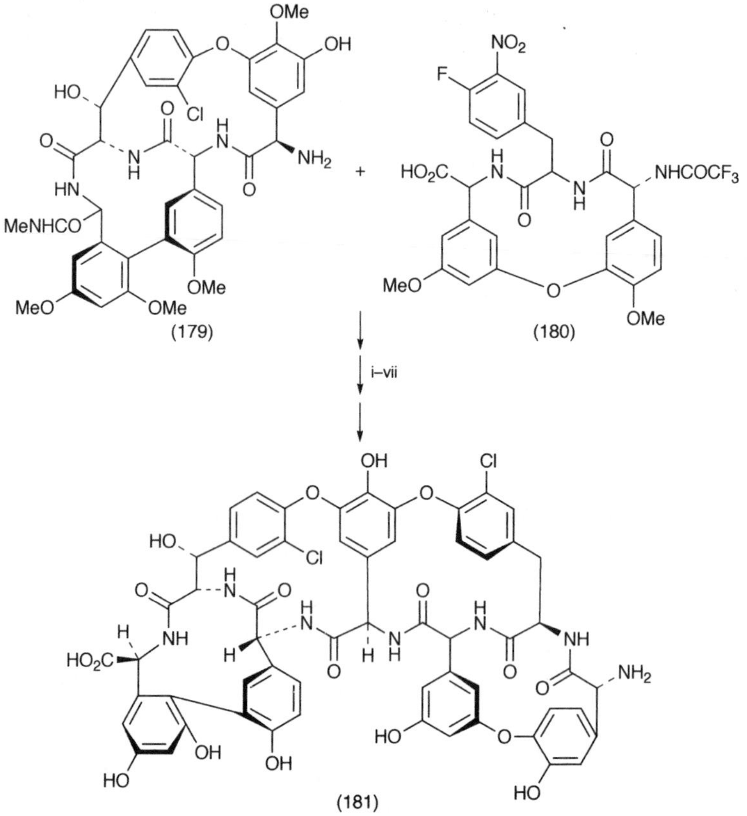

Reagents: i, DEPBT, −5 °C; ii, CsF, DMF, 10 °C; iii, H$_2$/10% Pd/C; iv, Bu^{t}ONO, HBF$_4$/CuCl,CuCl$_2$; v, N$_2$O$_4$, 0 °C; vi, DMF/H$_2$O/60 °C; vii, AlBr$_3$/EtSH, 0 °C

Scheme 15

achieved the total synthesis of teicoplanin aglycone (181) by a route whose last stages are summarised in Scheme 15. Key to their approach was the synthesis of subunits (179) and (180) in their correct oxidation states before further coupling. The fully functionalised DEFG ring system (182) of teicoplanin has also been synthesised[244] using an alternative methodology. Atropisomerism between the M and P forms of (182) was also studied. A novel selenium safety catch linker[245] has been useful in the solid phase semi synthesis of vancomycin aglycone(183) and in the synthesis of libraries. A number of useful antibacterial leads against van-comycin – resistant bacteria were obtained.

For some years now, 'reconnaissance' reports on the partial synthesis of difficult coupling fragments in this field have been periodically released. This year is no exception as exemplified by, (i) the ruthenium-promoted biaryl ether coupling between rings C and D to make[246] the BCDF ring system (184) of ristocetin A, (ii) the stereoselective synthesis[247] of the axially chiral A-B ring system of vancomycin aglycone (183) using a planar chiral tricarbonyl (aryl halide) chromium complex, (iv) a Ugi reaction incorporating the components illustrated in Scheme 16[248], (v) the thallium(III) trinitrate construction[249] of the

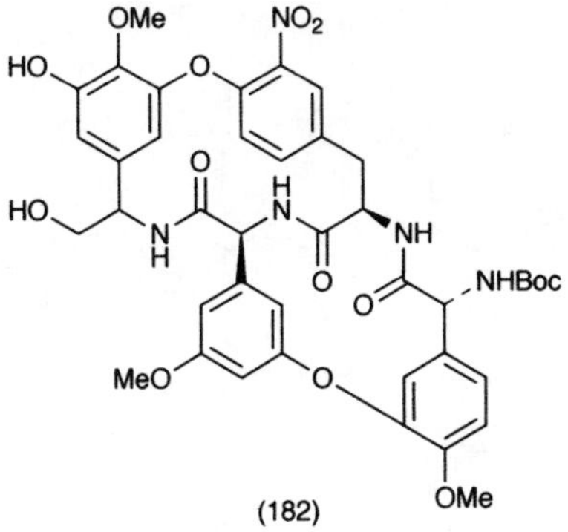

(182)

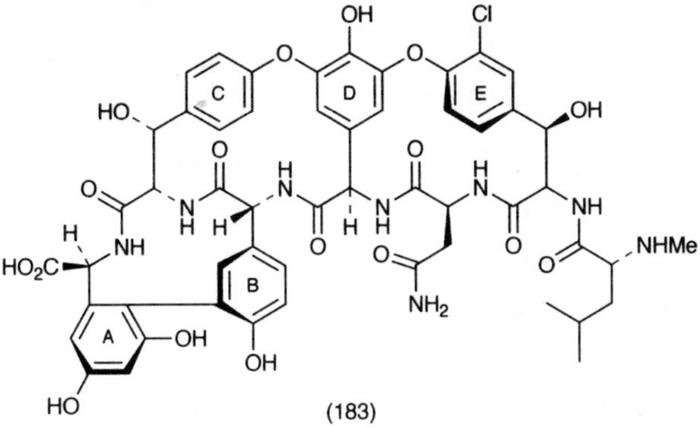

(183)

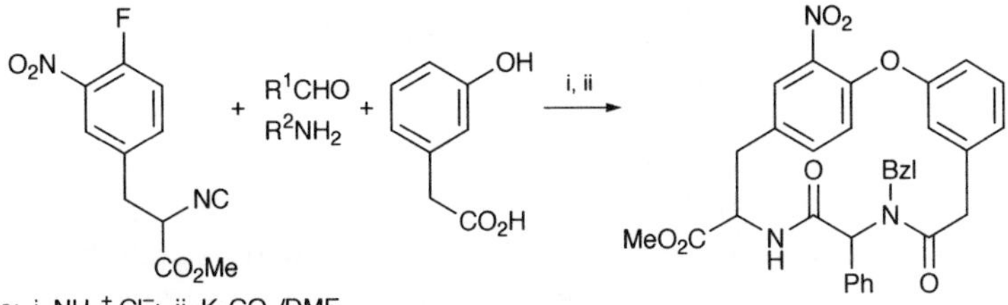

Reagents: i, NH₄⁺ Cl⁻; ii, K₂CO₃/DMF

Scheme 16

cyclic isodityrosine derivative (185), and (vi) the O-arylation of phenols with phenyl boronic acids to produce metalloproteinase inhibitors[250] such as (186).

In last year's Chapter in this series (Vol.33, p 269), details of the synthesis of dimers of vancomycin were revealed. Further syntheses of dimers using disulfide formation and olefine metathesis have now been adapted[251] for target-acceler-ated combinatorial synthesis, which on assay were found to have generated useful antibiotics against vancomycin-resistant bacteria. On allowing van-comycin, eremomycin and avoparcin to be exposed to neutral aqueous solutions containing traces of formaldehyde or acetaldehyde, tandem mass spectrometry has revealed[252] that derivatisation occurs at the N-terminus with formation of a 4-imidazolidine moiety. Hydrophobic N'-$C_{10}H_{21}$ and N'-p-(p-chlorophenyl)ben-zyl derivatives of eremomycin have been found[253] to be the most active against vancomycin-resistant enterococci. The role of the sugar residues in

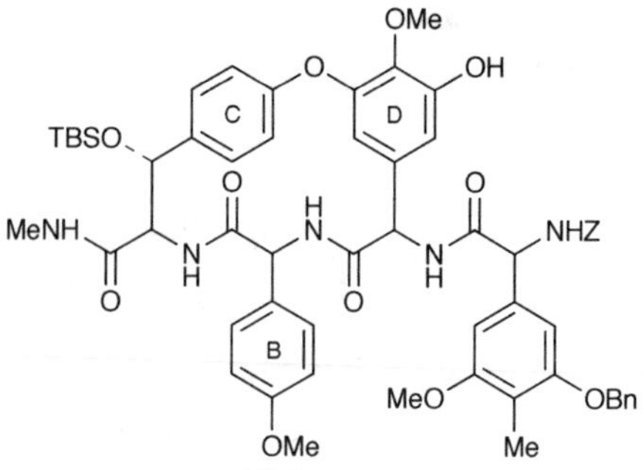

(184)

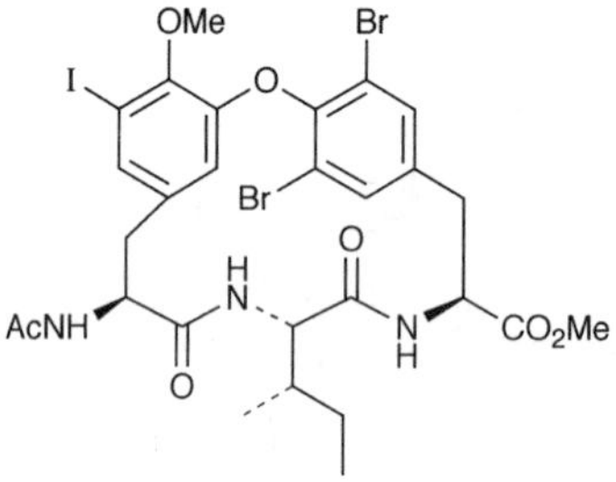

(185)

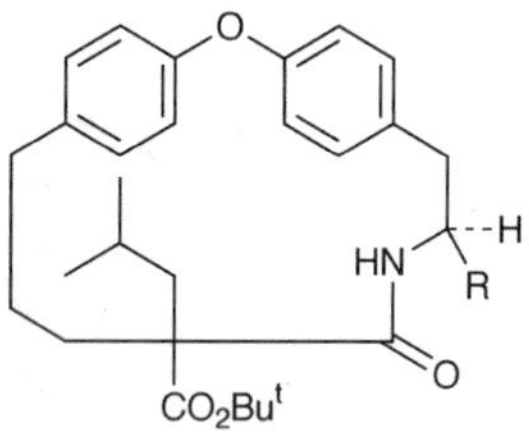

(186)

molecular recognition by vancomycin has been assessed[254] by comparing X-ray crystal structures of the aglycone (183) to structures of intact vancomycin:ligand complexes. It was shown that the aglycone binds acetate anions and forms back-to-back dimeric complexes similar to intact vancomycin, but it is argued that the sugar residues would enlarge and strengthen the dimer interface, and limit the molecule to a relatively small number of productive conformations. Cloned glycotransferases have been produced[255] to make novel carbohydrate derivatives of vancomycin and teicoplanin aglycones, and their substrate specificity is in the process of being ascertained. In order to be more compatible with lipid membrane environments, NMR techniques have been applied[256] to CD$_3$OD solutions of the aglycone of antibiotic ramoplanin (187) and show that it exists as a dimer, in contrast to its existence as a monomer in D$_2$O. The dimerised structure might provide the clefts necessary for binding to Lipid II. A triazacyclophane scaffold has been used to produce a library of receptor molecules that were tested against D-Ala-D-Ala and D-Ala-D-Lac ligands. After screening and validation, the bio-inspired mimic that came closest to vancomycin was (188).

With the long-term aim of making vancomycin analogues, studies on the biosynthesis of sections of the molecule continue. Thus the order of the oxidative bridging of the aromatic side chains has been studied[258] through the isolation of aglycone intermediates from culture filtrates. The observations support the hypothesis that β-hydroxylation and chlorination occur before the oxidative ring formation of the aglycone, with glycosylation and N-methylation occurring subsequently. O-Linked bridging between C and D rings is the first ring closure in the biosynthesis. The 3,5-dihydroxyphenyl acetate precursor of 3,5-dihydroxyphenylglycine has been shown[259] to be derived by catalysis by a type III

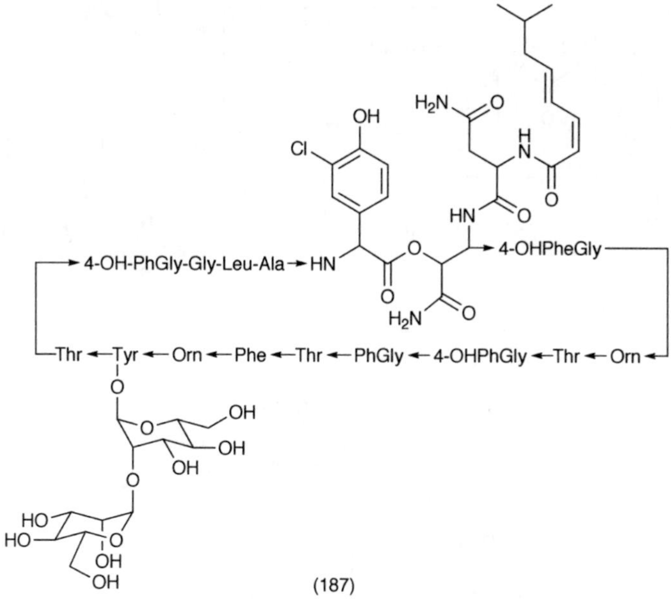

(187)

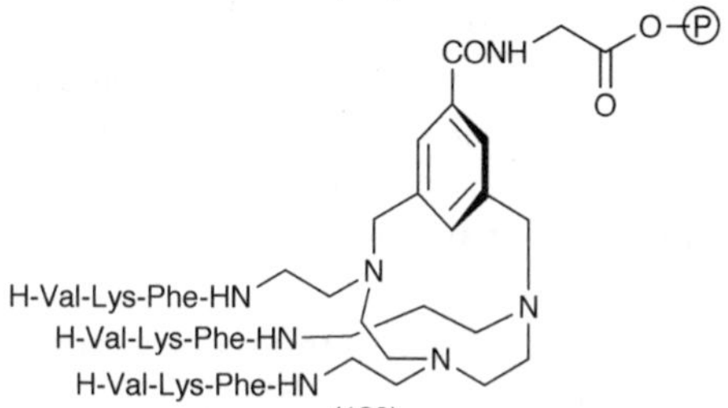

(188)

polyketide synthase using a malonyl-CoA starter unit. Of four proteins expressed[260] in *E.coli*, onlyDpgA, a type III polyketide synthase in association with protein DpgB, was able to catalyse the formation of 3,5-dihydroxyphenylglycine.

Of the eight conformationally rigid inserts that were investigated[261] as replacements for the methylvalerate (boxed) section of bleomycin (189) the isomer (A) (shown with 189) proved to have the best characteristics. Structurally related to the bleomycins are the tallysomycins, and an intermediate (190) on the synthetic pathway to tallysomycin has been assembled[262].

Chiral separations using glycopeptide antibiotics as stationary phases have recently gained popularity and this interest has been placed in context by a review[263]. Many of the antibiotics are complementary to one another, so that if a partial enantioresolution is obtained by one glycopeptide, there is a high probability that a baseline or better separation can be obtained with another. Retention and selectivity of teicoplanin stationary phases have been studied in a number of situations: in the separation of amino acids after copper complexation and isotopic exchange[264]; in separation of amino acids on silica gel modified with teicoplanin[265]; in LC-MS as a chiral stationary phase for amino acid separation

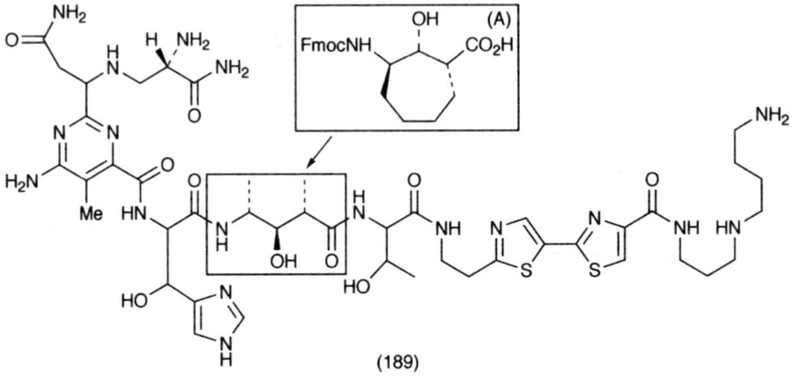

(189)

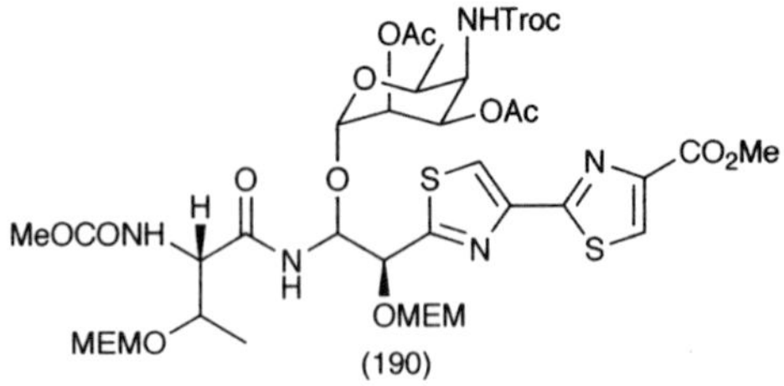

(190)

followed by ion spray mass spectrometry for ionisation and detection[266]; in the separation of dansyl amino acid enantiomers[267]. Several unnatural amino acids have been chirally separated using hplc separation of the underivatised acids on ristocetin A-bonded chiral stationary phase[268].

3.3 Glycopeptides. – The sub-division of this section into work associated with O-, N- and C-linked glycopeptides can again be justified. However a few of the lead reviews cover the whole breadth of the subject, as for example the coverage of recent trends in glycopeptide synthesis[269]. The link between complex carbohydrates and glycoproteins has been discussed[270], and the efficient synthesis of oligosaccharides and glycopeptides by chemoenzymatic approaches has been reviewed[271]. The synthesis of long-chain glycopeptides and the coupling of fragments by thioesters or by chemical ligation has been reviewed[272] in the context of the synthesis of diptericin and emmprin.

Dehydroalanine units in peptides can be the focal point for chemoselective conjugate addition of farnesyl thiolate and thioglycosides[273], while S-linked glycosyl amino acids are formed[274] from a reaction between protected serine and a sodium thiolate salt of a variety of unprotected 1-thiosugars. A combination of acid labile protecting groups, acid sensitive resins and two orthogonal chemoselective ligation reactions have been used[275] to create the model (191) of post translational tyrosine sulfation and O-glycosylation. A Tentagel solid support functionalised with a protected 2-amino sugar has been proposed[276] for the preparation of glycopeptides and nucleo-glycopeptides. 2-Bromoethyl glycosides corresponding to the Tn (GalNAc α Ser/Thr) and T [Gal (1→3) GalNAc α Ser/Thr] antigens have been prepared[277], starting from N-acetylgalactosamine. These were used to alkylate a homocysteine residue in a peptide that is able to

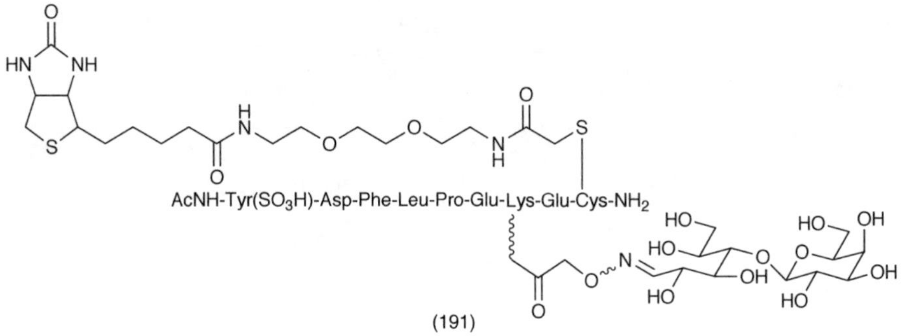

AcNH-Tyr(SO₃H)-Asp-Phe-Leu-Pro-Glu-Lys-Glu-Cys-NH₂

(191)

bind to class I MHC molecules on antigen-presenting cells, and further processed into glycopeptides which carry the sialyl-Tn and 2,3-sialyl-T antigens by using recombinant sialyltransferases.

3.3.1 *O-Glycopeptides*. Lipid II is the ultimate monomeric intermediate of bacterial cell wall peptidoglycan biosynthesis, and as part of the synthetic effort towards its synthesis, the core NAG-NAM disaccharide unit (192) has been assembled[278]. In the synthesis[279] of the hen egg lysozyme 52-61 sequence (193), it has been shown that when the R protecting groups are either 3-fluorobenzoyl or 2,5-difluorobenzoyl the extent of β-elimination during synthesis decreased from 80% to 10% and from 50% to 0% respectively. The first synthesised examples (e.g. 194) of natural-type glycoconjugates that simultaneously contain phosphate and sulfate as well as carboxylic acids have been reported[280]. Phosphate was introduced using $(Pr^i)_2NP(OBn)_2$/1H-tetrazole and sulfate using $SO_3.NMe_3$. A facile route to preparing homogeneous antifreeze glycoproteins (composed of repeating Ala-Thr-Ala units) has used[281] solid phase techniques using T_F antigen building blocks to form (195).

The chemoenzymatic approach overcomes some of the chemical problems and poor control of anomeric configuration in this demanding synthetic area. As discussed in an earlier reference above (Ref. 276), once the T and Tn antigens are obtained by chemical synthesis, further elaboration to sialyl-Tn and 2,3 sialyl-T can be catalysed by recombinant sialyltranferases[282] as applied to the tandem repeat domain of mucin MUC 1, and neoglycosylated derivatives of a T-cell stimulating viral peptide. As part of the effort to design inhibitors to target the polysaccharide backbone synthesis of bacterial cell walls, the precursor UDP-MurNAc (196) and its vancomycin-resistant analogue depsipeptide (197) have been synthesised by chemoenzymatic approaches[283]. The key enzyme-catalysed step was the coupling of the Thr side chain to the sugar using glucose dehydrogenase. Endo α-N-acetylgalactosaminidase from *Streptomyces* sp. has catalysed[284] the transfer of Gal β-(1→3)-GalNAc to the hydroxyl group of a serine within a hexapeptide using Gal β-(1→3)-GalNAcβ-*p*NP as donor. The yield[285] in transglycosylation to form the conjugate [Gal(β1-3) Gal(β1-4) Xyl(β1-O)-L-Ser] was much improved by using galactosyl fluoride as a donor.

A strategy to replace the critical β (1→3) link to the core GalNAc in mucin-type oligosaccharides with thioether analogues, has been brought about[286] via

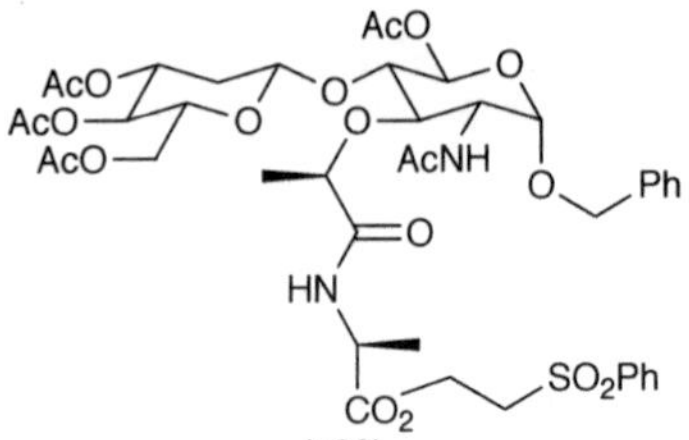

(192)

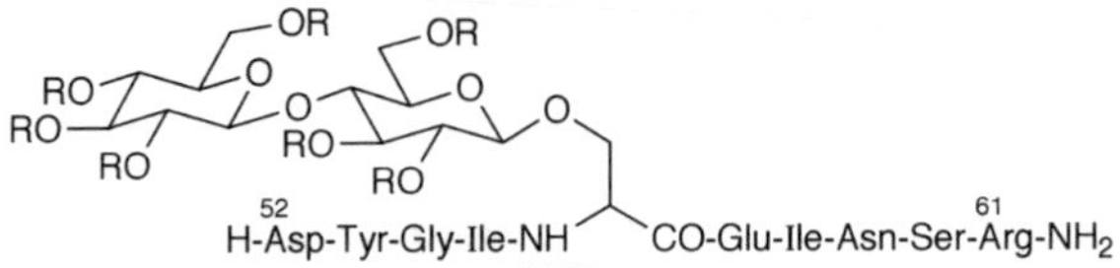

52
H-Asp-Tyr-Gly-Ile-NH   CO-Glu-Ile-Asn-Ser-Arg-NH₂
                                              61
(193)

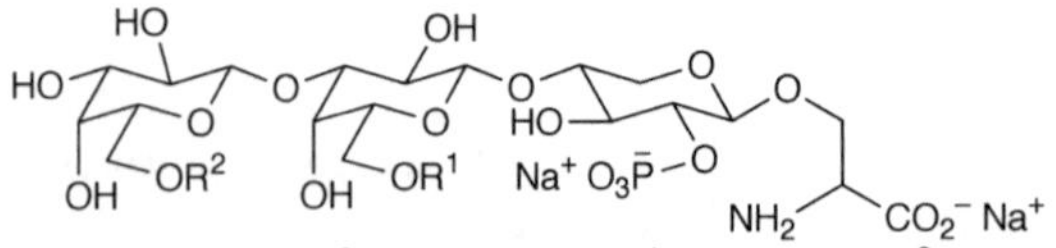

(194) R^1 = H, R^2 = SO₃⁻ Na⁺ or R^1 = SO₃⁻ Na⁺, R^2 = H

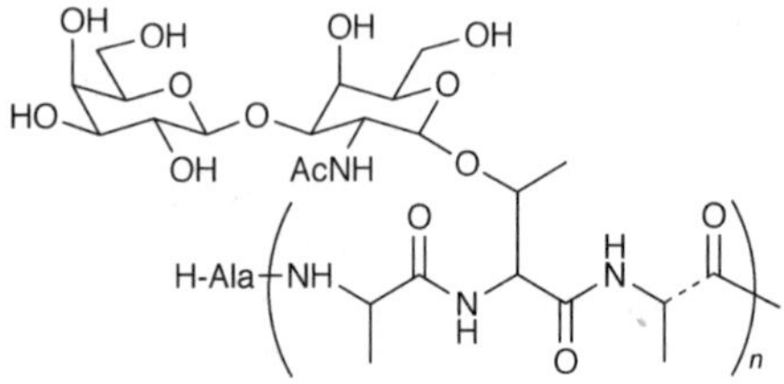

(195) *n* = 4, 8

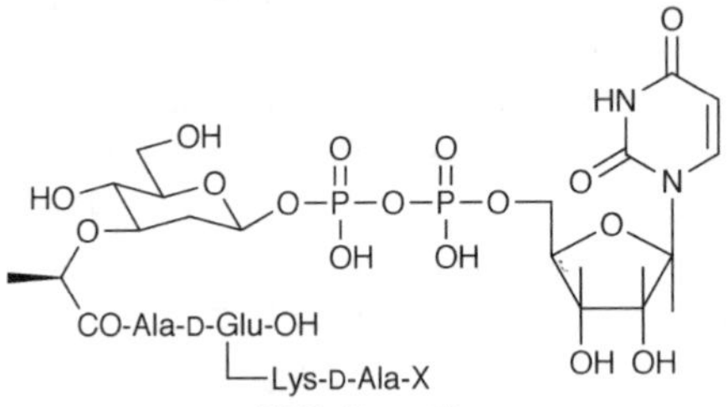

(196) X = D-Ala
(197) X = *R*-Lac

the key synthesis of (198), which was incorporated into the synthesis of (199). The key to synthesis[287] of antigen -bearing residues such as (200) was attachment to the amino acid residue via a Horner-Emmons olefination with a suitably pro-tected glycine-derived phosphonate, followed by catalytic asymmetric hydro-genation. Twenty-two enkephalin glycopeptide analogues of the general struc-ture (201) have been synthesised[288] using Fmoc-protected glycosylated amino acid building blocks. The glycosides tested were Xyl, Glc, Gal, Man, GlcNAc, GalNAc, lactose, cellobiose and melibiose, and each were linked to the amino acid via AgOTfl-catalysed condensation of acetobromosugars. The acetate-protected sugars gave poorer yields than their benzoate counterparts, but prod-

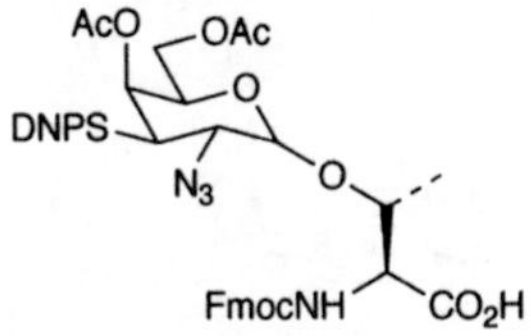

AcNH-[EYEYLDYDFLPETEPPE]-CONH$_2$
(199)

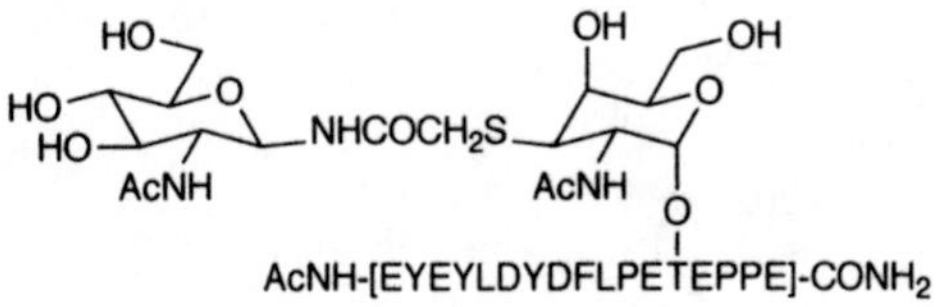

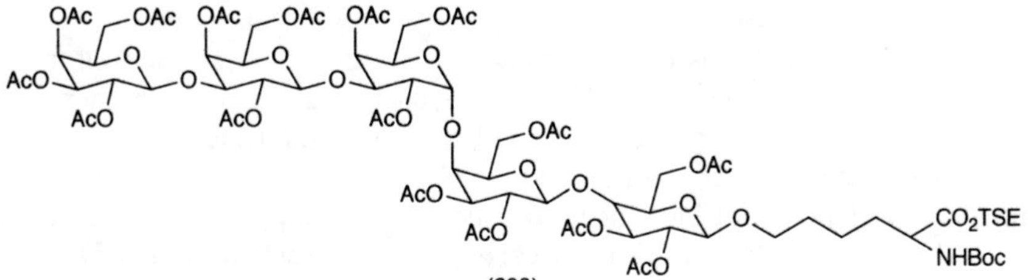

(198)

(200)

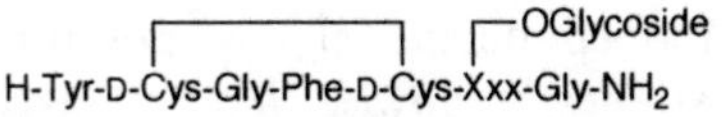

H-Tyr-D-Cys-Gly-Phe-D-Cys-Xxx-Gly-NH$_2$
(201)

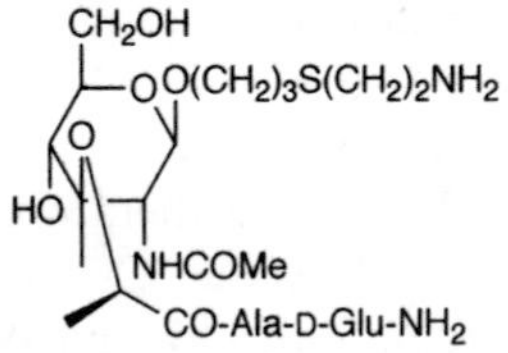

(202)

ucts were purer. The muramyl dipeptide (202) has been obtained via solid phase synthesis[289] and linked via an ethane thiol spacer to form neoglycopeptide polymers. The products were shown to stimulate TNF-α (tumour necrosis factor). An allyl ester type linker (cleaved by Pd(0) catalysis), used in combination with an acid labile Sieber amide resin, has enabled[290] glycopeptide blocks consisting of [O-(2,3,4-tri-O-acetyl-D-xylosyl)-L-seryl-glycine]$_n$ with n = 1-8, to be produced in good yields. These are prototypes for making serglycin a proteoglycan with repeating Ser-Gly sequences. Glycosyl trichloroacetimidate donors have achieved[291] high yields in the solid phase glycosylation of Ser and Thr residues. Also developed in this work was a photolabile linker to facilitate analysis by both MALDI TOF mass spectrometry and magic angle NMR.

Interest in glyco-amino acid building blocks continues, and includes (203) formed using isopropyl thioglycosides as donors. This derivative has been successfully used[292] in the preparation of a glycopeptide derivative related to *Lycium barbarum* L. Further application of the Michael addition of nitroglycals as summarised in scheme 16, has been reported[293] in the synthesis of neuraminic acid glycopeptide synthons. Three different routes[294] to building block (204) has allowed the incorporation of lipophilic side chains where R = (CH$_2$)$_5$Me or (CH$_2$)$_9$Me.

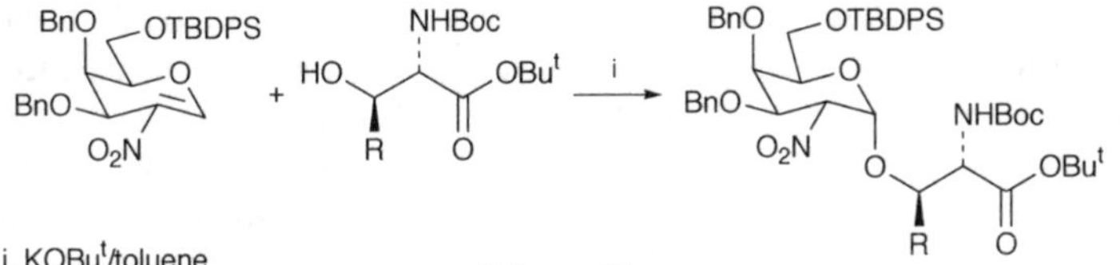

Reagent: i, KOBu^t/toluene

Scheme 16

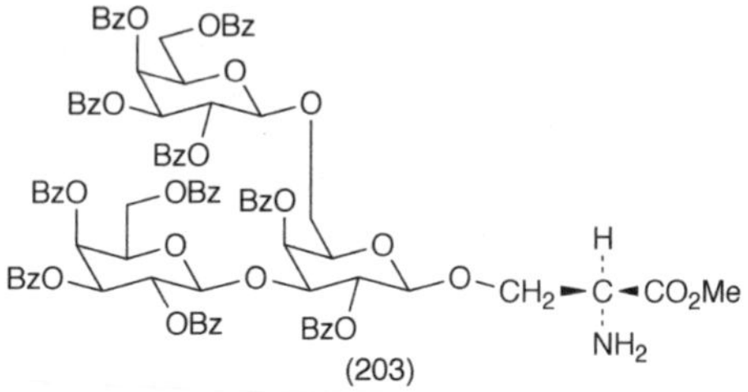

(203)

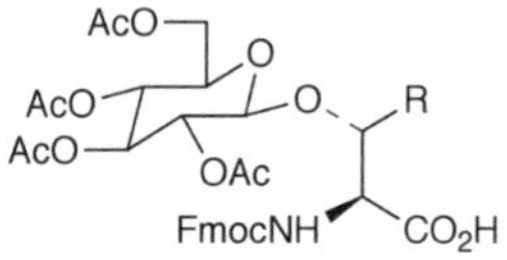

(204) R = Me

There is increasing discussion on the effect sugar units have on the conformation of the peptide or protein segments in glycoconjugates, and whether they have an influence on recognition and binding properties. Calcitonin has been used[295] as a model for such studies, and on N-glycosylation, the peptide backbone did not change, but biological activity was affected by the nature of the carbohydrate. On O-glycosylation, the carbohydrate affected both the 3D structure and the biological activity in a binding site-dependent manner. A detailed NMR investigation[296] in aqueous solution of the glycopeptide H-Val-[β-Gal (1→3)-α-GalNAc(1O)]-Thr-His-Pro-Gly-Tyr-OH showed that the 2-OH group of Gal had reduced contact with water due to interference from the 2-acetamido group of GalNAc, and there was restricted rotation around the sugar-peptide linkage. In a 17-residue O-linked glycopeptide model incorporating the α-mannosyl-serine link CD and NMR studies[297] showed that the glycopeptide retained its helicity in the presence of SDS micelles, whereas the native peptide lost secondary structure in the presence of micelles. A range of physical data[298] has also been accumulated on a series of O-glycosylated Thr peptides rich in Aib residues. The presence of glycosylated threonine residues tends to discourage formation of a 3_{10} – helical structure in Aib-rich peptides, but in antifreeze peptides, where the much less helicogenic Ala residues are found, the helical structure should still be regarded as a remote possibility.

The enzymic digestion of glycopeptides with endoglycosidases, prior to analysis by MALDI-MS has been modified[299] to take place in the presence of the MALDI matrix, which has reduced the digestion time to between 5 and 15 mins. The determination of glycosylation sites in the unusual post-translational O-fucosylation of proteins has been helped by the development[300] of nanoelectrospray quadrupole time of flight mass spectrometry (nano-ESI Q-TOF). In a single measurement of chemically untreated O-fucosylated peptides from thrombospondin–1, it was possible to determine the glycosylation sites and the glycan structure. Pre-treatment[301] with methylamine transforms glycosylated Ser and Thr residues into stable methylamine derivatives via β-elimination and addition of 13 Da to each residue. These changes can be picked up by CID-MS/MS to locate the original O-glycosylated sites.

3.3.2 N-Glycopeptides. With the aim of replacing the fucoside bond of sialyl Lewis[x] with β-D-arabino-pyranoside having the same, but unnatural, hydroxy configuration, the building block (205) was synthesised[302] in seven steps. This was then incorporated using solid phase techniques as the Asn(R) residue in Gly-Asn(R)-Leu-Thr-Glu-Leu-Glu-Ser-Glu-Asp, which corresponds to residues 672-

681 of ESL-1, and used as a synthetic E-selectin ligand stabilised against enzyme degradation. In *in vitro* tests of inhibition of adhesion of 32Dc3 cells to an E-selectin IgG-construct IC_{50} values of 0.12mM were obtained. Some demanding synthetic sequences[303] have been assembled for (206), which has a high mannose core but bears the full H-type 2 human blood group specificity. The carbohydrate structure was first put together before a final step of coupling to the Asn residue. Technology has been evolved[304] to make N-linked glycopeptides from glycosyl asparagines containing unprotected oligsaccharides, as in the key building block Fmoc-Asn(NGlnNAc)-OBu^t (207). Three N-linked glycopeptides, e.g. the N-terminal glycosylated sequence of CD52 of the immune system, Ac-Gly-Gln-Asn(NGlnNAc)-OH were made using this method which depended on precipitation of the products at critical stages.

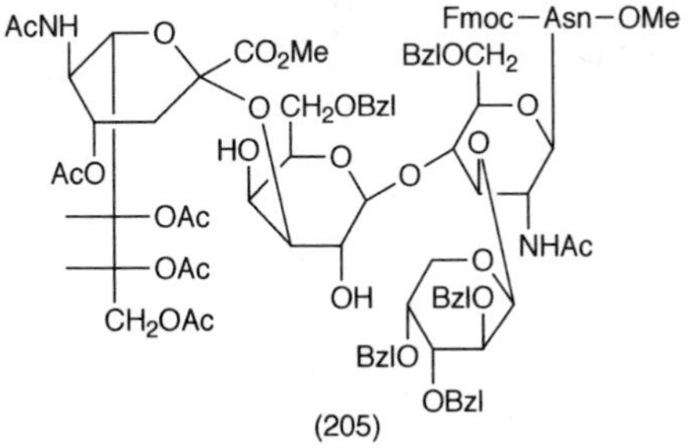

(205)

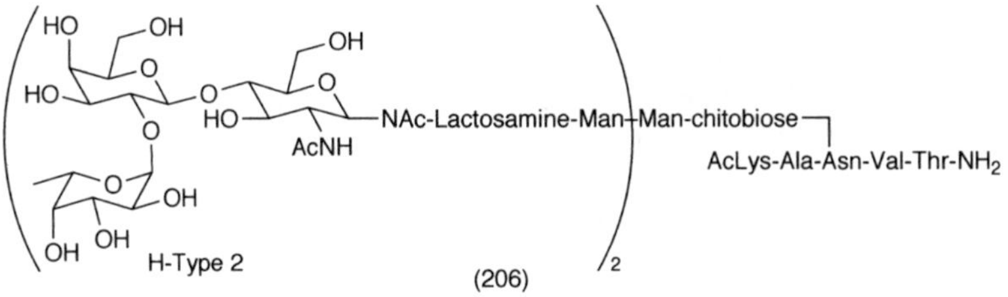

(206)

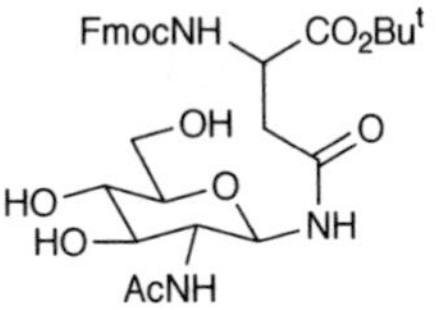

(207)

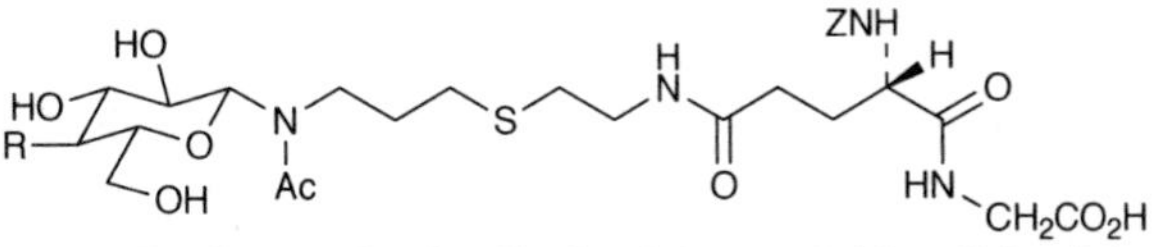

(208) R = OH, α-D-Glc, β-D-Glc, β-D-Gal or α-D-Gal (1→3)-β-D-Gal

N-Acetylmuramyl-L-alanyl-D-isoglutamine aryl β-glycosides have also been synthesised[305]. The starting point was the glycosylation of phenols with peracetylated α-glucosaminyl chloride, before condensation of these derivatives with a dipeptide.

The enzyme transglutaminase has been shown to use as substrate[306] a number of saccharides with amino-terminated thioether spacers, which can be transaminated on to Z-Gln-Gly at point (a) in structures (208). The trans-glycosylation activity of endo-β-N-acetylglucosaminidase of *Mucor hiemalis* (Endo-M) has been made use of in the further glycosylation[307] of N-acetylglucosaminyl peptides, e.g. addition of sialo-complex-type oligosaccharide to N-acetyl-glucosaminyl peptide T, glycosylation of calcitonin , and glycosylation of the glutamine residue of substance P. New derivatives of a muramoyl dipeptide with chromone aglycones have been described[308] following condensation of the methyl ester of N-acetylmuramoyl-Ala-D-isoglutamine with β-aryl-4,6-O-iso-propylidene-N-acetyl muramic acids. N-4-(β-GlcNAc)-2-substituted succinamic acid isomers have been synthesised[309] as analogues of N-4-(2-acetamido-2-deoxy-β-D-glucopyranosyl)-L-asparagine.

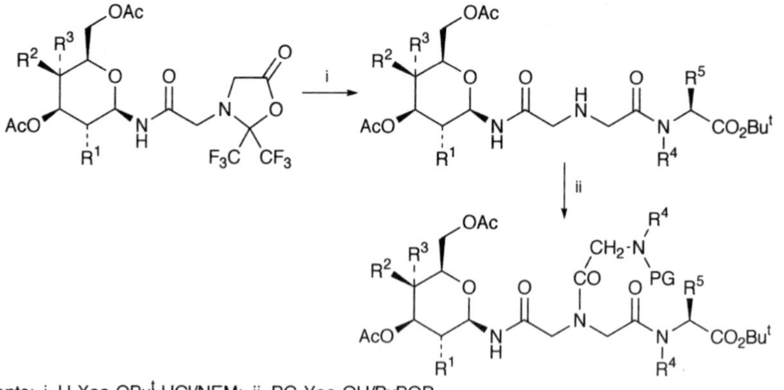

Reagents: i, H-Xaa-OBu^t-HCl/NEM; ii, PG-Yaa-OH/PyBOP

Scheme 17

A new, preparatively simple, access to N-linked glycopeptoids (N-substituted oligoglycines with alkylated amino groups) has been reported[310] using hexa-fluoroacetone as both protecting group and activating agent. A general summary of the stages appears in Scheme 17. The building block, Fmoc-Asn[β-chitobiose(TBDMS)$_5$]-OH (209) has been synthesised[311] and has been successfully incorporated into a glycosylated amyloidogenic peptide representing the 175-195 fragment of the mouse prion protein. Hydroxylamine and hydrazide links as shown in (210) have been developed[312] as asparagine surrogate groups in the assembly of N-linked glycopeptide mimetics for testing as substrates of oligosaccharyl transferase. 2-Nitrobenzenesulfonamides have been successfully used[313] to couple primary alcohol groups of saccharide and nucleoside deriva-tives under Mitsunobu reaction conditions, to form fully protected hybrids as exemplified by (211). Carbohydrate-based templates of the type represented by (212) have been produced[314] which are suitable for the preparation of dendrimers by solid-phase synthesis. The Ugi reaction has been used[315] to prepare the divalent galactose derivative (213) via the reaction summarised in Scheme 18.

3.3.3 C-Linked and Other Linked Glycopeptides. The hydrolytically stable C-glycoside analogue (214) of the Tn antigen has been synthesised[316] via a Wittig-Horner reaction of a C-linked aldehyde with Gly-derived phosphonate followed by asymmetric hydrogenation. Methylene isosteres of O-glycosyl serine have been obtained[317] using ethynyl ketoses as key intermediates. β-Linked ethynyl glycosides were transformed into N-Boc C-glycosyl α-aminobutyric acids by

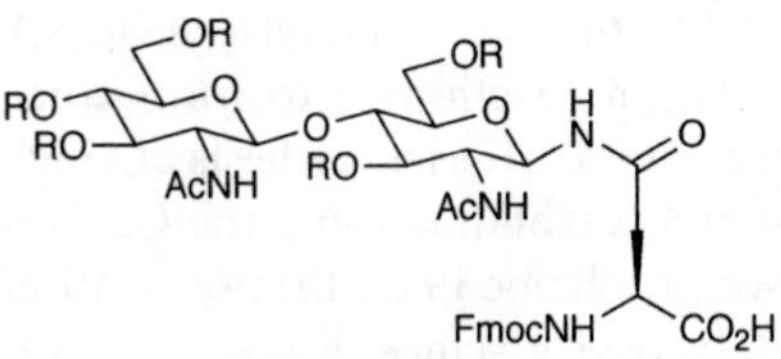

(209) R = TBDMS

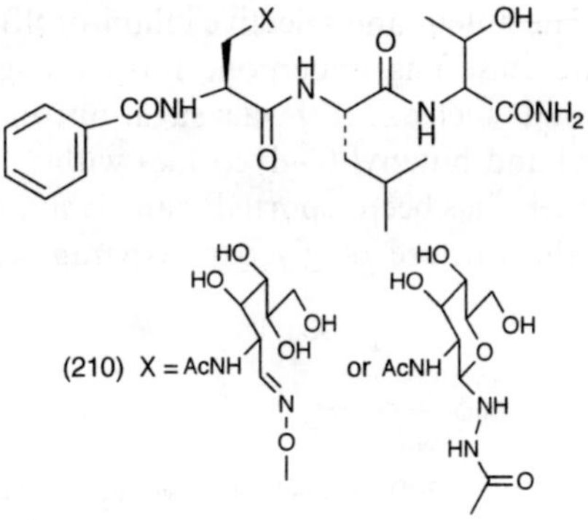

(210) X = AcNH ... or AcNH

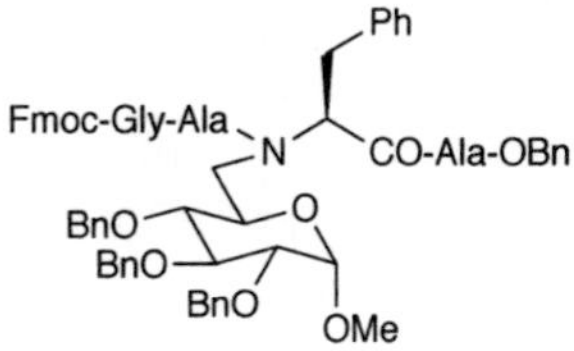

(211)

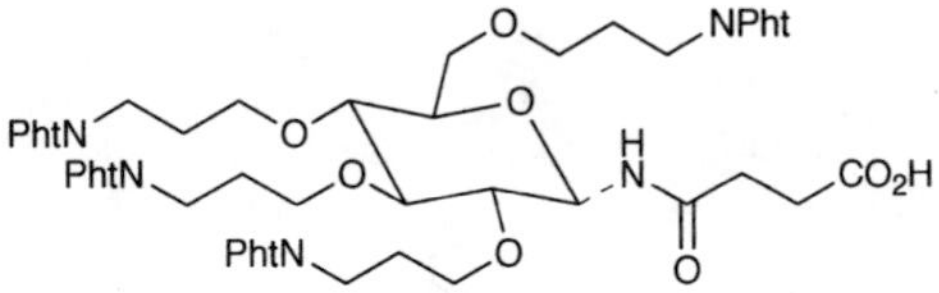

(212)

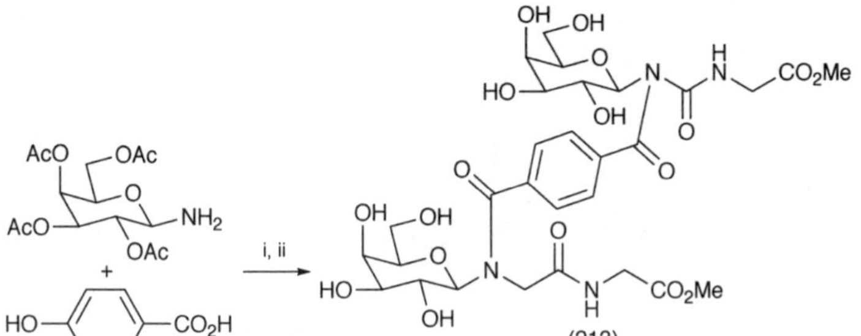

Reagents: i, HCHO + CNCH₂CO₂Me; ii, NaOMe

Scheme 18

reduction of the triple bond using $H_2/Pd(OH)_2$ and oxidative cleavage of the oxazolidine ring. A stereocontrolled synthesis[318] of α-C-mannosyltryptophan has been achieved by $Sc(ClO_4)_3$ mediated coupling between α-C-mannosylindole and 2-aziridine carboxylate as a key step. A one step procedure[319] has been worked out for the conversion of α-D-galacto-2-deoxy-oct-3-ulopyranosonic acid into unnatural glycopeptides based on the structure (215).

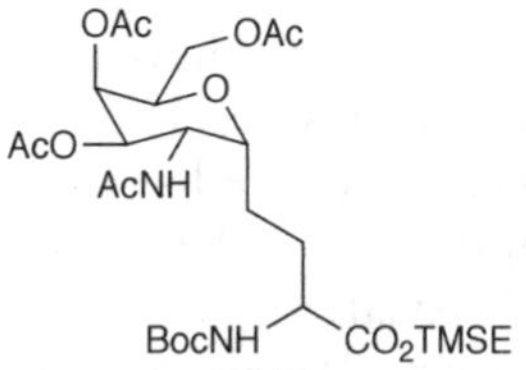

(214)

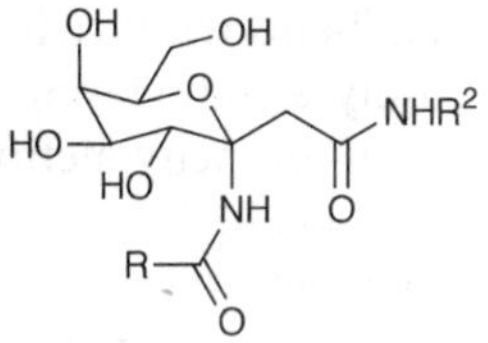

(215)

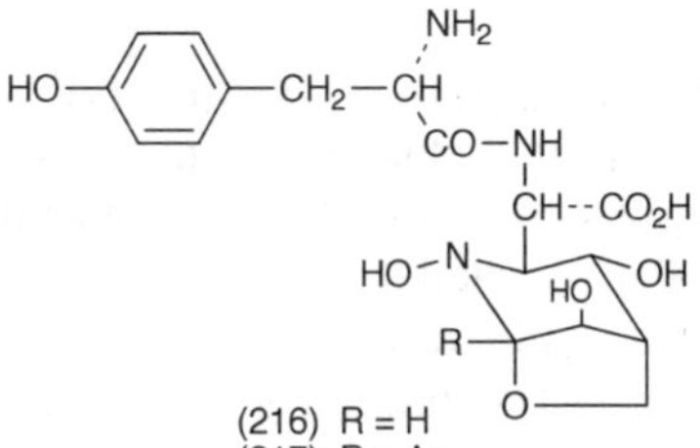

(216) R = H
(217) R = Ar

The potent and selective inhibitor SB-219383 (216) of bacterial tyrosyl t-RNA synthetase has undergone further C-glycosidation reactions[320] to form compounds such as (217) via a sugar nitrone intermediate. Cross metathesis of vinyl, allyl and butenyl C-glycosides with N-Boc-vinyloxazolidine using the Grubbs carbene has been reported[321] and is summarised in Scheme 19 for the preparation of the isostere of glycosyl asparagines. C-Glycosyl glycines, e.g 2-(4,6-di-O-

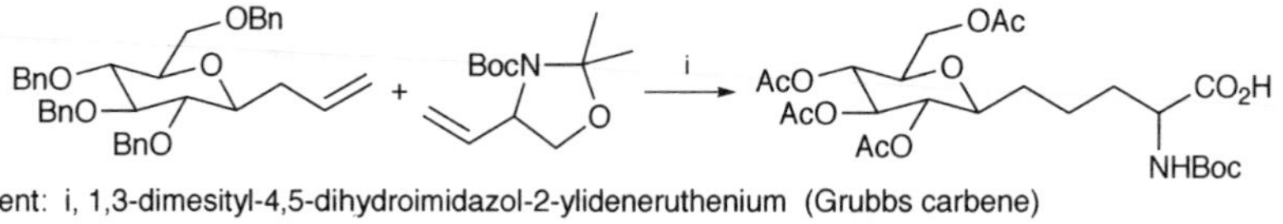

Reagent: i, 1,3-dimesityl-4,5-dihydroimidazol-2-ylideneruthenium (Grubbs carbene)

Scheme 19

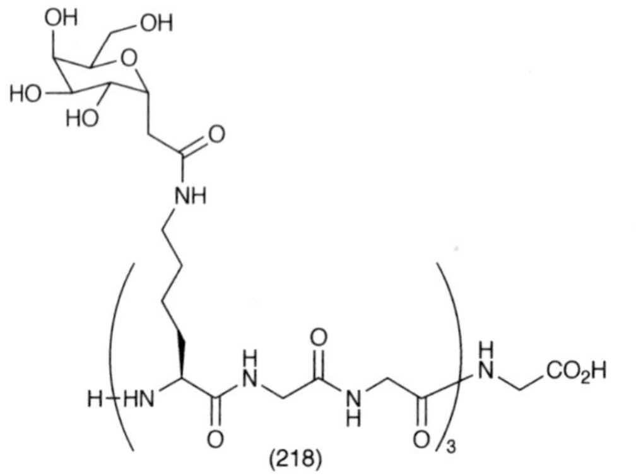

(218)

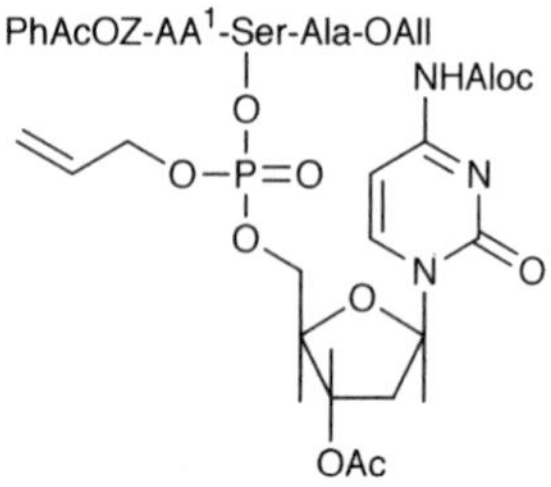

PhAcOZ-AA^1-Ser-Ala-OAll

(219)

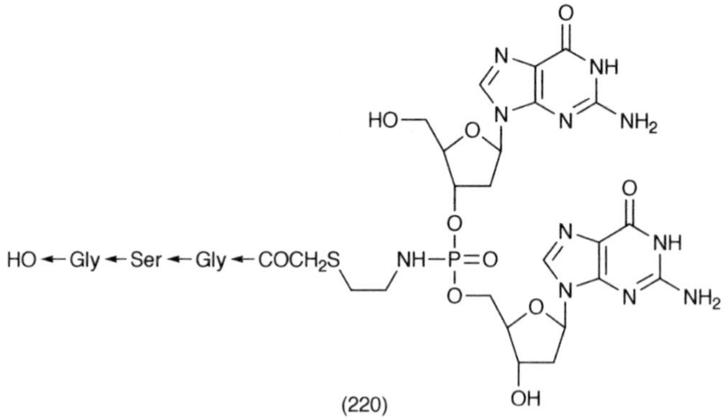

(220)

benzyl-2, 3-dideoxy-α-D-erythro-hex-2-enopyranosyl)-2 (R,S) glycine have been formed[322] starting from a Pd(0)-catalysed alkylation of 2,3-unsaturated aryl glycoside with ethyl nitroacetate or N-(diphenylmethylene)glycine ethyl ester. Perbenzylated amino α and β allyl glucosides have been shown[323] to be moderated substrates in the Sharpless asymmetric aminohydroxylation to form C-glycosyl- amino acids and -peptides. C-Linked mimetics e.g. (218) of the antifreeze glycoproteins have been synthesised[324] by introducing a methylene carboxylic acid unit on to the glycoside before coupling on to the side chain amino group on the peptide. Synthesis of new stable glycosyl amino esters has been achieved[325] by linking up the α-carbon of glycine moieties to C-6 position of pyranoses, or to C-5 positions of pentoses.

The reaction between amino acid vinyl ester acyl donors, and minimally or unprotected carbohydrate acyl acceptors, when catalysed by serine protease subtilisin at 45°C, yields amino acid esters of carbohydrates[326]. The enzyme labile phenylacetoxy benzyloxycarbonyl (PhAcOZ) has been useful for the synthesis of nucleopeptides based on (219). Coupling between nucleotide and peptide at the thioether link was used[328] to prepare analogues such as (220), of interest in the study of chromium-DNA-protein cross-links. The conjugation of oligonucleotide phosphorothioates with antennapedia peptide has been optimised[329] and can be used for large scale preparations.

3.4 Lipopeptides. – A secondary metabolite, antillatoxin B (221) has been characterised[330] from the marine cyanobacterium *Lyngbya majuscula*, and was shown to exhibit significant sodium channel-activating (EC_{50} = 1.77μM) and ichthyotoxic (LC_{50} = 1 μM) properties. The crystal structure of the lipoundecapeptide amphisin (222) from *Pseudomonas sp strain* DSS73 has been published[331]. The peptide, a close analogue of cyclic lipopeptides tensin and pholipeptin, is mainly 3_{10} helical and amphiphilic in character. The water soluble echinocandin-like lipopeptide FR131535 (223) was produced by cleaving off the lipophilic side chain of FR901379 and replacing it with the *p*-alkoxybenzoyl side chain[332]. The product inhibited 1,3 β-glucan synthase (IC_{50} = 2.8μg/mL) and had broad spectrum activity against a variety of fungal species.

A biotinylated, farnesylated Ca_1a_2X peptide [1-N-biotinyl-(13-N-succinimidyl(S-(E,E-farnesyl)-cysteinyl)-Val-Ile-Ala)-4,7,10-trioxatridecanediamine] containing a polyethylene glycol linker has been prepared by solid phase synthesis[333] on a Kaiser oxime resin. Radio-label was added in the penultimate step, and the product when tested on cloned yRce1p proved to be a good substrate (K_M = 1.3 ± 0.3 μM). The total synthesis of trunkamide A (224) has been reported[334]. Initial attempts to close the macrocycle from thiazole or thioamide-containing precursors proved to be a stereochemical burden, so cyclisation at point (a) in (224) was carried out on an all-amide precursor with the thiazole ring being generated after cyclisation. The epimer at C-45 was also synthesised. Hydrazino groups at the N-terminal position of peptides, e.g. H_2NNH-Lys-Val-Gly-Phe-Phe-Lys-Arg-NH_2 can be selectively acylated[335] at the N-terminus with Me$(CH_2)_{14}$COOSu in citrate buffer to form lipopeptides. N-Terminal modifications[336] at the fatty acid section of the potent antibacterial lipopeptide polymyxin B have shown that oligoalanyl substitutions do not affect antibacterial activity, but hydrophobic aromatic substitution generated high antibacterial activity and significantly reduced toxicity.

Conformational features of palmitoyl-Lys and palmitoyl-Gly-Lys analogues of bradykinin have been subjected to an NMR[337] study in zwitterionic lipoid environment. The results indicate that the palmitic acid and N-terminal residues were embedded into the micelles, while the rest of the polypeptide chain was closely associated with the water-micelle interface. Significant resolution enhan-

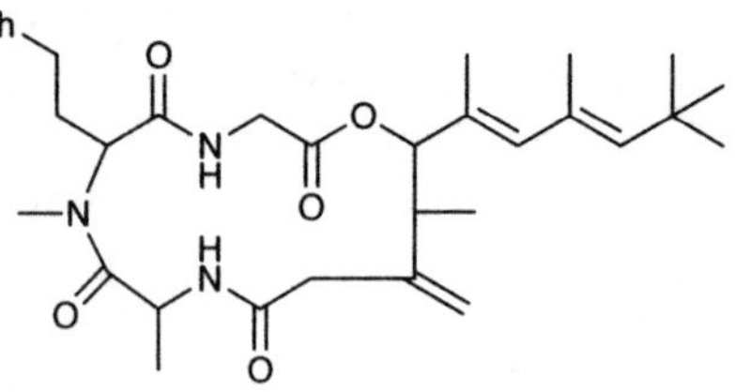

(221)

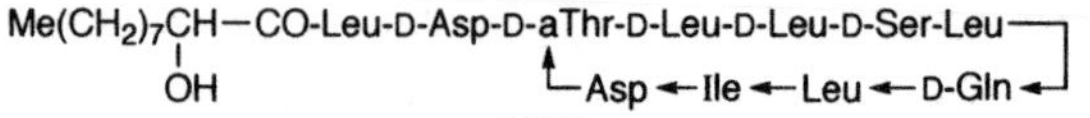

Me(CH₂)₇CH—CO-Leu-D-Asp-D-aThr-D-Leu-D-Leu-D-Ser-Leu⎤
       |
      OH   └Asp ◄— Ile ◄— Leu ◄— D-Gln ◄⎦

(222)

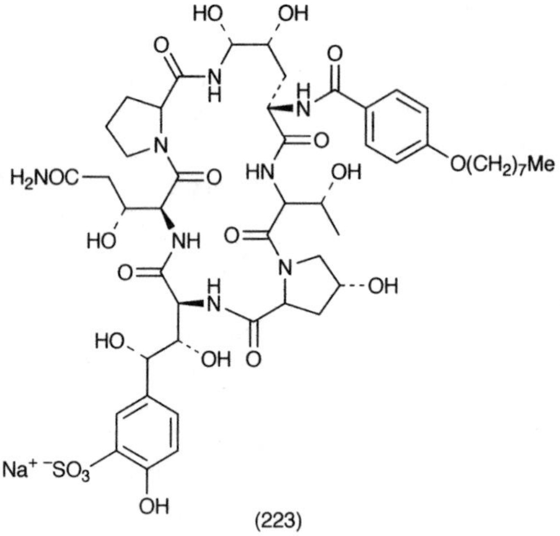

(223)

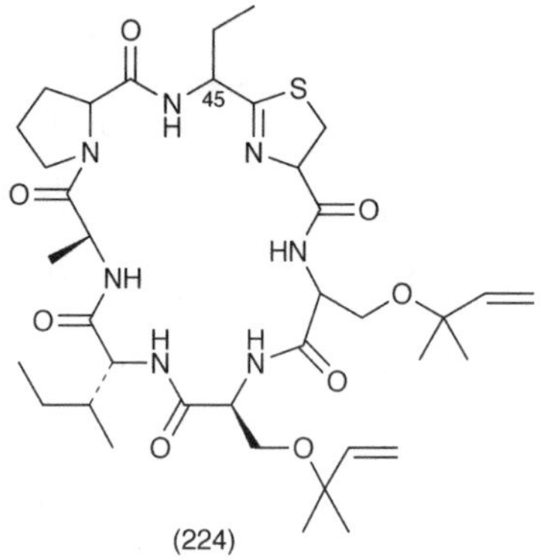

(224)

Vector / Cargo

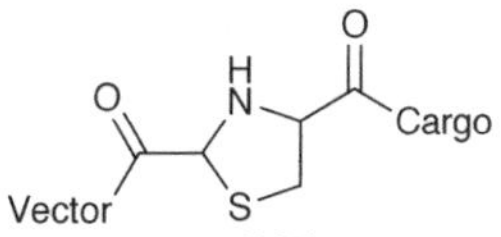

(225)

cement has been achieved[338] in the NMR spectra of a Ras lipopeptide [H-Gly-Cys (Pal)-Met-Gly-Leu-Pro-Cys (hexadecyl)-OMe] incorporated into 1,2-dimyristol-*sn*-glycero-3-phosphocholine, if the instrument is run using the magic angle spinning technique. A modular approach[339] has been used to produce a large series of cell-permeable lipopeptides. It involves linking a model peptide, the pseudo-substrate sequence of protein kinase C-zeta (labelled the cargo e.g.

X-SIYRRGARRWRKL-NH$_2$) to a shuttle system (labelled the vector, Gly-Asp-Gly-Arg-Lys-NH$_2$) via a series of ligands Optimal results were seen with the thiazolidine ligation product (225). Signal transduction pathways involved in leucocyte activation have been explored[340] using the synthetic lipopeptide N-palmitoyl-S-[2,3-bis(palmitoyloxy)-(2R,S)-propyl]-(R)-cysteinyl-seryl-(lysyl)3-ly-sine. This lipopeptide activated mitogen-activated protein kinases ERK1/2 and MAP kinase MEK1/2 in bone-marrow-derived macrophages, and in macrophage cell line RAW 264.7.

4   Miscellaneous Structures

Cyclopeptide alkaloids have found an honourable place in this sub-section over many years, and this year again nature has sourced a few more members to this family. Thus the bark of *Discaria Americana* has been shown[341] to contain discarenes C (226) and D (227), whose structures were elucidated by NMR and chiral chromatography. Paliurines G (228), H (229) and I (230) have been identified[342] together with six other known alkaloids from the stems of *Paliurus ramossisimus*.

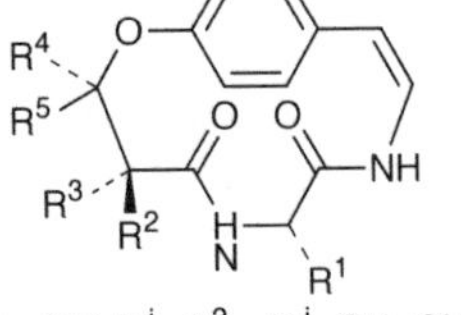

(226) R^1 = CH$_2$Pr^i, R^2 = Pr^i—CH=CHCONH,
      R^3 = R^4 = H, R^5 = Ph
(227) R^1 = CH$_2$Ph, R^2 = R^5 = H,
      R^3 = Pr^i—CH=CHCONH, R^4 = Ph

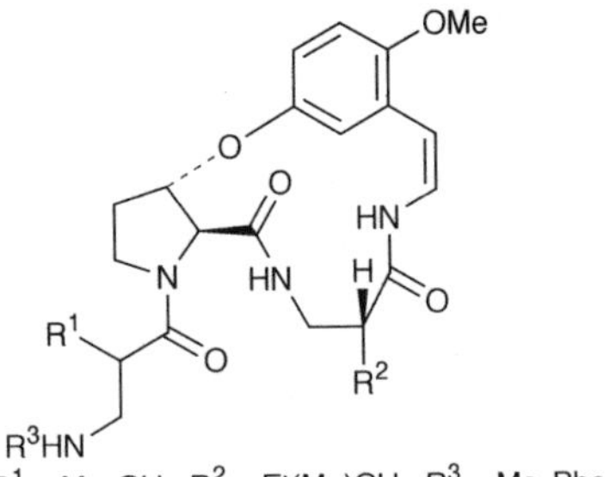

(228) R^1 = Me$_2$CH, R^2 = Et(Me)CH, R^3 = Me$_2$Phe
(229) R^1 = R^2 = Et(Me)CH, R^3 = MeLeu
(230) R^1 = Me$_2$CHCH$_2$, R^2 = PhCH$_2$, R^3 = MeIle

Quite large 22-32 membered macrocycles (231) have been obtained[343] by dimerisation of cysteine-based precursors, while coupling of Z-Ala-ONSu with 1,5-diamino-3-oxapentane and appropriate deprotection and reduction steps have yielded[344] (232) and (233). Molecular design and synthesis[345] of 'artificial peptides' containing rigid unnatural amino acids, such as 3-aminobenzoic acid provide opportunities for ion-channelling/chelation. The series of analogues shown under (234) form channels which are cation selective, these channels being formed from a dimer of the peptide. The peptide ring constructs the channel entrance and the alkanoyl chains line across the membrane to form the channel pore, and hence determine the rate of ion conduction. Acid catalysed cyclisation[346] of 4-[(MeO)$_2$CH]C$_6$H$_4$CO-Pro-Phe-NHNH$_2$ in the presence of LiI, generated a dynamic combinatorial library, with a peptide hydrazone trimer accounting for 98% of the library. This trimer bound to lithium in a 1:1 stoichiometry. Cyclic trimers such as (235) have been synthesised[347] and show supramolecular

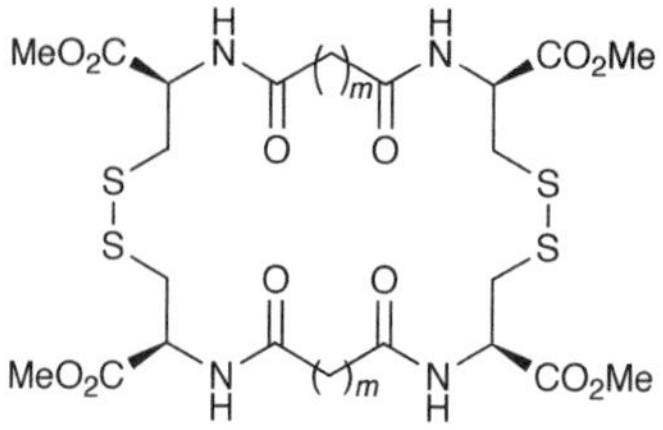

(231) *m* = 1–6

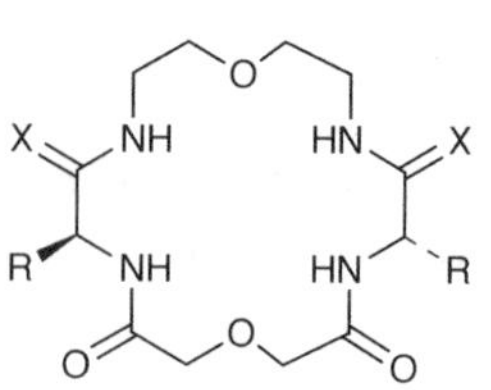

(232) X = H$_2$
(233) X = O

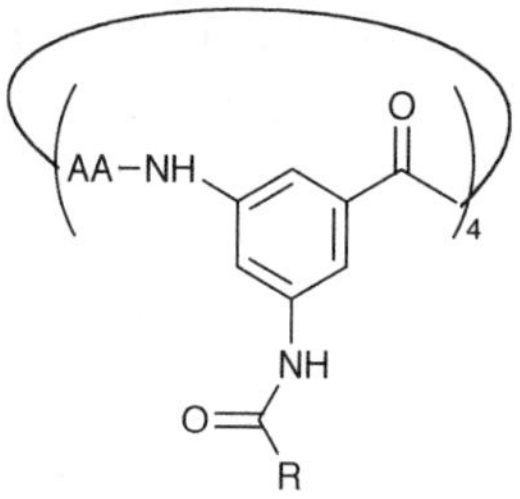

(234)
AA = Asp(OBzl), R = nC$_{15}$H$_{31}$
AA = Lys(ClZ), R = nC$_{17}$H$_{35}$
AA = Ser(Bzl), R = nC$_{17}$H$_{35}$

nanotubular stacked structures when crystallised. A flexible method[348] of synthesising cyclic peptides with unnatural aliphatic bridges has been proven through the synthesis of (236). A triply-orthogonal protecting strategy was used to incorporate the bridging unit. The constrained peptide, contrary to expectation, adopted a left-handed type II β-turn in aqueous media and a right-handed type I β-turn conformation in trifluoroethanol. Ester-linked glycopeptides have been synthesised[349] by linking the carboxy groups of simple peptides, as well as enkephalin, to the C-6 hydroxy group of D-glucose. Their properties indicate that some products formed from them may be similar to those formed as part of the Maillard reaction.

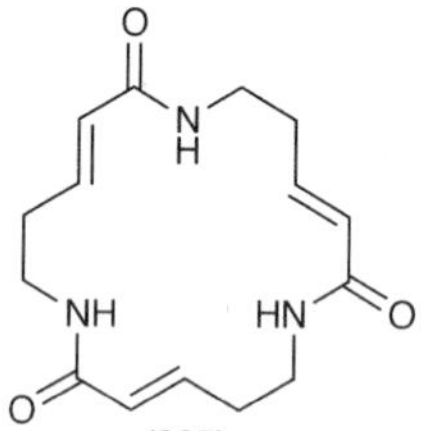

(235)

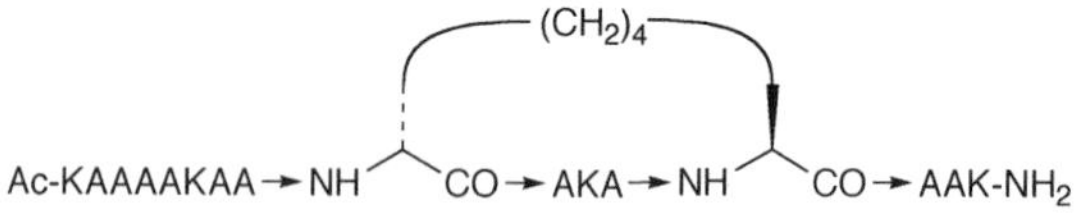

Ac-KAAAAKAA → NH⌒CO → AKA → NH⌒CO → AAK-NH$_2$

(236)

References

1. C. A. Selects on *Amino Acids, Peptides and Proteins*, published by the American Chemical Society and Chemical Abstracts Service, Columbus, Ohio.
2. *'The ISI Web of Science Service for UK Education'* on http://wos.mimas.ac.uk
3. *'Peptides 2000'* Proceedings of the 26th European Peptide Symposium, Montpellier, eds. J. Martinez and J-A.Fehrentz, Éditions EDK, Paris, 2001, 1054pp.
4. *'Peptides: Chemistry and Biology'* N. Sewald and H-D. Jakubke, Wiley-VCH Verlag GmbH, 2002 429pp.
5. J. N. Lambert, J. P. Mitchell and K. D. Roberts, *J. Chem. Soc. Perkin Trans 1*, 2001, 471.
6. F. Bordusa, *ChemBiochem.*, 2001, **2**, 405.
7. A. M. Burja, B. Banaigs, E. Abou-Mansour, J. G. Burgess and P. C. Wright, *Tetrahedron*, 2001, **57**, 9347.
8. M. MacDonald and J. Aube, *Curr. Org. Chem.*, 2001, **5**, 417.
9. M. Goodman, C. Zapf and Y. Rew, *Biopolymers*, 2001, **60**, 229.
10. D. Ranganathan, *Acc. Chem. Res.*, 2001, **34**, 919.

11. F. Albericio, R. Chinchilla, D. J. Dodsworth and C. Najera, *Org. Prep. Proced. Int.*, 2001, **33**, 203.
12. C. Rosenbaum and H. Waldmann, *Tetrahedron Lett.*, 2001, **42**, 5677.
13. R. M. Kohli, J. W. Trauger, D. Schwarzer, M. A. Marahiel and C. T. Walsh, *Biochemistry*, 2001, **40**, 7099.
14. T. Bayer, C. Riemer and H. Kessler, *J. Pept. Sci.*, 2001, **7**, 250.
15. M. A. Walker and T. Johnson, *Tetrahedron Lett.*, 2001, **42**, 5801.
16. T. Kimachi, *Farumashia*, 2001, **37**, 920.
17. H. E. Blackwell, J. D. Sadowsky, R. J. Howard, J. N. Sampson. J. A. Chao, W. E. Steinmetz, D. J. O'Leary and R. H. Grubbs, *J. Org. Chem.*, 2001, **66**, 5291.
18. K. Akaji, K. Teruya, M. Akaji and S. Aimoto, *Tetrahedron*, 2001, **57**, 2293.
19. W. D. Kohn, L. Zhang and J. A. Weigel, *Org. Lett.*, 2001, **3**, 971.
20. D. Scharn, L. Germeroth, L. Schneider-Mergener and H. Wenschuh, *J. Org. Chem.*, 2001, **66**, 507.
21. P. Grieco, P. M. Gitu and V. J. Hruby, *J. Pept. Res.*, 2001, **57**, 250.
22. G. Gellerman, A. Elgavi, Y. Salitra and M. Kramer, *J. Pept. Res.*, 2001, **57**, 277.
23. G. T. Bourne, S. W. Golding, W. D. F. Meutermans and M. L. Smythe, *Lett. Pept. Sci.*, 2000, **7**, 311.
24. G. T. Bourne, S. W. Golding, R. P. McGeary, W. D. F. Meutermans, A. Jones, G. R. Marshall, P. F. Alewood and M. L. Smythe, *J. Org. Chem.*, 2001, **66**, 7706.
25. L. M. A. McNamara, M. J. I. Andrews, F. Mitzel, G. Siligardi and A. B. Tabor, *Tetrahedron Lett.*, 2001 **42**, 1591.
26. Y. Sun, G. S. Lu and J. P. Tam, *Org. Lett.*, 2001, **3**, 1681.
27. C. P. Scott, E. Abel-Santos, A. D. Jones and S. J. Benkovic, *Chem. and Biol.*, 2001, **8**, 801.
28. T. O Larsen, B. O. Petersen and J. O. Duus, *J. Agric. Food Chem.*, 2001, **49**, 5081.
29. Y. Q. Tang, I Sattler, R. Thiericke, S. Grabley and X. Z. Feng, *Eur. J. Org. Chem.*, 2001, 261.
30. Y. Sugie, H. Hirai, T. Inagaki, M. Ishiguro, Y. J. Kim, Y. Kojima, T. Sakakibara, S. Sakemi, A. Suguria, Y. Suzuki, L. Brennan, J. Duignan, L. H. Huang, J. Sutcliffe and N. Kojima, *J. Antibiot.*, 2001, **54**, 911.
31. M. Gondry, S. Lautru, G. Fusai, G. Meunier, A. Menez and R. Genet, *Eur. J. Biochem.*, 2001, **268**, 1712.
32. M. Delaforge, C. Bouille, M. Jaouen, C. K. Jankowski, C. Lamouroux and C. Bensoussan, *Peptides* 2001, **22**, 557.
33. M. Graz, H. Jamie, C. Versluis and P. Milne, *Pharmazie*, 2001, **56**, 900.
34. L. E. Overman and D. V. Paone, *J. Am. Chem. Soc.*, 2001, **123**, 9465.
35. B. B. Snider and M. V. Busuyek, *Tetrahedron*, 2001, **57**, 3301.
36. N. Shibata, T. Tarui, Y. Doi and K. L. Kirk, *Angew. Chem. Int. Ed.*, 2001, **40**, 4461.
37. I. O. Donkor and M. L. Sanders, *Bioorg. Med. Chem. Lett.*, 2001, **11**, 2647.
38. A. Folkes, M. B. Roe, S. Sohal, J. Golec, R. Faint, T. Brooks and P. Charlton, *Bioorg. Med. Chem. Lett.*, 2001, **11**, 2589.
39. K. C. Huang, H. L.Gao, E. F. Yamasaki, D. R. Grabowski, S. J. Liu, L. L. Shen, K. K. Chan, R. Ganapathi and R. M. Snapka, *J. Biol. Chem.*, 2001, **276**, 44488.
40. M. L. Lopez-Rodriguez, M. J. Morcillo, E. Fernandez, E. Porras, L. Orensanz, M. E. Beneytez, J. Manzanares and J. A. Fuentes, *J. Med. Chem.*, 2001, **44**, 186.
41. H. Wennemers, M. Conza, M. Nold and P. Krattiger, *Chem.-Eur. J.*, 2001, **7**, 3342.
42. M. Giraud, N. Bernad, J. Martinez and F. Cavalier, *Tetrahedron Lett.*, 2001, **42**, 1895.
43. M. Royo, W. Van Den Nest, M. del Fresno, A. Frieden, D. Yahalom, M. Rosen-

blatt, M. Chorev and F. Albericio, *Tetrahedron Lett.*, 2001, **42**, 7387.

44.   A. N. Achayara, J. M. Ostresh and R. A. Houghten, *J. Comb. Chem.*, 2001, **3**, 612.

45.   G. Gerona-Navarro, M. A. Bonache, R. Herranz, M. T. Garcia-Lopez and R. Gonzalez-Muniz, *J. Org. Chem.*, 2001, **66**, 3538.

46.   B. T. Shireman and M. J. Miller, *J. Org. Chem.*, 2001, **66**, 4809.

47.   T. Kunisaki, K. Kawai, K. Hirohata, K. Minami and K. Kondo, *J. Polymer Sci.*, *Pt A – Polymer Chem.*, 2001, **39**, 927.

48.   T. Groth and M. Meldal, *J. Comb. Chem.*, 2001, **3**, 45.

49.   D. D. Long, R. J. Tennant-Eyles, J. C. Estevez, M. R. Wormald, R. A. Dwek, M. D. Smith and G. W. J. Fleet, *J. Chem. Soc.*, *Perkin Trans.1*, 2001, 807.

50.   S. D. Bull, S. G. Davies, A. C. Garner and N. Mujtaba, *Synlett.*, 2001, 781.

51.   S. D. Bull, S. G. Davies, A. C. Garner and M. D. O'Shea, *J. Chem. Soc.*, *Perkin Trans. 1*, 2001, 3281.

52.   M. MacDonald, D. Vander Velde and J. Aube, *J. Org. Chem.*, 2001, **66**, 2636.

53.   C. J. Creighton and A. B. Reitz, *Org. Lett.*, 2001, **3**, 893.

54.   A. Jehni, J. P. Lavergne, M. Rolland, J. Martinez and A. Hasnaoui, *Synth. Commun.*, 2001, **31**, 1707.

55.   E. Perrotta, M. Altamura, T. Barani, S. Bindi, D. Giannotti, N. J. S. Harmat, R. Nannicini and C. A. Maggi, *J. Comb. Chem.*, 2001, **3**, 453.

56.   M. A. Estiarte, A. Diez, M. Rubiralta and R. F. W. Jackson, *Tetrahedron*, 2001, **57**, 157.

57.   S. Pattanaargson, C. Chuapradit and S. Srisukphonraruk, *J. Food Sci.*, 2001, **66**, 808.

58.   S. Rastogi, M. Zakrzewski and R. Suryanarayanan, *Pharm. Res.*, 2001, **18**, 267.

59.   S. L. Wang, S. Y. Lin and T. F. Chen, *Chem. and Pharm. Bull.*, 2001, **49**, 402.

60.   B. Paizs and S. Suhai, *Rapid Comm.in Mass Spec.*, 2001, **15**, 2307.

61.   I. Jeric and S. Horvat, *Eur. J. Org. Chem.*, 2001, 1533.

62.   N. G. Vinokurova, N. F. Zelenkova, B. P. Baskunov and M. U. Arinbasarov, *J. Anal. Chem.*, 2001, **56**, 258.

63.   P. Gockel, R. Vogler, M. Gelinsky, A. Meissner, H. Albrich and H. Vahrenkamp, *Inorg. Chim. Acta*, 2001, **323**, 16.

64.   J. M. Farrugia, T. Taverner and R. A. J. O'Hair, *Int. J. Mass Spectrom.*, 2001, **209**, 99.

65.   K. Gademann and D. Seebach, *Helv. Chim. Acta*, 2001, **84**, 2924.

66.   P. Virta, J. Rosenberg, T. Karskela, P. Heinonen and H. Lonnberg, *Eur. J. Org. Chem.*, 2001, 3467.

67.   E. N. Prabhakaran, V. Rajesh, S. Dubey and J. Iqbal, *Tetrahedron Lett.*, 2001, **42**, 339.

68.   W. Kuriyama and T. Kitahara, *Heterocycles*, 2001, **55**, 1.

69.   L. Mou and G. Singh, *Tetrahedron Lett.*, 2001 **42**, 6603.

70.   S. B. Singh, D. L. Zink, J. M. Liesch, A. W. Dombrowski, S. J. Darkin-Rattray, D. M. Schmatz and M. A. Goetz, *Org. Lett.*, 2001, **3**, 2815.

71.   S. L. Colletti, R. W. Myers, S. J. Darkin-Rattray, A. M. Gurnett, P. M. Dulski, S. Galuska, J. J. Allocco, M. B. Ayer, C. S. Li, J. Lim, T. M. Crumley, C. Cannova, D. M. Schmatz, M. J. Wyvratt, M. H. Fisher and P. T. Meinke, *Bioorg. Med. Chem. Lett.*, 2001, **11**,107.

72.   S. L. Colletti, R. W. Myers, S. J. Darkin-Rattray, A. M. Gurnett, P. M. Dulski, S. Galuska, J. J. Allocco, M. B. Ayer, C. S. Li, J. Lim, T. M. Crumley, C. Cannova, D. M. Schmatz, M. J. Wyvratt, M. H. Fisher and P. T. Meinke, *Bioorg. Med. Chem. Lett.*, 2001, **11**, 113.

73. R. Furumai, Y. Komatsu, N. Nishino, S. Khochbin, M. Yoshida and S. Horinouchi, *Proc. Nat. Acad. Sci. U. S. A.*, 2001, **98**, 87.

74. J. Hlavacek, M. Budesinsky, B. Bennettova, J. Marik and R. Tykva, *Bioorg. Chem.*, 2001, **29**, 282.

75. C-T. Cheng, V. Lo, J. Chen, W-C. Chen, C-Y. Lin, H-C. Lin, C-H. Yang and L. Sheh, *Bioorg. Med. Chem.*, 2001, **9**, 1493.

76. S. Pohl, R. Goddard and S. Kubik, *Tetrahedron Lett.*, 2001, **42**, 7555.

77. Z. Kriz, P. H. J. Carlsen and J. Koca, *Theochem.*, 2001, **540**, 231.

78. S. H. Lin, S. Liehr, B. S. Cooperman and R. J. Cotter, *J. Mass Spectrom.*, 2001, **36**, 658.

79. T. Sano, T. Usui, K.Ueda, L. Osada and K.Kaya, *J. Nat. Prod.*, 2001, **64**, 1052.

80. K. L. Webster, A. B. Maude, M. E. O'Donnell, A. P. Mehrotra and D. Gani, *J. Chem. Soc. Perkin Trans 1*, 2001, 1673.

81. M. E. O'Donnell, J. Sanvoisin and D. Gani, *J. Chem. Soc. Perkin Trans 1*, 2001, 1696.

82. M. Haramura, A. Okamachi, K. Tsuzuki, K. Yogo, M. Ikuta, T. Kozono, H. Takanashi and E. Murayama, *Chem. and Pharm. Bull.*, 2001, **49**, 40.

83. M. Dettin L. Falcigno, T. Campanile, C. Scarnici, G. D'Auria, M. Cusin. L. Paolillo and C. DiBello, *J. Pept. Sci.*, 2001, **7**, 358.

84. E. S. Monteagudo, F. Calvani, F. Catrambone, C. I. Fincham, A. Madami, S. Meini and R. Terracciano, *J. Pept. Sci.*, 2001, **7**, 270.

85. P. Prusis, R. Muceniece, I. Mutule, F. Mutilis and J. E. S. Wikberg, *Eur. J. Med. Chem.*, 2001, **36**, 137.

86. J. Hlavacek, M. Budesinsky, B. Bennettova, J. Marik, and R. Tykva, *Bioorg. Chem.*, 2001, **29**, 282.

87. V. Marchi-Artzner, B. Lorz, U. Hellerer, M. Kantlehner, H. Kessler and E. Sackmann, *Chem. Eur. J.* 2001, **7**, 1095.

88. D. Boturyn and P. Dumy, *Tetrahedron Lett.*, 2001, **42**, 2787.

89. R. Haubner, H. J. Wester, F. Burkhart, R. Senekowitsch-Schmidtke, W. Weber, S. L. Goodman, H. Kessler and M. Schwaiger, *J. Nuclear Med.*, 2001, **42**, 326.

90. S. Liu, E. Cheung, M. Rajopadhye, N. E. Williams, K. L. Overoye and D. S. Edwards, *Bioconjugate Chem.*, 2001, **12**, 84.

91. S. Liu, E. Cheung, M. C. Ziegler, M. Rajopadhye and D. S. Edwards, *Bioconjugate Chem.*, 2001, **12**, 559.

92. S. Liu, D. S. Edwards, M. C. Ziegler, A. R. Harris, S. J. Hemingway and J. A. Barrett, *Bioconjugate Chem.*, 2001, **12**, 624.

93. J. Kobayashi, H. Suzuki, K. Shimbo, K. Takeya and H. Morita, *J. Org. Chem.*, 2001, **66**, 6626.

94. R. Sekizawa, I . Momose, N. Kinoshita, H. Naganawa, M. Hamada, Y. Muraoka, H. Iinuma and T. Takeuchi, *J. Antibiot.*, 2001, **54**, 874.

95. H. Kobayashi, K. Shin-Ya, K. Furihata, K. Nagai, K-I Suzuki, Y. Hayakawa, H. Seto, B-S. Yun, I-J. Ryoo, J-S. Kim, C-J. Kim and I-D. Yoo, *J. Antibiot.*, 2001, **54**, 1019.

96. P. Sonnet, L. Petit, D. Marty, J. Guillon, J. Rochette and J.-D. Brion, *Tetrahedron Lett.*, 2001, **42**, 1681.

97. Y. Hitotsuyanagi, Y. Matsumoto, S-i. Sasaki, K. Yamaguchi, H. Itokawa and K. Takeya, *Tetrahedron Lett.*, 2001, **42**, 1535.

98. O. Poupardin, F. Ferreira , J. P. Genet and C. Greck, *Tetrahedron Lett.*, 2001, **42**, 1523.

99. H. Eickhoff, G. Jung and A. Rieker, *Tetrahedron*, 2001, **57**, 353.

100. K. Alexopoulos, D. Panagiotopoulos, T. Mavromoustakos, P. Fatseas, M. C. Paredes-Carbajal, D. Mascher, S. Mihailescu and J. Matsoukas, *J. Med. Chem.*, 2001, **44**, 328.

101. H. Weisshoff, T. Nagel, A. Hansicke, A. Zschunke and C. Mugge, *FEBS Lett.*, 2001, **491**, 299.

102. J. Boer, D. Gottschling, A. Schuster, M. Semmrich, B. Holzmann and H. Kessler, *J. Med. Chem.*, 2001, **44**, 2586.

103. Z. Szewczuk, P. Buczek, P. Stefanowicz, K. Krajewski, Z. Wieczorek and I. Z. Siemion, *Acta Biochim. Pol.*, 2001, **48**, 121.

104. A. Basso and B. Ernst, *Tetrahedron Lett.*, 2001, **42**, 6687.

105. J. Bitta and S. Kubik, *Org. Lett.*, 2001, **3**, 2637.

106. S. Kubik, R. Goddard, R. Kirchner, D. Nolting and J. Seidel, *Angew. Chem. Int. Ed.*, 2001, **40**, 2648.

107. S. Kubik, J. Bitta, R. Goddard, D. Kubik, and S. Pohl, *Mater. Sci. Eng. C.*, 2001, **18**, 125; S. Kubil and R. Goddard, *Eur. J. Org. Chem.*, 2001, 311.

108. F. Palmer, R. Tunnemann, D. Leipert, C. Stingel, G. Jung and V. Hoffmann, *J. Mol. Struct.*, 2001, **563-564**, 153.

109. C. Marinzi, R. Longi and R. Consonni, *Electrophoresis*, 2001, **22**, 3257.

110. A. Katoh, Y. Hikita, M. Harata, J. Ohkanda, T. Tsubomura, A. Higuchi, R. Saito and K. Harada, *Heterocycles*, 2001, **55**, 2171.

111. L. J. Mu, H. Huang, J. Q. He, N. Zhang and J. P. Cheng, *Chin. Sci. Bull.*, 2001, **46**, 219.

112. P. Kedzierski and W. A. Sokalski, *J. Comput. Chem.*, 2001, **22**, 1082.

113. T. Golakoti, W. Y. Yoshida, S. Chaganty and R. E. Moore, *J. Nat. Prod.*, 2001, **64**, 54.

114. J. Tabudravu, L. A. Morris, J. J. Kettens-van den Bosch and M. Jaspars, *Tetrahedron Lett.*, 2001, **42**, 9273.

115. C. Baraguey, A. Blond, F. Cavalier, J-L. Pouset, B. Bodo and C. A. Auvin-Guette, *J. Chem. Soc.*, 2001, 2098.

116. C. Napolitano, I. Bruno, P. Rovero, R. Lucas, M. P. Peris, L. Gomez-Paloma and R. Riccio, *Tetrahedron*, 2001, **57**, 6249.

117. M. Hahn, D. Winkler, K. Welfle, R. Misselwitz, H. Welfle, H. Wessner, G. Hahn, C. Scholz, M. Seifert, R. Harkins, J. Schneider-Mergener, and W. G. Hohne, *J. Mol. Biol.*, 2001, **314**, 293.

118. N. Hue, L. Serani and O. Laprevote, *Rapid. Comm. in Mass Spect.*, 2001, **15**, 203.

119. L. A. Lawton and C. Edwards, *J. Chromat.*, 2001, **912**, 191.

120. Y. Cheng, J. Zhou and N. Tan, *Zhiwu Xuebao*, 2001, **43**, 760.

121. Y. X. Cheng, J. Zhou, N. H. Tan, T. Lu, X. Y. Liu and Q. T. Zheng, *Heterocycles*, 2001, **55**, 1943.

122. Z. T. Ding, J. Zhou, N. H. Tan, Y. X. Cheng and S. M. Deng, *Acta Botanica Sinica*, 2001, **43**, 541.

123. Q. Mu, R. W. Teng, C. M. Li, D. Z. Wang, Y. Wu, H. D. Sun and C. Q. Hu, *Chinese Chem. Lett.*, 2001, **12**, 607.

124. Y. Zhou, M. Wang, S. Peng and L. Ding, *Zhiwu Xuebao*, 2001, **43**, 431.

125. T. Matsumoto, A. Shishido and H. Morita, H. Itokawa and K. Takeya, *Phytochemistry*, 2001, **57**, 251.

126. G. R. Pettit, J. W. Lippert III, S. R. Taylor, R. Tan and M. D. Williams, *J. Nat. Prod.*, 2001, **64**, 883.

127. K. Krajewski, Z. Ciunik and I. Z. Siemion, *Tetrahedron: Asymmetry*, 2001, **12**, 455.

128. J. Leprince, H. Oulyadi, D. Vaudry, O. Masmoudi, P. Gandolfo, C. Patte, J.

Costentin, J. L. Fauchere, D. Davoust, H. Vaudry and M. C. Tonon, *Eur. J. Biochem.*, 2001, **268**, 6045.

129. V. Gut, V. Cerovosky, M. Zertova, E. Korblova, P. Malon, H. Stocker and E. Wunsch, *Amino Acids*, 2001, **21**, 255.

130. S. K. Jiang, S. Gazal, G. Gelerman, O. Ziv, O. Karpov, P. Litman, M. Bracha, M. Afargan, C. Gilon and M. Goodman, *J. Pept. Sci.*, 2001, **7**, 521.

131. V. Cavallaro, P. E. Thompson and M. T. W. Hearn, *J. Pept. Sci.*, 2001, **7**, 529.

132. D. T. Bong, and M. R. Ghadiri, *Angew. Chem. Int. Ed.*, 2001, **40**, 2163.

133. A. B. Pomilio, M. E. Battista and A. A. Vitale, *J. Mol. Struct.-Theochem.*, 2001, **536**, 243.

134. R. Oliva, L. Falcigno, G. D'Auria, G. Zanotti and L. Paolillo, *Biopolymers*, 200, **56**, 27.

135. A. Holtzel, R. W. Jack, G. J. Nicholson, G. Jung, K. Gebhart, H-P Fiedler and R. D. Sussmuth, *J. Antibiot.*, 2001, **54**, 434.

136. G. Saviano, F. Rossi, E. Benedetti, C. Pedone, D. F. Mierke, A. Maione, G. Zanotti, T. Tancredi and M. Saviano, *Chem. -Eur. J.*, 2001, **7**, 1176.

137. T. Niidome, H. Murakami, M. Kawazoe, T. Hatakeyama, Y. Kobashigawa, M. Matsushita, Y. Kumaki, M. Demura, K. Nitta and H. Aoyagi, *Bioorg. Med. Chem. Lett.*, 2001, **11**, 1893.

138. M. Doi, S. Fujita, Y. Katsuya, M. Sasaki, T. Taniguchi, and H. Hasegawa, *Arch. Biochem. Biophys.*, 2001, **395**, 85.

139. M. Jelokhani-Niaraki, E. J. Prenner, L. H. Kondejewski, C. M. Kay, R. N. McElhaney and R. S. Hodges, *J. Pept. Res.*, 2001, **58**, 293.

140. H. Sasaki, M. Makino, M. Sisido, T. A. Smith and K. P. Ghiggino, *J. Phys. Chem B*, 2001, **105**, 10416.

141. S. Peluso, T. Ruckle, C. Lehmann, M. Mutter, C. Peggion and M. Crisma, *ChemBiochem.*, 2001, **2**, 432.

142. Q. Zu, F. Borremans and B. Devreese, *Tetrahedron Lett.*, 2001, **42**, 7261.

143. T. Higa, J. Tanaka, I. I. Ohtani, M. Musman, M. C. Roy and I. Kuroda, *Pure and Appl. Chem.*, 2001, **73**, 589.

144. W. Zheng, J. Zhuo, Q. Weng and J. Zhang, *Zhongguo Kangshengsu Zazhi*, 2001, **26**, 417.

145. B. Kratochvil, M. Husak and A. Jegorov, *Chem. Listy* 2001, **95**, 9.

146. A. Jegorov and V. Havlicek, *J. Mass Spectrom.*, 2001, **36**, 633.

147. H. Jiang, J-B. Ma, X-M. Zhu, Y-Z. Hui, H-M Wu and H-B. Chen, *Huaxue Xuebao*, 2001, **59**, 1745.

148. A. M. Broussalis, U. Goransson, J. D. Coussio, G. Ferraro, V. Martino and P. Claeson, *Phytochemistry*, 2001, **58**, 47.

149. C. Jennings, J. West, C. Waine, D. Craik and M. Anderson, *Proc. Natl Acad. Sci. U. S. A.* 2001, **98**, 10614.

150. B. Jaki, O. Zerbe, J. Heilmann and O. Sticher, *J. Nat. Prod.*, 2001, **64**, 154.

151. E. Lioy, J. Suarez, F. Guzman, S. Siegrist, G. Pluschke, and M. E. Patarroyo, *Angew. Chem. Int. Ed.*, 2001, **40**, 2631.

152. S. Ono, M. Umezaki, N. Tojo, S. Hashimoto, H. Taniyama, T. Kaneko, T. Fujii, H. Moria, C. Shimasaki, I. Yamazaki, T. Yoshimura and T. Kato, *J. Biochem.*, 2001, **129**, 783.

153. Y. Kumaki, N. Matsushima, H. Yoshida, K. Nitta, and K. Hikichi, *Biochim. Biophys. Acta*, 2001, **1548**, 81.

154. D. Wang, L. Guo, J. Zhang, L. R. Jones, Z. Chen, C. Pritchard and R. W. Roeske, *J. Pept. Res.*, 2001, **57**, 301.

155. T. Shioiri and Y. Hamada, *Synlett.*, 2001, 184.
156. Y. H. Wang, S. J. Yan, J. Y. Su, L. M. Zheng and H. Li, *Chin. J. Org. Chem.*, 2001, **21**, 16.
157. J. R. Lewis, *Nat. Prod. Rep.*, 2001, **18**, 95.
158. L. L. Guan, Y. Sera, K. Adachi, F. Nishida and Y. Shizuri, *Biochem. Biophys. Res. Commun.*, 2001, **283**, 976.
159. F. Juttner, A. K. Todorova, N. Walch and W. von Philipsborn, *Phytochemistry*, 2001, **57**, 613.
160. X. Fu, J. Y. Su and L. M. Zeng, *Sci. China Series B-Chem.*, 2000, **43**, 643.
161. G. G. Harrigan, G. H. Goetz, H. Luesch, S. Yang and J. Likos, *J. Nat. Prod.*, 2001, **64**, 1133.
162. J. Li, S. Jeong, L. Esser and P. G. Harran, *Angew. Chem. Int. Ed.*, 2001, **40**, 4765.
163. J. Li, A. W. G. Burgett, L. Esser, C. Amezcua and P. G. Harran, *Angew. Chem. Int. Ed.*, 2001, **40**, 4770.
164. J. Li, X. Chin, A. W. G. Burgett and P. G. Harran, *Angew. Chem. Int. Ed.*, 2001, **40**, 2682
165. K. C. Nicolaou, X. Huang, N. Giuseppone, P. B. Rao, M. Bella, M. V. Reddy and S. A. Snyder, *Angew. Chem. Int. Ed.*, 2001, **40**, 4705.
166. P. Magnus and C. Lescop, *Tetrahedron Lett.*, 2001, **42**, 7193.
167. P. Wipf and J. L. Methot, *Org. Lett.*, 2001, **3**, 1261.
168. E. Vedejs and M. A. Zajac, *Org. Lett.*, 2001, **3**, 2451.
169. F. Yokokawa, T. Asano and T. Shioiri, *Tetrahedron*, 2001, **57**, 6311.
170. Z. Xia and C. D. Smith, *J. Org. Chem.*, 2001, **66**, 3459.
171. F. Yokokawa, H. Sameshima and T. Shioiri, *Synlett.*, 2001, 986.
172. F. Yokokawa, H. Sameshima and T. Shioiri, *Tetrahedron Lett.*, 2001, **42**, 4171.
173. J. M. Caba, I. M. Rodriguez, I. Manzanares, E. Giralt and F. Albericio, *J. Org. Chem.*, 2001, **66**, 7568.
174. A. Bertram and G. Pattenden, *Synlett.*, 2001, 1873.
175. L. Somogyi, G. Haberhauer and J. Rebek, *Tetrahedron*, 2001, **57**, 1699.
176. G. Zhao, X. W. Sun, H. Bienayme and J. P. Zhu, *J. Am. Chem. Soc.*, 2001, **123**, 6700.
177. T. Suzuki, A. Nagasaki, K. Okumura and C-G. Shin, *Heterocycles*, 2001, **55**, 835.
178. T. Yamada, K. Okumura, Y. Yonezawa and C-G. Shin, *Chem. Lett.*, 2001, 102.
179. G. Pattenden and T. Thompson, *Chem. Commun. (Cambridge)*, 2001, 717.
180. Y. Singh, N. Sokolenko, M. J. Kelso, L. R. Gahan, G. Abbenante and D.P. Fairlie, *J. Am. Chem. Soc.*, 2001, **123**, 333.
181. K. Kouda, *Farumashia*, 2001, **37**, 921.
182. L. A. Morris, B. F. Milne, M. Jaspars, J. J. Kettens-van den Bosch, K. Versluis, A. J. R. Heck, S. M. Kelly and N. C. Price, *Tetrahedron*, 2001, **57**, 3199.
183. L. A. Morris, M. Jaspars, J. J. Kettens-van den Bosch, K. Versluis, A. J. R. Heck, S. M. Kelly and N. C. Price, *Tetrahedron*, 2001, **57**, 3185.
184. B-S. Yun, K-I Fujita, K. Furihata and H. Seto, *Tetrahedron*, 2001, **57**, 9683.
185. A. Asano, M. Doi, K. Kobayashi, M. Arimoto, T. Ishida, Y. Katsuya. Y. Mezaki, H. Hasegawa, M. Nakai, M. Sasaki, T. Taniguchi and A. Terashima, *Biopolymers*, 2001, **58,** 295.
186. A. Asano, T. Taniguchi, M. Sasaki, H. Hasegawa, Y. Katsuya and M. Doi, *Acta Crystallograph. Section E- Structure Reports Online*, 2001, **57**, 834.
187. A. Randazzo, C. Debitus and L. Gomez-Paloma, *Tetrahedron*, 2001, **57**, 4443.
188. A. Randazzo, G. Bifulco, C. Giannini, M. Bucci, C. Debitus, G. Cirino and L. Gomez-Paloma, *J. Am. Chem. Soc.*, 2001, **123**, 10870.
189. M. A. Rashid, K. R. Gustafson, L. K. Cartner, N. Shigematsu, L. K. Pannell and M.

R. Boyd, *J. Nat. Prod.*, 2001, **64**, 117.

190. Y. Suzuki, M. Ojika, Y. Sakagami, K. Kaida, R. Fudou and T. Kameyama, *J. Antibiot.*, 2001, **54**, 22.

191. H. Luesch, W. Y. Yoshida, R. E. Moore, V. J. Paul and T. H. Corbett, *J. Am. Chem. Soc.*, 2001, **123**, 5418.

192. F. Wan and K. L. Erikson, *J. Nat. Prod.*, 2001, **64**, 143.

193. H. Maki, K. Miura and Y. Yamano, *Antimirob. Agents and Chemo.* 2001, **45**, 1823.

194. Y. Che, D. C. Swenson, J. B. Gloer, B. Koster and D. Malloch, *J. Nat. Prod.*, 2001, **64**, 555.

195. K. Takahashi, H. Koshino, Y. Esumi, E. Tsuda and K. Kurosawa, *J. Antibiot.*, 2001, **54**, 622.

196. U. Matern, L. Oberer, R. A. Falchetto, M. Erhard, W. A. Konig, M. Herdman and J. Weckesser, *Phytochemistry*, 2001, **58**, 1087.

197. B. Liang, D. J. Richard, P. S. Portnovo and M. M. Joullie, *J. Am. Chem. Soc.*, 2001, **123**, 9830.

198. D. Valognes, P. Belmont, N. Xi and M. A. Ciufolini, *Tetrahedron Lett.*, 2001, **42**, 1907.

199. E. Horikawa, M. Kodaka, Y. Nakahara, H. Okuno and K. Nakumara, *Tetrahedron Lett.*, 2001, **42**, 8337.

200. A. Lopez-Macia, J. C. Jimenez, M. Royo, E. Giralt and F. Albericio, *J. Am. Chem. Soc.*, 2001, **123**, 11398.

201. F. Yokokawa, A. Inaizumi and T. Shioiri, *Tetrahedron Lett.*, 2001, **42**, 5903.

202. T. Ast, E. Barron, L. Kinne, M. Schmidt, L. Germeroth. K. Simmons and H. Wenschuh, *J. Pept. Res.*, 2001, **58**, 1.

203. D. E. Ward, Y. Z. Gai, R. Lazny, M. Soledade and C. Pedras, *J. Org. Chem.*, 2001, **66**, 7832.

204. K. J. Hale, L. Lazarides and J. Cai, *Org. Lett.*, 2001, **3**, 2927.

205. C. Hermann, C. Giammasi, A. Geyer and M. E. Maier, *Tetrahedron*, 2001, **57**, 8999.

206. S. Nishiyma, *Yuki Gosei Kagaku Kyokaishi*, 2001, **59**, 938.

207. A. K. Ghosh and C. F. Liu, *Org. Lett.*, 2001, **3**, 635.

208. D. Barrett, A. Tanaka, A. Fujie, N. Shigematsu, M. Hashimoto and S. Hashimoto, *Tetrahedron Lett.*, 2001, **42**, 703.

209. A. Tanaka, D. Barrett, A. Fujie, N. Shigematsu, M. Hashimoto, S. Hashimoto and F. Ikeda, *J. Antibiot.*, 2001, **54**, 193; D. Barrett, A. Tanaka, E. Watabe, K. Maki and F. Ikida, *ibid.*, 2001, **54**, 844.

210. N. Murakami, S. Tamura, W. Wang, T. Takagi and M. Kobayashi, *Tetrahedron*, 2001, **57**, 4323.

211. R. Banteli, J. Wagner and G. Zenke, *Bioorg. Med. Chem. Lett.*, 2001, **11**, 1609.

212. T. K. Chakraborty, S. Ghosh and S. Dutta, *Tetrahedron Lett.*, 2001, **42**, 5085.

213. J. J. Duffield and G. R. Pettit, *J. Nat. Prod.*, 2001, **64**, 472.

214. S. Chandrasekhar, T. Ramachandar, and B. V. Rao *Tetrahedron:Asymmetry*, 2001, **12**, 2315.

215. N. Okamoto, O. Hara, K. Makino and Y. Hamada, *Tetrahedron:Asymmetry*, 2001, **12**, 1353.

216. C. M. Acevedo, E. F. Kogut and M. A. Lipton, *Tetrahedron*, 2001, **57**, 6353.

217. L. Trevisi, S. Bova, G. Cargnelli, D. Danieli-Betto, M. Floreani, E. Germinario, M. V. D'Auria and S. Luciani, *Biochem. Biophys. Res. Comm.*, 2000, **279**, 219.

218. X. Sun, M. Rodriguez, D. Zeckner, B. Sachs, W. Current, R. Boyer, J. Paschal, C. McMillian and S-H. Chen, *J. Med. Chem.*, 2001, **44**, 2671.

219. T. A. Mukhtar, K.P. Koteva, D. W. Hughes and G. D. Wright, *Biochemistry*, 2001,

40, 8877.

220. K. N. Koch, A. Linden and H. Heimgartner, *Tetrahedron*, 2001, **57**, 2311.

221. K. N. Koch, G. Hopp, A. Linden, K. Moehle and H. Heimgartner, *Helv. Chim. Acta*, 2001, **84**, 502.

222. F. Cardenas, M. Thormann, M. Feliz, J-M Caba, P. Lloyd-Williams and E. Giralt, *J. Org. Chem.*, 2001, **66**, 4580.

223. J. Kuroda, T. Fukai and T. Nomura, *J. Mass Spectrom.*, 2001, **36**, 30.

224. D. T. McLachlin and B. T. Chait, *Curr. Opinion in Chem. Biol.*, 2001, **5**, 591.

225. M. P. Molloy and P. C. Andrews, *Anal. Chem.*, 2001, **73**, 5387.

226. A. Beck, M. Deeg, K. Moeschel, E. K. Schmidt, E. D. Schleicher, W. Voelter, H.U.Haring and R. Lehmann, *Rapid Comm.Mass Spectrom.*, 2001, **15**, 2324.

227. M. Adamczyk, J. C. Gebler and J. Wu, *Rapid Commun. Mass Spectrom.*, 2001, **15**, 1481.

228. G. Fago, A. Filippi, A. Gardini, A. Lagana, A. Paladini and M. Speranza, *Angew. Chem. Int Ed.*, 2001, **40**, 4051.

229. A. Schlosser, R. Pipkorn, D. Bossemeyer and W.D. Lehmann, *Anal. Chem.*, 2001, **73**, 170.

230. J. W. Flora and D. C. Muddiman, *Anal. Chem.*, 2001, **73**, 3305.

231. R. S. Annan, M. J. Huddleston, R. Verma, R. J. Deshaires and A. Carr, *Anal. Chem.*, 2001, **73**, 393.

232. K. Janek, H. Wenschuh, M. Bienert and E. Krause, *Rapid Commun. Mass Spectrom.*, 2001, **15**, 1593.

233. H. Steen, B. Kuster, M. Fernandez, A. Pandey and M. Mann, *Anal. Chem.*, 2001, **73**, 1440.

234. H. Steen, B. Kuster and M. Mann, *J. Mass Spectrom.*, 2001, **36**, 782.

235. Y. L. Ma, Y. Lu, H. Q. Zeng, D. Ron, W. J. Mo and T. A. Neubert, *Rapid. Commun. Mass Spectrom.*, 2001, **15**, 1693.

236. J. S. McMurray, D. R. Coleman IV, W. Wang and M. L. Campbell, *Biopolymers*, 2001, **60**, 3.

237. S-B. Chen, Y-M Li, G. Zhang, S-Z. Luo and Y-F. Zhao, *Gaodeng Xuexiao Huaxue Xuebao*, 2001, **22**, 106.

238. Z. Kupihar, Z. Kele and G. K. Toth, *Org. Lett.*, 2001, **3**, 1033.

239. S. Z. Dong, H. Fu and Y. F. Zhao, *Synth. Commun.*, 2001, **31**, 2067.

240. C. Grison, C. Comoy and P. Coutrot, *Actualite Chimique*, 2001, 38.

241. D. L. Boger, *Med. Res. Rev.*, 2001, **21**, 356.

242. P. Lloyd-Williams and E. Giralt, *Chem. Soc.Rev.*, 2001, **30**, 145.

243. D. A. Evans, J. L. Katz, G. S. Peterson and T. Hintermann, *J. Am. Chem. Soc.*, 2001, **123**, 12411.

244. Y. Mori, J. J. McAtee, O. Rogel and D. L. Boger, *Tetrahedron Lett.*, 2001, **42**, 6061.

245. K. C. Nicolaou, S. Y. Cho, R. Hughes, N. Winssinger, C. Smethurst, H. Labis-chinski and R. Endermann, *Chemistry-Eur. J.*, 2001, **7**, 3798.

246. A. J. Pearson and S. Zigmantas, *Tetrahedron Lett.*, 2001, **42**, 8765.

247. K. Kamikawa, A. Tachibana, S. Sugimoto and M. Uemura, *Org. Lett.*, 2001, **3**, 2033.

248. P. Cristau, J-P. Vors and J. Zhu, *Org. Lett.*, 2001, **3**, 4079.

249. K. Nakamura, H. Nishiya and S. Nishiyama, *Tetrahedron Lett.*, 2001, **42**, 6311.

250. C. P. Decicco, Y. Song and D. A. Evans, *Org. Lett.*, 2001, **3**, 1029.

251. K. C. Nicolaou, R. Hughes, S. Y. Cho, N. Winssinger, H. Labishinski and R. Endermann, *Chem.-Eur. J.*, 2001, **7**, 3824.

252. A. J. R. Heck, P. J. Bonnici, E. Breukink, D. Morris and M. Wills, *Chem.-Eur. J.*,

2001, **7**, 910.
253. A. Y. Pavlov, O. V. Miroshnikova, S. S. Printsevskaya, E. N. Olsufyeva, M. N. Preobrazhenskaya, R. C. Goldman, A. A. Branstrom, E. R. Baizman and C. B. Longley, *J. Antibiot.*, 2001, **54**, 455.
254. J. Kaplan, B. D. Korty, P.H. Axelsen and P.J. Loll, *J. Med. Chem.*, 2001, **44**, 1837.
255. H. C. Losey, M. W. Peczuh, Z. Chen, U. S. Eggert, S. D. Dong, I. Pelczer, D. Kahne and C. T. Walsh, *Biochemistry*, 2001, **40**, 4745.
256. M-C. Lo, J. S. Helm, G. Samgadharan, I. Pelczer and S. Walker, *J. Am. Chem. Soc.*, 2001, **123**, 8640.
257. M. C. F. Monnee, A. J. Brouwer, L. M. Verbeek, A. M. A. van Wageningen, R. M. J. Liskamp, *Bioorg. Med. Chem. Lett.*, 2001, **11**, 1521.
258. D. Bischoff, S. Pelzer, H. Holtzel, G. J. Nicholson, S. Stockert, W. Wohlleben, G. Jung and R. D. Susmuth, *Angew. Chem. Int. Ed.*, 2001, **40**, 1693.
259. T. L. Li, O. W. Choroba, H. Hong, D. H. Williams and J. B. Spencer, *Chem. Commun.(Cambridge)*, 2001, 2156.
260. H. Chen, C. C. Tseng, B. K. Hubbard and C. T. Walsh, *Proc. Nat. Acad. Sci. U. S. A.*, 2001, **98**, 14901.
261. M. J. Rishel and S. M. Hecht, *Org. Lett.*, 2001, **3**, 2867.
262. M. L. Sznaidman and S. M. Hecht, *Org. Lett.*, 2001, **3**, 2811.
263. T. J. Ward and A. B. Farris, *J. Chromatog. A*, 2001, **906**, 73.
264. A. Berthot, A. Vallaix, V. Tizon, E. Leonce, C. Caussignac and D. W. Armstrong, *Anal. Chem.*, 2001, **73**, 5499.
265. I. A. Anan'eva, E. N. Shapovalova, O. A. Shpigun and D. W. Armstrong, *Vestnik Moskovskogo Universiteta Seriya 2: Khimiya*, 2001, **42**, 278.
266. K. Petritis, A. Valleix, C. Elfakir and M. Dreux, *J. Chromatog., A.*, 2001, **913**, 331.
267. E. Peyrin, A. Ravel, C. Grosset, A. Villet, C. Ravelet, E. Nicolle and J. Alary, *Chromatographia*, 2001, **53**, 645.
268. G. Torok, A. Peter, D. W. Armstrong, D. Tourwe, G. Toth and J. Sapi, *Chirality*, 2001, **13**, 648.
269. M. Mamoru, *Trends Glycosci. Glycotechnol.*, 2001, **13**, 11.
270. P. M. St. Hilaire, M. Meldal and K. Bock, *Special Publication -Royal Society of Chemistry*, 2001, **264**, 293.
271. K. Ajiska, *Trends Glycosci. Glycotechnol.*, 2001, **13**, 305.
272. H. Hojo, *Kagaku (Kyoto Japan)*, 2001, **56**, 66.
273. Y. Zhu and W. A. van der Donk, *Org. Lett.*, 2001, **3**, 1189.
274. S. B. Cohen and R. L. Halcomb, *Org. Lett.*, 2001, **3**, 405.
275. P. Durieux, J. Fernandez-Carneado and G. Tuchscherer, *Tetrahedron Lett.*, 2001, **42**, 2297.
276. G. Di Fabio, A. De Capua, L. De Napoli, D. Montesarchio, G. Piccialli, F. Rossi and E. Benedetti, *Synlett.*, 2001, 341.
277. S. K. George, B. Holm, C. A. Reis, T. Schwientek, H. Clausen and J. Kihlberg, *J. Chem. Soc. Perkin Trans. 1*, 2001, 880.
278. S. L. Saha, M. S. Van Nieuwenhze, W. J. Hornback, J. A. Aikins and L. C. Blaszcak, *Org. Lett.*, 2001, **3**, 3575.
279. P. Sjoelin and J. Kihlberg, *J. Org. Chem.*, 2001, **66**, 2957.
280. J-i. Tamura and J. Nishihara, *J. Org. Chem.*, 2001, **66**, 3074.
281. P-H. Tseng, W-T Jiaang, M-Y. Chang and S-T Chen, *Chem.-Eur. J.*, 2001, **7**, 585.
282. S. K. George, T. Schwientek, B. Holm, C. A. Reis, H. Clausen and J. Kihlberg, *J. Am. Chem. Soc.*, 2001, **123**, 11117.
283. H. Liu, R. Sadamato, P. S. Sears and C-H. Wong, *J. Am. Chem. Soc.*, 2001, **123**,

9916.

284. K. Ajisaka, M. Miyasato and I. Ishii-Karakasa, *Biosci. Biotechnol. Biochem.* 2001, **65**, 1240.

285. K. Fukase, T. Yasukochi and S. Kusumoto, *Bull.Chem.Soc. Jpn.*, 2001, **74**, 1123.

286. L. A. Marcaurelle and C. R. Bertozzi, *J. Am. Chem. Soc.*, 2001, **123**, 1587.

287. J. R. Allen, C. R. Harris and S. J. Danishefsky, *J. Am. Chem. Soc.*, 2001, **123**, 1890.

288. S. A. Mitchell, M. R. Pratt, V. J. Hruby and R. Polt, *J. Org. Chem.*, 2001, **66**, 2327.

289. A. Siriwardena, M. R. Jorgensen, M. A. Wolfert, M. L. Vandenplas, J. N. Moore and G-J. Boons, *J. Am. Chem. Soc.*, 2001, **123**, 8145.

290. Y. Nakahara, S. Ando, Y. Ito, J. H. Hojo and Y. Nakahara, *Biosci. Biotechnol. Biochem.*, 2001, **65**, 1358.

291. K. M. Halkes, C. H. Gotfredsen, M. Grotli, L. P. Miranda, J. O. Duus and M. Meldal, *Chem.-Eur. J.* 2001, **7**, 3584.

292. Y. Du, M. Zhang, F. Yang and G. Gu, *J. Chem. Soc. Perkin Trans. 1*, 2001, 3122.

293. G. A. Winterfeld and R. R. Schmidt, *Angew. Chem. Int. Ed.*, 2001, **40**, 2654.

294. M. M. Palian and R. Polt, *J. Org. Chem.*, 2001, **66**, 7178.

295. M. Tagashira, H. Iijima and K. Toma, *Trends Glycosci. Glycotech.*, 2001, **13**, 373.

296. L. Kindahl, C. Sandstrom, T. Norberg and L. Kenne, *Carbohydr. Res.*, 2001, **336**, 319.

297. M. M. Palian, N. E. Jacobsen and R. Polt, *J. Pept. Res.*, 2001, **58**, 180.

298. M. Gobbo, A. Nicotra, R. Rocchi, M. Crisma and C. Toniolo, *Tetrahedron*, 2001, **57**, 2433.

299. J. Colangelo and R. Orlando, *Rapid. Comm. Mass Spec.*, 2001, **15**, 2284.

300. B. Macek, J. Hofsteenge and J. Peter-Katalinic, *Rapid Comm. Mass Spec.*, 2001, **15**, 771.

301. E. Mirgorodskaya, H. Hassan, H. Clausen and P. Roepstroff, *Anal. Chem.*, 2001, **73**, 1263.

302. M. Rosch, H. Herzner, W. Dippold, M. Wild, D. Vestweber and H. Kunz, *Angew. Chem. Int. Ed.*, 2001, **40**, 3836.

303. Z-G. Wang, M. Visser, D. Live, A. Zatorski, U. Iserloh, K. O. Lloyd and S. J. Danishefsky, *Angew. Chem. Int. Ed.*, 2001, **40**, 1728.

304. S. Wen and Z. Guo, *Org. Lett.*, 2001, **3**, 3773.

305. A. E. Zemlyakov, V. V. Tsikalov, V. O. Kuryanov, V. Ya. Chirva and N. V. Bovin, *Russ. J. Bioorg. Chem.*, 2001, **27**, 390.

306. D. Ramos, P. Rollin and W. Klaffke, *J. Org. Chem.*, 2001, **66**, 2948.

307. K. Yamamoto, *J. Biosci. Bioeng.*, 2001, **92**, 493.

308. V. O. Kuryanov, T. A. Chupakhina, A. E. Zemlyakov, V. Y. Chirva, V. V. Ischenko and V. P. Khilya, *Chem. Nat. Prod.*, 2001, **37**, 39.

309. Y. Q. Xia and J. M. Risley, *J. Carbohyd. Chem.*, 2001, **20**, 45.

310. K. Burger, C. Bottcher, G. Radics and L. Hennig, *Tetrahedron Lett.*, 2001, **42**, 3061.

311. C. J. Bosques, V. W. F. Tai and B. Imperiali, *Tetrahedron Lett.*, 2001, **42**, 7207.

312. S. Peluso and B. Imperiali, *Tetrahedron Lett.*, 2001, **42**, 2085.

313. J. J. Turner, D. V. Filippov, M. Overhand, G. A. van der Marel and J. H. van Boom, *Tetrahedron Lett.*, 2001, **42**, 5763.

314. R. P. McGeary, I. Jablonkai and I. Toth, *Tetrahedron*, 2001, **57**, 8733.

315. H. Bradley, G. Fitzpatrick, W. K. Glass, H. Kunz and P. V. Murphy, *Org. Lett.*, 2001, **3**, 2629.

316. C. H. Rohrig, M. Takhi and R. R. Schmidt, *Synlett.*, 2001, 1170.

317. A. Dondoni, G. Mariotti, A. Marra and A. Massi, *Synthesis*, 2001, 2129.

318. T. Nishikawa, M. Ishikawa, K. Wada and M. Isobe, *Synlett.*, 2001, Spec Iss., 945.

319. F. Schweizer, A. Lohse, A. Otter and O. Hindsgaul, *Synlett.*, 2001, 1434.
320. J. M. Berge, C. S. V. Houge-Frydrych and R. L. Jarvest, *J. Chem. Soc. Perkin Trans. 1*, 2001, 2521.
321. A. Dondoni, P. P. Giovanni and A. Marra, *J. Chem. Soc. Perkin Trans. 1*, 2001, 2380.
322. A. Wernicke and D. Sinou, *J. Carbohyd. Chem.*, 2001, **20**, 181.
323. J. Xie and J-M. Valery, *J Carbohyd. Chem.*, 2001, **20**, 441.
324. A. Eniade, A. V. Murphy, G. Landreau and R. N. Ben, *Bioconjugate Chem.*, 2001, **12**, 817.
325. C. Grison, F. Coutrot and P. Coutrot, *Tetrahedron*, 2001, **57**, 6215.
326. V. Boyer, M. Stanchev. A. J. Fairbanks and B. G. Davis, *Chem. Commun (Cambridge)*, 2001, 1908.
327. D. A. Jeyaraj and H. Waldmann, *Tetrahedron Lett.*, 2001, **42**, 835.
328. E. R. Civitello, R. G. Leniek, K. A. Hossler, K. Haebe and D. M. Stearns, *Bioconjugate Chem.*, 2001, **12**, 459.
329. S. O. Doronina, A. P. Guzaev and M. Manoharan, *Nucleosides, Nucleotides Nucleic Acids*, 2001, **20**, 1007.
330. L. M. Nogle, T. Okino and W. H. Gerwick, *J. Nat. Prod.*, 2001, **64**, 983.
331. D. Sorensen, T. H. Nielsen, C. Christopherson, J. Sorensen and M. Gajhede, *Acta Crystallogr. Section C*, 2001, **57**, 1123.
332. A. Fujie, T. Iwamoto, B. Sato, H. Muramatsu, C. Kasahara, T. Furuta, Y. Hori, M. Hino and S. Hashimoto, *Bioorg. Med. Chem. Lett.*, 2001, **11**, 399.
333. E. K. Dolence, J. M. Dolence and C. D. Poulter, *Bioconjugate Chem.*, 2001, **12**, 35.
334. B. McKeever and G. Pattenden, *Tetrahedron Lett.*, 2001, **42**, 2573.
335. D. Bonnet, N. Ollivier, H. Gras-Masse and O. Melnyk, *J. Org. Chem.*, 2001, **66**, 443.
336. H. Tsubery, I. Ofek, S. Cohen and M. Fridkin, *Peptides*, 2001, **22**, 1675.
337. C. Giragossian, E. Nardi, C. Savery, M. Pellegrini, S. Meini, C. A. Maggi, A. M. Papini and D. F. Mierke, *Biopolymers*, 2001, **58**, 511.
338. D. Huster, K. Kuhn, D. Kadereit, H. Waldmann and K. Arnold, *Angew. Chemie Int. Ed.*, 2001, **40**, 1056.
339. D. Bonnet, K. Thiam, E. Loing, O. Melnyk and H. Gras-Masse, *J. Med. Chem.*, 2001, **44**, 468.
340. M. R. Muller, S. D. C. Pfannes, M. Ayoub, P. Hoffmann, W. G. Bessler and K. Mittenbuhler, *Immunology*, 2001, **103**, 49.
341. S. R. Giacomelli, F. C. Missau, M. A. Mostardeiro, U. F. Da Silva, I. I. Dalcol and A. F. Morel, *J. Nat. Prod.* 2001, **64**, 997.
342. S-S. Lee, W-C. Su, K. Liu and C. S. Chen, *Phytochemistry*, 2001, **58**, 1271.
343. H. Kudo, F. Sanda and T. Endo, *Tetrahedron Lett.*, 2001, **42**, 7847.
344. M. Achmatowicz, A. Szczepanska, D. T. Gryko, P. Salanski and J. Jurczak, *Supramolecular Chem.*, 2000, **12**, 93.
345. H. Ishida, Z. Qi, M. Sokabe, K. Donowaki and Y. Inoue, *J. Org. Chem.*, 2001, **66**, 2978; H. Ishida and Y. Inoue, *Biopolymers*, 2001, **55**, 469.
346. R. L. E. Furlan, Y-F. Ng, S. Otto and J. K. M. Sanders, *J. Am. Chem. Soc.*, 2001, **123**, 8876.
347. D. Gauthier, P. Baillargeon, M. Drouin and Y.L. Dory, *Angew. Chemie, Int. Ed.*, 2001, **40**, 4635.
348. L. M. A. McNamara, M. J. I. Andrews, F. Mitzel, G. Siligardi and A. B. Tabor, *J. Org. Chem.*, 2001, **66**, 4585.
349. S. Horvat, I. Jeric, L. Varga-Defterdarovic, M. Roscic and J. Horvat, *Croatia Chem. Acta*, 2001, **74**, 787.

4
Proteins

BY GRAHAM C. BARRETT

1   Introduction*

The aim of this Chapter has remained unchanged over the years: the coverage of a year's protein literature with emphasis on studies that uncover links between protein function and molecular structure. Routine biochemical studies are excluded, as are data compilations and reviews that do not reflect the aim of the Chapter. However, the contributions that can be made to descriptive biological topics, through the interpretation of molecular structures of proteins, have been illustrated regularly in this Chapter over the years. A striking example this year in this context, is the conclusion that the actin cytoskeleton had evolved before eukaryotes had developed (Ref. 42).

All too often, descriptions of studies deserving a place in this Chapter have had to be excluded. This is an inevitable restriction on the coverage that has had to be imposed, because protein science is now an area of phenomenal growth that this Chapter can only sample. But in the process of selection, papers have been chosen so that details of major current themes can be given, to provide a description of the frontiers of the topic. In other words, the general thrust of the year's protein literature is thought to be represented in the papers that have been cited here.

Graphics styles by which protein structures and mechanistic schemes are depicted in current research papers, are represented in this Chapter. It is hoped that this display of the extraordinary range of graphics techniques that have been developed for protein studies may be informative in its own right. Perhaps the display may help to introduce the techniques to new authors reporting results in the proteins field, and also to reveal the possibilities of these graphics techniques to authors reporting other areas of structural chemistry.

---

* Most of the protein structures reproduced in this Chapter are in colour in the original sources. Authors have usually chosen to use colour to help the reader to appreciate the discussion of their work, and the on-line version of this Chapter reproduces the colour versions. This can be found at http://pubs.rsc.org/ebooks/CONTENTS/AA034004figures.pdf.

---

2   Structure of this Chapter

Last year's Chapter set out a structure for the coverage on Proteins in this Specialist Periodical Report. This structure has been largely followed in this year's review. Of course, active topics that have been worked out to completion within the time covered by this review have been given only brief mention, and conversely, rapidly developing topics that were barely recognised last year have been given more space. So the structure for this Chapter will be understood to be flexible, in accordance with the ebb and flow of protein topics as determined by the research groups involved.

2.1   Cross-Referencing in This Chapter. – The papers discussed in this Chapter are frequently wide-ranging, and many of them could be located under more than one heading in the different Sections of this Chapter. The opportunity has been taken in each Section to alert the reader to the presence of relevant papers elsewhere in the Chapter. Reference back to an earlier Volume of this Specialist Periodical Report is made with, *e.g.*, 'Volume 33, p. 378'.

3   Textbooks and Monographs

In addition to a listing of textbooks and monographs published in 2001, a number of books that have escaped mention in this Chapter in earlier Volumes are cited here since they represent a resource of continuing usefulness. Although these titles will be familiar to regular readers of this Specialist Periodical Report, new researchers in protein science will welcome information about textbooks and monographs as a time-saving initiation to the field, and may therefore find it helpful to have a listing of recommended primary literature of Protein Science in this Specialist Periodical Report.

Titles added this year, to supplement listings in preceding Volumes, are: *Protein – Protein Interactions: A Molecular Cloning Manual,*[1] *Modern Protein Chemistry: Practical Aspects,*[2] *The Physical Foundation of Protein Architecture,*[3] *Perspectives in Amino Acid and Protein Geochemistry,*[4] *Protein Ligand Interactions (Practical Approach Series),*[5] *Guide to Cytochrome P450: Structure and Function,*[6] *Cloning, Gene Expression, and Protein Purification: Experimental Procedures and Process Rationale,*[7] *Nature's Robots: A History of Proteins,*[8] *Protein – Protein Recognition,*[9] and *Advances in Protein Chemistry: Protein Modules and Protein-Protein Interactions.*[10]

Books published in earlier years, that have escaped mention in preceding Volumes of this Specialist Periodical Report, are: *Biological Sequence Analysis: Probabilistic Models of Proteins and Nucleic Acids,* by R.Durbin, S.R.Eddy, A.Krogh and G.Mitchison, *The Protein Protocols Handbook,* ed. J.M.Walker, *The Chaperonins,* ed. R.J.Ellis, *Protein Folding,* eds. C.M.Dobson and A.R.Fersht, *Fundamentals of Enzymology,* by N.C.Price and L.Stevens, *Protein – Solvent Interactions,* by R.B.Gregory, *Biotechnology: Proteins to PCR: A Course in Strategies and Laboratory Techniques,* by D.Burden and D.Whitney, and *Protein Analysis and Purification: Benchtop Techniques,* by I Rosenberg.[11]

Later Sections of this Chapter provide titles of further recent textbooks.

3.1 Literature Searching in Protein Science. – The textbooks noted in the preceding section, and at the start of many other sections of this Chapter, are valuable as starting points for researching the status of a protein topic.

The online literature in protein science continues to expand, through the increased range of internet access to established Journals and the secondary literature. Background literature searching is well supported by this resource. Useful new portals have been set up (Ref. 90), which will help readers who seek more background on the topics described in this Chapter.

3.2 Protein Nomenclature. – No additional recommendations from international organisations have appeared in the period under review, and the primary literature reflects the continuing application of approved nomenclature that mostly has been in use for many years.

A glossary of abbreviations is usually lacking in reviews presenting protein results that involve many acronyms and other abbreviations,[12] which are a source of frustration for many readers. The usual freedom to introduce new names for individual proteins and families of proteins has been taken on board by authors of journal papers, as the opportunities arise, and especially the freedom to use more or less logical abbreviations and neologisms. There is some logic to many of the choices of abbreviations, for example the use of Protein Data Bank conventions, but some unravelling of abbreviations to provide the full protein name will be seen in the present Chapter. This seems especially necessary for achieving and maintaining communication with readers who are approaching protein science from the direction of classical organic chemistry.

4 Structure Determination of Proteins

Major techniques employed in structure determination of proteins are illustrated well in the studies selected for this Section. The most widely applicable technique continues to be X-ray crystallography, and this occupies the last place in this Section, partly because it is now a relatively routine technique, and generates little in the way of monographs and textbooks, or reviews. Also, copious cross-referencing would be needed between papers on X-ray crystallography in the text of this Chapter, if attention is given to the interests of some readers of this Chapter. However, routine use of the technique underpins most of the papers cited this year, and cross-referencing on this scale would be a tedious, space-consuming exercise. By being placed last in this Section, particularly interesting applications of X-ray crystallography are brought close to the main descriptive part of the Chapter.

4.1 Proteomics and Genomics. – The growth of the general field of protein structure determination, and links between structure and function has been supported by a large range of texbooks. Recent additions to the range are:

Proteome Research: New Frontiers in Functional Genomics,[13] *Proteome Research: Mass Spectrometry,*[14] *Post-Translational Modification of Proteins: Tools for Functional Proteomics (Methods in Molecular Biology, Vol. 194),*[15] *Proteomics in Practice: A Laboratory Manual of Proteome Analysis,*[16] *Protein Structure Prediction: Bioinformatic Approach,*[17] *Proteome Research: Two-Dimensional Gel Electrophoresis and Detection Methods (Principles and Practice),*[18] *Discovering Genomics, Proteomics, and Bioinformatics,*[19] *Proteomics,*[20] *Proteins and Proteomics: A Laboratory Manual,*[21] *The Proteome Revisited: Theory and Practice of All Relevant Electrophoretic Steps,*[22] *Proteomics Reviews 2001,*[23] *Proteins: Biochemistry and Biotechnology,*[24] *Introduction to Bioinformatics,*[25] *Bioinformatics, Sequence, Structure, and Databanks,*[26] and *Introduction to Bioinformatics.*[27]

4.2   Mass Spectrometry. – A recent textbook deals with *Interpreting Protein Mass Spectra,*[28] and *MS For Biotechnology*[29] deserves a repeat mention. Several textbooks listed in the preceding section cover the important role of mass spectrometry in protein structure determination.

Mass spectrometric methods are a major source of new results that assist structure determinations for proteins. The technique is now applied widely, and has not been given detailed coverage here. However, it must be said that the literature demonstrates that the contribution of mild ionization methods to the mass spectrometric analysis of large molecular complexes continues to undergo astonishing development. One new example is the complete amino acid sequence of the coat protein of Brome Mosaic Virus.[30] This is one of the first RNA plant viruses to have had its genome completely sequenced, its importance being its capacity to infect grasses and cereals, causing occasional crop destruction and therefore financial loss. The point is well made in this paper, that protein sequencing derived from interpreting the DNA sequence will not reveal post-translational changes that may take place. In the present example, the proteins studied were found to carry an N-terminal acetyl group, readily identified from the mass spectrum after the researchers had found difficulty in matching likely amino acid and peptide moieties with masses of fragments liberated in the mass spectrometer.

For other examples of detailed structure elucidation involving current mass spectrometric techniques see also Refs. 136, 141.

Complex protein structures are rarely solved by the use of only one physical method (unless that method is X-ray crystallography), and current methodology involving combinations of physical methods is featured in later sections of this Chapter.

4.3   NMR Spectroscopy. – Textbook coverage of the major role played by NMR spectroscopy is extended significantly by new or recent offerings: *Protein NMR for the Millennium (Biological Magnetic Resonance),*[31] and *Biomolecular NMR Spectroscopy.*[32]

High-pressure ^{15}N/^{1}H-NMR spectroscopy of ubiquitin (750 MHz for ^{1}H; the pressure range involved was 30 to 3500 bar)[33] has revealed the existence at equilibrium of two folded conformers at pH 4.5 at 20°C at 2000 bar. Sigmoid-

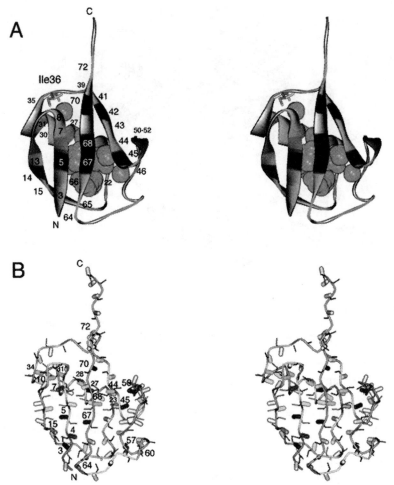

Figure 1   *(A) Location of residues showing anomalous amide ^{15}N pressure shifts in the three-dimensional X-ray structure of ubiquitin. The residues showing anomalous amide nitrogen shifts are marked by shading according to their deviation from the average pressure shift. (B) Location of residues showing anomalous amide ^{1}H pressure shifts on a representative NMR structure of ubiquitin. Cavities with a probe radius of 1.0 Å are represented by spheres of a darker shade. Isoleucine36, whose cross-peak preferentially broadens and disappears at 3000 bar, is drawn in stick format. The internally hydrogen-bonded amide groups that show anomalous ^{1}H pressure shifts are marked by shading according to their deviation from the average pressure shift. The NH and CO groups are represented in thick and thin stick format, respectively* (Reproduced with permission from Kitahara et al.[33])

shaped pressure-induced chemical shift relationships were diagnostic features from which this conclusion was reached. In Figure 1, (A) shows the location of residues showing anomalous amide ^{15}N pressure shifts in the three-dimensional X-ray structure of ubiquitin.

There are several points of interest, for example the cross-peak of isoleucine36

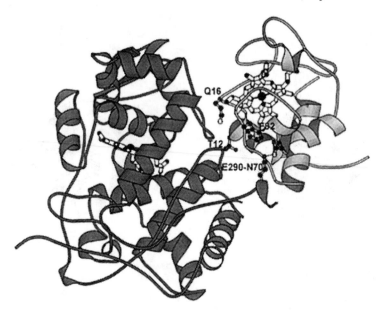

Figure 2   *The 1:1-complex formed between yeast iso-1-cytochrome c (light regions of this Figure) and cytochrome c peroxidase (black regions of this Figure)*
(Reproduced with permission from Worrall et al.[36])

preferentially broadens and disappears at high pressures (3000 bar), thus revealing a particular sensitivity of conformational parameters for this residue.

High-pressure $^{15}N/^{1}H$- NMR spectroscopy of the complex of a GTPase Rap1A/Ra1GDS (its effector) shows unusually large non-linear chemical shifts that occur as the pressure is changed, for the Ras-binding domain of Ra1 guanine dissociation stimulator. The data reveal the presence of an increasing proportion of a higher energy conformer at higher pressures.[34] Cdc42, a Rho-related member of the Ras superfamily, is a GTP-binding protein, functioning as a molecular switch to control a wide range of cellular processes. Its role in the slowing of Fas-induced apoptosis has now been established.[35]

1:1-Complex formation between yeast iso-1-cytochrome c and its physiological redox partner, cytochrome c peroxidase, has been probed by $^{15}N/^{1}H$-heteronuclear NMR spectroscopy.[36] Thirty-four amide bonds, corresponding to residues at the front face of the protein, are affected by complex formation with the diamagnetic ferrous cytochrome c; this agrees with the interface amino acids identified by X-ray crystallographic analysis (Figure 2).

However, for the paramagnetic ferric cytochrome c, 56 amide bonds are affected by complex formation. These are generally the same bonds that are affected for the ferrous form, but also include several amide bonds at the rear of the structure as viewed in Figure 2. The authors suggest that this is due either to a different binding mode for the two forms with the enzyme, or a consequence of the greater flexibility of the ferric protein compared with the ferrous analogue.

A combined $^{1}H/^{15}N$-NMR and X-ray study of methylglyoxal synthase (MGS) complexed with its competitive inhibitor phosphoglycolohydroxamic acid

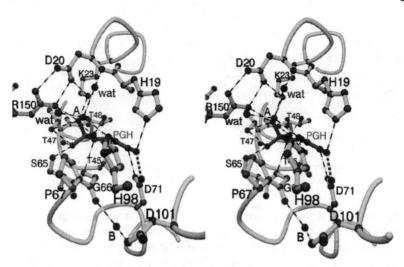

Figure 3 *Methylglyoxal synthase (MGS) complexed with its competitive inhibitor phos-phoglycolohydroxamic acid (PGH), showing water molecules involved in the catalytic mechanism*
(Reproduced with permission from Marks et al.[37])

(PGH) has been reported.[37] The structure established in this way reveals the locations of water molecules that are important in the catalysis mechanism (Figure 3). A point of interest is the fact that PGH is already known to be a competitive inhibitor of triosephosphate isomerase and therefore the new results link these two enzyme systems.

For an application of heteronuclear NMR spectroscopy to assess conformational mobility of localities in proteins, see Ref.133; Solid-state NMR results are discussed in a low temperature study with a similar objective (Ref. 63), and a combination of NMR with other physical measurements (see next Section) is illustrated in Ref. 200.

[129]Xe NMR focussed on an encrypted Xe atom bound to avidin (an egg white protein) has involved the biotin probe shown in Figure 4.[38] This effectively demonstrates the preparation of an NMR-based biosensor that exploits the enhanced signal-to-noise response of laser-polarized xenon leading to spectral simplicity and chemical-shift sensitivity. Detection of specific biomolecules at the level of tens of nanomoles is claimed to be potentially achievable by this approach.

4.4 Structures of Proteins Determined using Physical Methods in Combination with Structural Derivatization. – The role of NMR spectroscopy in protein structure determination continues to gain importance and competes with, and complements, the scope of X-ray crystallography in this respect. Taken with mass spectrometry studies that contribute to this endeavour, also often in a complementary role, the scenario of currently-used methods is nearly fully covered by just these three physical techniques.

However, significant progress continues to depend on other techniques, circu-

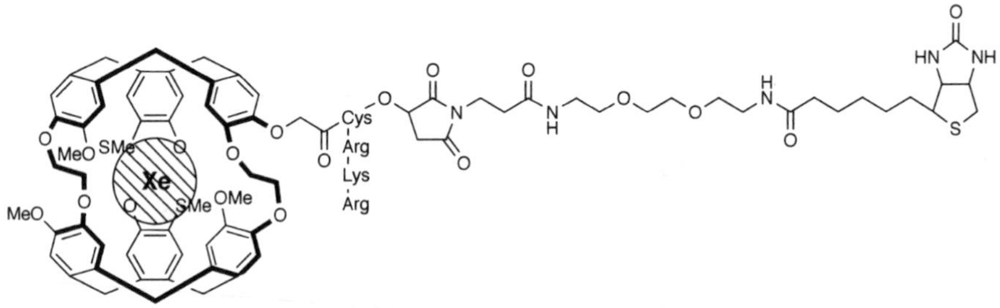

Figure 4   *A cryptophane-A xenon biosensor for ^{129}Xe protein NMR*
(After Ref. 38)

lar dichroism (CD) spectroscopy (Refs. 52, 53, 80, 200) and EPR measurements (Refs. 50, 149) in particular, as well as on routine spectrometric methods (UV – visible; Refs. 71, 149, 160), Fourier-transform IR spectroscopy (Ref. 76), and on various subdivisions of fluorescence techniques (Refs. 68, 71, 79, 96, 107, 194, 200).

4.5   Molecular Modelling. – The role of molecular modelling is becoming more widely accepted as contributing usefully to structure determination and interpretation. It is a relatively routine exercise undertaken with the help of 'off the shelf' packages, and is mentioned in connection with studies throughout this Chapter where it is a subsidiary element of the investigations (Refs. 81, 144, 148, 181, 198).

4.6   Other Structural Features Established for Proteins by Physical Methods: Crosslinks. – There are post-translational modifications to proteins to be sought out by standard methods, including mass spectrometry (Ref. 30), as soon as the overall structure of the native protein has been established.

Discovery of the location of intramolecular crosslinks has often been a demanding enterprise in the past. The determination of other modifications to amino acid side-chain functional groups is generally more routine. This year's literature provides examples of both familiar crosslinks (disulfide bonds) and less familiar, sometimes unique, covalent bonding patterns.

4.6.1   The Disulfide Bond. The determination of the location of disulfide bonds is featured in several current studies, concentrating on the distinction between multiple locations in proteins containing several cysteine residues and the role of this crosslink in determining conformation (Refs. 53, 177).

4.6.2   Other Crosslinks. Lysine-derived pyrrole crosslinks found in collagen some years ago have received attention for the obvious interest in their physiological roles. Uncertainty concerning their precise structures and genesis has lingered for the intervening time, but a detailed mass spectrometry-supported study of model collagen derivatives using a biotinylated Ehrlich's reagent to generate the crosslink (Scheme 1) has put the topic into clearer focus. The outcome of this study supports one of the structures proposed earlier.[39]

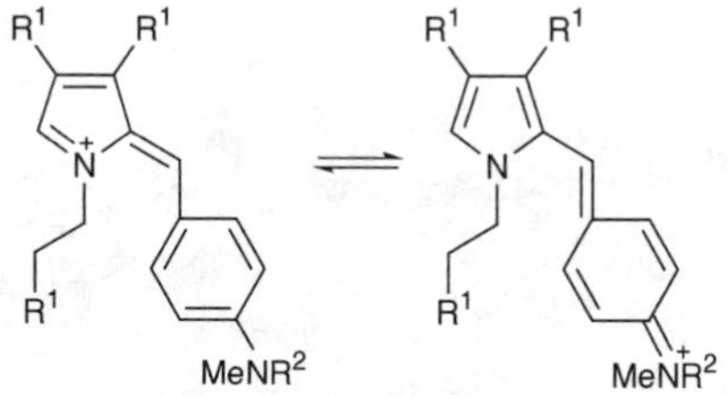

Scheme 1

4.7 X-Ray Crystallographic Studies of Proteins. – Many of the new interpretations of protein behaviour that have been cited in this Chapter, have depended on detailed molecular structures derived by X-ray crystal analysis. X-Ray data underpinning studies described in other sections of this Chapter are covered in many of the references cited.

An overview of the year's literature suggests that classical single X-ray crystal structure determinations of proteins are now less common than corresponding studies of complexes between proteins and their physiological partners. This is a sign of the increasing sophistication of current physical techniques, and the ease of their implementation, although interpretation of the data is less easy in some respects (for example, the mis-identification of alkali metal cations in structures on the assumption that the data revealed water molecules has been discussed in Volume 33, p. 378). It is also a sign of inspiration received by researchers seeking novel problems for X-ray study that can lead to new insights into protein behaviour.

4.7.1 Case Studies: Photosystems I and II. As an indication of the impressive scope of some current studies, the picture that has emerged of multiply-interacting components in photosystems I and II (Figures 5 – 9) can hardly be bettered as an example of the achievements of current physical techniques.[40]

The crystal structure of photosystem I from the thermophilic cyanobacterium *Synechococcus elongatus* reveals 12 protein subunits and 127 cofactors (comprising 96 chlorophylls, 2 phylloquinones, 3 Fe_4S_4 clusters, 22 carotenoids, 4 lipids, a putative Ca^{2+} ion, and 201 water molecules). The structural information provides a basis for understanding how the efficiency of photosystem I in light capturing and electron transfer is achieved.

In Figure 5, the structure of photosystem I trimer obtained at 2.5Å resolution is shown in schematic form. Monomers are labelled I, II, and III, and different structural elements are shown in each of the three monomers. Figure 6 (view parallel to the membrane plane) shows the cofactors of the electron transfer chain (ETC) and of PsaC. Figures 7 – 9 give further relevant details.

Novel modes of chlorophyll binding are revealed in this study, which is also likely to account for the role of the carotenoids in this system. Not least, a better picture of exciton transfer and electron transport in plants and cyanobacteria can now be contemplated.

Work on photosystem II is reported by the same research group, with interpretation of an X-ray analysis at 3.8Å resolution (Figures 10, 11).[41] This descrip-

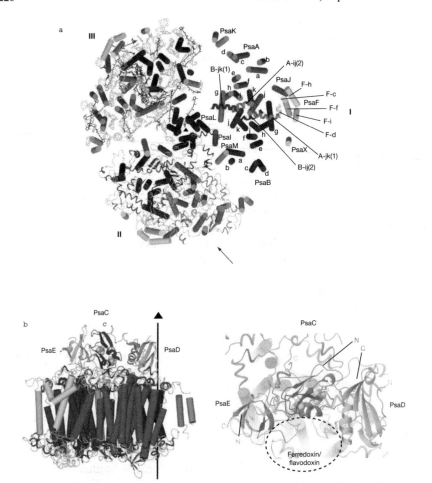

Figure 5   *Structure of photosystem I trimer at 2.5 Å resolution. a, View along the membrane normal from the stromal side. In I, the arrangement of the transmembrane α-helices (cylinders) is shown (six helices in extra-membranous loop regions are drawn as spirals); II shows membrane-intrinsic subunits, and III shows the complete set of cofactors with the transmembrane α-helices (the side chains of the antenna Chla molecules have been omitted). Electron transfer chain (ETC): quinones and chlorophylls are shown in grey-black, and iron and sulfur atoms of the three Fe₄S₄ clusters as dark grey and light grey spheres, respectively. Antenna system: chlorophylls in light grey, carotenoids in black, lipids in medium grey. b, Side view of the arrangement of all proteins in one monomer of photosystem I (shading as in a), including the stromal subunits PsaC (light grey), PsaD (medium grey), PsaE (dark grey) and the Fe₄S₄ clusters. The view direction is indicated by the arrow at monomer II in (a). The vertical line (right) shows the crystallographic C₃ axis. c, View as in (a) showing stromal subunits PsaC, PsaD and PsaE. These cover some of the loop regions and helices of PsaA and PsaB (light grey). The dashed ellipse is the putative docking site of ferredoxin, which covers loops of PsaA*
(Reproduced with permission from Jordan et al.[40] and *Nature*, copyright 2001, Macmillan Magazines Ltd)

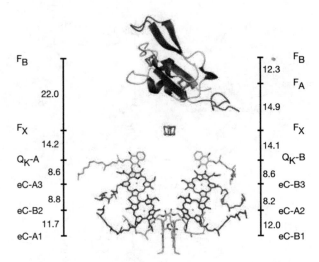

F_B		F_B	
	22.0	12.3	F_A
		14.9	
F_X		F_X	
	14.2	14.1	
Q_K-A		Q_K-B	
	8.6	8.6	
eC-A3		eC-B3	
	8.8	8.2	
eC-B2		eC-A2	
	11.7	12.0	
eC-A1		eC-B1	

Figure 6 *View parallel to the membrane plane showing the electron transfer chain (ETC). The chlorophyll pairs are arranged in two branches A and B. Phylloquinones are labelled QK-A and QK-B and Fe_4S_4 clusters are labelled FX, FA, and FB*
(Reproduced with permission from Jordan et al.[40] and *Nature*, copyright 2001, Macmillan Magazines Ltd)

tion complements the earlier report in a number of respects, giving special attention to the mechanism of water oxidation and the location, for the first time, of the manganese cluster within the complex.

4.7.2 Other Protein Studies. Bacterial MrcB protein assembles into filaments with a subunit repeat similar to that of F-actin, the physiological polymer of eukaryotic actin, part of the cytoskeleton.[42] A point of interest arising from this result lies in discovering something about the distant past; it shows that the actin cytoskeleton had evolved before eukaryotes had developed.

A 'quorum sensing protein' is a constituent of Gram-positive bacteria that shows altered behaviour in response to autoinducer compounds operating in the same cell.[43] The explanation of such behaviour based on molecular structure is clearly an attractive challenge. LuxS is an enzyme of this type, that has been shown to possess a catalytic metal site, possessing structural similarities to the active site of well-studied metalloproteases such as thermolysin (*cf* Ref. 161). The zinc atom in LuxS, located by X-ray crystal analysis, is shown in the structures depicted in Figures 12-14.

The interest in this system lies not only in the role of protein ligands His54, His58, and Cys126 (Figure 12) but also in the remarkably polar active site cavity. This includes an oxidized cysteine residue, Cys84, near the zinc atom. This is a sulfonic acid moiety (or a mixture of sulfonic and sulfinic acids), based on interpretation of the X-ray data. The significance of this for the catalytic mechanism is not clear, but the discovery of oxidized cysteine side-chains in catalytic sites of enzymes is not new (cysteine sulfenic acid is a particularly interesting example, involving a functional group considered to be fleeting and unstable

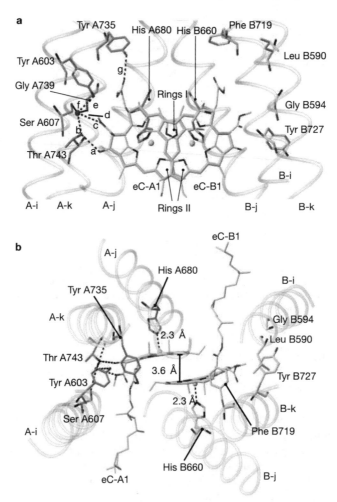

Figure 7   *Local environment of P700 (view perpendicular to the chlorin planes). A, eC-A1 (Chla') and eC-B1(Chla), with surrounding transmembrane α-helices (grey). All possible hydrogen bonds between eC-A1, water (dark grey sphere) and amino acids of PsaA (medium grey) are shown as dashed lines. These residues are strictly conserved between subunits PsaA of many organisms, but none of them is conserved between PsaA and PsaB. The corresponding amino-acid side chains in PsaB (light grey) cannot form hydrogen bonds to eC-B1. His A680 and His B660 coordinating eC-A1and eC-B1, respectively, are also shown. B, As in A, but viewed along the pseudo-C_2 axis*
(Reproduced with permission from Jordan et al.[40] and *Nature*, copyright 2001, Macmillan Magazines Ltd)

until recently).[44] LuxS possesses the Cys-His pair that is found in the active site of papain and some amidotransferases.

The N-terminal two-domain fragment of human CD4 coreceptor (hCD4), which is expressed on the surface of all helper T cells, has been probed through X-ray crystal analysis of the complex of this protein with murine I-A^k class II

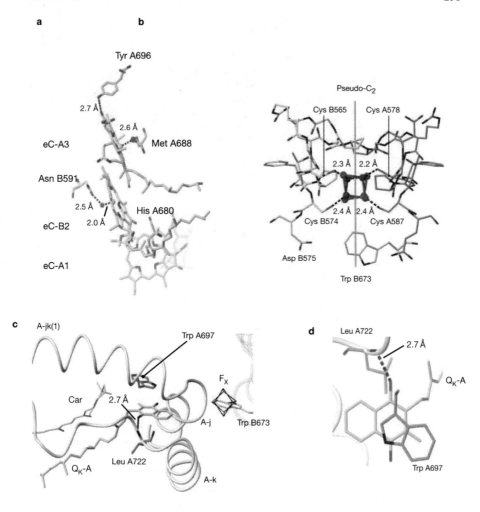

Figure 8 *Local environment of membrane-intrinsic cofactors of the electron transfer chain. (A), Chlorophylls of the A branch of the ETC, with axial ligands and hydrogen bonds involving chlorophylls eC-B2 and eC-A3. In the B branch, eC-A2 and eC-B3 show similar interactions, as these amino acids are conserved between PsaA and PsaB. (B), Loops A-hi of PsaA (medium grey) and B-hi of PsaB (light grey) surround the inter-polypeptide iron sulfur cluster FX, coordinated by cysteine residues A578, A587 and B565, B574. Pseudo-C_2 axis shown as grey vertical line. (C), Phylloquinone QK-A (light grey) in its binding pocket. QK-B bound similarly by PsaB. Main chain of PsaA as ribbon and a carotenoid (car) in grey. Trp A697 stacks with QK-A. Only one hydrogen bond, Leu A722 NH to carbonyl-oxygen 4 of QK-A, is formed. A carotenoid (car) is located in the vicinity of QK-A. (D), View of the quinone binding site normal to the quinone plane*

(Reproduced with permission from Jordan et al.[40] and *Nature*, copyright 2001, Macmillan Magazines Ltd)

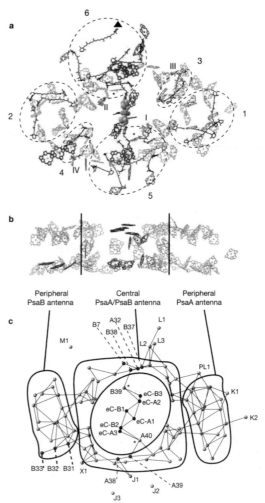

Figure 9   *Spatial organization of the cofactors of the ETC and the antenna system in one monomer of PSI. (A), View from the stromal side onto the membrane plane. The C_3 axis is indicated (triangle). Ring substituents of all chlorophylls omitted for clarity, putative nine excitonically coupled 'dark grey' chlorophylls are dark grey, chlorophylls coordinated by PsaA/PsaB in light grey, those bound to peripheral subunits in the shade of the coordinating subunits (Figure 5), chlorophylls of the ETC are black. Asterisks denote 'connecting chlorophylls' (see text). The 22 carotenoids are arranged in clusters 1 to 6. Four lipids (dark grey) labelled with roman numerals. (B), Side view of PSI antenna rotated $90°$ about the horizontal axis with respect to (A). Two vertical lines indicate separation of the antenna Chlas coordinated by PsaA/B into a central part and two peripheral parts. (C), Schematic view of the chlorophyll positions represented by their central Mg^{2+} ions; view direction as in (A). Chlas belonging to the ETC in dark grey, those of the antenna in light grey and black. Some of the chlorophylls referred to in the text are named by a capital letter indicating the subunits involved in axial liganding and numbered from N to C; PL1 is the single Chla bound to a phospholipid*

(Reproduced with permission from Jordan et al.[40] and *Nature*, copyright 2001, Macmillan Magazines Ltd)

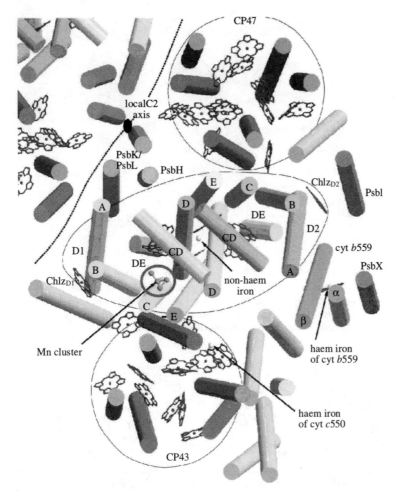

Figure 10   *The Photosystem II complex showing the location of the manganese cluster. The chlorophyll units are represented by porphyrin ring systems, and the Fe atoms as grey spheres. Other aspects can be appreciated by referring to information given with Figure 5*
(Reproduced with permission from Fromme et al.[41])

major histocompatibility complex (pMHCII). Structural information about the CD4 receptor – ligand interaction involved in such events is of prime importance in understanding immune processes. Particularly, development of methods to block CD4 – pMHCII interactions might lead to valuable new classes of immunosuppressive compounds. Results shown in Figures 15 and 16 reveal specific new details of contacts involving CD4.[45]

The structure of a human γδ T-cell antigen receptor (γδTCR) from a T-cell clone has been determined at 3.1Å resolution.[46] The background to this work lies in the need to account for the different behaviour of γδTCRs, αβTCRs, and antibodies that constitute the main types of immune system receptors. The first

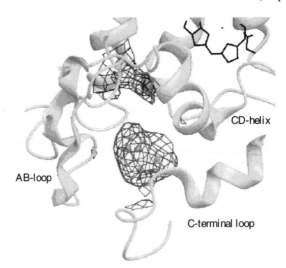

Figure 11   *A region of the Photosystem II complex in the vicinity of the manganese cluster (indicated as an electron density globule near the C-terminal loop), also showing a tyrosine residue used to bind a water molecule, from which it is hypothesized that the manganese cluster extracts a proton*
(Reproduced with permission from Fromme et al.[41])

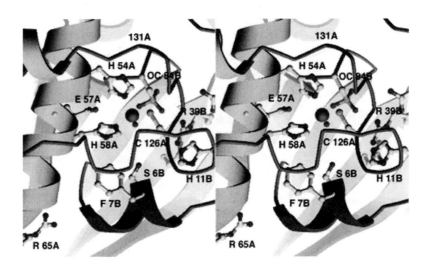

Figure 12   *The zinc-ligand cluster and active site of LuxS, with the helix carrying the HXXE57H54 motif at the left. The chain from Ala118 to Ala132 underpins the zinc-binding site in this view. The N-terminal 3_{10} helix that may control entry to the active site is shown in dark grey*
(Reproduced with permission from Hilgers and Ludwig,[43] copyright 2000, National Academy of Sciences, USA)

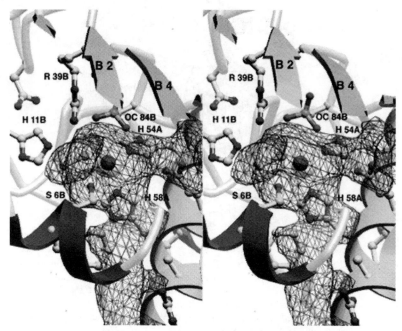

Figure 13 *As in Figure 12 but rotated aproximately 180° about the vertical. This shows better the cavity and the channel to solvent, outlined in the dark grey mesh* (Reproduced with permission from Hilgers and Ludwig,[43] copyright 2000, National Academy of Sciences, USA)

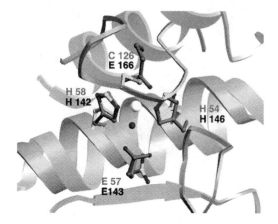

Figure 14 *Comparison of the LuxS binding site (medium grey) with that of thermolysin (lighter grey)* (Reproduced with permission from Hilgers and Ludwig,[43] copyright 2000, National Academy of Sciences, USA)

two types differ in their constitution, the former carrying γ- and δ-polypeptide chains, the latter carrying α- and β-polypeptide chains.

Figures 17-19 show views of the two receptors, and the interest in γδTCRs

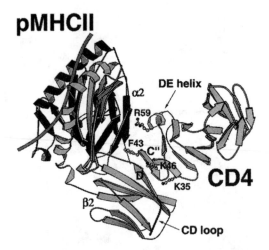

Figure 15   *Ribbon diagram of the CD4-pMHCII complex. Interaction involves the α2
(dark grey) and β2 (light grey) domains of pMHCII and domain 1 of hCD4
through residues Lys35, Phe43, Lys46, and Arg59 on CD4. The CD loop on the
b2 domain has no direct interactions with CD4*
(Reproduced with permission from Wang et al.,[45] copyright 2001, National
Academy of Sciences, USA)

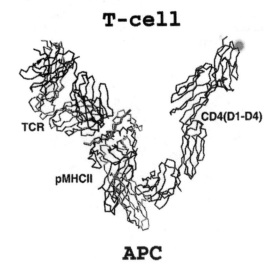

Figure 16   *The ternary CD4-pMHCII-TCR complex has a V-shaped conformation. The
complex is depicted with the T-cell membrane at the top, and the antigen-
presenting cell membrane (APC) at the bottom of the model*
(Reproduced with permission from Wang et al.,[45] copyright 2001, National
Academy of Sciences, USA)

(carried by about 5% of peripheral blood T-cells) lies in their recognition of
intact proteins or non-peptide phosphorylated antigens (in contrast to αβTCRs).
These antigens are rather simple organic compounds; an example is 3-formyl-1-
butylpyrophosphate, produced by *Mycobacterium tuberculosis*, and responsible

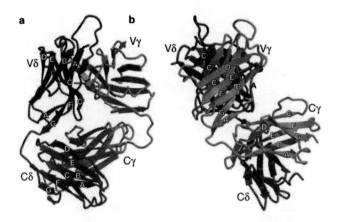

Figure 17   *Views of the G115 Vγ9Vγ2 TCR. (a), View of Vγ and Cδ (black) and Vγ and Cγ (dark grey) domains. CDR loops are at the top. (b), View after a 90° rotation around the vertical axis*
(Reproduced with permission from Allison et al.[46] and *Nature*, copyright 2001, Macmillan Magazines Ltd)

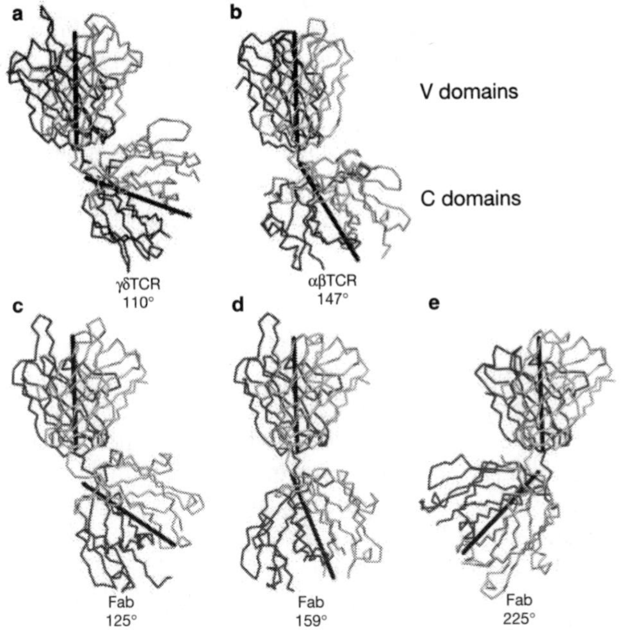

Figure 18   *Overall structure of γδTCR, αβTCR and Fabs. (a), γδTCR G115 (black); (b), αβTCR (dark grey); and (c) – (e), three Fabs (medium grey) aligned using both V domains. Note the atypical position of the γδ C domains. The pseudo two-fold symmetry axes (dyads) within the V and C domains are shown (solid lines). The γ-, β- and H chains and the δ-, α- and L chains are in lighter and darker shades, respectively*
(Reproduced with permission from Allison et al.[46] and *Nature*, copyright 2001, Macmillan Magazines Ltd)

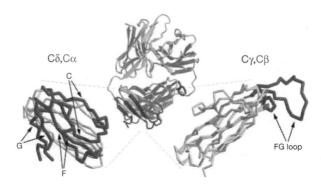

Figure 19   *Comparison of the γδTCR and αβTCR C domains. Superpositions of Cδ with*
Cα (left) and Cγ with Cβ (right) are shown. Differences between the domains are
highlighted; the locations of these differences within the G115 structure are also
shown. Cδ and Cγ are shown in black. Cα and Cβ from a particular receptor are
shown in dark grey. The outer face of Cδ contains β-strands C, F and G, which
form a regular β-sheet. Cγ has a much shorter FG loop
(Reproduced with permission from Allison et al.[46] and *Nature*, copyright
2001, Macmillan Magazines Ltd)

for T-cell activation in the human host.

As a result of this study, a special role for γδTCRs in immunity is suggested.
The fact that the C-domains of the two receptor types are very different (Figures
17-19) while the V-domains are rather similar, indicates a structural basis for the
differing response of the two classes of receptor.

In Figure 17, the nine β-strands of the V domains are labelled A-G, including
A9, C9, and C0. The seven β-strands of the C domains are labelled A-G. In each
domain, strands A, B, E, and D form one β-sheet, and strands G, F and C
(including A9, C9 and C0 in the V domains) form the other β-sheet. The Vγ-Vδ
interface involves the A9GFCC9C0 β-sheets. The Cγ-Cδ interface involves the
ABED β-sheets.

The B7-1/CTLA-4 complex that inhibits human immune responses (Figures
20-23) continues to be the target of structural studies, by X-ray crystallography,[47]
and by details of the CTLA-4/B7-2 binding sites established in parallel work
(Figure 24).[48] These papers give an excellent illustration of current research
activity in the general protein – protein interaction field (covered mainly in
Section 6) with key references to the general accumulated knowledge of in-
teratomic distances and interacting amino acid side-chains. In particular, the
opportunities that are offered through knowledge of T-cell – antigen interactions
offer considerable scope for immunotherapeutic applications. Figure 20 displays
β-sheets involved in the receptor-ligand interaction. Glycosylation sites on
sCTLA-4 (Asn78 and Asn110) and on sB7-1 (Asn19, Asn55, Asn152, Asn173 and
Asn192) are all surface exposed. None of the ordered glycosides is involved in the
receptor-ligand recognition. Two hydrogen bonds formed at the elbow of the
homodimer are indicated as dashed lines. Spatial proximity of the C termini
(Glu120) would enable a formation of disulfide bond between cysteines at
positions 122. Asn110 is the only strictly conserved glycosylation site in CTLA-4

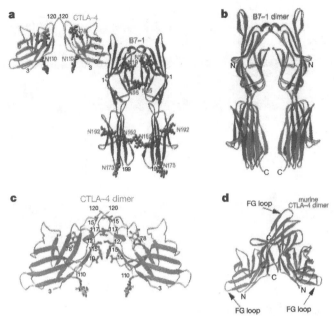

Figure 20   *Structural comparison of sCTLA-24/sB7-1 complex with uncomplexed forms of sCTLA-4 and sB7-1. (a), Ribbon diagram of the sCTLA-4/sB7-1 complex showing two sB7-1 (black) and two sCTLA-4 (medium grey) molecules in the asymmetric unit. Disulfide bonds (medium grey shade) and sugar moieties (light grey) are also shown. (b), Superposition of the ligand-binding V-set domains of complexed (dark ribbons) and uncomplexed (light grey ribbons) forms of sB7-1. (c), Detailed view of the sCTLA-4 homodimer interface. Residues involved in hydrophobic interactions between the two monomers are shown in dark grey. (d), Superposition of human sCTLA-4 dimer (medium grey) on murine sCTLA-4 dimer (black) reveals the contradiction between the two dimerization modes. The FG loop of sCTLA-4 in the complex is displaced by 2.5Å towards sB7-1* (Reproduced with permission from Stamper et al.[47] and *Nature*, copyright 2001, Macmillan Magazines Ltd)

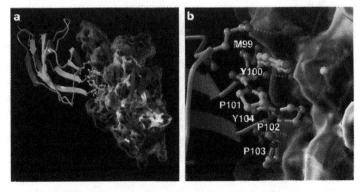

Figure 21   *Receptor – ligand system discussed in Ref. 47: T-cell – antigen interactions* (Reproduced with permission from Stamper et al.[47] and *Nature*, copyright 2001, Macmillan Magazines Ltd)

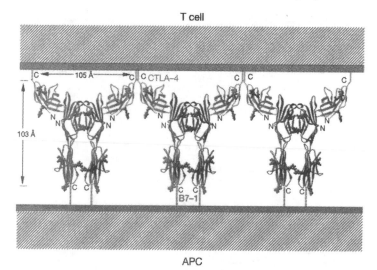

Figure 22   *Crystal arrangement of the antigen – receptor pairs. The complexes are evenly*
spaced along the membrane surfaces with a separation of 105 Å. APC = antigen
presenting cell
(Reproduced with permission from Stamper et al.[47] and *Nature*, copyright
2001, Macmillan Magazines Ltd)

and CD28. Located between the two monomers, N-glycans in these positions
might stabilize the homodimer.

For CTLA-4, the strands are labelled A-E: A (residues 4-8), A9 (residues 11-13),
B (residues 19-26), C (residues 33-42), C9 (residues 45-54), C0 (residues 56-60), D
(residues 67-72), E (residues 75-81), F (residues 90-99) and G (residues 105-109;
112-115). For B7-2, the strands are labelled A (residues 3-7), B (residues 9-11), C
(residues 27-35), C9 (residues 38-45), C0 (residues 47-49), D (residues 60-64), E
(residues 67-73), F (residues 81-91) and G (residues 94-108).

4.8   Structural Information for Proteins from Other Physical Techniques. –
Textbook coverage has been extended recently by the publication of *Protein-*
Ligand Interactions: Structure and Spectroscopy (A Practical Approach).[49]
The redox state dependence of rotamer distributions in tyrosine and the neutral
tyrosyl radical generated in proteins has been investigated by electron paramag-
netic resonance (EPR) techniques applied to ^{2}H-labelled substrates.[50] At 170K,
UV photolysis generates trapped ^{2}H$_4$-tyrosyl side-chains which adopt one of a
range of C1-Cβ and other rotamers (e.g., the amine-gauche Cβ-Cγ rotamer;
Figure 25) in proportions that are influenced by the structural environment of
these residues.

Recombinant and native soybean β-conglycinin β homotrimers (Figure 26)
have been determined by X-ray crystal analysis. Glycinin constitutes nearly half
the seed protein of soybean, and clearly the role of this plant in food presentation
(good gel-forming properties; useful emulsifier) and health terms (hypocholes-
terolemic properties) justifies the extensive literature on this protein.

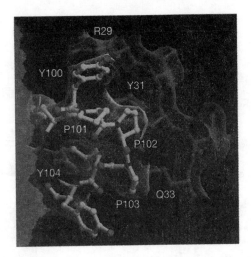

Figure 23 *Electron density map of the receptor-ligand binding site. Shades of grey as in Figures 20, 21*
(Reproduced with permission from Stamper et al.[47] and *Nature*, copyright 2001, Macmillan Magazines Ltd)

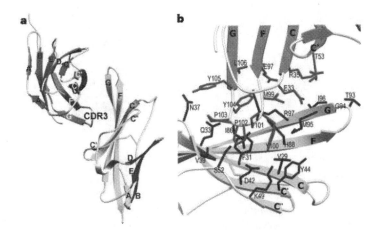

Figure 24 *The CTLA-4/B7-2-binding interface. (a), Ribbon diagram of the CTLA-4/B7-2 monomers that form the binding interface. The A9GFCC9 (front) and ABEDC0 (back) sheets are shaded light grey and black, respectively. The CDR3-like loop of CTLA-4 is labelled. The AGFCC9C0 (front) and BED (back) sheets are shaded black and dark grey, respectively. All inter-sheet disulfide bonds are shown in light grey. (b), Detailed view of the human CTLA-4/B7-2 interface. The interface is formed by residues from the front sheets of CTLA-4 (CDR3 and the C and C9 strands) and a concave surface on B7-2 (the G, F, C, C9 and C0 strands and the CC9, C0D and FG loops)*
(Reproduced with permission from Schwartz et al.[48] and *Nature*, copyright 2001, Macmillan Magazines Ltd)

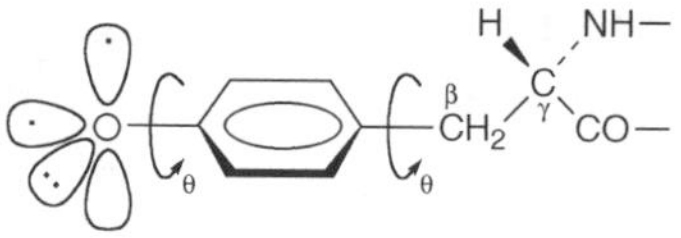

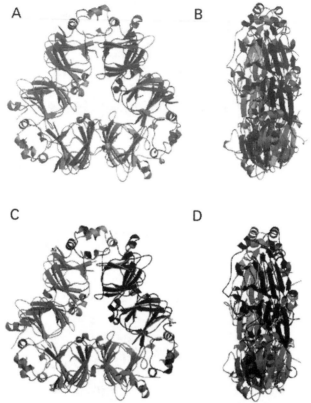

Figure 25   *The amine-gauche Cβ-Cγ rotamer of the side-chain tyrosyl radical*
(After Ref. 50)

Figure 26   *Recombinant (A and B) and native (C and D) β-homotrimers of β-conglycinin.*
The three monomers in the recombinant and native β-homotrimers are coloured
differently. The carbohydrate moieties are shown in light grey as a ball-and-
stick model. A and C are at right angles to the threefold symmetry axis and B
and D are side-on views
(Reproduced with permission from Maruyama et al.,[51] *Eur.J.Biochem.,*
Blackwell Science)

The results reveal structures consistent with those of canavalin and phaseolin,
which are similar proteins within the vicilin class.[51] Thus, it can be concluded that
the presence of N-linked glycans does not have an effect on the structure of the
β-homotrimer, which resembles canavalin rather than phaseolin, and that
canavalin and phaseolin differ the most amongst the members of this class.

4.9 Circular Dichroism Spectroscopy. – Circular dichroism (CD) spectroscopy continues to be exploited for the provision of information on the secondary structure of proteins and polypeptide fragments. The contributions made in this way are often not feasible with any other technique, but the information does not generally bring local structural features of proteins into sharp focus.

Limited proteolysis of horse heart cytochrome c′, and CD spectroscopic data for the resulting fragments, reveal interactions within the intact protein deduced from the ease of proteolysis of the protein and properties of its fragments.[52] [Figure 28 (Ref. 61) shows the structure of cytochrome c′, and Figure 2 shows the structure of the cytochrome c/peroxidase complex].

Bovine α-lactalbumin fragment 53-103 carries two disulfide bonds within the helix that covers residues 86 – 98, released intact by pepsin-catalysed hydrolysis of the protein at low pH. This polypeptide has been shown to undergo 30% aqueous trifluoroethanol-assisted recovery of its original structure, in almost quantitative yield, with correctly-located disulfide bonds, after reduction and re-oxidation.

CD spectroscopy reveals such details by unambiguous interpretation of easily-obtained data. The reduced polypeptide is largely unfolded in aqueous solution, but becomes helical in aqueous trifluoroethanol.[53]

5 Folding and Conformational Studies

Textbook coverage of this topic continues to expand: *Protein Folding in the Cell (Advances in Protein Chemistry, Vol. 59),*[54] *Protein Flexibility and Folding (Biological Modelling Series),*[55] *Protein Stability and Folding: A Collection of Thermodynamic Data,*[56] and *Protein Stability and Folding: A Collection of Thermodynamic Data: Supplement.*[57]

5.1 Background to Protein Folding Studies. – The significance of this topic continues to be more widely appreciated, and examples in this year's literature give good examples of the ways the topic has infiltrated the traditional subdivisions of protein science.

The contortions that a protein molecule is called upon to undergo, for transport through a membrane (Ref. 73) is a newer thought-provoking topic; more mainstream are the studies aiming to reveal the folding pathways available to a protein molecule, on the way from genesis to maturity, essential for the development of the protein's function.

5.1.1 General Considerations and Theoretical Studies. The assignment of secondary structure to a protein often focusses on the identification of hydrogen-bond donor/hydrogen-bond acceptor frameworks. A new predictive approach has been proposed that will assist the process of working from primary structures towards the prediction of three-dimensional proximities of amino acid residues within proteins.[58]

A thoughtful study of the pathways of protein folding stresses the need to start from an accurate description of the denatured state.[59] Although this may seem obvious, there is real scope for building a sequence of steps to assign an incorrect pathway, if starting from a false foundation.

5.1.2   Protein Design. A 'search of sequence space for protein catalysts' is the description used by some of those researching the *de novo* design of catalytic proteins. A successful outcome can only be achieved if folding steps that lead to a uniquely tailored structure are available to the protein under consideration.

The creation of proteins that are tailored to function as enzymes is far from being a routine task. Knowledge of the relationship between sequence, structure and function is incomplete, and the demands of optimized active site dimensions are severe constraints. The accumulation of knowledge, from site-directed mutagenesis in particular, shows that misplacement of catalytic residues by even a few tenths of an angstrom can mean the difference between full activity, and none at all.

The opportunity exists, to create pools of random polypeptides, and select catalysts out of these pools, as has been done for the creation of RNA catalysts for a variety of chemical reactions. A moment's reflection on the nature of this opportunity, given that a 100-residue protein has $20^{100}(1.3 \times 10^{130})$ possible sequences, leads to the conclusion that a way of cutting down the global number of members of such a library would be needed. It is concluded[60] that this sort of narrowing down can be achieved by using basic stuctural knowledge, such as the sequence preferences of helices and sheets, and the tendency of hydrophobic residues to be buried in the protein interior.

Binary patterning of polar and nonpolar amino acids has successfully led to structures that fold as four-helix bundle proteins. In this work, a large library of binary patterned structures, using the AroQ-class chorismate mutase from *Methanococcus jannaschii* (Figure 27) as a model, has been shown to contain a number of different sequences that are compatible with a catalytically active helical bundle fold. The enzyme catalyses the Claisen rearrangement of chorismate to prephenate with rate accelerations greater than 10^6, and is shaped as three intertwined helices that are separated by two loops.

Mixing the two fragments of proteolysed horse heart cytochrome c (Figure 28; the heme contains N-terminal sequence, residues 1 – 56, and the C-terminal fragment, residues 57 – 104), gives a 1:1-complex with a compact conformation, in which the α-helical structure and the native Met80-Fe(III) axial bond are recovered.[61] The fragments themselves are substantially unstructured in solution at neutral pH. This work provides another simple example to add to many reports over the years, illustrating the control by primary sequence on three-dimensional structure.

5.1.3   Mechanics of Protein Folding. The 50th anniversary of the recognition of the right-handed α-helix as a feature of molecular structures has been celebrated with a review of the history of the topic.[62] It was actually drawn by Pauling in the left-handed configuration at first, an arbitrary choice at the time, though this configuration was later discovered to be represented in polypeptides.

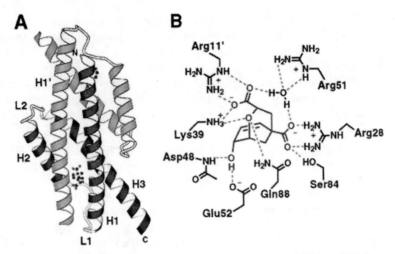

Figure 27   *(A) The AroQ active site carrying a transition state inhibitor. (B) An array of polar active site residues (black) provides extensive hydrogen bonding and electrostatic interactions with the bound inhibitor (dark grey)*
(Reproduced with permission from Taylor et al.,[60] copyright 2001, National Academy of Sciences, USA)

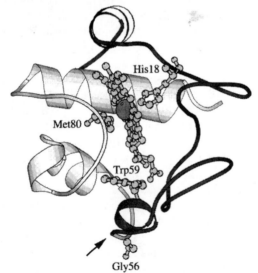

Figure 28   *Fragment of horse cytochrome c with N-terminal fragment shown as dark grey and the C-terminal fragment in light grey. The cleavage site targetted in Ref.61 is shown with an arrow*
(Reproduced with permission from Sinibaldi et al.,[61] *Eur.J.Biochem.*, Blackwell Science)

All proteins undergo a structural change near 200K that has become known as a 'glass transition'. Consistent with this is the fact that below about 220K, enzyme function ceases[63] (but see Vol.33, p.409, for reports of substantial enzymic activity at 173K involving beef liver catalase and calf intestine alkaline phosphatase). The important link between these simple observations, and any explanation that brings in structural factors, must be presumed to be the embedded solvent. A role probably exists for solvent to freeze internal motion within protein structures at low temperatures. Previous NMR studies of ^{2}H-water in lysozyme, ribonuclease, and crambin, showed the existence of mobile – stationary transition at about 180K. Particularly, crambin was an ideal testbed for this work since it possesses a remarkably ordered set of water molecules within its structure.

An amino acid code for protein folding has been developed, in which a native structure is envisaged to generate its own folding pathway. Not only does the amino acid sequence of a protein determine its final folded state, but it also determines the existence of a small set of discrete partially folded intermediates, that between them bear features of the final native state. The conclusions are drawn mainly from hydrogen exchange-based methods, and other kinetic studies that determine the first step to be a free-energy uphill conformational search followed by a folding regime that is free-energy downhill. Of course, this establishes the existence of an initial transition state; the time taken to find this sets the maximum possible overall folding rate.[64]

5.1.4   T- and R-Allosteric States. Tetramethyl orthosilicate sol-gels have been shown to trap T and R allosteric states of pig kidney fructose-1,6-biphosphatase, so reducing the rate of allosteric transition of the enzyme.[65] Further recent work giving attention to allosteric states is covered in Refs. 82, and 152-154.

5.1.5   Misfolding and Unfolding of Proteins. High-pressure unfolding of the 33-KDa protein from the spinach photosystem II particle is promoted by elevated temperatures; added sodium chloride tends to protect the protein from unfolding at elevated pressures.[66] Other recent work has addressed the effect of elevated pressures on protein structure (Refs. 33, 34), and consequences of temperature change on conformation are usually minor details within general protein studies (but see Ref. 177 for unusual effects of site-directed mutagenesis on thermodynamic stability).

Stepwise proteolytic removal of the β-subdomain of α-lactalbumin leaves the protein folded and capable of adopting the molten globule state. This is one of several reports during 2001, of protein studies depending on circular dichroism spectroscopic data, emerging from this research group (see also Refs. 52, 53).[67]

Many proteins adopt a partially folded conformation, termed the molten globule state, under physiological or mildly denaturing conditions. This possibly represents an intermediate state in protein folding generally. This has been a feature of the structure of α-lactalbumin, studied by time-resolved fluorescence anisotropy decay, which gives information on the relative dynamic properties of the three tryptophan residues in this protein. On the basis of fluorescence arising from Trp104 (the signals for Trp60 and Trp118 are significantly quenched by adjacent disulfide bonds) the order of solvent access to these residues is Trp104 > Trp60 > Trp118 in the native state, but Trp60 > Trp104 > Trp118 in the molten globule state (Figure 29).[68]

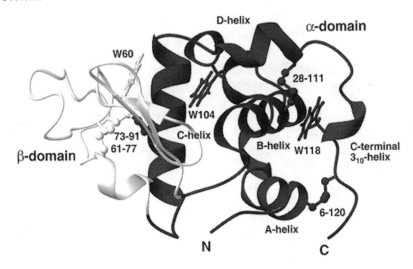

Figure 29 *Human α-lactalbumin showing the location of the three tryptophan residues Trp60, Trp104, and Trp118*
(Reproduced with permission from Chakraborty et al.[68])

An insulin analogue lacking the A7-B7 disulfide linkage has been shown to undergo asymmetric unfolding (the B-subdomain provides a template to guide the folding of the A-chain).[69] The analogue is one of a set prepared by pairwise substitution of the cysteine residues of insulin by serine, the result of which being substantial disordering and loss of biological activity of the protein. The loss of the intrachain A6-A11 disulfide bridge through the replacement of cysteine residues results in a much smaller loss of thermodynamic stability than that which occurs through the loss of the inter-chain disulfide bridge.

5.2 Effects of Metal Complexation on Protein Structure. – Metal ions contribute to the conformational stability of ribonuclease T_1, which was known for many years to show remarkably enhanced stability in salt solutions of high ionic strength. However, the search for an explanation for the contribution of divalent metal ions to the stabilization of this enzyme was not initiated until recently. This property is now shown by X-ray crystal analysis to be the result of unusually efficient binding at Asp49, particularly of Mg^{2+}, Ca^{2+} and Sr^{2+} ions. Figure 30 shows calcium ions located in the vicinity of acidic side-chains (Asp15, Asp29, and Asp49).[70]

Acquisition of Mn^{2+} by apo-fosmycin-resistance protein (FosA) is shown by UV – intrinsic fluorescence spectroscopy, to be a multi-step process.[71] The protein catalyses Mn^{2+}-dependent addition of glutathione to the antibiotic fosmycin, (1R,2S)-epoxypropylphosphonic acid, rendering it inactive. A possible coordination structure is shown in Scheme 2, and other possibilities depicted in the cited paper are obvious variants of this structure.

The cadmium-sequestering metallothionein from *Helix pomatia* (Roman snail) has been concluded to possess the same Cd_3Cys_9 cluster (Scheme 3) as other

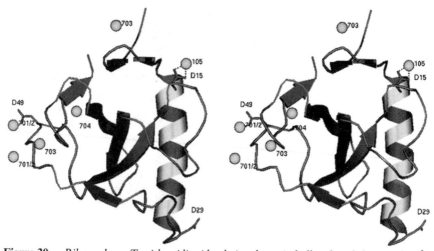

Figure 30   *Ribonuclease T_1 with acidic side-chains shown in ball and stick form, and Ca^{2+} ions as dark grey spheres*
(Reproduced with permission from Deswarte et al.,[70] *Eur.J.Biochem.*, Blackwell Science)

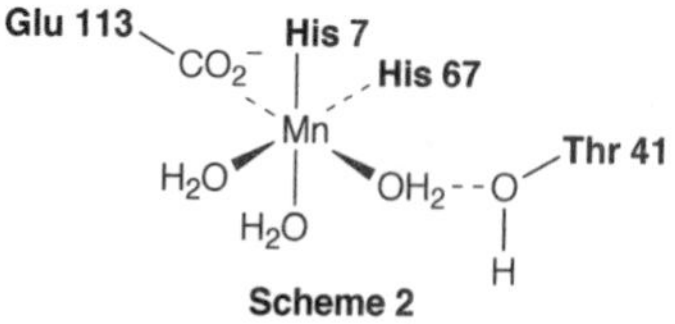

Scheme 2
The FosA – Mn²⁺ complex
(After Ref. 50)

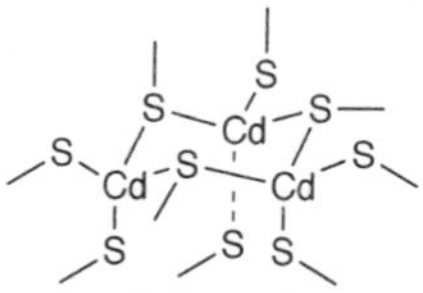

Scheme 3

metallothioneins.[72] Circular dichroism data were crucial in reaching this conclusion.

Results elsewhere in this Chapter deal with metal – protein interactions (Refs. 82, 84, 91, 92, 96, 142, 187) including metal-dependent enzymes (Refs. 168, 187, 195).

5.3   Membrane Proteins and Channel Proteins. – The importance of the cell membrane in controlling the integrity of an organism can hardly be overstated. This control must be exerted in obvious ways, permitting the release through channels of molecules, both large and small, at a crucial stage in each individual case, after a function of the molecule has been completed, or after the synthesis of a functional protein ready for export to a site of action elsewhere. The signalling involved in this process is much less obvious and is a topic of continuing study.

The TolC family of proteins embedded in the outer membrane of *E.coli* are involved in the export of various molecules, ranging from large protein toxins, such as α-hemolysin, to small toxic compounds, such as antibiotics. The TolC proteins therefore play important roles in conferring pathogenic bacteria with both virulence and multidrug resistance.[73]

Studies dealing with membrane proteins are also located elsewhere in this Chapter (Refs. 40, 47, 129, 133, 189).

5.3.1   The Principal Transport Processes Involved in Secretion Through the Cell Membrane. The general schemes by which proteins move out of the cell after biosynthesis, are shown in Figures 31-36. This topic may seem to be only for the specialist's enjoyment, but clearly there are contributions to be made to this topic by those with an understanding of protein structure at the molecular level. Conversely, knowledge of membrane structure and of transport mechanisms is clearly able to suggest topics for study in mainstream protein science.

Figure 31 illustrates 'type I secretion', in which proteins travel directly from the cytoplasm to the extracellular medium. The complexity of this simple-sounding process is clear from the example of α-haemolysin in *E.coli* (A in Figure 31). Transport of this protein is mediated by an inner membrane transport ATPase (HlyB), the outer membrane protein TolC, and a membrane fusion protein HlyD that is anchored to the inner membrane. In 'type II secretion' (B in Figure 31), proteins sharing a particular N-terminal sequence are first translocated across the inner membrane into the peroplasm by the SecA and SecYEG proteins (yellow).

Folding may occur in the periplasm and then the proteins are translocated across the outer membrane. There are 12 to 15 proteins involved in this process, most of which are located in the inner membrane.

In 'type III secretion' (C in Figure 31), activation signals are required to initiate the process, and this requires about 20 proteins to function. Of particular interest, is the conclusion that a protein forms a bridge across the periplasmic space that connects protein secretion channels in the inner and outer membranes.

In 'type IV secretion' (D in Figure 31), at least five membrane-spanning or membrane-associated proteins are involved.

In 'type V secretion' (E in Figure 31), the protein that is secreted organises its own transport. Proteins of this type share a particular N-terminal signal sequence required for transport across the inner membrane (as in 'type II secretion') and a β-domain located at the C-terminus that, upon translocation across the periplasmic space, inserts into the outer membrane and mediates the transport of the pasenger domain.

Figures 31 – 36 show other aspects of the criteria for membrane-crossing mechanisms.

This work[73] proposes new constraints on previously-adopted principles, e.g. that large proteins can only be handled by the 'type I secretion' pathway if the periplasmic component has an α-helical barrel that envelopes the bottom portion of the outer membrane protein TolC as part of the transport sequence.

The newly-discovered protein CaT1 is a constituent of the ion-conducting

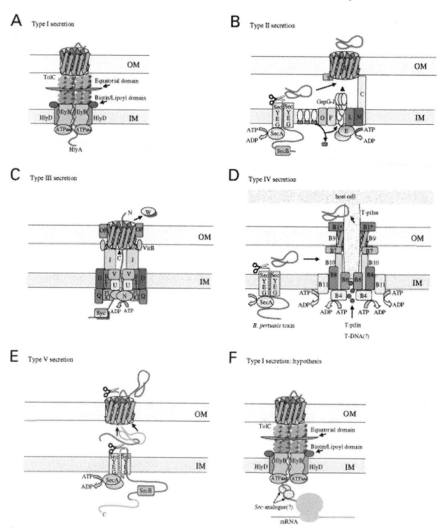

Figure 31   (*A*), *Type I secretion. The individual components are differently shaded. (B),*
Type II secretion. (C), Type III secretion. (D), Type IV secretion. (E), Type
V secretion. (F), A new proposal to account for 'Type I secretion' of large
proteins
(Reproduced with permission from Sharff et al.,[73] *Eur.J.Biochem.*, Blackwell
Science)

pore of a capacitative calcium-entry channel. The structure emerging from this
study may lead to a better understanding of the cellular and molecular mechanism by which this channel is controlled.[74]

The structure of the gating domain of a Ca^{2+}-activated K^+ channel complexed
with Ca^{2+}/calmodulin originating in the rat has been determined. The detail that
has been derived in this work can be appreciated from Figures 37-39, and is fully
delineated in the text of this paper.[75]

The structure provides an improved view of both calcium-dependent and

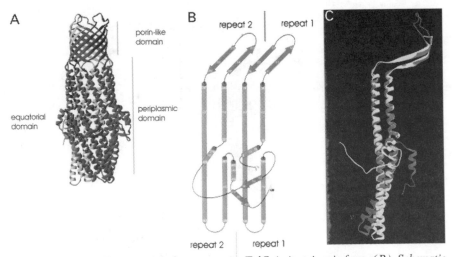

Figure 32 *(A), The outer membrane protein, TolC, in its trimeric form. (B), Schematic depiction of secondary structural elements of the protomer. (C), Ribbon structure showing structural repeats illustrated in (B)*
(Reproduced with permission from Sharff et al.,[73] *Eur.J.Biochem.*, Blackwell Science)

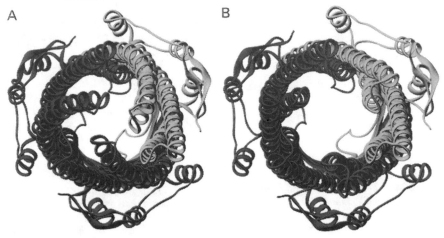

Figure 33 *(A), The TolC channel opening viewed along the three-fold axis from the cell interior; it is sealed in the resting state. (B), The open state of the channel*
(Reproduced with permission from Sharff et al.,[73] *Eur.J.Biochem.*, Blackwell Science)

calcium-independent calmodulin – protein interactions. In the context of this study, the calmodulin-binding domain (CaMBD) is concluded to be crucial in eliciting channel opening and channel closing, through the transmission of conformational changes in CaMBD to the sixth transmembrane helix (S6) of the channel. This conclusion is based on the finding that S6 is immediately adjacent to the calmodulin-binding domain.

Calcium binding to calcineurin B, a calcium- and calmodulin-dependent

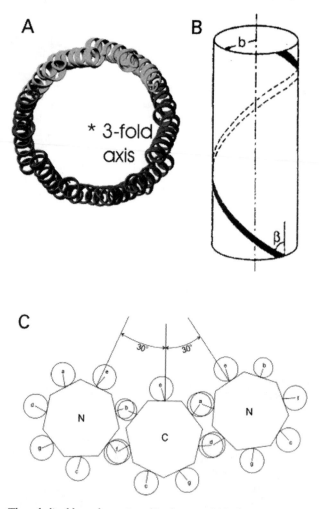

Figure 34   *The α-helical barrel mentioned in the text. (A), the top portion of the periplasmic domain, (B) the path of α-helices embedded in the surface of the barrel. (C) packing of the α-helical barrels*
(Reproduced with permission from Sharff et al.,[73] *Eur.J.Biochem.*, Blackwell Science)

serine/threonine protein phosphatase with a crucial role for T-cell activation in all eukaryotic cells, has been studied in detail.[76] All four calcium binding sites in the enzyme are in communication; what is meant by this rather cryptic statement, is that mutation of glutamic acid residues at these sites into glutamine causes changes in the affinity for calcium ions at the the other binding sites. Fourier-transform IR spectroscopy was crucially helpful in this study.

5.4   Prion Proteins. – Textbook coverage of this topic, e.g., *Prion Proteins (Advances in Protein Chemistry)*,[77] is starting to appear.

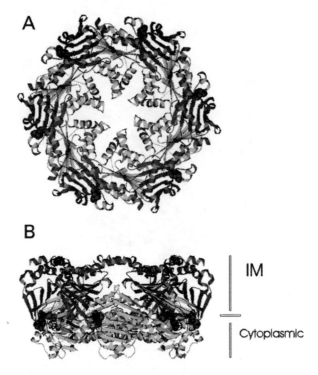

Figure 35   *Crystal structure of the ATPase of the* Helicobacter pylori *type IV secretion system (A) viewed down the sixfold axis, (B) viewed in the perpendicular direction. Bound ADP is shown in grey space-filling representation*
(Reproduced with permission from Sharff et al.,[73] *Eur.J.Biochem.*, Blackwell Science)

Oxidative folding of the fragment of murine prion fragment (residues 23-231) involves formation of a single disulfide bond (Cys179 – Cys214). This is only one of several folding features identified from this study for the reduced form of this protein. The optimum pH for folding in the absence of denaturant was 4 – 5, and folding is almost absent at pH 8, conditions under which disulfide bond formation is not favoured. An important finding is that the prion protein is capable of adopting various stable conformations separated by energy barriers, a unique property of a prion protein that has been predicted for some time, and is now verified.[78]

However, a peptide that has sequence 195-213 of human prion protein (PrP) forms fibrils without passing through a stable intermediate folding stage.[79] This is a region of Prp within which a cluster of mutations is represented in samples of the protein from patients with Geistmann-Straussler-Scheinker disease, and those with the well-publicized familial CJD disease. A study of a small part of a protein, especially this one, does not necessarily give a conclusion that can be extended to the total protein, but this section-by-section approach can be useful in showing the constraints offered by various parts of the molecule. Fluorescence resonance energy transfer methods were shown to be well suited to the monitor-

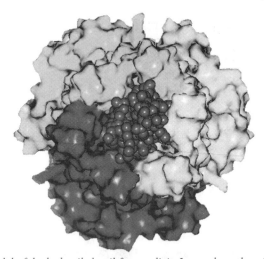

Figure 36   *Model of docked coiled-coil from colicin Ia, to show the width of the channel.*
The size of the channel must force larger proteins to adopt elongated helical
conformations, and to restrict proteins from entering into a globular structure
during the folding process within the periplasmic space
(Reproduced with permission from Sharff et al.,[73] *Eur.J.Biochem.*, Blackwell
Science)

ing of fibrillogenesis of the peptide in this study.

Folding of recombinant mouse prion protein PrP^C has been a test-bed for current investigations in this area. The β-sheet-rich conformation is thermodynamically more stable, but CD spectroscopic and size exclusion chromatographic studies demonstrate its refolding into its native α-helical conformation, a process that is therefore a clear example of kinetic control.[80]

Molecular modelling approaches that design a series of mutations in all regions of the β-amyloid protein (Aβ) have proved useful. It has been assumed, and supported in this study, that the C-domain (residues 29-40), the median region (residues 17-22) and the N-domain (residues 1-16) are all crucial as origins of Aβ - Aβ interactions in the initial stages leading to fibril formation. This study's conclusion, in other words, is that sequence specific Aβ - Aβ interactions are the initiating steps in β-amyloid peptide nucleation.[81]

Binding of copper and zinc by the β-amyloid peptide leads to an allosterically-ordered membrane-penetrating structure containing superoxide dismutase-like sub-units.[82] Thus, the general picture displaying the properties of this peptide, currently at the centre of intensive research, is filling with detail. Clearly, some of this detail is found to be quite unexpected, when considered against prior assumptions.

A novel β-amyloid peptide-binding protein (BBP) that contains a G-protein-coupling module adds to the two other proteins containing this conserved structure.[83] The significance of this is the induction of caspase-dependent vulnerability to β-amyloid peptide toxicity and therefore, BBP is a target of the neurotoxic β-amyloid peptide. Perhaps this is an important insight into the molecular pathophysiology of Alzheimer's disease.

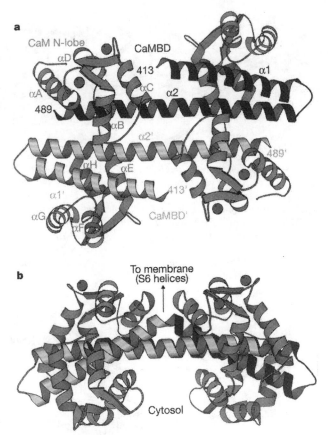

Figure 37 *(a) Ribbon diagram of the dimeric calmodulin-binding domain (CaMBD) complexed with Ca^{2+} and calmodulin (CaM). CaMBD subunits are in grey-black and light grey, CaM molecules are in dark grey, and the Ca^{2+} ions are black. (b) View as in (a) but rotated through 90°, showing the orientation of the complex relative to the membrane*
(Reproduced with permission from Schumacher et al.[75] and *Nature*, copyright 2001, Macmillan Magazines Ltd)

5.4.1 Prion Proteins in Alzheimer's Disease. A role for amyloid precursor protein and presenilins 1 and 2 in Alzheimer's disease continues to be sought. A number of current studies give an indication of good progress that is being made in this area.

Calsenilin is a member of the recoverin family of neuronal calcium-binding proteins, previously shown to interact with presenilin 1 (PS1) and presenilin 2 (PS2) holoproteins. The expression of calsenilin can regulate the levels of a proteolysis product of PS2 and reverse the presenilin-mediated enhancement of calcium signalling.[84] Calsenilin is now shown to have the ability to interact with endogenous 25-kDa C-terminal fragment (CTF) that is a product of regulated endoproteolytic cleavage of PS2, and that the presence of the N141I PS2 mutation does not significantly alter the interaction of calsenilin with PS2.

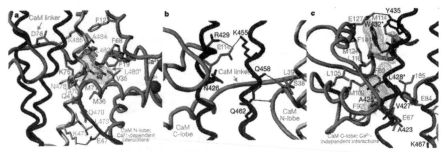

Figure 38   (a) Ca^{2+} dependent interactions between CaMBD and the CaM N-lobe (shades as in Figure 37). Hydrogen bonding between the constituents is shown as medium thickness lines. (b) Contacts between the CaM N-lobe and the second CaMBD dimer (dark grey). (c) Ca^{2+} independent interactions between the CaMBD and the CaM N-lobe. The van der Waals surface (medium grey) is shown for the three distinct prongs, Ala425, Leu428, and Trp432, that interact with this lobe

(Reproduced with permission from Schumacher et al.[75] and *Nature*, copyright 2001, Macmillan Magazines Ltd)

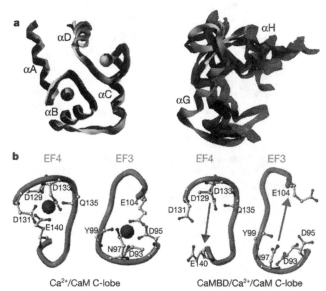

Figure 39   The CaM lobes on the in the CaMBD/Ca^{2+}/CaM complex. Left in (a): Cα superimposition of N-lobe residues 10-67 of the CaMBD/Ca^{2+}/CaM complex (dark grey) on to those of the Ca^{2+}/CaM complex (light grey). Ca^{2+} ions are shown as spheres. Right in (a): Cα superimposition of C-lobe residues 81-146 of the CaMBD/Ca^{2+}/CaM complex (dark grey) and the apoCAM C-lobe structure (medium grey) on to that of the Ca^{2+}/CaM complex (light grey). (b) Comparison of the C-lobe EF hand regions of the Ca^{2+}/CaM complex and the CaMBD/Ca^{2+}/CaM complex. Ca^{2+}-coordinating residues are shown as sticks. Residues Asp95, Asp131, and Glu104 are displaced out of the Ca^{2+}-coordination sphere of the CaMBD complex

(Reproduced with permission from Schumacher et al.[75] and *Nature*, copyright 2001, Macmillan Magazines Ltd)

When the 25-kDa PS2 CTF and the 20-kDa PS2 CTF are both present, calsenilin preferentially interacts with the 20-kDa CTF.

Since increases in amounts of the 20-kDa fragment are associated with the presence of familial Alzheimer's disease-associated mutations, a start is being made in this type of study, towards a better understanding of the problems presented by this condition. However, the finding that the production of the 20-kDa fragment is regulated by the phosphorylation of PS2[85] suggests that it is a regulated physiological event that also occurs in the absence of the familial Alzheimer's disease-associated mutations in PS2. The results give a good indication of points at which the pathways to development of Alzheimer's disease might be blocked.

Calsenilin is a substrate for caspase-3, and site-directed mutagenesis has been used to locate the caspase-3 cleavage site next to the calcium binding domain of calsenilin.[86]

Further results have been reported on the α-helix – β-sheet transition PrPc – PrPsc of recombinant prion protein, showing that partially-unfolded intermediates may be involved in the early stage in which the monomer, PrPc, changes towards the oligomeric scrapie-like form.[87]

Interactions of viral nucleic acids with ovine and human PrP, and with an HIV-1-encoded nucleocapsid protein, show that the two proteins have chaperone properties with respect to annealing of complementary nucleic acid strands. PrP's natural role has still to be identified, but the results of this study suggest that PrP is a multifunctional protein that is a possible participant in nucleic acid metabolism.[88]

5.5 Rare Folding Motifs within Proteins. This sub-heading was introduced in a preceding Volume to cover unusual structural motifs within folded proteins. In essence, what was unusual even relatively recently, has become more common. Such things as four-helix bundles (Refs.60, 122) and unusual turns (Ref. 43) can be mentioned without further comment since they are more familiar now, and discussion in this Chapter does not dwell on these characteristics.

6 Adhesion and Binding Studies

6.1 Textbooks and Monographs. – Binding and complex formation involving proteins has a long series of supporting textbooks going back over the years, and is now supplemented by *RNA Binding Proteins: New Concepts in Gene Regulation (Endocrine Updates)*.[89]

6.2 General Results. – *In vivo* protein – protein interaction assays that follow the general 'yeast two-hybrid assay' of Fields and Song (1989) have been significantly improved in technical aspects, and the modified versions are now used in numerous laboratories. Interacting proteins are listed at http://portal.curagen.com, and this site can also be searched for DNA- and RNA-protein interactions and small molecule – protein interactions (including chemical inducers of dimerization).[90]

Metal-binding properties of the Cys-Cys-Cys-His motif of 50S ribosomal protein L36 from *Thermus thermophilus* have been studied. This is an otherwise rare sequence, present in this protein, consisting of three cysteine residues and a histidine residue, separated by other amino acid residues, $C(Xaa)_2C(Xaa)_{12}X(Xaa)_4H$. It is a zinc finger motif which is known to be crucial, in other zinc finger proteins, for RNA hairpin cleavage. A core 26-mer peptide containing this 22-residue peptide motif was synthesised, and found to bind metal ions with decreasing potency Co(II) > Hg(II) > Zn(II).[91]

A mini-review of zinc metallohydrolases related to Class B β-lactamase has appeared.[92]

6.3   New Results from Binding Studies. – Studies under this heading can be based on quite simple resources, and give relatively limited but still useful factual results; explanations for the results often call for a little more experimental rigour. Thus, proline-rich proteins (histatins) in saliva have a strong affinity for tannins in particular, and polyphenols in general, and some link in the *in vivo* delivery of antioxidants to sites of action could be postulated as an outcome of such a result.[93]

Binding of the peptide Glu-Asp-Thr-Arg-Leu to the PDZ1 domain of the sodium ion – proton exchanger regulatory factor (NHERF) provides insight into interactions involving the cystic fibrosis transmembrane conductance regulator.[94] The peptide is the carboxy-terminal sequence of this regulator, which plays a central role in the cellular localization and physiological regulation of the chloride channel. This study reports the crystal structure of human NHERF PDZ1 bound to Glu-Asp-Thr-Arg-Leu and reveals the specificity and affinity determinants of the interaction, underlining the significance of C-terminal leucine as a recognition site.

Whereas this section would be expected to deal primarily with interactions of *in vivo* importance, affinity chromatography isolation and purification of enzymes also falls into this general category. An interesting development has been described in which knowledge derived from X-ray crystallography, NMR and homology structures has been combined with combinatorial chemical synthesis for the rational design of affinity ligands. The products were attached to agarose and shown to be superior affinity chromatography media for a target enzyme, *Pseudomonas fluorescens* β-galactose dehydrogenase.[95]

The calcium-dependent calmodulin-binding domain (Ser76-Ser92) of the 135kDa human protein 4.1 isoform has been characterized earlier by various techniques (fluorescence spectroscopy and comparison with non-phosphorylated or serine-phosphorylated peptides synthesised in the laboratory). It is now shown that phosphorylation of two serine residues within this 17-residue peptide alters the ability of the peptide to adopt a helical conformation in a position-dependent manner (Figure 40).[96]

6.4   Protein – Protein Interactions Involving Chaperones. – The topic is supported by a recent textbook: *Molecular Chaperones in the Cell (Frontiers in*

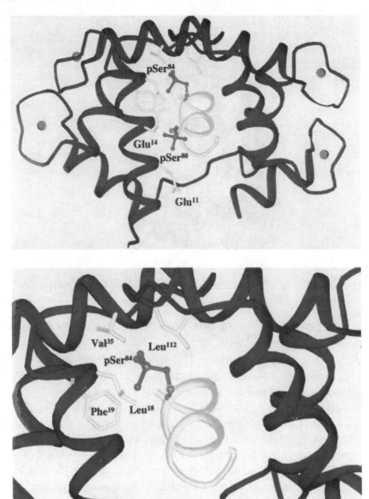

Figure 40   *The complex of the CaM binding domain of CaM kinase II and CaM used to model the possible binding mode of the P4.1 peptides to CaM. The backbone ribbons of the peptide (medium grey) and of CaM (black) are shown, and the two phosphoserine residues are shown as ball and stick models. The upper figure shows the orientation of the peptide in the complex and CaM residues (medium grey) in proximity to the phosphoserine residues. The lower figure shows the side-chain of phosphoserine84 pointing into the hydrophobic pocket formed by Leu18, Phe19, Val35, and Leu112*
(Reproduced with permission from Vetter and Leclerc,[96] *Eur.J.Biochem.*, Blackwell Science)

Molecular Biology).[97] Several papers mentioned elsewhere in this Chapter (Refs. 88, 99) report new chaperone research.

Transforming growth factor-β receptor-associated protein 1 is the first chaperone of Smad4 (a MAD-related protein) and part of a widely-important signal transduction pathway, to be identified. Its role is to bring the Smad 4 protein into

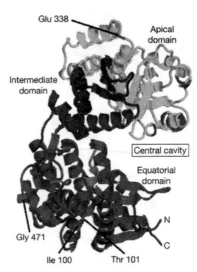

Figure 41   *A subunit of a 14-mer of the protein GroEL from* E.coli, *showing Glu338, known to be crucial in GroEL toxicity, and the residues Ile100, Thr101, and Gly471, also involved in the toxicity*
(Reproduced with permission from Yoshida et al.[99] and *Nature*, copyright 2001, Macmillan Magazines Ltd)

the vicinity of the receptor complex. This work, and many similar studies from other laboratories, graphically demonstrates the social heirarchy that exists in the world of interacting proteins assisted by chaperones.[98]

The paralysing toxin produced by bacterial endosymbionts in the saliva of antlions (larvae of the *Myrmeleontidae*) is a homologue of the protein GroEL, previously discovered to be a protective heat shock protein chaperone.[99] The amino acid residues that are critical for the toxicity of this protein are located away from the regions essential to its protein-folding activity (Figure 41), indicating that the dual function of this GroEL homologue may benefit both the antlion and the symbiotic bacteria. GroEL was originally of interest as an *E.coli* mutant that inhibits bacteriophage growth, and the example of a chaperone that is also a toxin will stimulate studies that will surely lead to the discovery of proteins categorised as polyfunctional.

The p35 protein is an effective broad spectrum inhibitor acting against all three groups of caspases from mammals and other metazoans, fulfilling a role in rescuing cells from apoptosis so as to revive their normal cellular functions. A slow-binding inhibition is characteristic of this system, now explained by rate-determining structural transitions occurring after covalent bonding between the catalytic residue Cys360 of caspase-8, and Asp87 of the bound protein (Figures 42, 43).[100]

X-Linked inhibitor-of-apoptosis protein (XIAP) interacts with caspase-9, and Smac (*alias* DIABLO) relieves this inhibition.[101] XIAP associates *via* the active caspase-9 – Apaf-1 holoenzyme complex through binding with the N-terminus of the linker peptide on the small subunit of caspase-9, which becomes exposed after proteolysis of procaspase-9 at Asp315.

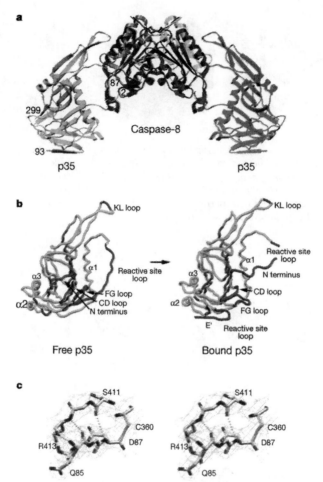

Figure 42 *(a) The dimeric p35/caspase-8 complex with the two-fold axis in the vertical orientation. p35 is medium grey and dark grey, α-subunit of caspase-is medium grey and dark grey, β-subunit is medium grey and light grey. (b) Conformational transition of p35 resulting from cleavage, Larger shifts are shown as darker regions: N-terminal residues 2-12, the CD loop (residues 35-40) the caspase recognition sequence (residues 85-87), the reactive site loop after the cleavage site (residues 93-101), the FG loop (residues 157-165), and the KL loop (residues 254-255). (c) The complex near the active site of caspase-8 overlaid with an electron density map, with hydrogen bonds indicated by dotted lines. Side chains of Met86 of p35 and of Tyr412 of caspase-8 are omitted for clarity*
(Reproduced with permission from Xu et al.[100] and *Nature*, copyright 2001, Macmillan Magazines Ltd)

Hydrogen peroxide is known to both induce and inhibit apoptosis; a curious pattern of behaviour for this familiar compound, common in laboratory, household and industrial environments. These properties have now been traced to effects exerted on recombinant caspase-3 and caspase-8, effects that can be

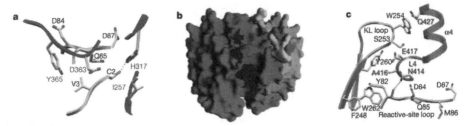

Figure 43   *(a) Interaction at N-terminus of p35 with the active site of caspase-8 (p35, light grey; α-subunit of caspase-8, dark grey). (b) p35 on the surface of caspase-8 (grey and light grey for each of the αβ-units). (c) Interactions near the KL loop of p35 (p35, light grey; β-subunit of caspase-8, dark grey)*
(Reproduced with permission from Xu et al.[100] and *Nature*, copyright 2001, Macmillan Magazines Ltd)

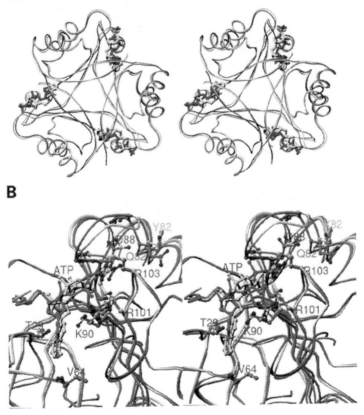

Figure 44   *Comparison of P_{II}/ATP (form II) complex (light grey/medium grey) with GlnK/ATP complex (off-white/light grey). (A) Superimposition of trimer with three ATP molecules. (B) ATP Binding site showing side-chains of residues that interact with ATP*
(Reproduced with permission from Xu et al.,[103] *Eur.J.Biochem.*, Blackwell Science)

prevented and reversed by dithiothreitol, while glutathione has little protective effect.[102]

6.5   Proteins Complexed with Non-protein Species. – The P_{II}-ATP complex, involving a signal transduction protein that is part of the cellular machinery used by many bacteria to regulate the activity of glutamine synthetase (GlnK) and the transcription of its gene, has been crystallized in two forms, and each has been subjected to X-ray crystal analysis. Surprising differences in secondary structure were seen in these structures in comparison with that of GlnK, a functional homologue (Figure 44).[103]

Imidazole glycerol phosphate synthase is one of the glutamine amidotransferases that link histidine and purine biosynthesis in bacteria (*Thermatoga maritima* in this study).[104] It is a two-compartment system with the glutaminase (HisH) producing ammonia at the active site, which is used by synthase subunit HisF and N'-[(5'-phosphoribulosyl)formimino]-5-aminoimidazole-4-carboxamide-ribonucleotide to yield imidazolylglycerol phosphate and 5-aminoimidazole-4-carboxamide ribotide. A conventional reaction kinetics study of the pure system including site-directed mutagenesis revealed Asp11 to be essential for the synthase component and Asp130 also crucial (but replacement by glutamate was permitted). General acid/base catalysis appears to account for the catalytic mechanism.

The Msx-1 homeodomain, a sequence of 60 amino acids, forms a complex with DNA that is revealed through X-ray crystallographic analysis to bind to DNA adjacent to the core TAAT sequence through water-mediated interactions at Gln50 (Figures 45 – 47).[105] The protein is crucial in human craniofacial development, and growth of limbs and nervous system.

A 2.5Å X-ray structure of the abundant nuclear protein Ku bound to DNA gives some insight into the role of the protein in repair of DNA double-strand breaks. A core set of proteins mediates this repair through bringing about non-homologous end-joining operations.[106]

Interaction of human adenovirus proteinase (AVP) and DNAs was monitored by changes in enzyme activity or by fluorescence anisotropy. Although other proteinases have been shown to bind to DNA, the stimulation of proteinase activity by DNA discovered in this study is unprecedented.[107]

Actin can act as a cofactor for human adenovirus proteinase (AVP) when it acts to cleave the cytokeratin network of viruses, as well as acting as substrate to the enzyme leading to its own destruction.[108] Interaction of AVP and AVP-DNA complexes with the undecapeptide cofactor pVIc (Gly-Val-Glu-Ser-Leu-Lys-Arg-Arg-Arg-Cys-Phe) has been characterised, and a contribution of individual pVIc side chains in the binding and stimulation of AVP has been discerned (Figures 48, 49). Binding does not seem to alter greatly the secondary structure of the enzyme, as judged from circular dichroism (CD) data.[109] The merit in this study from the point of view of clinical applications, lies in inhibition of infectious virus when added to Hep-2 cells infected with adenovirus serotype 5. This is presumably the consequence of prematurely activating the proteinase so that it cleaves virion precursor proteins before virion assembly, thereby aborting the infection.

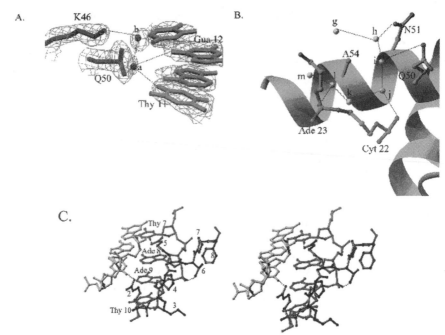

Figure 45   *(A) Gln50 – water – DNA interaction. (B) Water surrounding Ala54 and filling the cavity between the protein and the DNA backbone in this region. (C) Stereoview of the N-terminal sequence of Msx-1 with hydrogen bonding shown (dotted lines)*
(Reproduced with permission from Hovde et al.[105])

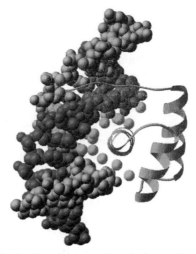

Figure 46   *Water (dark spheres) surrounding the homeodomain-DNA complex*
(Reproduced with permission from Hovde et al.[105])

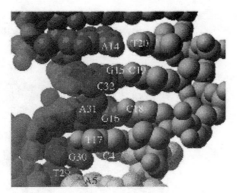

Figure 47   *(A) Another view of the homeodomain-DNA complex*
(Reproduced with permission from Hovde et al.[105])

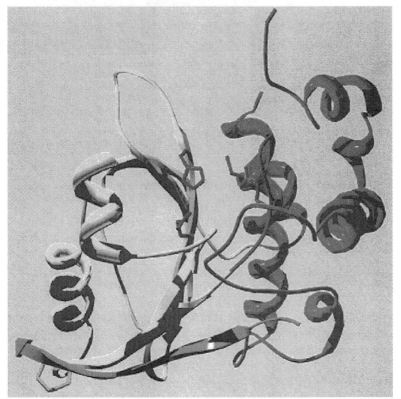

Figure 48   *The two domains of AVP are light grey and dark grey, and pVIc is coloured*
medium grey. The catalytic groupings in the active site are shown in stick form.
A disulfide bond, connecting AVP (Cys104) and pVIc (Cys10), is present in
this picture
(Reproduced with permission from Baniecki et al.[109])

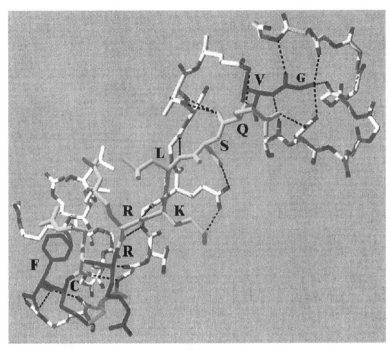

Figure 49   *The complex of AVP with pVIc shown in molecular structural form for the region of AVP that binds to the peptide. The AVP backbone is shown: carbon atoms are white, nitrogen atoms are medium grey, oxygen atoms dark grey. The peptide pVIc is shown in dark grey and light grey*
(Reproduced with permission from Baniecki et al.[109])

Two cysteine residues Cys104 and Cys10 are present in the 11-amino acid peptide activator pVIc but formation of a disulfide bond between these residues is not necessary for maximal stimulation of enzyme activity.[110] This conclusion was based on a kinetics study focussing on the effects of formation of disulfide bonds involving pVIc on its properties as enzyme cofactor.

A complex formed between *Burkholderia cepacia* lipase (see also Section 7.2.21) and a transition state analogue of R-(+)-1-phenoxy-2-acetoxybutane has been subjected to X-ray crystal analysis, to explain the high preference of the enzyme for acylation leading to the particular stereoisomer (Figures 50 – 52). The boomerang-shaped cleft of the enzyme accommodates well the transition state analogue, (R_P,S_P)-O-(2R)-(1-phenoxybut-2-yl)-methyphosphonic acid chloride, and molecular modelling data are in good agreement with the topography of the active site that is revealed in the X-ray data.[111]

Binding interactions between barley thaumatin-like proteins and (1,3)-β-D-glucans have been studied. These are PR5 proteins, a group within the larger family of pathogen resistant proteins (including osmotin), that may bind to nascent (1,3)-β-D-glucans during fungal cell wall synthesis, so explaining the antifungal activity of these proteins (Figure 53).[112]

A glia-derived soluble acetylcholine-binding protein (AChBP) has been de-

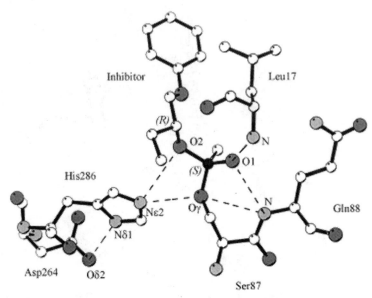

Figure 50 Burkholderia cepacia *lipase with* (R_P, S_P)-*O-(2R)-(1-phenoxybut-2-yl)-methyphosphonate at the active site. Hydrogen bonds that are shown, are those needed for productive binding*
(Reproduced with permission from Luic et al.,[111] *Eur.J.Biochem.,* Blackwell Science)

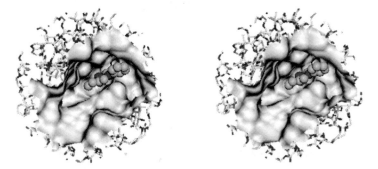

Figure 51 Burkholderia cepacia *lipase with the inhibitor in the boomerang-shaped cleft of the protein*
(Reproduced with permission from Luic et al.,[111] *Eur.J.Biochem.,* Blackwell Science)

scribed, a naturally occurring analogue of the ligand-binding domains of nicotinic acetylcholine receptors. Unlike the nAChRs, it lacks the domains to form a transmission ion channel.[113] This work allows a mechanism to be described by which glial cells release AChBP in the synaptic cleft, and to explain how the cells modulate synaptic cholinergic transmission.

X-Ray crystal analysis of the protein AchBP at 2.7Å gives important structural information that will assist drug design aimed at the ligand-gated ion channels (LGIC) superfamily. This is the area that is perhaps becoming well-known to the

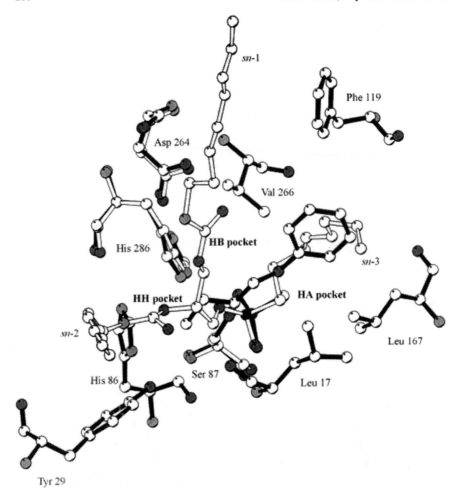

Figure 52   *Superimposition of the triacylglycerol-like, and secondary alcohol-like, inhibi-tors in the active site of* Burkholderia cepacia *lipase*
(Reproduced with permission from Luic et al.,[111] *Eur.J.Biochem.*, Blackwell Science)

general public, including GABA receptors, serotonin (5-HT3) receptors, and glycine receptors. Understanding the ligand-binding characteristics of these ion channels could lead to new antiemetics aimed at the 5HT3 receptors and mood-defining drugs that target the GABA receptors.

Figures 54 – 57 give the results of the study in graphic terms. An immediate benefit flows from the fact that the structures revealed are representative of the N-terminal domain of an α-subunit of nAChRs, which are of central interest in drug design, especially in the context of Alzheimer's disease, epilepsy, and addiction to tobacco smoking.[114]

Three-dimensional structures of avian H5-subtype and swine H9-subtype influenza virus hemagglutinins bound to avian and human receptor analogues

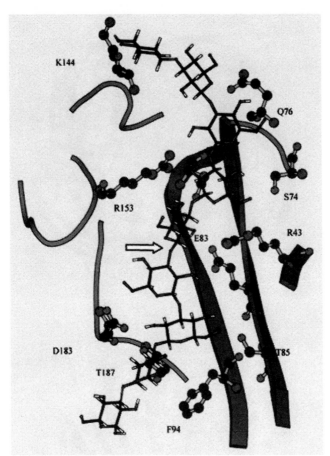

Figure 53   *Amino acid residues in the cleft of HvPR5c that could interact with the docked (1,3)-β-D-glucan. Ser74, Glu83, Phe94, and Asp101 are part of the β-barrel, and the other amino acids are in loops of region II. The arrow shows the distorted glycosidic linkage*
(Reproduced with permission from Osmond et al.,[112] *Eur.J.Biochem.*, Blackwell Science)

have been studied.[115] These were obtained from viruses that were closely related to those that caused outbreaks of human influenza in Hong Kong in 1997 and 1999. The interest arises from the explanation of the origins of pandemics, as a change of initial preference of avian viruses for sialic acid receptors in α2,3-linkage to a preference for human viruses with α2,6 linkages.

The four new hemagglutinin – receptor structures developed in this X-ray study show that the binding sites specific to human receptors are wider than those preferring avian receptors (Figure 58). Also, the configurations and other structural details at the glycosidic link, differ for the two species, connected with the role of Gln226, an avian-specific residue. Contacts with Leu226, a human-specific residue, are facilitated by a cis-configured glycosidic link.

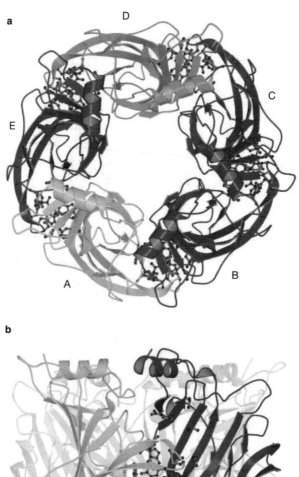

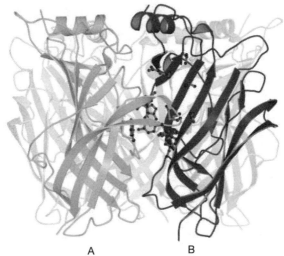

Figure 54   *The pentameric structure of AchBP. (a) Each protomer has a different colour. Subunits are labelled anti-clockwise, with A-B, B-C, C-D, D-E, and E-A forming the plus and minus interface side, with the principal and complementary ligand-binding sites, respectively (ball-and-stick representation). (b) View of the AchBP pentamer perpendicular to the five-fold axis. The equatorially located ligand-binding site (ball-and-stick representation) is highlighted only in the A (light grey)-B (dark grey) interface*
(Reproduced with permission from Brejc et al.[114] and *Nature*, copyright 2001, Macmillan Magazines Ltd)

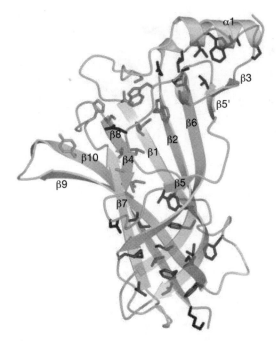

Figure 55 *Conservation in the pentameric LGIC superfamily; very few conserved residues are at the surface. Residues conserved between pentameric LGICs but not AchBP are indicated by a light grey main chain*
(Reproduced with permission from Brejc et al.[114] and *Nature*, copyright 2001, Macmillan Magazines Ltd)

7 Enzyme Studies

7.1 Textbooks and Monographs; Reviews. – This topic is well-served by secondary literature and continues to be supplemented with new material: *From Protein Folding to New Enzymes*,[116] and *Proteolytic Enzymes: A Practical Approach*.[117]

A Volume of invited papers, covering 'Protein Kinase A and Human Disease', has appeared.[118] The topic is important because inherited disruption of cAMP signalling results in cancer. The PKA system is an important pathway that mediates cyclic nucleotide signalling, appearing to have a crucial role in the integration of many messages from a variety of senders and centres.

A detailed review of the structures and specificities of α-amylases has been published.[119]

7.2 Mechanistic Studies. – This topic is covered in this Chapter at a number of different sub-headings, and within this Section there is some lack of organisation of the discussion. However, the common theme of work selected for discussion in this Chapter is new mechanistic information flowing from the interpretation of structural and kinetics data.

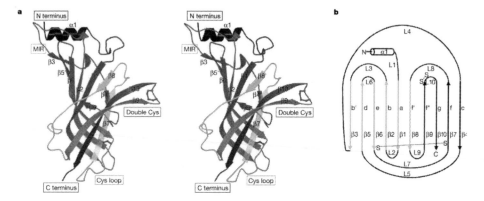

Figure 56   *Overview of the AchBP protomer structure. (A), the AChBP protomer as viewed from outside the pentameric ring, coloured as a shading gradient, from medium grey (N terminus) through intermediate shades to dark grey (C terminus). Disulfide bridges are indicated in ball-and-stick representation. In a complete ion channel the N terminus would point towards the synaptic cleft and the C terminus would enter the membrane at the bottom, continuing into the first transmembrane domain. (B), topology diagram of the AchBP protomer. For comparison with Ig-folds the strands have been labelled a–g, showing the additional strand (b9) and hairpin (f9-f0). In this structure, strands have been labelled b1–b10 with loops (or turns) L1–L10 preceding each strand with the same number. The b5 strand is broken (b5-b59) with internal loop L59; b6 also has a small break, but it is shown continuously (see Figure 54). The topology seen here will be highly conserved across the entire family of pentameric LGICs. S = disulfide bridge*
(Reproduced with permission from Brejc et al.[114] and *Nature*, copyright 2001, Macmillan Magazines Ltd)

7.2.1   Motor Proteins. A textbook, *Mechanics of Motor Proteins and the Cytoskeleton*, has appeared covering aspects of this topic.[120]

Kinesin motors are enzymes that use the hydrolysis of ATP to generate force and movement along their cellular tracks, the microtubules. A switch-based mechanism has been depicted as a contribution to picturing the components of this machine.[121] Large conformational transitions are the only plausible mechanical movements, but switches that control these are more subtle. Myosin motors (see also Ref. 129) and roles for G proteins are also well-studied in the context of protein mechanics.

Mitochondrial protein kinases are molecular switches for enzymes that are involved in the oxidative processing of branched-chain α-ketoacids and pyruvate. Elevated levels of these metabolites are implicated in several common diseases, notably insulin-resistant Type II diabetes, and ketoaciduria. X-ray analysis of rat branched-chain α-ketoacid dehydrogenase kinase (BCKD kinase), which features a characteristic nucleotide-binding domain (requiring potassium and magnesium) and a four-helix bundle domain, has been reported. Proper interaction of these domains is shown in this study to be crucial for the integrity of the binding site, and suggests that a search should be made for

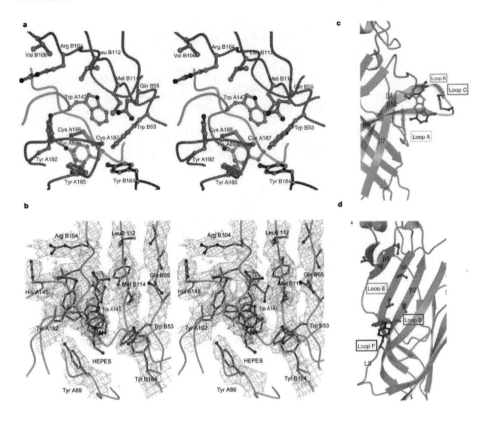

Figure 57   *a. Stereo representation of the ligand-binding site, in ball-and stick representation, showing the principal loops A—E (increasingly darker shades of grey,
respectively). b. Electron density map displaying a HEPES buffer molecule in
the ligand-binding site. c. Location of the principal ligand-binding residues
(grey shades as in a, orientation as in Figure 54, part b). d. Location of the
complementary ligand-binding residues (grey shades as in a, orientation as in
Figure 54, part b).*

disruption mechanisms that may help to find therapeutic agents for the intrac-
table diseases that BCKD kinases apparently initiate. Binding of ATP induces
conformational changes in a loop region of the nucleotide-binding site, disrupt-
ing the interface with the four-helix bundle domain, and causing a lateral shift of
the top portions of two helices. Phosphorylation effectively traps ADP and
explains product inhibition of mPKs.[122]

7.2.2   Synthetases. A newly identified cysteinyl tRNA synthetase (CysRS) in
Methanococcus jannaschii and *Deinococcus radiodurans*, members of the *Archae*,
is one of the first to have been uncovered with sequences that are distinct from
those of every other known aminoacyl tRNA synthetase. This contradicts an

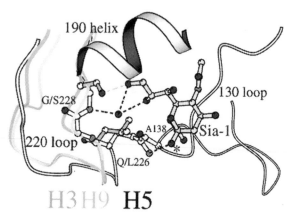

earlier recent proposal that aminoacylation by cysteine is performed by a prolyl tRNA synthetase.[123]

Studies discussed elsewhere in this Chapter cover the behaviour of synthetases, a broad range of functions being covered by this general class of enzyme (Refs. 103, 191, 193).

7.2.3 Peroxidases. Extended X-ray absorption fine structure analysis has been reported, of the environment of the heme iron of catalase-peroxidases from *Mycobacterium tuberculosis* and that from *E.coli*.[124] The objective of the study was to determine whether structural differences in this region were associated with differences in reaction profiles of the two enzymes. The former enzyme is a more effective activator of the antibiotic isoniazid, compared with the latter. However, overall similarity of the heme iron environment leads to the conclusion that the explanation for the properties lies elsewhere.

7.2.4 β-Ketoacyl-carrier Protein Synthase. X-ray study of *Mycobacterium tuberculosis* β-ketoacyl-carrier protein (ACP) synthase III (*Mycobacterium tuberculosis* FabH) has an important role in the synthesis of mycolic acids, α-alkyl-β-hydroxy acids that cover the surface of mycobacteria. Inhibition of their biosynthesis is an established mechanism of action for several current anti-tuberculosis drugs. In that context, the enzyme studied in this work, which catalyses a Claisen-type condensation between long-chain acyl-coenzyme-A substrates (myristoyl-CoA) and malonyl-ACP has been compared with the *E.coli* FabH (Figures 59 – 61).[125]

The *Mycobacterium tuberculosis* FabH is evidently structured to allow binding of long acyl chains, whereas the *E.coli* analogue can only admit smaller

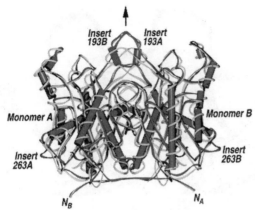

Figure 59   *Comparison of secondary structures of dimers of ecFabH (dark grey) and Mycobacterium tuberculosis FabH (medium grey). The two-fold axis is indicated by the arrow*
(Reproduced with permission from Scarsdale et al.[125])

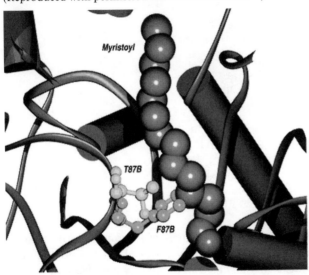

Figure 60   *Position of the myristoyl group in binding channel 2 showing the position of residue 87B, which is a threonine residue in Mycobacterium tuberculosis FabH (dark grey) and a phenylalanine residue in ccFabH (medium grey)*
(Reproduced with permission from Scarsdale et al.[125])

structures to its active site, explaining why it is restricted to operate on acetyl-CoA.

7.2.5 *Amino Acid Oxidases.* Kinetic isotope and pH effects have been interpreted to reveal small differences in mechanism for the Schiff base-forming step for D-amino acid oxidase, from that applicable at pH below 8 (A in Scheme 4) to the higher pH alternative (B in Scheme 4).[126]

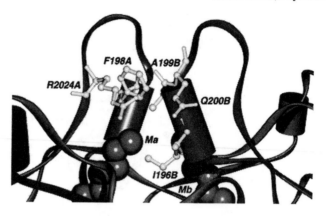

Figure 61   *Distal end of the myristoyl binding site in* Mycobacterium tuberculosis *FabH at the junction of the inserts of each monomer*
(Reproduced with permission from Scarsdale et al.[125])

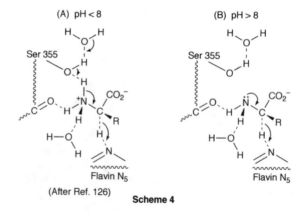

Scheme 4

The structure of the glycan content of L-amino acid from the venom of the Malayan pit viper (*Calloselasma rhodostoma*) has been determined. The major oligosaccharide, a bis-sialylated bi-antennary core-fucosylated dodecasaccharide, accounts for approximately 90% of the glycan content (Figure 62).[127]

Other oxidase topics are discussed in Section 7.2.24).

7.2.6   Superoxide Dismutases (se also Ref. 82). A crucial role has been demonstrated for a tryptophyl residue, Trp161, in human manganese superoxide dismutase. The enzyme protects mitochondria from oxidative damage associated with electron transport, exerting this role through catalysing the disproportionation of the superoxide radical anion.

This tryptophyl residue presents a hydrophobic side to the active site cavity. Its replacement by Ala or Phe causes significant conformational changes at adjacent residues near the active site, and reaction rates are some 100 times slower (Figures 63, 64).[128]

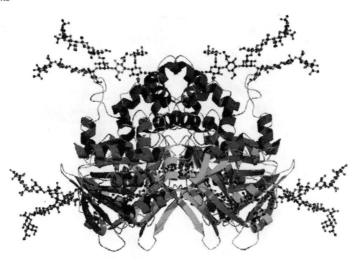

Figure 62   *The dimeric glycoprotein LAAO, with a dodecasaccharide linked to the glycosylation sites Asn172 and Asn361. In aqueous solution the carbohydrate chains populate rotation cones due to the movement dynamics around the side chain torsions of Asn and around glycosidic torsions. Both (1-6) arms probably populate several rotamers of which only one is shown here*
(Reproduced with permission from Geyer et al.,[127] *Eur.J.Biochem.*, Blackwell Science)

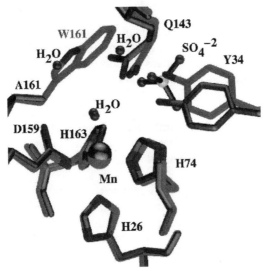

Figure 63   *Active site of mutant human manganese superoxide dismutase (MnSOD; Trp161 replaced by Ala; grey) superimposed on that of wild-type human MnSOD (medium grey)*
(Reproduced with permission from Hearn et al.[128])

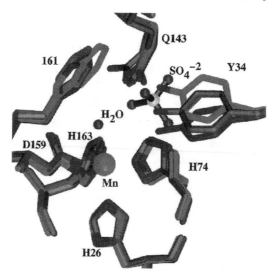

Figure 64   *Active site of wild-type human MnSOD (medium grey) superimposed on that of mutant MnSOD (Trp161 replaced by Phe; sulfate-free; light grey), of mutant MnSOD (Trp161 replaced by Phe; with sulfate; dark grey), and of mutant MnSOD (Trp161 replaced by Ala; with sulfate as described in the text; light grey)*
(Reproduced with permission from Hearn et al.[128])

A 2.7Å crystal structure analysis of the activated FERM domain of moesin (common to several enzymes, e.g. protein phosphatases, and myosin) has provided details of structural changes that occur on activation The FERM domain is the first 300 amino acid residues along the sequence from the N-terminus, and is now shown to consist of three subdomains. Residues 260-264 are in a loop that is shifted by removing the C-terminal domain, and the helix 166-170 is also relocated within the three-dimensional framework by this treatment (Figure 65).[129] The ERM (ezrin/radixin/moesin) family of proteins forms one of the physical links between the plasma membrane and the actin cytoskeleton.

7.2.7   Dehydrogenases. Acyl CoA dehydrogenase is rapidly inhibited by racemic 3,4-dienoyl-CoA thioester derivatives with a stoichiometry of two molecules of racemate per enzyme flavin, the (R)-enantiomer being the active inhibitory stereoisomer (Figures 66 – 68).[130] The role of flavin is depicted in Scheme 5, and an enolate intermediate is revealed through the battery of physical methods, including X-ray crystallographic analysis to 3.4Å, employed in this study.

In Figure 68, the distances (dotted lines), ranging from 1.5 to 2.2Å, indicate severe steric hindrance between the S-isomer and the side chain of Tyr375. A model of the S-isomer (dark blue) in the active site shows the substrate binding mode, whereas the R-isomer (purple) depicts the inhibitor binding mode. The carbonyl thioester moiety of the S-isomer binds about 1.3Å 'deeper' into the active site of the enzyme and forms a hydrogen bond with the 2-OH of the FAD moiety.

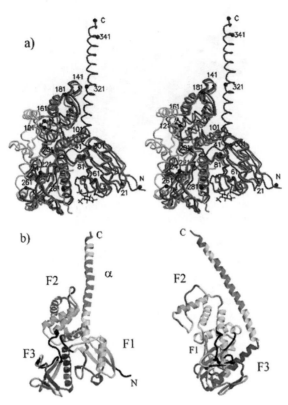

Figure 65   *(a) The run of Cα atoms through the activated FERM domain. (b) The activated moesin structure; the dark grey regions are those where shifts greater than 3 Å occur as a result of activation. Lighter shades of grey represent smaller structural shifts.*
(Reproduced with permission from Edwards and Keep[129])

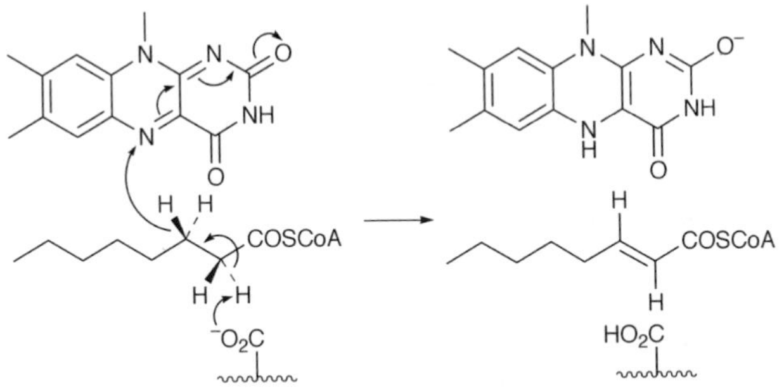

Scheme 5

An L-lactate dehydrogenase mutant (from *Bacillus stearothermophilus* involving replacement by Ala of Lys245, expressed in *E.coli*), has been used to demonstrate a new aspect of reaction mechanism for the native enzyme. The rate of formation of the complex of enzyme + NADH + pyruvate is rate-limiting in the mutant, and the ensuing hydride transfer is the first slow step; this is the reverse

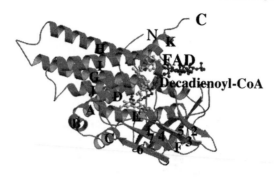

Figure 66   *Overall polypeptide folding of a monomer of the complex between medium chain acyl-CoA dehydrogenase and R(-)-3,4-decadienoyl-CoA. The FAD and the inhibitor portion of the adduct are shown in light grey and dark grey, respectively*
(Reproduced with permission from Wang et al.[130])

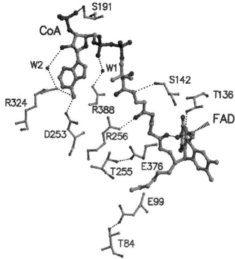

Figure 67   *Residues in the vicinity of the inhibitor in the 3,4-decadienoyl-CoA adduct. Potential hydrogen bonds are shown by the dotted lines. Carbon atoms are depicted in light grey, oxygen in medium grey, nitrogen in black, and sulfur in dark grey. Most of the side chain conformations are very similar to those found in the octenoyl-CoA complex*
(Reproduced with permission from Wang et al.[130])

of the relative rates for the native enzyme (Figure 69).[131]

The role proposed for quantum tunnelling, that accounts for facilitated hydride transfer in enzyme-catalysed reactions, has now had several years in which to receive considerable support. A further example in the 2001 literature develops the idea, to give a role to side chains that have no functional groups, by which they can contribute to a particular stage of the mechanism through conventional reaction schemes. Val292 is thought to assist the hydride transfer step, by virtue of its bulk and mobility near the active site of liver alcohol dehydrogenase. This

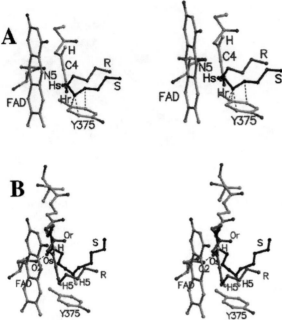

Figure 68   *Models of R- and S-3,4-decadienoyl-pantetheine in the active site of MCAD. The common portion of the inhibitor molecules is shown in atom color (carbon, light grey; oxygen, medium grey; nitrogen, black; and sulfur, dark grey) while C6-C10 atoms are depicted in different shades of dark grey, for the R- and S-isomer, respectively. (A) R- and S-isomers docked into the active site in the 'substrate-binding' mode. The H-atom on C-5 is marked as HR and HS for R- and S-isomers, respectively. (B) Comparison of the two different ligand binding modes in the active site of MCAD: the 'inhibitor-binding' mode and the 'substrate-binding' mode.*
(Reproduced with permission from Wang et al.[130])

conclusion was derived from comparison (Figure 70) of the structure of the natural enzyme (apo-enzyme) with the structure derived for the mutant enzyme V292S (in which Val292 was substituted by serine).[132]

Dihydrolipoyl dehydrogenase P64K from the pathogenic bacterium *Neisseria meningitidis* is found in the outer membrane of the cell, and attention has been given to the structure of its lipoyl domain by heteronuclear NMR spectroscopy (Figure 71). The enzyme was found to be a flattened β-barrel composed of two four-stranded antiparallel β sheets. The lysine residue that becomes lipoylated is in an exposed β turn that appears to be freely mobile, judging from a ^{1}H-^{15}N heteronuclear Overhauser effect NMR experiment.[133]

7.2.8 Orotodine Monophosphate Decarboxylase. Recently reported crystal structures for free orotodine monophosphate decarboxylase, and for the enzyme complexed with inhibitors, have been interpreted to bestow unusual mechanistic characteristics on the mode of catalysis. The enzyme has an essential role in nucleic acid biosynthesis, and its fascination is bolstered by the claim that it is the

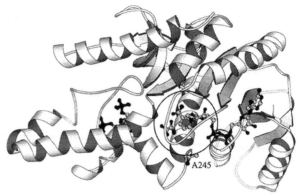

Figure 69   *One subunit of* Bacillus stearothermophilus *L-lactate dehydrogenase. The interface of the subunits in a dimer is on the bottom side of this picture, and the interface of two dimers forming a tetramer is on the back side. NADH, pyruvate, and the alanine moiety at the mutated position 245 are ball-and-stick models. The active site is marked with a circle*
(Reproduced with permission from Kedzierski et al.[131])

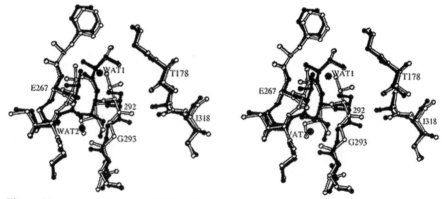

Figure 70   *Comparison of V292S (black) and apo-enzyme (unfilled tubes) showing changes in conformation of residues 290-294, 263-268, 178, and 317-318*
(Reproduced with permission from Rubach et al.[132])

most proficient enzyme known so far, with $(k_{cat}/k_m)/k_{uncat} = 2 \times 10^{23}$. Another source of interest lies in the mechanistic curiosity, that although all other known biochemical decarboxylation reactions involve resonance stabilization of a carbanion formed by loss of CO_2 from a carboxylate anion, other mechanisms are plausible in this case and deserve further investigation (Scheme 6).[134]

7.2.9   Glycosidases. In a series of significant studies, finely-tuned structures for active sites, optimally configured for transition state stabilization, have been characterized.[135] This has led to enhanced understanding of the mechanism of action of glycosidases. Structural analysis of various enzyme complexes that represent stable intermediates along the reaction coordinate in conjunction with detailed mechanistic studies on wild-type and mutant enzymes, have delineated

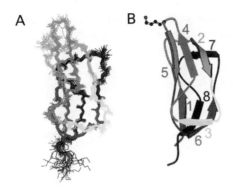

Figure 71   *Solution structure of the P64K lipoyl domain. (A), Superposition of backbones from 21 structures. (B), The overall fold showing the lysine residue (Lys42 at the top left) that receives the lipoyl group. Strands are shown in shades of grey, with grey-black at the N-terminus and medium grey at the C-terminus* (Reproduced with permission from Tozawa et al.,[133] *Eur.J.Biochem.,* Blackwell Science)

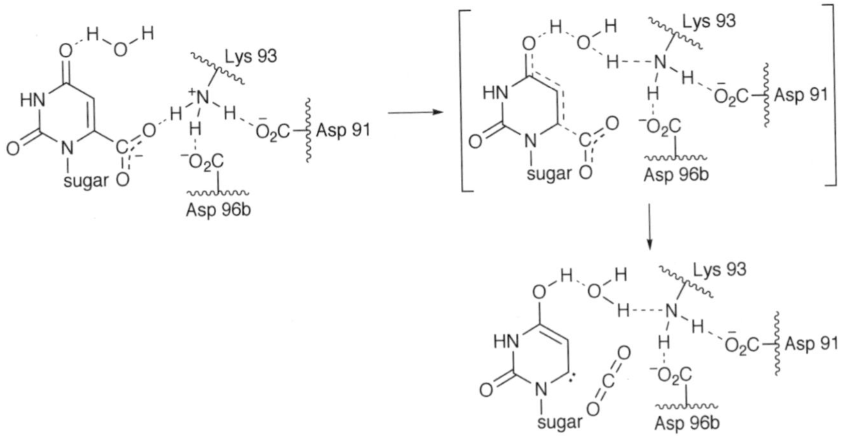

Scheme 6

the contribution of nucleophilic and general acid/base catalysis, as well as the roles of noncovalent interactions, to account for the 10^{17}-fold acceleration of the hydrolysis of glycosidic bonds by enzymes of this class.

Hen egg-white lysozyme is a β-glycosidase, the first enzyme to have its three-dimensional structure determined by X-ray diffraction, and Phillips' catalytic mechanism deriving from this work involves a long-lived oxocarbenium ion intermediate.

This was thought to be a settled matter for enzymes that cleave glycosidic linkages with retention of configuration. Now, a replacement for this, a common mechanism for these enzymes that involves a covalent glycosyl-enzyme intermediate, has been verified by electrospray ionization mass spectrometry and X-ray crystal analysis.[136] The uncertainty until now has lain in the timing of bond-making at the glycosidic centre, and the stereochemical outcome – bond-making with retention of configuration – is now accommodated with a glycosyl-Asp52 ion-pair intermediate (Scheme 7).

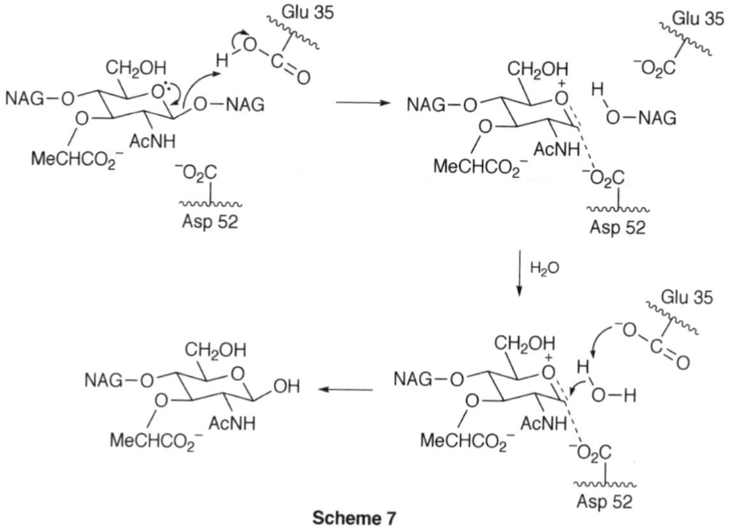

Scheme 7

A mutant β-mannosidase has been demonstrated to be effective in the synthesis of β-mannosides through a mechanism involving Glu519 and nearby Glu429, stereochemical details being revealed by the use of α-D-mannosyl fluoride as substrate.[137] The carboxylate anion of Glu519 performs an S_N2 displacement of the anomeric C1 substituent of the mannoside, with Glu429 providing the proton to complete the synchronized sequence. Reversal of the sequence with a secondary alcohol (the C_4-OH group of another glycoside) generates a different mannoside.

The feruloyl esterase domain of cellulosomal xylanase Z from *Clostridium thermocellum* has been searched to find the characteristic catalytic triad (the side-chains of His260, Ser172, and Asp230) at the heart of the active site.[138] The ferulic acid moieties in the substrate are cleaved and seen in a hydrophobic binding pocket in the X-ray structure. The carbohydrate binding pocket was identified from the position of the ferulic acid moieties (Figure 72).

7.2.10 Flavoprotein Phenol Hydroxylase. Structural information derived from site-directed mutagenesis studies of flavoprotein phenol hydroxylase shows that Arg281 and Asp54 are part of the active site environment.[139] An important part of this study is the demonstration that eleven amino acid residues in the published sequence of this enzyme are incorrect.

7.2.11 UDP-N-Acetylmuramyl-L-alanine Lignase. This enzyme, from *E.coli*, is the catalyst for the third step in Phase I of bacterial cell wall biosynthesis. Its role can be inhibited by a substrate analogue.[140] A transition state attained by nucleophilic attack on the acyl phosphate group has been presumed, and has

Figure 72 α/β *Domain of FAE–Xyn2 and the catalytic triad of this enzyme* (Reproduced with permission from Schubot et al.[138])

guided the choice of potential inhibitors; the large number of antibacterial agents that effectively target cell wall biosynthesis (55% of the antibacterials market in 1998) is a driving force for research in this area.

7.2.12 Glucose 4,6-Dehydratase. Dehydration involving *E.coli* dTDP-glucose 4,6-dehydratase is catalysed by Glu136 and Asp135 (Scheme 8 and Figure 73).[141] MALDI-TOF Mass spectrometry was used to establish the kinetics of solvent ^1H – ^2H exchange at C_5 of the deuterated product, dTDP-4-keto-6-deoxyglucose-$^2H_7$.

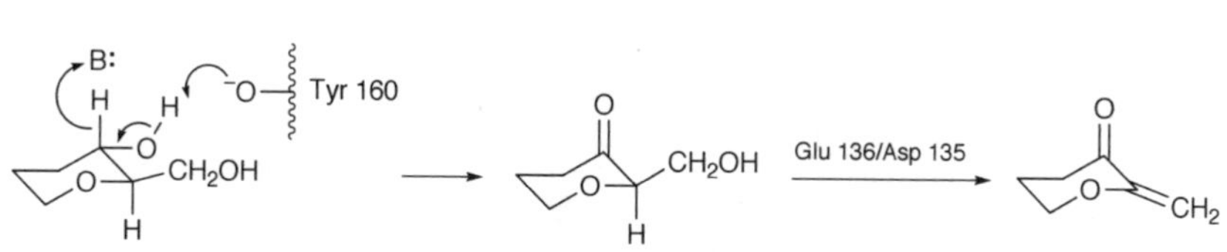

Scheme 8

7.2.13 Carbonic Anhydrases. The marine diatom *Thalassiosira weissflogii* synthesises a cadmium carbonic anhydrase in place of the normal zinc-liganded enzyme, when stressed with a deficiency of zinc (through addition of cadmium to zinc-depleted growing media).[142] Since this was found to enhance the growth rate of the organism, it was concluded that cadmium assimilation was a valid alternative for the organism.

This is the first example of a cadmium enzyme; this element has not been assumed to have a biological function, but its oceanic distribution follows closely that of major algal constituents such as phosphorus, so a strong indication is being conveyed that searching through marine microorganisms would be profit-

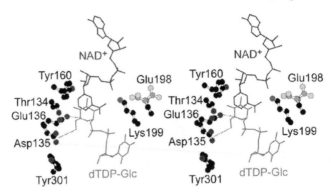

Figure 73   *Active site of* E.coli *dTDP-glucose 4,6-dehydratase*
(Reproduced with permission from Gross et al.[141])

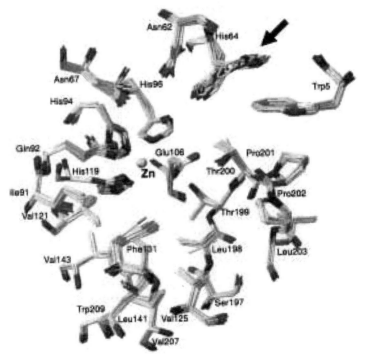

Figure 74   *Superposition of 24 protein + ligand complexes of hcaII (ligands not shown).*
With the exception of His 64 (arrow), the binding pocket is rather rigid
(Reproduced with permission from Gruneberg et al.[144])

able in identifying new metal-concentrating species.

Aggregation of bovine carbonic anhydrase can be achieved by α-cyclodextrins with modified OH groups. This indicates a way forward in devising inhibitors of protein aggregation, with cyclic polyhydroxylic structures as lead compounds.[143]

Human carbonic anhydrase II (hcaII), of established structure, features in a search for inhibitors through analysis of the binding pocket structure (Figures 74

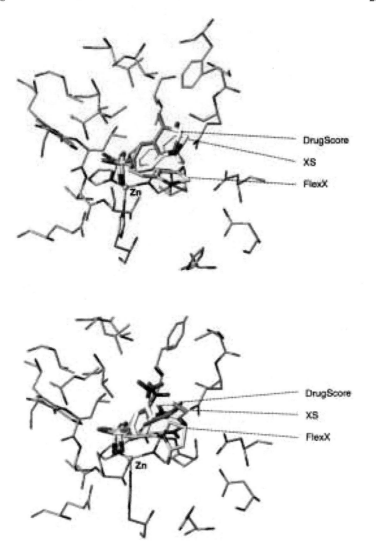

Figure 75   *Superposition of the X-ray structure (XS) and docked models for inhibitors detailed in Ref. 144, in hcaII*
(Reproduced with permission from Gruneberg et al.[144])

– 76).[144] The importance of a better understanding of the role of computer modelling, in competition with traditional pharmaceutical testing of intuitively derived lead compounds, was the driving force behind this research, and the results can be better appreciated from the Figures reproduced here (Figures 74 – 76) rather than a detailed description in words of the conclusions offered.

An approach to detect functional similarity among proteins independent from sequence and fold homology has been described.[145] The context of this is centred on drug discovery in the human carbonic anhydrase II field, and features a thoughtful discourse on many familiar aspects of the design of lead compounds

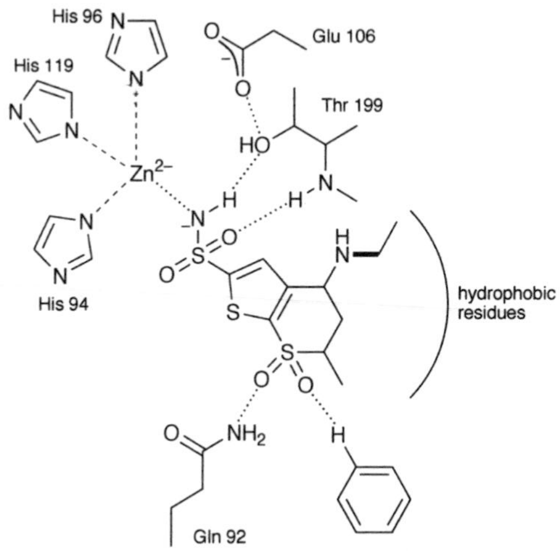

Figure 76   *Schematic drawing of the binding mode of dorzolamide to hcaII*
(Reproduced with permission from Gruneberg et al.[144])

in pharmaceutical research, where the compounds capable of controlling functional proteins are sought.

·7.2.14   *Reductases and Dehydrogenases.* The neighbourhood of the key active site acid/base arginine residue, in a soluble fumarate reductase (closely related to succinate dehydrogenase) from *Shewanella frigidimarina*, has been mapped by kinetic and crystallographic analysis of mutants in which Arg402 has been substituted (Figure 77 and Scheme 9).[146]

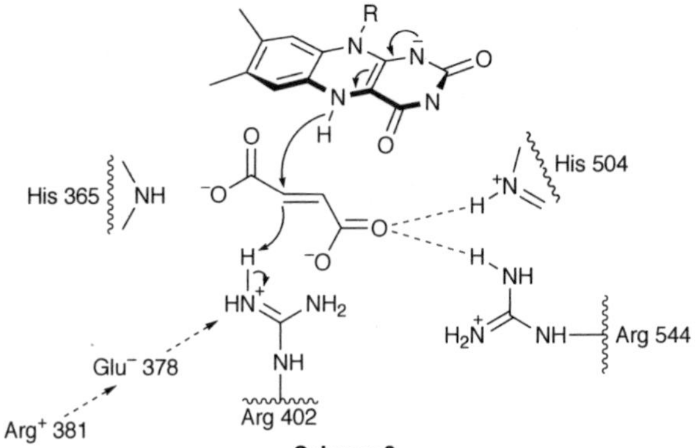

Scheme 9

(Adapted from Ref. 146 and from sources cited there)

NADH-Dependent cytochrome b5 reductase has provided an enigma until now; the two active site residues Lys83 and Cys245 were remote from the active site as revealed in an X-ray structure of the pig enzyme.[147] Considerable structural differences between rat cytochrome b5 reductase and pig cytochrome b5

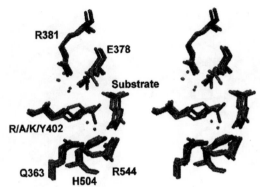

Figure 77 *Active sites of wild-type (dark grey), and mutants with Arg402 replaced by Ala, Glu, and Tyr (shown using different grey shades) showing the proximity of proton source to bound substrate molecules. The water molecules enclosed in the active site are shown*
(Reproduced with permission from Mowat et al.[146])

reductase have now been found, including these lysine and cysteine locations (Figures 78, 79), and it is concluded that the pig enzyme structure needs correction.

Inhibitors of dihydropicolinate reductase, a key enzyme of the diaminopimelate pathway of *Mycobacterium tuberculosis*, have emerged through a molecular modelling approach that used the established crystal structure of the enzyme (Figures 80, 81).[148] The candidate inhibitors were established as those that fitted best within the active site. A conventional screening approach to finding new inhibitors is also a feature of this study.

7.2.15 Hydrogenases. Cytochrome c_3 from sulfate-reducing bacteria *Desulfovibrio desulfuricans* Essex 6 provides the electron carrier that supports periplasmic hydrogenase function.[149] Heme IV of the cytochrome appears to dock with the distal [4Fe-4S] cluster of the hydrogenase, a conclusion supported through X-ray crystal analysis, and EPR and UV/Vis spectroscopic study (Figure 82). The spectroscopic methods concentrated on the role of the metal centres, and showed that heme IV is the entry point for electrons in this system.

7.2.16 Transaldolases. Transaldolase B of *E.coli* depends on Asn35 and Thr156 for the positioning of a catalytic water molecule at the active site. The side chain of Glu96 participates in concert with this water molecule in proton transfer during catalysis. Site-directed mutagenesis studies gave these details.[150] Furthermore, substitution of Ser176 by alanine resulted in a mutant enzyme with only 2.5% residual activity. Other data point to a function for Ser176 in the binding of the C_1 hydroxy group of the donor substrate. Figure 83 shows the Schiff base generated with Lys132. Figure 84 shows the water molecule involved in these events.

7.2.17 Endonucleases. Catalysis by TaqI endonuclease involves two metal ions and the Lys157 – Lys158 pair (Scheme 10),[151] as concluded from the results of

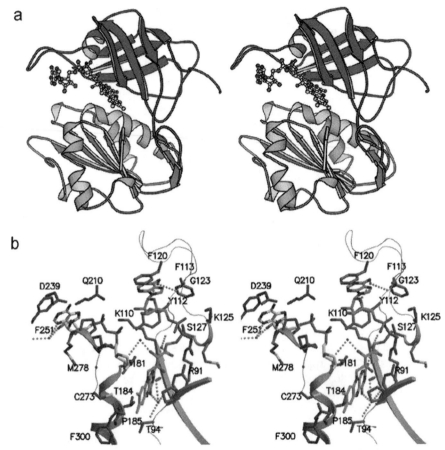

Figure 78   *(a) Elements of each domain of rat cb5r. The FAD binding domain is medium grey, the NADH-binding domain is dark grey, and the FAD molecule is shown in ball-and-stick representation. (b) Interactions of FAD with NAD. Water-mediated hydrogen bonds are shown as dotted lines*
(Reproduced with permission from Bewley et al.[147])

Asp, His, and Glu substitutions at Lys158 in native and in the mutant with Lys157 substituted by serine. DNA cleavage activity was partly rescued for these mutants by Mn^{2+} salts, but not by other divalent metal cations, and this has led to the proposal of a variant of the mechanism in Scheme 10. The place of the bridging water molecule is taken by Mn^{2+} in this proposed variant.

7.2.18 Sulfurylases. The crystal structure of ATP sulfurylase of *Penicillium chrysogenum*, crystallised from a medium containing a high concentration of adenosine 5′-phosphosulfate (APS; adenylyl sulfate) gives insights into the allosteric regulation of sulfate assimilation.[152] The active site reveals specific hydrogen bonds between adenine nucleotides and the side-chains of adjacent amino acid residues. APS is bound to the allosteric site less firmly than to the catalytic site, so

a

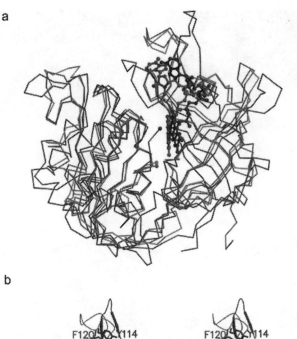

b

Figure 79 *Movement of the FAD-binding domain may facilitate binding of NAD⁺ in a productive manner in cb5r. (a) Superposition of rat cb5r (medium grey) with maize nitrate reductase (dark grey) and the Y308S mutant of pea FNR (medium grey). (b) Active sites of Y308S mutant of pea FNR and wild-type cb5r in complex with NADP⁺ and NAD⁺ respectively*
(Reproduced with permission from Bewley et al.[147])

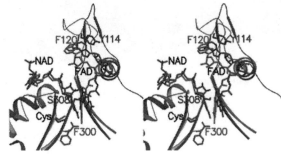

(After Ref. 151)

Scheme 10

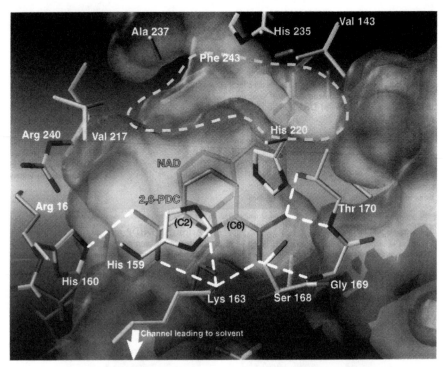

Figure 80   *Active site of dihydrodipicolinate reductase. The competitive inhibitor 2,6-PDC (see original paper) bound in the active site of DHPR is stacked with the nicotinamide ring of NAD⁺. The active site cavity is represented by the background*
(Reproduced with permission from Paiva et al.,[148] and Elsevier)

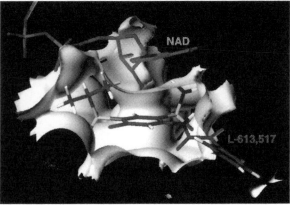

Figure 81   *Sulfonamide docking in the active site of dihydrodipicolinate reductase. The surface of the active site is represented as the white area; this is only part of the active site; enough has been cut away to be able to see the inhibitor. An alkyl sulfone makes a good fit within the left side whereas on the right side there is room only for an unsubstituted sulfonamide*
(Reproduced with permission from Paiva et al.[148] and Elsevier)

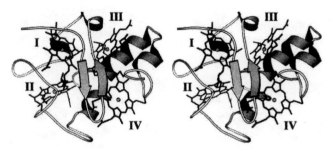

Figure 82 *Cytochrome c₃ hydrogenase complex from* Desulfovibrio desulfuricans *Essex 6*
(Reproduced with permission from Einsle et al.,[149] *Eur.J.Biochem.*, Blackwell Science)

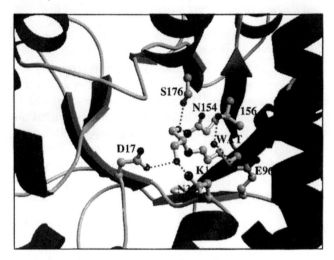

Figure 83 *Active site of transaldolase showing the reduced Schiff base intermediate linked to Lys132 and its interactions with other residues. Hydrogen bonds are shown as dotted lines*
(Reproduced with permission from Schorken et al.,[150] *Eur.J.Biochem.*, Black-well Science)

the structure that was determined in this study is assumed to be the active R-state (Figures 85- 88; for discussion of T- and R-allosteric states in other contexts, see Refs. 65, 82, 152-154, and Section 5.1.4).

An X-ray structure of the enzyme bound to the allosteric inhibitor 3′-phosphoadenosine 5′-phosphosulfate (Figure 89) is featured in this paper, in which it is concluded that there is a low affinity for substrate resulting from large rearrangement of domains that represents the R-to-T transition.[153] Allosteric inhibition is pictured in Figure 90.

X-Ray structure determination of another ATP sulfurylase from yeast (*Saccharomyces cerevisae*) has been reported recently, the enzyme showing no allosteric properties but possessing very similar multidomain features to those for the *Penicillium* enzyme.[154]

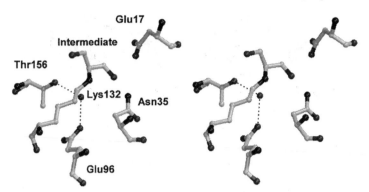

Figure 84   *The catalytic water molecule (dark sphere) in the active site of transaldolase,
and the reduced Schiff base intermediate with dashed lines showing hydrogen
bonds*
(Reproduced with permission from Schorken et al.,[150] *Eur.J.Biochem.*, Blackwell Science)

7.2.19   Nucleoside Diphosphate Kinase. Phosphoryl transfer by nucleoside
diphosphate kinase shows little specificity for either the nucleoside base or for
the monosaccharide component.[155] Several mutants modified at Lys16, Tyr56,
and Asn119 show that this enzyme belongs to the 'substrate assistance' category,
since nucleotide substrates lacking the 3'-OH group are phosphorylated some
three to four times slower than the natural analogues. Figure 91 shows the
structure of the NDP kinase from *Dictyostelium* in the vicinity of the active site.

The important topic of roles for kinases is represented in several sections of
this Chapter [Refs. 96 (Fig. 40), Refs. 118, 122, 152 (Fig.88), 190].

7.2.20   Phosphatases. Kinetic data have been re-evaluated and extended for
phosphoryl transfer by *Yersinia* protein tyrosine phosphatase, to show that the
first step involves attack by the cysteine403 anion on the di-anion of the substrate, p-nitrophenyl phosphate.[156] The active site structure is shown in Figure
92.

Other results for phosphatases are described in Refs. 65, 76, 129 and 156.

7.2.21   Lipases (see also Ref. 111). *Bacillus subtilis* secretes two lipolytic enzymes, LipA and LipB, which are 74% identical in terms of their primary
structure.[157] X-ray structure determination of LipA allows a structure to be
proposed for LipB, and shows that a catalytic triad (Ser78, His157, and Asp134)
constitutes the working parts of the active site (Figure 93).[158]

7.2.22   Proteases. A computer-generated docking program, Eudoc, devised for
demonstrating the most energetically stable complexes formed between papain
and 2,4,5,7-tetranitrofluorenone (TNFN) or human adenovirus cysteine protease (hAVCP), reveals shape complementarity between TNFN and the enzyme
active site.[159] This work assists a search for effective antiviral agent, and in the
case of hAVCP, reversible binding is followed by irreversible inhibition, suggesting a new approach to the search.

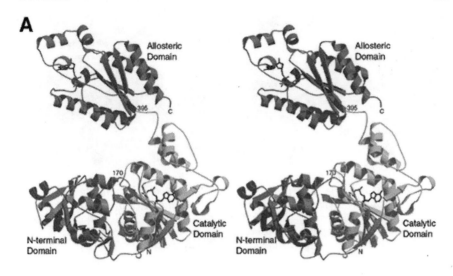

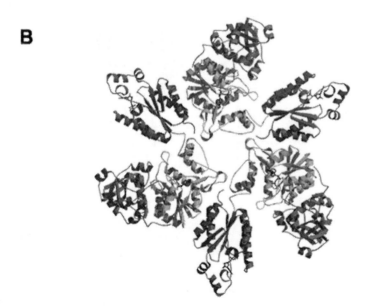

Figure 85 *(A) A monomer of ATP sulfurylase of* Penicillium chrysogenum, *showing three distinct domains: N-terminal (light grey), catalytic (medium grey), and allosteric (dark grey). (B) The three-fold relationship of the three subunits of ATP sulfurylase, showing points of interaction between catalytic and allosteric domains*
(Reproduced with permission from MacRae et al.[152])

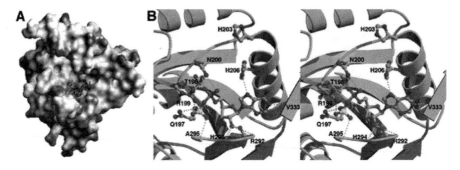

Figure 86   *Views of the catalytic domain of ATP sulfurylase showing the bound APS (adenylyl sulfate). (A) Electrostatic surface diagram showing the cavity which is accessible only from the external bulk solution. (B) Residues that interact with bound substrates. Dashed lines are potential hydrogen bonds*
(Reproduced with permission from MacRae et al.[152])

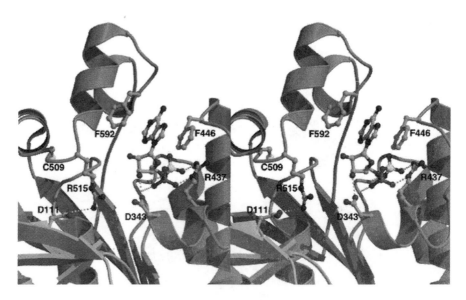

Figure 87   *Allosteric site showing the bound APS (adenylyl sulfate) molecule. Amino acid side-chains that make specific contacts with the substrate are shown. Arg515 is in a suitable position to bind the 3'-hydroxy group of PAPS, and is involved in a salt linkage to Asp111 of the N-terminal domain of the adjacent subunit. Dashed lines are potential hydrogen bonds*
(Reproduced with permission from MacRae et al.[152])

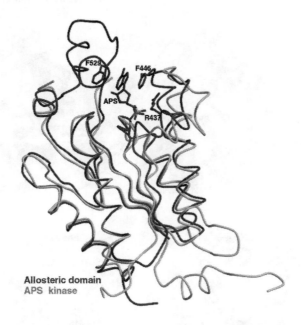

Figure 88 *Superimposition of the ATP sulfurylase allosteric domain (dark grey) and APS kinase (light grey) Cα backbones. APS (adenylyl sulfate) substrate is shown interacting with allosteric domain side-chains Arg437, Phe446, and Phe529 (Reproduced with permission from MacRae et al.[152])*

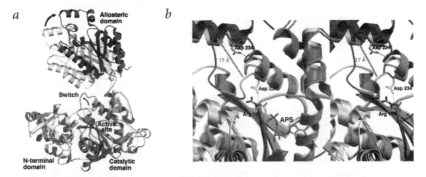

Figure 89 *(a) Conformational changes involved in the R- to T-transition of the allosteric domain of ATP sulfurylase. The R-state of the allosteric domain is shown partly transparent to distinguish it from the T-state (shown in a darker shade). The consequent active site switch (shown with an arrow) involves residues 228-238. The N-terminal and catalytic domains do not move significantly relative to each other during the R- to T-transition. (b) Detail of the active site switch, a 17Å movement. The active site region of the T-state is shown superimposed on the R-state active site, and the latter, as well as the bound APS, is shown partially transparent to distinguish it from the T-state structure. In the R-state, Asp234 forms a salt link to Arg199; in the T-state, Asp234 forms an N-cap with the adjacent helix*
(Reproduced with permission from MacRae et al.[153] and *Nature Struct.Biol.*, copyright 2002, Macmillan Magazines Ltd)

a

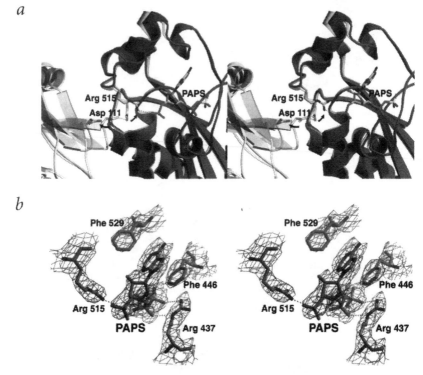

b

Figure 90   *(a) Allosteric inhibitor-binding site involving PAPS bound to the allosteric domain (the dark grey region). Arrows indicate the direction of movement in the R- to T-transition. (b) Electron density map of the final PAPS-binding site. The adenine ring of PAPS stacks between Phe446 and Phe592. Arg437 forms a bidentate salt linkage with the phosphosulfate group and Arg525 forms a salt linkage with the 3'-phosphate of PAPS. Except for the Arg515 salt linkage, these interactions are identical with those observed for the binding of APS, which is not an allosteric inhibitor of this enzyme*
(Reproduced with permission from MacRae et al.[153] and *Nature Struct.Biol.*, copyright 2002, Macmillan Magazines Ltd)

Nitric oxide donors inhibit cysteine proteinase from promastigotes of *Leishmania infantum* on a 1:1-basis, generating a new λ_{max} 330-350 nm.[160] Nitric oxide is already well-known for its reactions with proteins (e.g. myoglobin, Ref. 196), and in the context of targetting cysteine residues in proteins. The catalytic activity of the enzymes is restored by treatment with dithiothreitol or L-ascorbic acid.

The hydrophobic S1' subsite of thermolysin-like proteases (Figure 94) can be modified, and effects on substrate specificity then are considerable.[161] Replacement of Leu202 by Gly, Ala, Val in one group of experiments, and by Phe or Tyr in another, unexpectedly increased the catalysed cleavage of peptide substrates. Further studies are needed to account for these findings, though the well-known exquisite dependence of catalytic rates on separations of catalytic side-chains (Section 5.1.2) is a likely underlying cause.

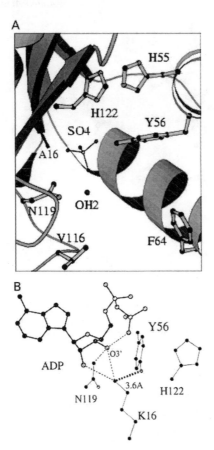

Figure 91   *(A) Active site of a mutant NDP kinase (Lys16 replaced by Ala) from* Dictyostelium. *(B) Active side NDP kinase from* Dictyostelium, *showing bound ADP-AlF3 and the pyrophosphate moiety of ADP, the bound Mg^{2+} ion, and surrounding protein side-chains*
(Reproduced with permission from Schneider et al.,[155] *Eur.J.Biochem.*, Blackwell Science)

The thrombin-like serine protease (LM-TL) from *Lachesis muta* preferentially cleaves Arg-Gly bonds in fibringen A-chains.[162] Its clotting function involves the release of fibrinopeptide A and fibrin, analogous to thrombin, but it also catalyses the hydrolysis of synthetic peptide substrates, with specificity similar to that shown by trypsin.

Structure comparisons such as these for several of the best-known proteases are discussed in this paper. Figures 95 and 96 show structural features of LM-TL with emphasis on the catalytic amino acid residues.

Human pigment epithelium-derived factor (PEDF) is a fundamentally important regulatory protein. The factor is a component of the retina as well as of vitreous humour and aqueous humour in the adult eye, and its mRNA derivative is found in most human tissues.[163] It is a non-inhibitory member of the serpin

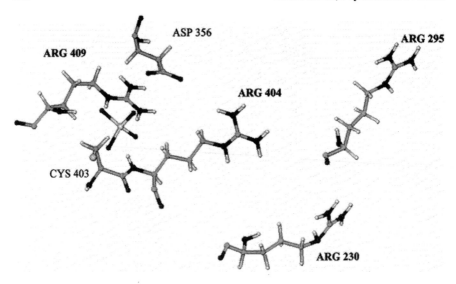

Figure 92   *Positions of Arg residues providing positive charge in the proximity of the active site of* Yersinia *protein tyrosine-phosphatase*
(Reproduced with permission from Czyryca and Hengge,[156] and Elsevier)

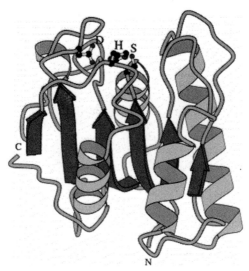

Figure 93   *Model of LipB from* Bacillus subtilis *showing amino and carboxy termini of the polypeptide main chain (N and C) and the three crucial amino acid residues at the active site (S, H, D)*
(Reproduced with permission from Eggert et al.,[158] FEBS Lett., Blackwell Science)

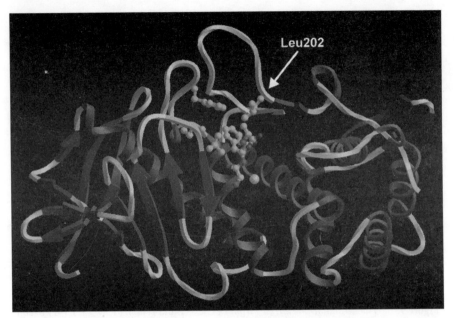

Figure 94   *Thermolysin with a substrate (the tripeptide Gly-Phe-Ala) in the active site cleft. The left side shows the N-terminal domain consists predominantly of β-sheet; the right side the predominantly α-helical C-terminal domain. The central α-helix, at the bottom of the active site cleft, contains several residues important for catalysis, including those coordinating the Zn^{2+} ion (shown as a dark sphere). The substrate occupies the S1 to S2' positions; the S1' side chains shown are (counterclockwise, starting at Leu202: Leu202, Phe130, Leu133, and Val139)*
(Reproduced with permission from de Kreij et al.,[161] *Eur.J.Biochem.,* Blackwell Science)

family, and is the most potent inhibitor of angiogenesis in the mammalian ocular compartment. Neurotrophic activity is a feature of the protein, both in the retina and in central nervous system. The elucidation of the structure of the serpin (Figure 97) could explain its effects (including the fact that it does not inhibit proteinases), and might lead to the development of therapeutic agents against uncontrolled angiogenesis.

The structure of PEDF derived from X-ray crystal analysis reveals details of possible receptor-binding sites and N-acetylglucosamine (heparin) binding sites, but also reveals that unlike any other previously characterized serpin, PEDF has an unusual asymmetric charge distribution that may be important in explaining its function.

In vivo regulation of protease activity may be mediated by small peptides as well as by proteins, judging from studies with human β-casomorphin-7 (Tyr-Pro-Phe-Val-Glu-Pro-Ile).[164] This is a natural peptide inhibitor of celestase that forms an acyl enzyme complex that is stable enough to survive X-ray structure analysis.

Tryptase-4, the fourth member of the mouse tryptase family of serine pro-

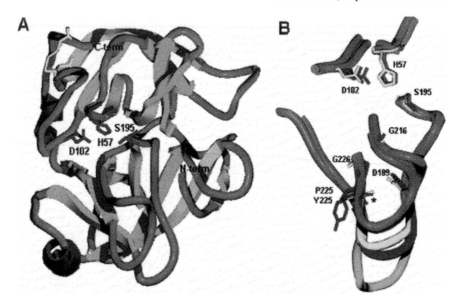

Figure 95   *The thrombin-like enzyme from* Lachesis muta *venom (LM-TL). (A) Secondary structure of LM-TL showing the catalytic triad His57, Asp102, and Ser195. The extended region, helix and β-strand are shown in differing shades of grey. The LM-TL additional disulfide bridge characteristic of snake serine protease is indicated in light grey. (B) Tube plot representation of the superposed catalytic domains of LM-TL, trypsin, and thrombin. The catalytic triad, as well as S1 (Asp189), S2 (Gly216), and S3 (Gly226) of LM-TL, trypsin and thrombin (grey-black) are shown in stick form. The variable S1 loops are highlighted revealing a LM-TL S1-loop shown in light grey, similar to trypsin being less deep and less narrow than that of thrombin (medium grey). Residue 225 (Pro or Tyr) and the orientation of the carbonyl O atom of residue 224, shifted by the presence of Pro225 (*) are also indicated. Other loops are omitted, for clarity*
(Reproduced with permission from Castro et al.[162] and Elsevier)

teases, has been characterised by standard genomic blot analysis methods. It seems that this new entity has importance in exerting a convertase-like role in the maturation of certain families of proteins.[165]

7.2.23   *Epimerases.* Human UDP-galactose epimerase is the third enzyme in the galactose metabolism pathway.[166] It is important in the context of treatment of galactosemia, an inherited disorder characterised by an inability to metabolize galactose. An X-ray structure determination now gives a more complete understanding of the three-dimensional consequences imparted by substitution of Val94 by methionine in human epimerase.

The crystal structure of native chicken fibrinogen at 2.7Å resolution complexed with two synthetic peptides H-Gly-Pro-Arg-Pro-NH$_2$ and H-Gly-His-Arg-Pro-NH$_2$ reveals the polypeptide chain arrangement in the central domain where the two halves of the molecule are joined, as well as a putative thrombin-

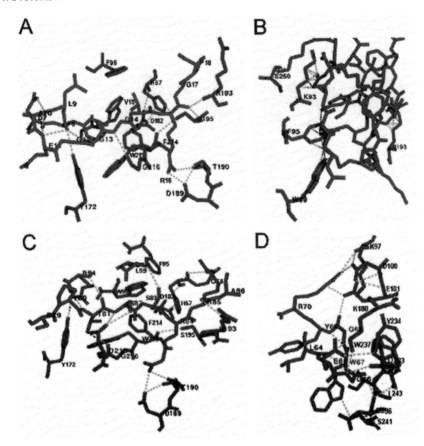

Figure 96   *Docking complexes of LM-TL with fibrinopeptide. A (A), BPTI (B) and ecotin-TSR R/R using primary (C) and secondary (D) binding sites. The enzyme is shown in medium grey, substrate and inhibitors in grey-black, hydrogen bonds are dotted lines (A, C, and D) and steric hindrance by grey dotted lines (B). (A) Environment of fibrinopeptide A (grey-black) in the LM-TL active site cleft (medium grey) shows the R16-V15 bond of fibrinopeptide lying proximal to the catalytic triad (His57, Asp102, and Ser195). (B) Part of the hypothetical docking complex formed between LM-TL (medium grey) and BPTI (grey-black). The BPTI docking is as in the trypsin-Arg15-BPTI complex. Only the cleft region of LM-TL is shown as well as the binding site of BPTI revealing the collision of Phe95 and Arg193 of LM-TL with the BPTI structure (grey dotted lines). (C) The ecotin-TSR R/R primary binding site presents most of its mutated residues involved in hydrogen bond interactions with the catalytic site of LM-TL. The Arg84 of ecotin-TSR R/R makes a salt bridge with the LM-TL site (Asp189), while Arg193 and Tyr172 are involved in hydrogen bond interactions, similar to fibrinogen. The hydrophobic site (Fp5, Trp99, Phe214, Trp215) of LM-TL partially interacts with Leu59 of ecotin-TSR R/R, and Ser195 of the catalytic triad interact with Phe214, not with His57. (D) The ecotin-TSR R/R secondary binding site's most important mutated position (Arg70) is involved in interactions with Asp100, Lys180, and Lys97. The α-helix C-terminal portion of LM-TL composed of Asp233, Tyr234, Ser241, and Ile243 also interacts with Phe65, Gly66, Trp67, and Gly68* (Reproduced with permission from Castro et al.[162] and Elsevier)

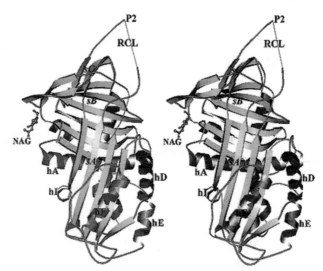

Figure 97   *Glycosylated PEDF in the typical serpin orientation. β-Strands, α-helices, coils and turns are shown in different shades of grey, and the carbohydrate residue is in ball-and-stick form. The reactive centre loop is RCL and N-acetyl-glucosamine is NAG*
(Reproduced with permission from Simonovic et al.,[163] copyright 2001, National Academy of Sciences, USA)

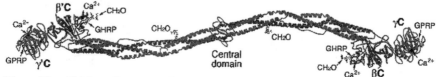

Figure 98   *Chicken fibrinogen showing distal globular domains connected by three-stranded coiled coils that meet at a central focus*
(Reproduced with permission from Yang et al.[167])

binding site.[167] A notable feature is the high degree of disorder in amino-terminal a and b chains, as well as in the aC domain, shown in Figures 98 and 99.

7.2.24 *Oxygenases and Oxidases* (*see also Ref.* 126). Four of the six classes of copper oxidase or mono-oxygenase enzymes have a chemically modified amino acid residue bonded to, or near, the active-site copper centre.[168] These are a topaquinone residue (S-cysteinyl-Tyr), and N^r-histidinyl-Tyr, lysyl-tyrosine-quinone and (S-cysteinyl-His) residues. The mechanism of action that has been proposed calls for an explanation of the puzzling fact that nucleophilic phenol and imidazole rings are coupled with non-electrophiles (SH groups and imidazolyl rings) in the process that generates these modified amino acid residues.

7.2.25 *Transformylases.* The active site of glycinamide ribonucleotide trans-formylase (GAR Tfase; Figures 100 – 103) has been shown to accumulate an enzyme-assembled multisubstrate adduct epoxide. This leads to a potential

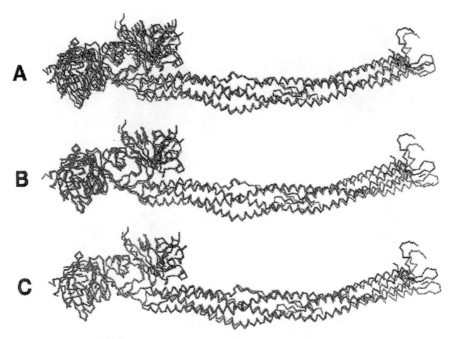

Figure 99   *Central domain of chicken fibrinogen (three different views) showing associ-*
ation of the two halves by three short β-sheets and penetration of chain from
opposite sides (α-chains, medium grey; β-chains, light grey; γ-chains, dark grey).
The γ-γ 'bow tie' is on the dorsal side of the molecule. The N-terminal sections of
α- and β-chains both emerge from the ventral side
(Reproduced with permission from Yang et al.[167])

application for epoxides, and perhaps aziridines, as potential GAR Tfase inhibitors.[169] 10-Bromo-10-bromomethyl-5,8,10-trideazafolic acid (10-formyl-TDAF)
exhibits time-dependent inhibition of the enzyme, now ascribed to the accumulation of the adduct in the active site.

7.2.26 Synthases. This is a loosely-defined class of enzyme that operates in
cellular construction machinery, not only to create small aliphatic structures, but
also larger species, as is the case with prostaglandin H synthase-I. This enzyme,
operating a key step in the conversion of arachidonic acid into prostaglandin
G^2/H^2, is of considerable mechanistic interest in undergoing irreversible self-
inactivation.[170]

1-Aminocyclopropane-1-carboxylate synthase has Glu47 as a major specificity determinant, positioned near the sulfonium grouping of (S,S)-S-adenosyl-L-
methionine (SAM) in the enzyme/SAM complex (Figure 104).[171] This result
follows from a study of models of the SAM-PLP quinonoid intermediate. The
possible elimination mechanisms leading to the amino acid (either with, or
without, a transient ketimine intermediate), are shown in Scheme 11. The
Scheme accommodates the role of Lys273 as proton source and proton abstrac-
tor.

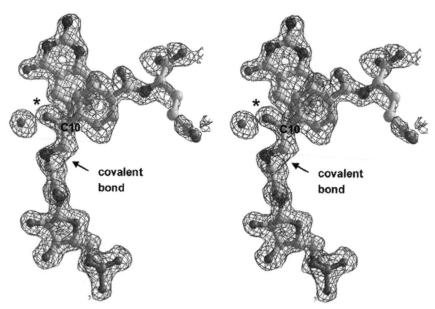

Figure 100 *Electron density for the multisubstrate adduct formed between β-GAR and the dibromide (10-formyl-TDAF), clearly showing connecting density between the substrate and inhibitor and insufficient electron density for a tertiary bromide. The C_{10} carbon has been labelled for clarity*
(Reproduced with permission from Greasley et al.[169])

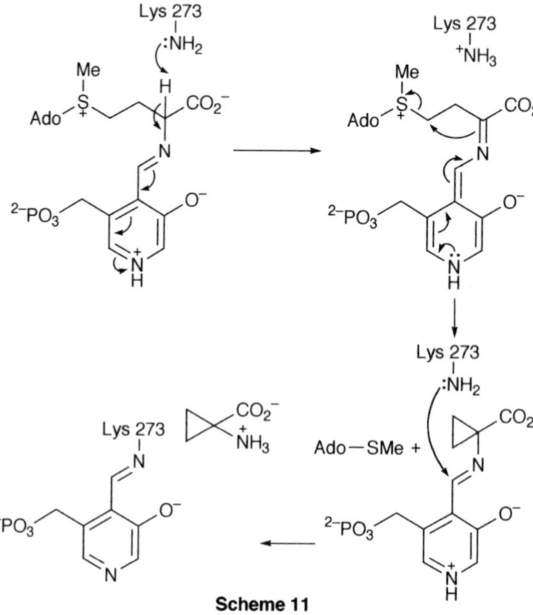

Scheme 11

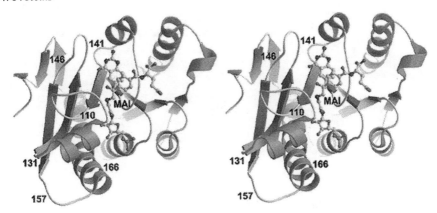

Figure 101 *Stereoview of GAR Tfase showing the multisubstrate adduct (MAI) formed between β-GAR and 10-formyl-TDAF. The labelled regions correspond to loops that contain the more disordered regions of the structure. Helices are shown in medium grey, β-strands in dark grey, and connecting loops in light grey. The MAI is shown as a ball-and-stick representation*
(Reproduced with permission from Greasley et al.[169])

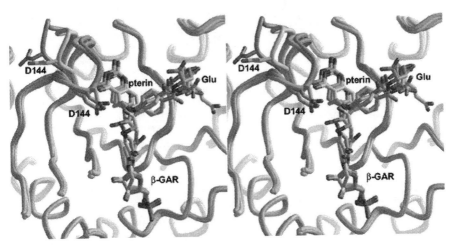

Figure 102 *Active site of GAR Tfase illustrating the different conformations of the flexible 'folate dependent' loop 141-146, and substrate/inhibitor structures. The structure in light grey is the current structure while all other models are dark grey, except for the flexible loop (Asp144 is also shown) which is medium grey (10-formyl-5,8,10-trideazafolic acid), and other shades for various other inhibitors. The pterin and glutamate moieties of the folate analogues are highlighted, as is the substrate, β-GAR*
(Reproduced with permission from Greasley et al.[169])

7.2.27 Heme-binding Enzymes. Site directed mutagenesis studies in which Phe393 is substituted by Ala, His, or Tyr have been conducted for flavocytochrome P450 BM3. The residue is close to the heme moiety, and its ligand, Cys40. The heme is influenced by Phe393, to the extent of controlling the delicate

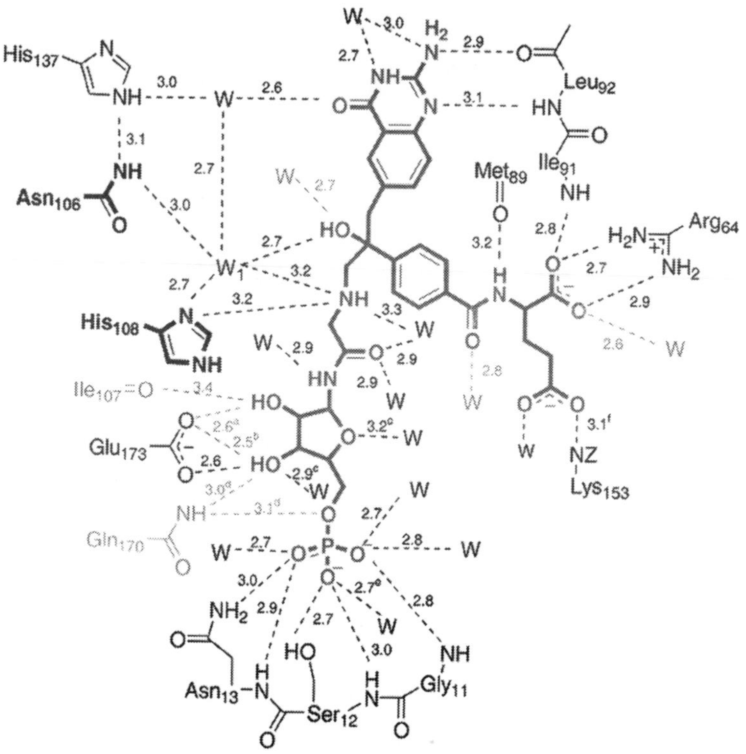

Figure 103 *Potential hydrogen bond interactions between GAR Tfase and the enzyme-assembled multisubstrate adduct (dark grey). Hydrogen bonds are shown in medium grey or light grey to emphasize that they can occur in some molecules and not others. Catalytic residues Asn106 and His108 are shown in bold (Reproduced with permission from Greasley et al.[169])*

equilibrium involving rate of heme reduction and the rate at which ferrous heme can bind and then reduce molecular oxygen.[172]

Protein engineering of *Bacillus megaterium* CYP102 provides cytochrome P450 (CYP) enzymes for study. These enzymes are involved in activating the carcinogenicity of polycyclic aromatic hydrocarbons, but are also utilized in this microorganism for the degradation of these compounds.[173] Clearly, there are potential environmental benefits possible through exploitation of this microorganism, adding to the recent example (Vol. 33, p.408) reporting the enzymic degradation of polychlorinated biphenyls.

The cytochrome P450 monooxygenase family includes species such as CYP11B1 (P450₁₁β) from yeast, which performs 11β-hydroxylation of steroids (conversion of 11-deoxycortisol into hydrocortisone).[174] The presence in mitochondria of a cluster of similar proteins that deal with steroid hormone production has long been a focus of interest in view of the physiological importance of their function.

Contexts for cytochrome P450 are discussed also in Ref. 6.

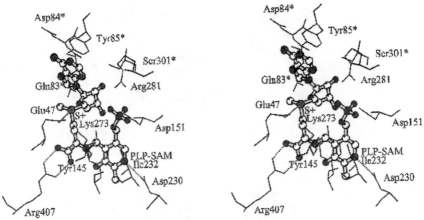

Figure 104 *Quinonoid intermediate of the reaction of ACC synthase with (S,S)-SAM. Several important residues around the adduct, which is depicted in ball-and-stick mode, are represented in wire-frame mode. Some important hydrogen bonds are represented as dashed lines. A solid line connects the S atom of the adduct and the side chain carboxylate of Glu47 (distance of 4.5 Å). Asterisks denote the residues that belong to the neighbouring subunit* (Reproduced with permission from McCarthy et al.[171])

7.2.28 Peptidases. Extracellular proteolysis that is needed in the adult central nervous system for cell migration and neurite outgrowth is believed to be conducted by tissue plasminogen activator (t-PA). This enzyme is expressed abundantly in neural tissues, both during development and nerve regeneration.[175] The structure (Figure 105) contains an arginine residue near the crucial Asn448 residue that mediates N-glycosylation, and multiply sulfated glycans in the final structure are present to the extent of about 20 novel bi- and triantennary N-glycans.

An atypical papain family peptidase, dipeptidyl peptidase I (DPPI; *alias* cathepsin C), appears to serve an essential function in the activation of a number of granule serine peptidases involved in cell-mediated apoptosis, inflammation, and connective tissue remodelling. These are just a few examples among several other roles taken by this enzyme.[176] Damaged biosynthesis of the enzyme leads to the autosomal recessive disorder 'Papillon-Lefevre syndrome', and clinical benefits should accrue from better understanding of its biochemistry. Crystal structure analysis providing more detail on the unique tetrameric structure (Figures 106, 107) can only promote this aim, and substrate binding and activation have also become better understood from this detailed study.

Mesophilic pyroglutamyl peptidase from *Bacillus amyloliquefaciens* and the corresponding thermophilic enzyme from *Thermococcus litoralis* have been compared through standard procedures. Site-directed mutagenesis of the cysteine replacement for Ser185 was performed, guided by the fact that the serine residue in the former enzyme is aligned with Cys190 of the thermophilic enzyme.[177] Since the catalytic efficiency of the enzyme was not impaired by this change, but temperatures some 30°C higher were tolerated by the mutant, it is clear that the

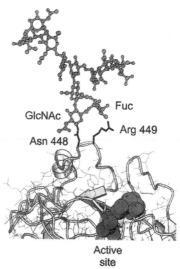

GlcNAc

Fuc

Arg 449

Asn 448

Active
site

Figure 105   *The serine protease domain of human t-PA, showing the proximity of Arg449 to the core residues of the glycan bonded at Asn448. The glycan shown is sulfated on either the first GlcNAc or the core fucose residue (Fuc)*
(Reproduced with permission from Zamze et al.[175])

interunit disulfide bond is an important factor in accounting for the thermal stability.

The essential S_2 pocket subsite of trypanosomal cathepsin L-like cysteine proteases has been probed with substrates constructed with phenylalanine-based amino acids, from the point of view of establishing structural features of viable peptidyl substrates. A model of one enzyme of this group, congopain, was constructed to study the fit of phenylalanine analogues within the S_2 pocket.[178] Stuctures of congopain and other structures for comparison, are shown in Figures 108 and 109, including the structure of cruzain derivatized with a dipeptidyl-fluoromethylketone (Z-Tyr-Ala-FMK as reagent).

These studies have led to the discovery of novel reversible inhibitors of cruzain and congopain, tetrapeptides of the general structure Cha-X^1-X^2-Pro (where Cha is a cyclohexylalanine residue, and X^1 and X^2 are phenylalanine analogues).[179]

The use of non-aqueous media for X-ray crystal analysis of proteins is showing some value in giving cleaner pictures of disordered regions on the molecule. The crystal structure of bovine β-trypsin complexed with an artificial mung bean inhibitor (a 22-residue synthetic peptide) has been described in this context.[180] The complex was soaked in neat cyclohexane to reduce conformational mobility, giving an electron density map carrying all residues of the inhibitor within the protein architecture.

Further examples of protein studies aimed at potential clinical applications include the determination of the crystal structure of Fab198, the rat monoclonal antibody 198 that protects the main immunogenic region of the human muscle acetylcholine receptor against the destructive action of *Myasthenia gravis* antibodies (Figures 110 – 112). Loops within the structure that make a major

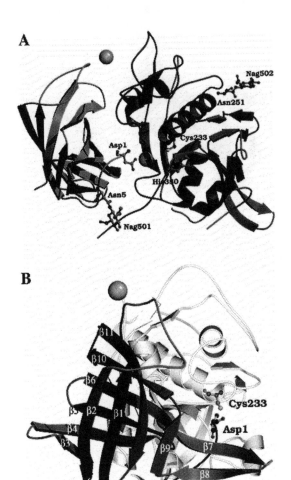

Figure 106 *Fold and domain structure of DPPI. (A) The three polypeptide chains of the subunit in the activated enzyme include (to the right) the papain fold domain. The light grey spheres are chloride ions. Active site residues Cys233 and His380 as well as N-acetylglucosamine residues and their associated Asn moieties are shown in ball-and-stick representation. (B) As in (A), rotated to show the β-barrel and the active site. The eight-stranded β-barrel pro-protein, the chloride ions, and the active site nucleophile Cys233 are shown.*
(Reproduced with permission from Olsen et al.,[176] FEBS Lett., Elsevier)

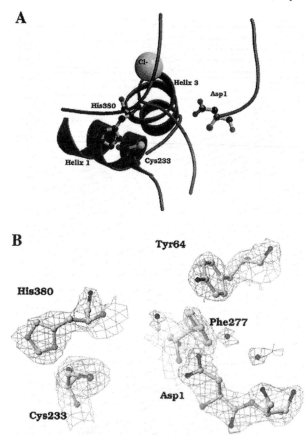

Figure 107 *(A) The DPPI active site with catalytic residues Cys233 and His380, showing residue Asp1 responsible for substrate binding. (B) Electron density map featuring the active site/substrate binding cleft contours*
(Reproduced with permission from Olsen et al.,[176] FEBS Lett., Elsevier)

contribution to binding, and residues of potential importance for antigen binding, were identified by computer modelling.[181]

7.2.29 S-Transferases. Theta class maize glutathione S-transferase has Asn49 in the Type 1 β-turn formed by residues 49-52, and this residue is involved in extensive hydrogen-bonding interactions between α-helix 2 and the rest of the N-terminal domain.[182] Site-directed mutagenesis from Asn49 to alanine, and physical studies of the resulting enzyme (Figures 113, 114), reveals the importance of this asparagine residue in modulating substrate binding, catalysis and intersubunit communication.

α-Class maize glutathione S-transferases regulate the intracellular concentrations of lipid peroxidation products that may be involved in signalling mechanisms of apoptosis.[183] Important properties of these enzymes are now shown to include protection against lipid peroxidation, protection against hydrogen per-

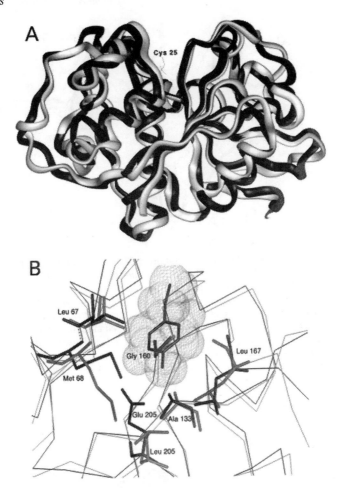

Figure 108   *(A) Superimposition of congopain (catalytic domain; grey) on cruzain (black).
(B) Structure of the S_2 subsite of congopain (grey). The S_2 pocket of cruzain,
corresponding to the cruzain/Z-Tyr-Ala-FMK complex, is superimposed
(dark). For the sake of clarity, only the tyrosyl side chain (P_2 position) of the
peptidyl-FMK is represented here. Residue numbers correspond to papain
numbering*
(Reproduced with permission from Lecaille et al.,[178] *Eur.J.Biochem.*, Blackwell Science)

oxide-induced apoptosis and inhibition of JNK and caspase-3 activation.

Rat glutathione S-transferase variants (GST T1) from *Methylobacterium di-
chloromethanicum* are of considerable interest as a part of the metabolic machinery of this bacterium. The fascination lies in the ability of the organism to live
with dichloromethane as sole carbon source, from which it generates formaldehyde. The production of CH_2Cl_2-resistant transconjugants, that still express the
GST T1 protein, has allowed a routine study to be undertaken, that establishes
the involvement of Glu102 and Arg107 in the catalytic mechanism for GST.[184]

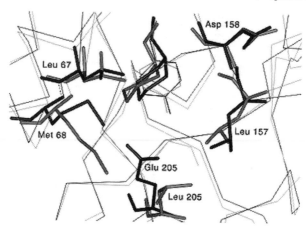

Figure 109   *Conformation of the cyclohexylalanyl (Cha) side chain inside the S2 subsite of trypanosomal cysteine proteases. Residues of the S2 subsite are shown in grey (thick line) for congopain and in dark (thick line) for cruzain. Cha inside the S2 pocket of trypanosomatid cysteine proteases is represented in grey (thick line) for congopain or in dark (thick line) for cruzain. Tyr residue (dark, thin line), issued from the crystal structure of complexed cruzain which has been used as control, has been superimposed to Cha at P2 and compared to the predicted position of the tyrosyl side chain (grey, thin line)*
(Reproduced with permission from Lecaille et al.,[178] *Eur.J.Biochem.*, Blackwell Science)

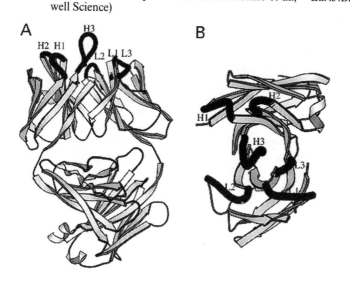

Figure 110   *Fab198 shows the typical immunoglobulin fold. Complementarity-determining regions (CDRs) are shown in black. (A) is a representation viewed to show the variable chains at the top and the constant features at the bottom. (B), As (A), viewed from the top, showing access to the binding crevice*
(Reproduced with permission from Poulas et al.,[181] *Eur.J.Biochem.*, Blackwell Science)

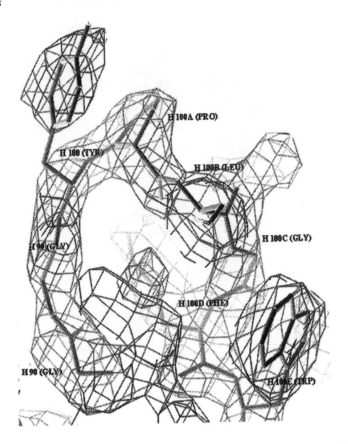

Figure 111 *View centred at the CDR H3 loop (residues H98-H100F) with a water molecule in the middle*
(Reproduced with permission from Poulas et al.,[181] *Eur.J.Biochem.,* Blackwell Science)

8   Proteins as Enzyme Inhibitors

A 322-residue protein (BJ46a) from *Bothrops jararaca* contains four N-glycosylation sites. It is a snake venom constituent that has the property of inhibiting metalloproteinases, but more detail needs to be teased out from the structure to account for its mode of action.[185] The extraordinary feature, that one inhibitor molecule forms a non-covalent complex with two molecules of enzyme, suggests an unusual mechanism of inhibition.

9   Enzyme Inhibition by Non-Protein Species

Insect silk from *Galleria mellonella* consists of a fibrous protein (fibroin) and a sticky protein (sericin). There are also some peptides, two of which are shown to

CDR's

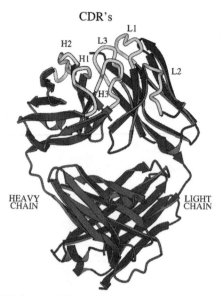

Figure 112   *Fab192 showing CDR loops, to be compared with Fab198*
(Reproduced with permission from Poulas et al.,[181] *Eur.J.Biochem.,* Blackwell
Science)

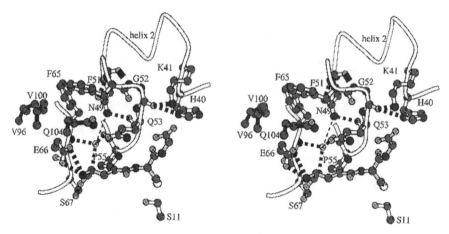

Figure 113   *The Asn49 residue and other important nearby residues of maize glutathione
S-transferase-I. An isolated water molecule is also important, represented as a
sphere (near Pro55), and hydrogen bonds are signalled by dotted lines*
(Reproduced with permission from Labrou et al.,[182] *Eur.J.Biochem.,* Blackwell Science)

be bacterial and fungal proteinase inhibitors; one is a typical Kunitz-type inhibitor, and the other is a Kazak-type inhibitor.[186] Their natural role is presumably
to protect the silk against microbial destruction.

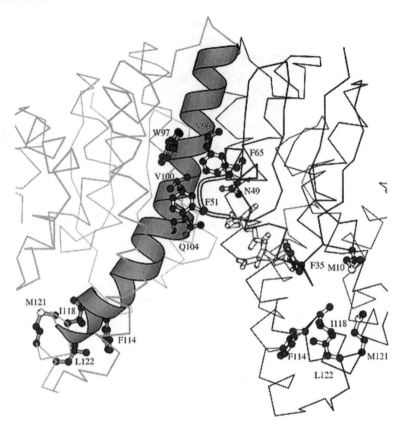

Figure 114 *Possible mode of communication between H-sites in the Asn49 – Ala mutant. The β-turn containing Asn49 is adjacent to the long helix of the alternate subunit, depicted as a ribbon. Glutathione is shown as an all-white ball-and-stick model*
(Reproduced with permission from Labrou et al.,[182] *Eur.J.Biochem.*, Blackwell Science)

Tetrathiomolybdate reacts readily with copper ions, and inhibits *Enterococcus hirae* CopB copper ATPase.[187] Inhibition is reversible through the addition of copper or silver salts. Copper is an essential component of more than 30 currently known enzymes, notably superoxide dismutase (clearly a vital enzyme, because it combats cellular radical formation).

Molybdenum uptake and utilization in bacteria is regulated by the molybdate-dependent transcriptional regulator known as ModE, from *E.coli*. The topic has been studied by several research groups in the past, and new X-ray studies show that oxyanion binding alters the conformation and quaternary structure of the protein (Figure 115).[188] Although there is still uncertainty as to how oxyanions are bound in proteins generally, there is evidence for important roles for tryptophan residues in this context.

Antagonists of α₄-integrins have considerable potential as anti-inflammatory

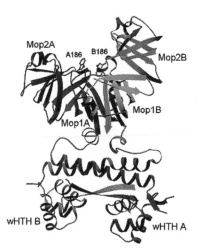

Figure 115 *Apo-ModE, in which Trp18 is shown (twice; labelled as A186 and B186 to indicate the subunit) in ball-and-stick representation. N-Terminal DNA-binding domains are dark grey and medium grey, and the helix-turn-helix motifs are labelled wHTHA and wHTHB. The molybdate-binding domains are labelled Mop1A, Mop2A, Mop1B, Mop2B*
(Reproduced with permission from Gourley et al.[188])

agents, and the osteopontin motif Ser-Val-Val-Tyr-Gly-Leu-Arg within the 158-168 residue region Gly-Arg-Gly-Asp-Ser-Val-Val-Tyr-Gly-Leu-Arg is important as a site for a binding interaction between these species.[189] The integrins are a large family of transmembrane receptors that mediate both cell-cell and cel-matrix interactions. The important elements in the interaction in the present example are Leu167 and the C-terminal residue Arg168 that is exposed by thrombin cleavage of the protein.

10   Other Biological Functions for Proteins

Levels of cyclin E are higher than normal in archipeligo mutant cells and in some tumour cell lines. Several research groups have identified this protein to be as a crucial factor that ensures the degradation of a regulator of the cell-division cycle, by controlling cyclin-dependent kinases.[190]

11   Other Protein Studies

11.1 Assisted Biosynthesis of Proteins. – The biosynthesis of proteins is accomplished efficiently, and for a diverse range of structures, but operates within the relatively small number of amino acids specified in the genetic code. A study leading to the incorporation of a versatile set of phenylalanine analogues (p-iodo-, p-cyano-, p-ethynyl-, p-azido-, and 4-pyridylalanine) into proteins has been described, using a mutant form of the *E.coli* phenylalanyl-tRNA synthetase in which Ala294 was substituted by glycine, in bacterial hosts.[191]

Hexafluoroleucine has been incorporated in a similar way, to lead to a novel highly stable coiled-coil protein.[192]

Several new methionine analogues (cis-crotylglycine, 2-aminoheptanoic acid, norvaline, 2-butynylglycine and allylglycine) have been incorporated into proteins of *E.coli*, with the help of a methionyl-tRNA synthetase.[193] This work reveals the more flexible limits of the protein synthesis machinery in the context of amino acids that can be involved. The boundaries were being shown to extend beyond the classic coded amino acids in recent years, and methods used in the present work are going to bring many other new substrates into novel biosynthesized proteins.

It is usually difficult to incorporate amino acids carrying fluorophores into proteins. Progress has now been made in this endeavour, and highly fluorescent β-anthraniloyl-L-αβ-diaminopropionic acid (λ_{em} beyond 400 nm) has been incorporated into various sites in streptavidin, as directed by a CGGG/CCCG four-base codon/anticodon pair. Position 120 of the resulting protein turns out to be a useful test-bed for binding studies – e.g. biotin binding in the vicinity of the fluorophore is revealed through changes in fluorescence wavelength and intensity.[194]

11.2 Disulfide Interchange. – Glutathione undergoes disulfide interchange with metallothionein (a zinc protein) catalysed by organoselenium compounds. It is suggested that the pharmacological actions of organoselenium compounds undergoing study for cancer prevention, and other antiviral and anti-inflammatory therapeutic applications, may relate to their role in coupling the thiol redox state and the zinc state.[195] A reversible cycle is generated in this system that either mobilizes zinc from the metallothionein, or releases thionein, which is thus made available to bind zinc.

12 Other Protein Properties

Myoglobin has been suggested to be a nitric oxide scavenger, preventing cellular respiration and explaining the preferential colonization of heart and skeletal muscle by *Trypanosoma cruzi*.[196] This pathogen is the cause of Chagas disease, a progressive fatal cardiomyopathy that is widespread in Central and South America, and myoglobin is perhaps capable of protecting from the antiparasitic effects of nitric oxide.

Recombinant tyrosine aminotransferase from *Trypanosoma cruzi* exhibits identical properties to the native enzyme. It resists limited tryptic cleavage and contains no disulfide bond. Its broad substrate specificity is notable (valine, isoleucine, aspartic acid and glutamic acid are poor substrates; polar aliphatic amino acids are not transaminated; several 2-oxo-acids are accepted in addition to the expected pyruvate, oxaloacetate and 2-oxoglutarate). Variants of the enzyme were prepared for this study, with native residues replaced with Arg389

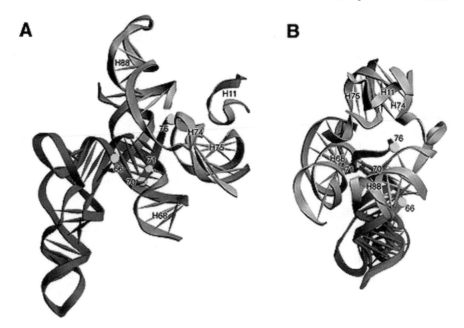

Figure 116   *Interaction of E-site tRNA with the 50S subunit. (A). View from the side with TψC-loop in front and the D-loop behind, so that positions 66 and 76 are in front, and 70 and 71 are behind (light grey spheres). (B). View from the top of the acceptor stem-TψC-helix with the anticodon stem pointing into the paper. Positions 66 and 76 are on the surface to the right, and positions 70 and 71 to the left*
(Reproduced with permission from Feinberg and Joseph,[199] copyright 2001, National Academy of Sciences, USA)

and Arg283. The former mutant was completely inactive, and the latter was fully active, thus confirming proposals for the molecular basis of action of this enzyme.[197]

Chagasin from *Trypanosoma cruzi* is the first member of a new family of cysteine protease inhibitors. A modelling study of chagasin leads to the conclusion that the protein exemplifies a new class of protease inhibitor, enclosing an immunoglobulin-like domain that is unprecedented in protein inhibitors. The presence of the sequence His-Asn-Gly-Ala in positions 8-11 is essential for inhibition by peptides of the cathepsin L-like protease, cruzipain.[198]

In protein synthesis, there is a stepwise movement (translocation) as each amino acid residue is built into the growing polypeptide chain, and work aiming at describing the essential molecular interactions that lead to this motion has been published.[199] Translocation of the tRNA-mRNA complex is a fundamental step, and it is now shown that the ribosome can translocate a P-site-bound tRNA^Met with a break in the phosphodiester backbone between positions 56 and 57 in the TψC-loop (Figure 116).

Monellin is an intensely sweet protein, in the native form consisting of two

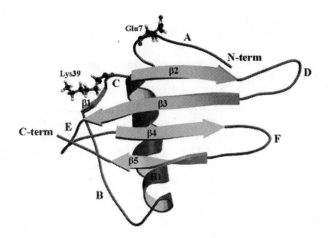

Figure 117 *Ordered structural elements and relative orientations of secondary structures for single chain monellin, also showing mutated residues with side-chains as ball-and-stick representations*
(Reproduced with permission from Sung et al.[200])

chains. Single-chain monellin, an engineered 94-residue polypeptide, is equally sweet; its solution structure and backbone dynamics have been assessed through a suite of techniques, notably circular dichroism, fluorescence spectroscopy and NMR.[200] Clearly, an objective for molecular structural studies of such proteins is the interpretation of structure in terms of perceived sweetness. Mutant single-chain monellins were prepared in this study, with changes at Glu2, Asp7, and Arg39, the structures and properties of the proteins indicating an important role for these particular residues (Figure 117).

References

1. *Protein – Protein Interactions: A Molecular Cloning Manual*, E.A.Golemis, ed., Cold Spring Harbor Laboratory Press, Cold Spring Harbor, New York, 2001.
2. *Modern Protein Chemistry: Practical Aspects*, G.C.Howard and W.E.Brown, eds., CRC Press, Boca Raton, Florida, 2001.
3. *The Physical Foundation of Protein Architecture*, N.Saito and Y.Kobayashi, World Scientific Publishing, Tokyo, 2001.
4. *Perspectives in Amino Acid and Protein Geochemistry*, G. A. Goodfriend, M. J. Collins, M. L. Fogel, S. A. Macko, and J. F. Wehmiller, eds., Oxford University Press, Oxford, 2001.
5. *Protein Ligand Interactions (Practical Approach Series)*, S. E. Harding and B. Chowdhry, eds., Oxford University Press, Oxford, 2001.
6. *Guide to Cytochrome P450: Structure and Function*, D. F. V. Lewis, Taylor & Francis, Andover, 2001.
7. *Cloning, Gene Expression, and Protein Purification: Experimental Procedures and Process Rationale*, C. C. Hardin, ed., Oxford University Press, Oxford, 2001.
8. *Nature's Robots: A History of Proteins*, C. Tanford and J. Reynolds, Oxford

University Press, Oxford, 2001.

9. *Protein – Protein Recognition*, C. Kleantous, ed., Oxford University Press, Oxford, 2001.

10. *Advances in Protein Chemistry: Protein Modules and Protein-Protein Interactions*, Janin, Wodsak, Academic Press, New York, 2002.

11. *Biological Sequence Analysis: Probabilistic Models of Proteins and Nucleic Acids*, R. Durbin, S. R. Eddy, A. Krogh and G. Mitchison, Cambridge University Press, Cambridge, 1998; *The Protein Protocols Handbook*, J. M. Walker, ed., Humana Press, Totowa, NJ, 1996; *The Chaperonins*, R. J. Ellis, ed., San Diego Press, CA, 1996; *Protein Folding*, C. M. Dobson and A. R. Fersht, eds., Cambridge University Press, Cambridge, UK, 1995; *Fundamentals of Enzymology (Third Edition)*, N. C. Price and L. Stevens, Oxford University Press, Oxford, 1999; *Protein – Solvent Interaction*, R. B. Gregory, Dekker, New York, 1995; *Biotechnology: Proteins to PCR: A Course in Strategies and Laboratory Techniques*, D. Burden and D. Whitney, Birkhauser Verlag, Zurich, 1995; *Protein Analysis and Purification: Benchtop Techniques*, I. Rosenberg, Birkhauser Verlag, Zurich, 1996.

12. e. g., M. Gross, *Chem. Brit.*, 2002, Vol. 38, No. 12, p. 24.

13. *Proteome Research: New Frontiers in Functional Genomics*, M. R. Wilkins, ed., Springer-Verlag, Berlin and Heidelberg, 1997. ; see also 'A Trends Guide to Proteomics', W. Blackstock and M. Mann, eds., *Trends Biotechnol.*, 2001, **19**(Suppl.).

14. *Proteome Research: Mass Spectrometry*, P. James, ed., Springer-Verlag, Berlin and Heidelberg, 2001.

15. *Post-Translational Modification of Proteins: Tools for Functional Proteomics (Methods of Molecular Biology*, Vol. 194), C. Kannicht, ed., Humana Press, Totowa, NJ, 2002.

16. *Proteomics in Practice, A Laboratory Handbook of Proteome Analysis*, R. Westermeier and T. Naven, Wiley, New York, 2002.

17. *Protein Structure Prediction: Bioinformatic Approach*, I. F. Tsigelny, International University Line, New York, 2002.

18. *Proteome Research: Two-Dimensional Gel Electrophoresis and Detection Methods (Principles and Practice)*, T. Rabilloud, ed., Springer-Verlag, Berlin and Heidelberg, 1999.

19. *Discovering Genomics, Proteomics, and Bioinformatics*, A. M. Campbell and L. J. Heyer, Addison Wesley, New York, 2002.

20. *Proteomics*, T. Palzkill, Kluwer Academic Publishers, Dordrecht, 2002.

21. *Proteins and Proteomics: A Laboratory Manual*, R. Simpson and J. L. Hotchkiss, eds., Cold Spring Harbor Laboratory Press, Cold Spring Harbor, New York, 2002.

22. *The Proteome Revisited: Theory and Practice of All Relevant Electrophoretic Steps* (Journal of Chromatography Library), P. G. Righetti, A. Stoyanov and M. Y. Zhukov, eds., Elsevier, Amsterdam, 2001.

23. *Proteomics Reviews 2001*, M. J. Dunn, Wiley-VCH, New York, 2001.

24. *Proteins: Biochemistry and Biotechnology*, G. Walsh, Wiley, New York, 2001.

25. *Introduction to Bioinformatics*, A. M. Lesk, Oxford University Press, Oxford, 002.

26. *Bioinformatics, Sequence, Structure, and Databanks*, D. Higgins and W. Tylor, Oxford University Press, Oxford, 2000.

27. *Introduction to Bioinformatics*, T. K. Attwood and D. J. Parry-Smith, Pearson, Harlow, 1999.

28. *Interpreting Protein Mass Spectra: A Comprehensive Resource*, P. Snyder, American Chemical Society, 2000.

29. *MS For Biotechnology*, G. Siuzdak, Academic Press, New York, 1996.

30. Y.-M. She, S. Haber, D. L. Seifers, A. Loboda, I. Chernushevich, H. Perreault, W. Ens and K. G. Standing, *J. Biol. Chem.*, 2001, **276**, 20039.

31. *Protein NMR for the Millennium (Biological Magnetic Resonance)*, N. Rama Krishna and L. J. Berliner, eds., Kluwer Academic/Plenum Publishers, Dordrecht, 2002.

32. *Biomolecular NMR Spectroscopy*, J. N. S. Evans, Oxford University Press, Oxford, 1995.

33. R. Kitahara, H. Yamada and K. Akasaka, *Biochemistry*, 2001, **40**, 13556.

34. K. Inoue, T. Maurer, H. Yamada, C. Herrmann, G. Horn, H. R. Kalbitzer and K. Akasaka, *FEBS Lett.*, 2001, **506**, 180.

35. S. Tu and R. A. Cerione, *J. Biol. Chem.*, 2001, **276**, 19656.

36. J. A. R. Worrall, U. Kolczak, G. W. Canters and M. Ubbink, *Biochemistry*, 2001, **40**, 7069.

37. G. T. Marks, T. K. Harris, M. A. Massiah, A. S. Mildvan and D. H. T. Harrison, *Biochemistry*, 2001, **40**, 6805.

38. M. M. Spence, S. M. Rubin, I. E. Dimitrov, E. J. Ruiz, D. E. Wemmer, A. Pines, S. Q. Yao, F. Tian and P. G. Schultz, *Proc. Nat. Acad. Sci. USA*, 2001, **98**, 10654.

39. J. D. Brady and S. P. Robins, *J. Biol. Chem.*, 2001, **276**, 18812.

40. P. Jordan, P. Fromme, H. T. Witt, O. Klukas, W. Saenger and N. Krauss, *Nature*, 2001, **411**, 909.

41. P. Fromme, J. Kern, B. Loll, J. Biesiadka, W. Saenger, H. T. Witt, N. Krauss and A. Zouni, *Phil. Trans. R. Soc. Lond. B*, 2002, **357**, 1337.

42. F. van den Ent, L. A. Amos and J. Lowe, *Nature*, 2001, **413**, 39 (for a commentary on this paper and its background, see H. P. Erickson, *Idem*, p. 30).

43. M. T. Hilgers and M. L. Ludwig, *Proc. Nat. Acad. Sci. USA*, 2001, **98**, 11169.

44. G. C. Barrett, in *The Chemistry of Sulfenic Acids and Their Derivatives*, S. Patai, ed., Wiley, New York, 1990, p. 1.

45. J. Wang, R. Meijers, Y. Xiong, J. Liu, T. Sakihama, R. Zhang, A. Joachimiak and E. L. Reinherz, *Proc. Nat. Acad. Sci. USA*, 2001, **98**, 10799.

46. T. J. Allison, C. C. Winter, J.-J. Fournie, M. Boneville and D. N. Garboczi, *Nature*, 2001, **411**, 820.

47. C. C. Stamper, Y. Zhang, J. F. Tobin, D. V. Erbe, S. Ikemizu, S. J. Davis, M. L. Stahl, J. Seehra, W. S. Somers and L. Mosyak, *Nature*, 2001, **410**, 608.

48. J.-C. D. Schwartz, X. Zhang, A. A. Fedorov, S. G. Nathenson and S. C. Almo, *Nature*, 2001, **410**, 604.

49. *Protein-Ligand Interactions: Structure and Spectroscopy: A Practical Approach*, S. E. Harding and B. Chowdhry, eds., Oxford University Press, Oxford, 2000.

50. K. Warncke and M. S. Perry, *Biochim. Biophys. Acta*, 2001, **1545**, 1.

51. N. Maruyama, M. Adachi, K. Takahashi, K. Yagasaki, M. Kohno, Y. Takenaka, E. Okuda, S. Nakagawa, B. Mikami and S. Utsumi, *Eur. J. Biochem.*, 2001, **268**, 3595.

52. B. Spolaore, R. Bermejo, M. Zambonin and A. Fontana, *Biochemistry*, 2001, **40**, 9460.

53. P. Polverino de Laureto, M. Donadi, E. Scaramella, E. Frare and A. Fontana, *Biochim. Biophys. Acta*, 2001, **1548**, 29.

54. *Protein Folding in the Cell* (Advances in Protein Chemistry, Vol. 59), A. Richard, A. Richards, and A. Horwich, eds., Academic Press, New York, 2001.

55. *Protein Flexibility and Folding* (Biological Modelling Series), L. A. Kuhn and M. F. Thorpe, Elsevier Science, Amsterdam, 2001.

56. *Protein Stability and Folding: A Collection of Thermodynamic Data*, W. Pfeil, Springer-Verlag, 2001.

57. *Protein Stability and Folding: A Collection of Thermodynamic Data*: Supplement, W. Pfeil, Springer-Verlag, 2001.

58. C. A. Andersen, H. Bohr and S. Brunak, *FEBS Lett.*, 2001, **507**, 6.

59. W. F. van Gunsteren, R. Burgi, C. Peter and X. Daura, *Angew. Chem. Int. Ed.*, 2001, **40**, 352.

60. S. V. Taylor, K. U. Walter, P. Kast and D. Hilvert, *Proc. Nat. Acad. Sci. USA*, 2001, **98**, 10596.

61. F. Sinibaldi, L. Fiiorucci, G. Mei, T. Ferri, A. Desideri, F. Ascoli and R. Santucci, *Eur. J. Biochem.*, 2001, **268**, 4537.

62. J. D. Dunitz, *Angew. Chem. Int. Ed.*, 2001, **40**, 4167.

63. M. M. Teeter, A. Yamano, B. Stec and U. Mohanty, *Proc. Nat. Acad. Sci. USA*, 2001, **98**, 11242.

64. J. Rumbley, L. Hoang, L. Mayne and S. W. Englander, *Proc. Nat. Acad. Sci. USA*, 2001, **98**, 105.

65. J. K. McIninch and E. R. Kantrowitz, *Biochim. Biophys. Acta*, 2001, **1547**, 320.

66. K. Ruan, C. Xu, Y. Yu, J. Li, R. Lange, N. Bec and C. Balny, *Eur. J. Biochem.*, 2001, **268**, 2742.

67. P. Polverino de Laureto, D. Vinante, E. Scaramella, E. Frare and A. Fontana, *Eur. J. Biochem.*, 2001, **268**, 4324.

68. S. Chakraborty, V. Ittah, P. Bai, L. Luo, E. Haas and Z. Peng, *Biochemistry*, 2001, **40**, 7228.

69. Q.-X. Hua, S. H. Nakagawa, W. Jia, S.-Q. Hu, Y.-C. Chu, P. G. Katsoyannis and M. A. Weiss, *Biochemistry*, 2001, **40**, 12299.

70. J. Deswarte, S. De Vos, U. Langhorst, J. Steyaert and R. Loris, *Eur. J. Biochem.*, 2001, **268**, 3993.

71. B. A. Bernat and R. N. Armstrong, *Biochemistry*, 2001, **40**, 12712.

72. R. Dallinger, Y. Wang, B. Berger, E. A. Mackay and J. H. R. Kagi, *Eur. J. Biochem.*, 2001, **268**, 4126.

73. A. Sharff, C. Fanutti, J. Shi, C. Calladine and B. Luisi, *Eur. J. Biochem.*, 2001, **268**, 5011.

74. L. Yue, J.-B. Peng, M. A. Hediger and D. E. Clapham, *Nature*, 2001, **410**, 705; see also J. W. Putney, *Nature*, 2001, **410**, 648, for a commentary on this paper.

75. M. A. Schumacher, A. F. Rivard, H. P. Bachinger and J. P. Adelman, *Nature*, 2001, **410**, 1120.

76. S. C. Gallagher, Z.-H. Gao, S. Li, R. B. Dyer, J. Trewhella and C. B. Klee, *Biochemistry*, 2001, **40**, 12094.

77. *Prion Proteins* (Advances in Protein Chemistry, Vol. 57), B. Caughey and B. Caughey, eds., Academic Press, New York, 2001.

78. B.-Y. Lu, P. J. Beck and J.-Y. Chang, *Eur. J. Biochem.*, 2001, **268**, 3767.

79. W.-Q. Zou, D.-S. Yang, P. E. Fraser, N. R. Cashman and A. Chakrabartty, *Eur. J. Biochem.*, 2001, **268**, 4885.

80. I. V. Baskakov, G. Legname, S. B. Prusiner and F. E. Cohen, *J. Biol. Chem.*, 2001, **276**, 19687.

81. E. Festy, L. Lins, G. Peranzi, J. N. Octave, R. Brasseur and A. Thomas, *Biochim. Biophys. Acta*, 2001, **1546**, 356.

82. C. C. Curtain, F. Ali, I. Volitakis, R. A. Cherny, R. S. Norton, K. Beyreuther, C. J. Barrow, C. L. Masters, A. I. Bush and K. J. Barnham, *J. Biol. Chem.*, 2001, **276**, 20466.

83. E. M. Kajkowski, C. F. Lo, X. Ning, S. Walker, H. J. Sofia, W. Wang, W,Edris, P. Chanda, E. Wagner, S. Vile, K. Ryan, B. McHendry-Rinde, S. C. Smith, A. Wood,

K. J. Rhodes, J. D. Kennedy, J. Bard, J. S. Jacobsen and B. A. Ozenberger, *J. Biol. Chem.*, 2001, **276**, 18748.

84. M. A. Leissring, T. R. Yamasaki, W. Wasco, J. D. Buxbaum, I. Parker and F. M. LaFerla, *Proc. Natl. Acad. Sci. U. S. A.*, 2000, **97**, 8590.

85. J. Walter, A. Schindzielorz, J. Grunberg and C. Haass, *Proc. Natl. Acad. Sci. U. S. A.*, 1999, **96**, 1391.

86. E.-K. Choi, N. F. Zaidi, J. S. Miller, A. C. Crowley, D. E. Merriam, C. Lilliehook, J. D. Buxbaum and W. Wasco, *J. Biol. Chem*, 2001, **276**, 19197.

87. M. Morillas, D. L. Vanik and W. K. Surewicz, *Biochemistry*, 2001, **40**, 6982.

88. C. Gabus, E. Derrington, P. Leblanc, J. Chnaiderman, D. Dormont, W. Swietnicki, M. Morillas, W. K. Surewicz, D. Marc, P. Nandi and J.-L. Darlix, *J. Biol. Chem.*, 2001, **276**, 19301.

89. *RNA Binding Proteins: New Concepts in Gene Regulation (Endocrine Updates)*, K. Sandberg and S. E. Mulroney, eds. Kluwer Academic Publishers, Dordrecht, 2001.

90. H. Lin and V. W. Cornish, *Angew. Chem. Int. Ed.* , 2001, **40**, 871.

91. R. I. Boysen and M. T. W. Hearn, *J. Peptide Res.*, 2001, **57**, 1.

92. H. Daiyasu, K. Osaka, Y. Ishimo and H. Toh, *FEBS Lett.* , 2001, **503**, 1.

93. K. Wroblewski, R. Muhandiram, A. Chakrabartty and A. Bennick, *Eur. J. Biochem.*, 2001, **268**, 4384.

94. S. Karthikeyan, T. Leung and J. A. A. Ladias, *J. Biol. Chem.*, 2001, **276**, 19683.

95. C. F. Mazitsos, D. J. Rigden, P. G. Tsoungas and Y. D. Clonis, *Eur. J. Biochem.* , 2002, **269**, 5391.

96. S. W. Vetter and E. Leclerc, *Eur. J. Biochem.* , 2001, **268**, 4292.

97. *Molecular Chaperones in the Cell (Frontiers in Molecular Biology)*, Ed. P. Lund, Oxford University Press, Oxford, 2001.

98. J. U. Wurthner, D. B. Frank, A. Felici, H. M. Green, Z. Cao, M. D. Schneider, J. G. McNally, R. J. Lechleider and A. B. Roberts, *J. Biol. Chem.*, 2001, **276**, 19495.

99. N. Yoshida, K. Oeda, E. Watanabe, T. Mikami, Y. Fukita, K. Nishimura, K. Komai and K. Matsuda, *Nature*, 2001, **411**, 44.

100. G. Xu, M. Cirilli, Y. Huang, R. L. Rich, D. G. Myszka and H. Wu, *Nature*, 2001, **410**, 494.

101. S. M. Srinivasula, R. Hegde, A. Saleh, P. Datta, E. Shiozaki, J. Chai, R.-A. Lee, P. D. Robbins, T. Fernandez-Alnemri, Y. Shi and E. S. Alnemri, *Nature*, 2001, **410**, 112.

102. V. Borutaite and G. C. Brown, *FEBS Lett.* , 2001, **500**, 114.

103. Y. Xu, P. D. Carr, T. Huber, A. G. Vasudevan and D. L. Ollis, *Eur. J. Biochem.* , 2001, **268**, 2028.

104. S. Beismann-Driemeyer and R. Sterner, *J. Biol. Chem.*, 2001, **276**, 20387.

105. S. Hovde, C. Abate-Shen and J. H. Geiger, *Biochemistry*, 2001, **40**, 12013.

106. J. R. Walker, R. A. Corpina and J. Goldberg, *Nature*, 2001, **412**, 607.

107. W. J. McGrath, M. L. Baniecki, C. Li, S. M. McWhirter, M. T. Brown, E. Peters, D. L. Toledo and W. F. Mangel, *Biochemistry*, 2001, **40**, 13237.

108. M. T. Brown, K. M. McBride, M. L. Baniecki, N. C. Reich, G. Marriott and W. F. Mangel, *J. Biol. Chem.* , 2002, **277**, 46298.

109. M. L. Baniecki, W. J. McGrath, S. M. McWhirter, C. Li, D. L. Toledo, P. Pellicena, D. L. Barnard, K. S. Thorn and W. F. Mangel, *Biochemistry*, 2001, **40**, 12349.

110. W. J. McGrath, M. L. Baniecki, E. Peters, D. T. Green and W. F. Mangel, *Biochemistry*, 2001, **40**, 14468.

111. M. Luic, S. Tomic, I. Lescic, E. Ljubovic, D. Sepac, V. Sunjic, L. Vitale, W. Saenger and B. Kojic-Prodic, *Eur. J. Biochem.* , 2001, **268**, 3964.

112. R. I. W. Osmond, M. Hrmova, F. Fontaine, A. Imberty and G. B. Fincher, *Eur. J.*

Biochem. , 2001, **268**, 4190.

113. A. B. Smit, N. I. Syed, D. Schaap, J. van Minnen, J. Klumperman, K. S. Kits, H. Lodder, R. C. van der Schors, R. van Elk, B. Sorgedrager, K. Brejc, T. K. Sixma and W. P. M. Geraerts, *Nature*, 2001, **411**, 261.

114. K. Brejc, W. J. van Dijk, R. V. Klaassen, M. Schuurmans, J. van der Oost, A. B. Smit and T. K. Sixma, *Nature*, 2001, **411**, 269.

115. Y. Ha, D. J. Stevens, J. J. Skehel and D. C. Wiley, *Proc. Nat. Acad. Sci. USA*, 2001, **98**, 11181.

116. *From Protein Folding to New Enzymes* (Biochemical Society Symposium 68), A. Berry and S. E. Radford, eds., Portland Press, London, 2001.

117. *Proteolytic Enzymes: A Practical Approach (Practical Approach Series)*, R. Beynon and J. S. Bond, eds., Oxford University Press, Oxford, 2001.

118. *Protein Kinase A and Human Disease, Ann. N. Y. Acad. Sci.*, 2002, Vol. 968, C. A. Stratakis and Y. S. Cho-Chung, eds.

119. E. A. MacGregor, S. Janacek, and B. Svensson, *Biochim. Biophys. Acta*, 2001, **1546**, 1.

120. *Mechanics of Motor Proteins and the Cytoskeleton*, J. Howard, Palgrave Macmillan, New York, 2001.

121. M. Kikkawa, E. S. Sablin, Y. Okade, H. Yajima, R. L. Fletteruch and N. Hirokawa, *Nature*, 2001, **411**, 439.

122. M. Machius, J. L. Chuang, R. M. Wynn, D. R. Tomchick and D. T. Chuang, *Proc. Nat. Acad. Sci. USA*, 2001, **98**, 11218.

123. C. Fabrega, M. A. Farrow, B. Mukhopadhyay, V. de Crecy-Laagard, A. R. Ortiz and P. Schimmel, *Nature*, 2001, **411**, 110.

124. L. Powers, A. Hillar and P. C. Loewen, *Biochim. Biophys. Acta*, 2001, **1546**, 44.

125. J. N. Scarsdale, G. Kazanina, X. He, K. A. Reynolds and H. T. Wright, *J. Biol. Chem.*, 2001, **276**, 20516.

126. C. M. Harris, L. Pollegioni and S. Ghisla, *Eur. J. Biochem.* , 2001, **268**, 5504.

127. A. Geyer, T. B. Fitzpatrick, P. D. Pawelek, K. Kitzing, A. Vrielink, S. Ghisla and P. Macheroux, *Eur. J. Biochem.* , 2001, **268**, 4044.

128. A. S. Hearn, M. E. Stroupe, D. E. Cabelli, J. R. Lepock, J. A. Tainer, H. S. Nick and D. N. Silverman, *Biochemistry*, 2001, **40**, 12051.

129. S. D. Edwards and N. H. Keep, *Biochemistry*, 2001, **40**, 7061.

130. W. Wang, Z. Fu, J. Z. Zhou, J.-J. P. Kim and C. Thorpe, *Biochemistry*, 2001, **40**, 12266.

131. P. Kedzierski, K. Moreton, A. R. Clarke and J. J. Holbrook, *Biochemistry*, 2001, **40**, 7247.

132. J. K. Rubach, S. Ramaswamy and B. V. Plapp, *Biochemistry*, 2001, **40**, 12686.

133. K. Tozawa, R. W. Broadhurst, A. R. C. Raine, C. Fuller, A. Alvarez, G. Guillen, G. Padron and R. N. Perham, *Eur. J. Biochem.* , 2001, **268**, 4908.

134. K. N. Houk, J. K. Lee, D. J. Tantillo, S. Bahmanyar and B. N. Hietbrink, *Chem. Biochem.* , 2001, **2**, 113.

135. D. L. Zechel and S. G. Withers, *Accounts Chem. Res.* , 2000, **33**, 11.

136. D. J. Vocadio, G. J. Davies, R. Laine and S. G. Withers, *Nature*, 2001, **412**, 835. For a discussion, see A. J. Kirby, *Nature Struct. Biol.* , 2001, **8**, 737.

137. O. Nashiru, D. C. Zechel, D. Stoll, T. Mohammadzadeh, R. A. J. Warren and S. G. Withers, *Angew. Chem. Int. Ed.* , 2001, **40**, 417.

138. F. D. Schubot, I. A. Kataeva, D. L. Blum, A. K. Shah, L. G. Ljungdahl, J. P. Rose and B.-C. Wang, *Biochemistry*, 2001, **40**, 12524.

139. D. Xu, D. P. Ballou and V. Massey, *Biochemistry*, 2001, **40**, 12369.

140. S. Marmor, C. P. Petersen, F. Reck, W. Yang, N. Gao and S. L. Fisher, *Biochemistry*, 2001, **40**, 12207.

141. J. W. Gross, A. D. Hegeman, B. Gerratana and P. A. Frey, *Biochemistry*, 2001, **40**, 12497.

142. H. Strasdeit, *Angew. Chem. Int. Ed.*, 2001, **40**, 707. T. W. Lane and F. M. M. Morel, *Proc. Nat. Acad. Sci. USA*, 2000, **97**, 4627.

143. L. Sharma and A. Sharma, *Eur. J. Biochem.*, 2001, **268**, 2456.

144. S. Gruneberg, B. Wendt and G. Klebe, *Angew. Chem. Int. Edit.*, 2001, **40**, 389; S. Gruneberg, M. T. Stubbs and G. Klebe, *J. Med. Chem.*, 2002, **45**, 3588.

145. S. Schmitt, M. Hendlich and G. Klebe, *Angew. Chem. Int. Edit.*, 2001, **40**, 3141.

146. C. G. Mowat, R. Moysey, C. S. Miles, D. Leys, M. K. Doherty, P. Taylor, M. D. Walkinshaw, G. A. Reid and S. K. Chapman, *Biochemistry*, 2001, **40**, 12292.

147. M. C. Bewley, C. C. Marohnic and M. J. Barber, *Biochemistry*, 2001, **40**, 13574.

148. A. M. Paiva, D. E. Vanderwall, J. S. Blanchard, J. W. Kozarich, J. W. Williamson and T. M. Kelly, *Biochim. Biophys. Acta*, 2001, **1545**, 67.

149. O. Einsle, S. Foerster, K. Mann, G. Fritz, A. Messerschmidt and P. M. H. Kroneck, *Eur. J. Biochem.*, 2001, **268**, 3028.

150. U. Schorken, S. Thorell, M. Schurmann, J. Jia, G. A. Sprenger and G. Schneider, *Eur. J. Biochem.*, 2001, **268**, 2408.

151. W. Cao and J. Lu, *Biochim. Biophys. Acta*, 2001, **1546**, 253.

152. I. J. MacRae, I. H. Segel and A. J. Fisher, *Biochemistry*, 2001, **40**, 6795.

153. I. J. MacRae, I. H. Segel and A. J. Fisher, *Nature Struct. Biol.*, 2002, **9**, 945.

154. T. C. Ullrich, M. Blaesse and R. Huber, *EMBO J.*, 2001, **20**, 316.

155. B. Schneider, M. Babolat, Y. W. Xu, J. Janin, M. Veron and D. Deville-Bonne, *Eur. J. Biochem.*, 2001, **268**, 1964.

156. P. G. Czyryca and A. C. Hengge, *Biochim. Biophys. Acta*, 2001, **1547**, 245.

157. G. van Pouderoyen, T. Eggert, K.-E. Jaeger and B. W. Dijkstra, *J. Mol. Biol.*, 2001, **309**, 215.

158. T. Eggert, G. van Pouderoyen, B. W. Dijkstra and K.-E. Jaeger, *FEBS Lett.*, 2001, **502**, 89.

159. Y.-P. Pang, K. Xu, T. M. Kollmeyer, E. Perola, W. J. McGrath, D. T. Green and W. F. Mangel, *FEBS Lett.*, 2001, **502**, 93.

160. P. Ascenzi, L. Salvati, M. Bolognesi, M. Colasanti, F. Polticelli and G. Venturini, *Curr. Protein Pept. Sci.*, 2001, **2**, 137.

161. A. de Kreij, B. van der Burg, O. R. Veltman, G. Vriend, G. Venema and V. G. H. Eljsink, *Eur. J. Biochem.*, 2001, **268**, 4985.

162. H. C. Castro, D. M. Silva, C. Craik and R. B. Zingali, *Biochim. Biophys. Acta*, 2001, **1547**, 183.

163. M. Simonovic, P. G. W. Gettins and K. Volz, *Proc. Nat. Acad. Sci. USA*, 2001, **98**, 11131.

164. P. A. Wright, R. C. Wilmouth, I. J. Clifton and C. J. Schofield, *Eur. J. Biochem.*, 2001, **268**, 2969.

165. G. W. Wong, L. Li, M. S. Madhusudhan, S. A. Krilis, M. F. Gurish, M. E. Rothenberg, A. Sali and R. L. Stevens, *J. Biol. Chem.*, 2001, **276**, 20648.

166. J. B. Thoden, T. M. Wohlers, J. L. Fridovich-Keil and H. M. Holden, *J. Biol. Chem.*, 2001, **276**, 20617.

167. Z. Yang, J. M. Kollman, L. Pandi and R. F. Doolittle, *Biochemistry*, 2001, **40**, 12515

168. M. A. Halcrow, *Angew. Chem. Int. Edit.*, 2001, **40**, 346

169. S. E. Greasley, H. Marsilje, H. Cai, S. Baker, S. J. Benkovic, D. L. Boger and I. A. Wilson, *Biochemistry*, 2001, **40**, 13538.

170. G. Wu, J. L. Vuletich, R. J. Kulmacz, Y. Osawa and A.-L. Tsai, *J. Biol. Chem.*, 2001, **276**, 19879.
171. D. L. McCarthy, G. Capitani, L. Feng, M. G. Gruetter and J. F. Kirsch, *Biochemistry*, 2001, **40**, 12276.
172. T. W. B. Ost, C. S. Miles, A. W. Munro, J. Murdoch, G. A. Reid and S. K. Chapman, *Biochemistry*, 2001, **40**, 13421.
173. A. B. Carmichael and L. L.-Wong, *Eur. J. Biochem.*, 2001, **268**, 3117.
174. G. Canet, D. Balbuena, T. Achstetter and B. Dumas, *Eur. J. Biochem.*, 2001, **268**, 4054.
175. S. Zamze, D. R. Wing, M. R. Wormald, A. P. Hunter, R. A. Dwek and D. J. Harvey, *Eur. J. Biochem.*, 2001, **268**, 4063.
176. J. G. Olsen, A. Kadziola, C. Lauritzen, J. Pedersen, S. Larsen and S. W. Dahl, *FEBS Lett.*, 2001, **506**, 201.
177. T. Kabashima, Y. Li, N. Kanada, K. Ito and T. Yoshimoto, *Biochim. Biophys. Acta*, 2001, **1547**, 214.
178. F. Lecaille, E. Authie, T. Moreau, C. Serveau, F. Gauthier and G. Lalmanach, *Eur. J. Biochem.*, 2001, **268**, 2733.
179. F. Lecaille, J. Cotton, J. H. McKerrow, M. Ferrer-Di Martino, E. Boll-Bataille, F. Gauthier and G. Lalmanach, *FEBS Lett.*, 2001, **507**, 362.
180. G. Zhu, Q. Huang, Y. Zhu, Y. Li, C. Chi and Y. Tang, *Biochim. Biophys. Acta*, 2001, **1546**, 98.
181. K. Poulas, E. Eliopoulos, E. Vatzaki, J. Navaza, M. Kontou, N. Oikonomakos, K. R. Achaya and S. J. Tzartos, *Eur. J. Biochem.*, 2001, **268**, 3685. This continues a sequence of studies in this area by this research group: M. Kontou, D. D. Leonidas, E. H. Vatzaki, P. Tsantili, A. Mamalaki, N. Oikonomakos, K. R. Achaya and S. J. Tzartos, *Eur. J. Biochem.*, 2000, **267**, 2389.
182. N. E. Labrou, L. V. Mello and Y. D. Clonis, *Eur. J. Biochem.*, 2001, **268**, 3950.
183. Y. Yang, J.-Z. Cheng, S. S. Singhal, M. Saini, U. Pandya, S. Awasthi and Y. C. Awasthi, *J. Biol. Chem.*, 2001, **276**, 19220.
184. D. Gisi, J. Maillard, J. U. Flanagan, J. Rossjohn, G. Chelvanayagam, P. G. Board, M. W. Parker, T. Leisinger and S. Vuilleumier, *Eur. J. Biochem.*, 2001, **268**, 4001.
185. R. H. Valente, B. Dragulev, J. Perales, J. W. Fox and G. B. Domont, *Eur. J. Biochem.*, 2001, **268**, 3042.
186. V. Nirmala, D. Kodrik, M. Zurovec and F. Sehntal, *Eur. J. Biochem.*, 2001, **268**, 2064.
187. K.-D. Bissig, T. C. Voegelin and M. Solioz, *FEBS Lett.*, 2001, **507**, 367.
188. D. G. Gourley, A. W. Schüttelkopf, L. A. Anderson, N. C. Price, D. H. Boxer and W. N. Hunter, *J. Biol. Chem.*, 2001, **276**, 20641.
189. P. M. Green, S. B. Ludbrook, D. D. Miller, C. M. T. Horgan and S. T. Barry, *FEBS Lett.*, 2001, **503**, 75.
190. K. H. Moberg, D. W. Bell, D. C. R. Wahrer, D. A. Haber and I. K. Jariharan, *Nature*, 2001, **413**, 311; H. Strohmaier, C. H. Spruck, P. Kaiser, K.-A. Won, O. Sangfelt and S. I. Reed, *Nature*, 2001, **413**, 311; see M. Schwab and M. Tyers, *Nature*, 2001, **413**, 268, for a discussion of this work and its background.
191. K. Kirshenbaum, I. S. Carrico and D. A. Tirrell, *Chem. BioChem.*, 2002, **2**, 235.
192. Y. Tang and D. A. Tirrell, *J. Am. Chem. Soc.*, 2001, **123**, 11089.
193. K. L. Kiick, R. Weberskirch and D. A. Tirrell, *FEBS Lett.*, 2001, **502**, 25 (corrigendum: *Idem*, **505**, p. 465).
194. M. Taki, T. Hohsaka, M. Murakami, K. Taira and M. Sisido, *FEBS Lett.*, 2001, **507**, 35.

195. Y. Chen and W. Maret, *Eur. J. Biochem.*, 2001, **268**, 3346.
196. P. Ascenzi, L. Salvati and M. Brunori, *FEBS Lett.*, 2001, **501**, 103.
197. C. Nowicki, G. R. Hunter, M. Montemartini-Kalisz, W. Blankenfeldt, H.-J. Hecht and H. M. Kalisz, *Biochim. Biophys. Acta*, 2001, **1546**, 268.
198. D. J. Rigden, A. C. S. Monteiro and M. F. G. de Sa, *FEBS Lett.*, 2001, **504**, 41; corrigendum, see *Idem*, p. 296.
199. J. S. Feinberg and S. Joseph, *Proc. Nat. Acad. Sci. USA*, 2001, **98**, 11120.
200. Y.-H. Sung, J. Shin, H.-J. Chang, J. M. Cho and W. Lee, *J. Biol. Chem.*, 2001, **276**, 19624.